Contents

Preface ix
Acknowledgments xvii
About the Authors xix

1 Overview of Corrections and Criminal Justice 1

Introduction 2
Criminal Justice as a Social and Governmental Institution 3
Criminal Justice as an Academic Field of Study 6
Overview of the Evolution of Criminal Justice 7
The Contemporary Criminal Justice System 11
Criminal Justice Process 14
Models and Images of Crime and Justice 19
Chapter Resources 24

2 The Picture of Crime and Punishment in the United States 31

Introduction 32
Sources of Crime Data in the United States 33
Patterns of Crime 42
Sources of Data for Adult Corrections 48
Conclusion 54
Chapter Resources 55

3 Crime and the Law 61

Introduction 62
Law and Government Structure 62
Law and Discretion 63
The Evolution of Law 64
Civil and Criminal Law 64
Substantive and Procedural Criminal Law 66
Sources of Law 67
Principles of Substantive Law 70
Principles of Procedural Law: The Ideal and the Real 78
Defining and Categorizing Crime 88

Application to Corrections 97
Chapter Resources 100

4 **Policing and Corrections** **109**

Introduction 111
History and Evolution of Policing 111
Structure and Organization of Policing in the United States 117
Roles, Functions, and Activities of Police 123
Police Management 125
Working Personality of a Police Officer 132
Police Subculture 133
Legal Aspects of Policing 133
Discretion 141
Implications for Corrections 142
Conclusion 145
Chapter Resources 146

5 **Courts and Corrections** **157**

Introduction 158
Philosophy of the U.S. Court System 159
History of the U.S. Court System 159
Structure and Organization of Twenty-First Century Courts in the United States 162
The Courtroom Work Group 169
Nonprofessional Courtroom Participants 176
Pretrial Activities 179
Steps of a Criminal Trial 183
Implications for Corrections 187
Conclusion 189
Chapter Resources 190

6 **Sentencing and Corrections** **203**

Introduction 204
Sentencing Options 204
Philosophies of Punishment 206

CRIMINAL JUSTICE ILLUMINATED

Corrections and the Criminal Justice System

David C. May, PhD
Professor
Kentucky Center for School Safety Research Fellow
Department of Safety, Security, and Emergency Management
Eastern Kentucky University

Kevin I. Minor, PhD
Professor and Chair
Department of Correctional and Juvenile Justice Studies
Eastern Kentucky University

Rick Ruddell, PhD
Graduate Program Director
Department of Correctional and Juvenile Justice Studies
Eastern Kentucky University

Betsy A. Matthews, PhD
Associate Professor
Department of Correctional and Juvenile Justice Studies
Eastern Kentucky University

JONES AND BARTLETT PUBLISHERS
Sudbury, Massachusetts
BOSTON TORONTO LONDON SINGAPORE

World Headquarters
Jones and Bartlett Publishers
40 Tall Pine Drive
Sudbury, MA 01776
978-443-5000
info@jbpub.com
www.jbpub.com

Jones and Bartlett Publishers
Canada
6339 Ormindale Way
Mississauga, Ontario L5V 1J2
Canada

Jones and Bartlett Publishers
International
Barb House, Barb Mews
London W6 7PA
United Kingdom

Jones and Bartlett's books and products are available through most bookstores and online booksellers. To contact Jones and Bartlett Publishers directly, call 800-832-0034, fax 978-443-8000, or visit our website www.jbpub.com.

Substantial discounts on bulk quantities of Jones and Bartlett's publications are available to corporations, professional associations, and other qualified organizations. For details and specific discount information, contact the special sales department at Jones and Bartlett via the above contact information or send an email to specialsales@jbpub.com.

Production Credits
Acquisitions Editor: Jeremy Spiegel
Editorial Assistant: Lisa Gordon
Production Director: Amy Rose
Production Editor: Renée Sekerak
Production Assistant: Julia Waugaman
Marketing Manager: Wendy Thayer
Manufacturing and Inventory Control Supervisor: Amy Bacus
Composition: Auburn Associates, Inc.
Interior Design: Anne Spencer
Cover Design: Anne Spencer
Cover Image: © Kimber Rey Solana/ShutterStock, Inc.
Chapter Opener Image: © Masterfile
Photo Research Manager and Photographer: Kimberly Potvin
Photo Researcher: Timothy Renzi
Text Printing and Binding: Malloy Incorporated
Cover Printing: Malloy Incorporated

Library of Congress Cataloging-in-Publication Data

Corrections and the criminal justice system / by David C. May ... [et al.]. — 1st ed.
p. cm.
Includes bibliographical references and index.
ISBN-13: 978-0-7637-3500-5 (pbk.)
1. Corrections—United States. 2. Criminal justice, Administration of—United States.
3. Punishment—United States. I. May, David C., 1966-
HV9304.C6754 2008
364.973—dc22

2007032441

6048

Printed in the United States of America
11 10 09 08 07 10 9 8 7 6 5 4 3 2 1

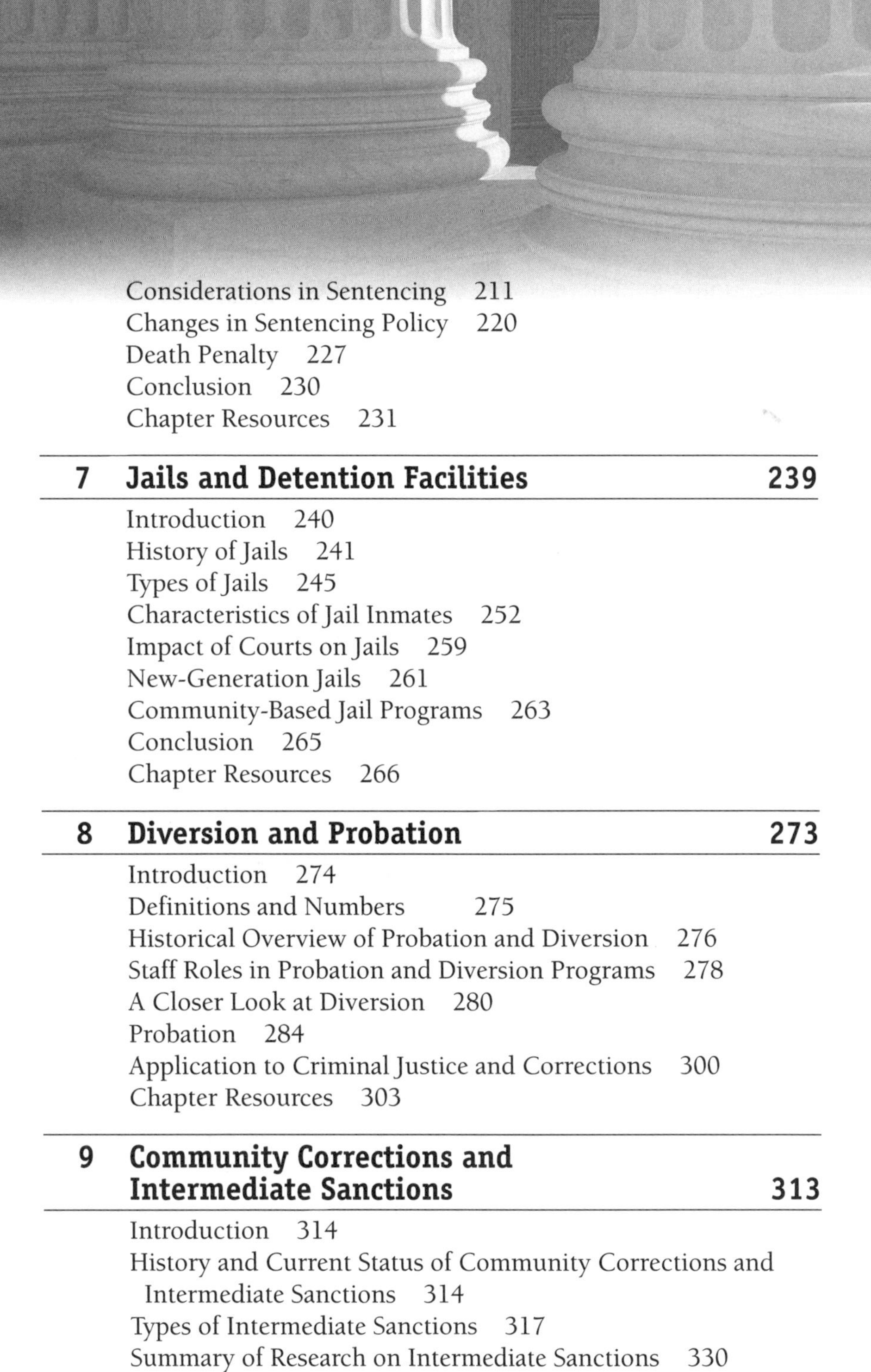

Considerations in Sentencing 211
Changes in Sentencing Policy 220
Death Penalty 227
Conclusion 230
Chapter Resources 231

7 **Jails and Detention Facilities** **239**

Introduction 240
History of Jails 241
Types of Jails 245
Characteristics of Jail Inmates 252
Impact of Courts on Jails 259
New-Generation Jails 261
Community-Based Jail Programs 263
Conclusion 265
Chapter Resources 266

8 **Diversion and Probation** **273**

Introduction 274
Definitions and Numbers 275
Historical Overview of Probation and Diversion 276
Staff Roles in Probation and Diversion Programs 278
A Closer Look at Diversion 280
Probation 284
Application to Criminal Justice and Corrections 300
Chapter Resources 303

9 **Community Corrections and Intermediate Sanctions** **313**

Introduction 314
History and Current Status of Community Corrections and Intermediate Sanctions 314
Types of Intermediate Sanctions 317
Summary of Research on Intermediate Sanctions 330
Future Directions for Intermediate Sanctions 332

Impact of Intermediate Sanctions on Corrections 334
Chapter Resources 335

10 Prison Systems and Staff 347

Introduction 348
History of the Prison 348
Modern Prison Systems 356
Prison Staff 364
Conclusion 371
Chapter Resources 372

11 Prisoners, Prison Life, and Inmate Rights 379

Introduction 380
Prison Society, Culture, and Experience 394
Inmate Rights and Case Law 398
Conclusion 403
Chapter Resources 404

12 Prison Security, Programs, and Services 413

Introduction 414
Inmate Classification 414
Institutional Security 418
Institutional Programs 427
Institutional Services and Health Care 433
Conclusion 434
Chapter Resources 436

13 Release from Prison, Parole, and Prisoner Reentry 443

Introduction 445
Release from Incarceration 446
Parole 449
Recidivism and Reentry 460
Application to Criminal Justice and Corrections 473
Chapter Resources 475

14 Juvenile Justice and Corrections 487

Introduction 488
Defining Delinquency 489
The Nature and Extent of Juvenile Crime and Victimization 490
Causes of Delinquency 497
Controlling Delinquency Throughout History 505
Contemporary Juvenile Courts 508
Processing Delinquency Cases in Juvenile Court 509
Juvenile Corrections 515
Effective Correctional Interventions for Youth 520
Contemporary Issues in Juvenile Justice 522
Chapter Resources 526

15 The Future of Corrections and Juvenile Justice 539

Introduction 540
Trends and Challenges Facing Corrections in the Twenty-First Century 540
Correctional Futurism 555
Conclusion 559
Chapter Resources 561

Index 569

Preface

About two weeks after the tragic events of September 11, 2001, one of the authors, David May, was deployed to Diego Garcia, an island about 13 miles long, 1,000 miles off the southern coast of India in the Indian Ocean. Although he had been in the military for more than 17 years, this experience was both exciting and somewhat scary at the same time, primarily because it was unfamiliar. Most of his military experience had been in the United States Army Infantry. For this deployment, he would be a United States Air Force Security Forces officer (or the Air Force version of the military police).

One of the most troubling aspects of David May's deployment with the Air Force was that there were a number of terms and concepts with which he was unfamiliar but were needed to perform daily duties. Terms such as in-flight emergencies (a suspected malfunction on a plane), services squadron (the group responsible for food, clothing, and shelter of Air Force personnel), and PERSCO (an acronym for Personnel Services Company) were regularly used by the vast majority of the people with whom he came into contact; however, having never heard these terms, he had no idea of their meaning or significance. You may have that same experience in this book. Terms such as PSI (presentence investigation reports), probation, segregated housing units, and discretion are all terms that you will regularly hear your instructor discuss in class and read in this book. At first, you may not be familiar with these terms and may experience the same frustrations that May experienced at the beginning of his first Air Force deployment. Like him, however, with enough reading and studying (and help from the authors of this book) these terms will become familiar to you.

During that deployment, May also experienced a brand new set of expectations. His Army experience had not prepared him for driving on a flight line full of taxiing aircraft, dealing with civilians moving in and out of "secure" areas, and being forced to confiscate cameras from military personnel for taking photographs of classified materials. In other words, he was introduced to a brand new paradigm.

As you read this book, you will also experience a brand new paradigm. You have heard that murderers either receive the death penalty or go to jail; in fact, both of those statements are rarely true. Most people convicted of murder do not receive a death sentence (only 128 convicted murderers were sentenced to death in 2005) or go to jail (in fact, practically all go to prison for a long time, which is quite different from going to jail). You may also have been told by your relatives and friends that crime policies are too lenient in the United States; in fact, the United States is the most punitive industrialized country in the world. Finally, if you are like most others, you think that the vast majority of people who go to prison each year are murderers, rapists, robbers, or other violent criminals; in fact, more than half of the offenders sentenced to federal prison each year are sentenced for drug-related crimes, and the majority of offenders sentenced to state prisons each year have committed nonviolent crimes. You will learn, as David May learned, that your opinions are not always based on fact; by the end of your journey through this textbook, you will have the facts about corrections and juvenile justice and will never again have the same opinion about these matters.

The unfamiliar terms and concepts and the differing set of expectations May experienced during his deployment were troubling at first; but, over time, he developed a friendship with another noncommissioned officer who helped him understand the puzzles and unravel the mysteries of the Air Force Security Forces. You, on the other hand, do not have to worry about finding that friend. We (the authors of this textbook) are those friends. This textbook has been designed so that whenever there is an unfamiliar term, the definition is provided. Whenever there are troubling questions or concepts, a solution has been provided. With the authors' help, you will navigate this textbook successfully. It cannot guarantee you will receive a good grade in your course (that is really up to you), but the textbook has been designed so that achieving a good grade is easier.

Chapter Resources

Each chapter in this book has a number of distinctive sections to assist you in your journey through the world of criminal justice, corrections, and juvenile justice. Each chapter begins with a set of chapter objectives. These objectives let you know what you need to learn from the chapter. After completing each chapter, go back and test your knowledge by re-reading those objectives.

Each chapter includes a Case Study that presents a "real-world" scenario describing a problem likely to be faced by a practitioner. Each Case Study includes two or three questions. As you read the chapter, keep those questions in the back of your mind; by the end of the chapter, you should be able to answer the questions by using knowledge you have gained. Answers to those questions have been included at the end of each chapter so that the reader can check their answers.

To help you both (1) understand the material in the chapter and (2) prepare for the exams that your instructor has planned for you, two sections have been included. The first is "Key Terms." In this section, the most important terms in the chapter are identified and are marked in bold within the chapter text. When you see a term in bold, you can go to the end of the chapter to find that term's definition and make a note of that term and its definition for future study and review.

At the end of each chapter, a section entitled "Ready for Review" is included. This section consists of a concise summary of the central points included within each chapter. After reading the chapter, the reader should be able to review each bullet and derive a good overview of the most important information in the chapter. As you study, we encourage you to write an expanded summary of each bulleted point.

The hope is that these tools included in the chapters will make your journey through the world of corrections easier and more structured. You should consider these items as *information booths*; use them just as you would an information booth in an airport. Whenever you discover some concept or idea that you do not understand, check these sections to see if we have included a tool to help you understand that concept.

Chapter Overviews

This textbook is designed to provide a comprehensive understanding of corrections within the larger, more complex system of criminal justice.

Corrections policies and processes are influenced heavily by the criminal processes that precede them (e.g., arrest, sentencing, etc.). Thus, without a good understanding of how all the various components of the criminal justice system operate, it is difficult to completely understand the correctional and juvenile justice components of the system. This complex system and its associated processes are discussed one step at a time.

The journey of understanding begins in Chapter 1, Overview of Corrections and Criminal Justice. This chapter describes the evolution of criminal justice as an academic field and discusses its importance in the larger arena of social institution of government. Later, the components of the criminal justice system and the phases in the criminal justice process are described. The chapter closes by emphasizing the importance of the media and distinguishing between the due process and crime control models of criminal justice.

The second stage of your journey is Chapter 2, The Picture of Crime and Punishment in the United States. In this chapter, the various sources of criminal justice, corrections, and juvenile justice data in the United States are discussed. The authors talk about the benefits and limitations of each and then close by discussing the demographic patterns that these data reveal. Although not directly included at any point in the criminal justice process, each phase of the system collects data that inform decisions made by that stage and the other stages following it. By the end of this chapter, you should have a much better understanding of what is known about crime and corrections in the United States.

The next stage on your journey is Chapter 3, Crime and the Law. This chapter discusses the origins of law (both civil and criminal), and how both substantive and procedural law impact criminal justice and corrections in the United States. Also discussed are the various approaches to classifying crimes by using law (e.g., *mala in se* v. *mala prohibita* crimes, felonies v. misdemeanors). After completing this chapter, you should understand how the various aspects of the law impact not just criminal justice data, but corrections data as well.

The next stop on your journey occurs in Chapter 4, Policing and Corrections. Policing is most important in the first stages of the criminal justice process. In this chapter, the authors discuss the history of policing in the United States and throughout the world, the structure and organization of policing in the 21st century, and the important role that policing plays in both corrections and juvenile justice. After you have completed this chapter, you should understand how decisions by police to enforce (or not to enforce) the law impact not just individuals involved in criminal justice, but corrections as well.

In Chapter 5, Courts and Corrections, we discuss the history and evolution of courts in the United States and go into detail about the federal and state court systems and the similarities and differences between the two systems. The courts are heavily involved in the second and third phases of the criminal justice flowchart. Discussed later in the chapter are the various stages of the criminal trial and the actors that work in the "courtroom work group" to ensure that the process of justice flows as smoothly as possible. Upon completion of this chapter, you will have a detailed knowledge of the court systems in the United States and the impact of those court systems on corrections.

The next stop on the journey is Chapter 6, Sentencing and Corrections. This chapter focuses on the decisions made by the judges and magistrates in the sentencing phase of the system. Also described are the different philosophies of punishment that drive sentencing decisions and provide a detailed description of the sentencing process, from plea bargaining to conviction. Finally discussed are the changes that have occurred in sentencing strategies over time. Some of the philosophical shifts that have made sentencing harsher at the beginning of the 21st century than any time in the 20th century have been outlined. After finishing this chapter, you will have a good understanding of why the United States continues to be the most punitive industrialized society in the world.

Chapter 7, Jails and Detention Facilities, begins a detailed discussion of the field of corrections with one of the most poorly-researched areas in criminal justice, the jail. In this chapter, you will learn about the history and evolution of jails, both throughout the world and in the United States. Also discussed are the various types of jails in the United States, along with the wide variety of people that pass through jails each day. The impacts that local decisions made by the courts and the police have on jails and their operations will also be learned. Upon completion of this chapter, you will have a much better understanding of how local jails are so different from one another and the reasons for those differences.

The next stop along the journey to the understanding of corrections and juvenile justice occurs in Chapter 8, Diversion and Probation. An introduction to the historical development of both of these concepts and exposure to the various stages in each of these processes will be covered. You will then learn about the various types of diversion and the agencies that administer diversion programs, along with the advantages and disadvantages of diversion programs. Finally, you will read about the various types of probation and the processes through which offenders are placed on probation, supervised, and released. By the end of Chapter 8, you will understand the important roles that diversion and probation play in the larger field of corrections, and you will leave that chapter understanding that without diversion and probation, the picture of corrections in the United States would be vastly different.

The journey then continues into Chapter 9, Community Corrections and Intermediate Sanctions. This chapter introduces and discusses the concepts of community corrections and alternative sanctions and the historical development of these "intermediate sanctions." The goals of intermediate sanctions are described, as well as the major types of intermediate sanctions available today. This chapter concludes with discussing the research available on the effectiveness of intermediate sanctions and their key roles in the field of corrections. As you finish Chapter 9, you will realize that your idea of what constitutes corrections was probably different than what corrections really involves.

Beginning in Chapter 10, Prison Systems and Staff, you finally enter the world of institutional corrections, one of the most misunderstood entities in criminal justice. Discussed in this chapter is the origin of prisons and their development throughout the history of the United States. We discuss the different types of prisons, including levels of security and specialized prisons such as the supermax. The focus is then turned to the correctional officers, their daily activities and training, and the challenges that they face. By the end of this chapter, you will have a good understanding of how prisons developed and of the daily lives and challenges of those who operate prisons in the 21st century.

Chapter 11, Prisoners, Prison Life, and Inmate Rights, describes the "special needs" populations of prisons, the types of litigation that have arisen from activities within the prison setting, and the harsh realities of prison life, including violence and gang membership. At the conclusion of this chapter, you will probably realize that the world of prison is much more complex than you had originally believed.

You then continue your journey and exit the prison system in Chapter 12, Prison Security, Programs, and Services. This chapter describes the various elements of an effective security program in the institutional setting and the threats to security in those settings, along with the various health-care issues faced by correctional managers. This chapter concludes by exploring the different elements of correctional rehabilitation and discussing the importance of evidence-based corrections for effective rehabilitation in the correctional setting. Upon finishing this chapter, you will have a good idea of the challenges that face correctional managers in terms of security, health care, and rehabilitative programming.

In Chapter 13, Release from Prison, Parole, and Prisoner Reentry, we discuss what happens to prisoners after they are released from prison and describe the various types of release from prison. Later we distinguish parole from probation and describe the historical development of parole. We cover the various phases of the parole process and critically analyze the major issues in parole. We close the chapter by critically analyzing the controversies surrounding prisoner reentry and discuss the issues that so often make reentry difficult for those offenders. By the end of this chapter, you will realize that "getting out" of prison is not always the end of the difficulties that prisoners face.

In Chapter 14, Juvenile Justice and Corrections, you are introduced to the world of juvenile justice, perhaps for the first time. If you look at the chart describing the criminal justice process in Chapter 1 (page 15), notice the separate pipeline presented at the bottom of the flowchart that represents the path of juvenile offenders. In this chapter, you will learn about that path and how the juvenile justice system differs from the adult justice system. You will read about the unique challenges faced by juveniles and those responsible for their care. You will learn about the historical development of the juvenile court and will develop a new vocabulary of terms specific to the juvenile justice system. By the end of the chapter, you will realize that the juvenile justice system is a system of which few people are aware.

You close your journey with Chapter 15, The Future of Corrections and Juvenile Justice. In this chapter, you will learn about the challenges facing corrections and juvenile justice in the 21st century and how our current systems are dealing with those challenges. At the conclusion of this chapter, you will be well aware that the future of corrections and juvenile justice is fraught with challenges that may completely reshape both fields by the end of the century.

Boxed Features

Throughout the various chapters, you will find boxes designed to demonstrate the applicability of the chapter to the field. The four types of boxes include:

- *Corrections in the Real World*, where a concept discussed in the chapter is applied to an issue that could be encountered by a corrections professional.
- *Corrections and Policy*, where consideration is given to how some type of policy affects corrections.
- *Race and Gender in Corrections*, where attention is given to race and gender issues confronting the field.
- *Controversies*, where current controversies in the field are applied to chapter content.

Upon completion of your journey through the corrections topics covered in this book, you will have both a better understanding of those individuals involved in the correctional system and a newfound respect for the professionals who work in these fields. Both areas are challenging fields of study and work, and the rewards experienced by the students, researchers, and professionals in those fields are unparalleled. As the Chinese philosopher Lao-tzu once said, "A journey of a thousand miles begins with a single step." Let's begin the journey to an understanding of corrections with that first step now.

Acknowledgments

This textbook would have never been finished without the help of a wide variety of people. First, we would like to thank Chambers Moore, David Cella, and Jeremy Spiegel with Jones and Bartlett Publishers. Their encouragement throughout the long process of creating this book kept us on task when we often strayed, and on time when we delayed. We would particularly like to thank Lisa Gordon and Renée Sekerak with Jones and Bartlett, who were instrumental in the final phases of manuscript development by ensuring that all the needed pieces of the text were present and in order.

We would also like to thank our family members, who understood when we said we had to work on "the book," as well as our colleagues in the College of Justice and Safety at Eastern Kentucky University (EKU), who understood when we delayed other projects to work on this textbook and provided critical ears and eyes when asked their opinions about the text. At the risk of omitting someone who should have been included, the list of EKU staff and students (present and former) who assisted us includes the following, and we are grateful for your help: Janice Marcum, Adam Cart, Greg Coomes, Tiffany Coots, Elizabeth Donnelly, Amy Eades, Paula Elliston, Tonya Jordan, Mitchell King, Lena Kline, Graham Rutherford, and Trisha Zeek.

About the Authors

David C. May is a Professor and Kentucky Center for School Safety Research Fellow in the Department of Safety, Security, and Emergency Management at Eastern Kentucky University. He received his PhD in sociology with emphasis in criminology from Mississippi State University in 1997. He has published numerous articles in the areas of responses to school violence, perceptions of the severity of correctional punishments, adolescent fear of crime and weapon possession, and two books examining the antecedents of gun ownership and possession among male delinquents.

Kevin I. Minor is Professor and Chair in the Department of Correctional and Juvenile Justice Studies at Eastern Kentucky University. His specializations include institutional and community corrections and juvenile delinquency and justice, as well as evaluation and applied research. Dr. Minor has published various books, anthology chapters, and articles in these areas, with articles appearing in such journals as *Crime & Delinquency, Federal Probation, Journal of Criminal Justice,* and *Youth Violence and Juvenile Justice.* He has also performed consultant services for the United States Department of Justice (including the Bureau of Justice Statistics, the Office of Juvenile Justice and Delinquency Prevention, the National Institute of Corrections, and the National Institute of Justice) as well as for numerous state level and private agencies. Earlier in his career, Dr. Minor worked as a practitioner in both juvenile and adult correctional institutions.

Rick Ruddell is the Graduate Program Director in the Department of Correctional and Juvenile Justice Studies at Eastern Kentucky University. Dr. Ruddell has experience as a supervisor and manager within the Department of Corrections and Public Safety in Saskatchewan, Canada, and served as Associate Professor of Political Science at the California State University, Chico. His research has been published in over 30 scholarly articles and focuses on corrections and criminal justice policy. Rick Ruddell is the co-author of *Making Sense of Criminal Justice*, author of *America Behind Bars: Trends in Imprisonment, 1950 to 2000*, and co-editor of *Issues in Correctional Health.*

Betsy A. Matthews has worked as a youth worker, an adult probation officer, and a research associate for the American Probation and Parole Association. She earned her PhD from the University of Cincinnati in 2004. She has published articles on various topics including intensive supervision, restorative justice, performance-based measures in probation and parole, faith-based programming, and working effectively with girls in the juvenile justice system. Dr. Matthews is currently an Associate Professor in the Department of Correctional and Juvenile Justice Studies at Eastern Kentucky University.

1

Overview of Corrections and Criminal Justice

Chapter Objectives

- Define criminal justice as part of the social institution of government. Include understanding of branches and levels of government as these apply to criminal justice.
- Describe criminal justice as an academic field.
- Summarize the evolution of criminal justice, demonstrating understanding of its growth.
- Identify and describe the components of the criminal justice system, and state why a system analogy can be misleading.
- Identify and describe the phases in the criminal justice process. Discuss how this process can be compared to a funnel.
- Describe the image of crime and justice portrayed by the media, and critically assess this image.
- Distinguish between the due process and crime control models of criminal justice.

CASE STUDY

After graduating from the local university, you were hired as a juvenile probation officer in the town in which you grew up. After about a year in your agency, you have been asked by the local Kiwanis Club, of which your dad is a member, to provide a 20-minute overview of corrections and juvenile justice at a public forum on crime and delinquency at the upcoming civic fair. You are honored by the request and want to do a professional job that both (1) demonstrates your knowledge of the fields of corrections and juvenile justice and impresses your father and his friends with how knowledgeable you are about your profession and (2) challenges any distorted images the audience might have based on exposure to the media.

1. Because you have only 20 minutes allotted for the presentation, what are the most important topics that you should cover to provide the audience the widest breadth of knowledge regarding corrections and juvenile justice?
2. Most of your audience probably has at least a cursory knowledge of policing and the courts, two of the "three Cs" of the criminal justice system. Nevertheless, few probably have much accurate knowledge of corrections or juvenile justice. How will you use that existing knowledge to inform them about corrections and juvenile justice and how the various aspects of the criminal justice system are interrelated and dependent upon one another?

Introduction

When most people hear the term *criminal justice*, they are likely to think of police officers, lawyers (both prosecuting and defense attorneys), and perhaps even elected officials who pass laws. They might think of officials such as wardens or correctional officers who staff jails and prisons, especially if asked to consider the term *corrections*. If asked to consider the term *juvenile justice*, they might think of counselors or social workers. They might also think about individuals who break the law and the various things that happen when lawbreakers get caught, such as arrest, trial, sentencing, and incarceration. In essence, when people hear *criminal justice* or related terms, they think about officials who work in legislative, police, court, and correctional agencies—what these officials do and how these agencies operate.

As the distinction drawn later in this chapter between the criminal justice *system* and *process* will show, these are all correct ways of thinking about the topic. But there is a broader way to conceptualize criminal justice, and that is to think of it as part of the social institution of government. Doing so helps us understand how criminal justice is organized and connected to the wider society of which it is a part, a society that both affects and is affected by criminal justice.[1]

Criminal Justice as a Social and Governmental Institution

All societies have **social institutions**. These are established groupings of people, beliefs, and practices with three main characteristics. First, social institutions are relatively stable and lasting. They are built on custom, so they tend to change slowly and gradually over time. Second, each institution is set up to address a particular set of societal needs, such as the need to rear children in the case of the family institution. Besides the family, other important institutions are religion, education, the medical field, the economy, and the political or governmental system. Third, rather than being isolated from each other, social institutions are linked to one another; what takes place in one (e.g., the family) will usually affect another (e.g., education).[2] (See **Figure 1–1**.)

In the United States, it is helpful to think of criminal justice as a component of government established to meet three broad needs of our society. The first is the need to control crime, thereby preserving social order and the safety of citizens. Citizens want to feel secure and protected in their homes and in public places. The second is the need for those who obey the law to feel that those who break it are getting what they deserve. The public often feels unjustly treated when punishment is not seen as "fitting" the crime. The third is the need to ensure that offenders are treated fairly and justly, and in the case of our society, this means according to the rights set forth in the U.S. Constitution. In short, then, criminal justice serves the needs society has for both justice and safety. These needs are evident in the discussion later in this chapter of the crime control and due process models of criminal justice.

Criminal justice is linked closely to other social institutions, but the most obvious linkage is to the political institution of government. In the United States, government has primary responsibility for operating criminal justice agencies. The private sector performs many functions such as private security or sometimes providing treatment for offenders addicted to drugs. But these functions are typically performed under contractual arrangements with government and are subject to government oversight.

Branches of Government

In civics and government classes, students are taught that U.S. government is divided into three branches (the legislative, executive, and judicial branches), such that there is a division of power and a system of checks and balances. Students also learn that government exists at three levels, including the federal, state, and local (county and town/city) levels. All three governmental branches are involved in the operation of criminal justice, and criminal justice agencies and activities operate at all three levels. This point is illustrated in simplified form in **Table 1–1**.

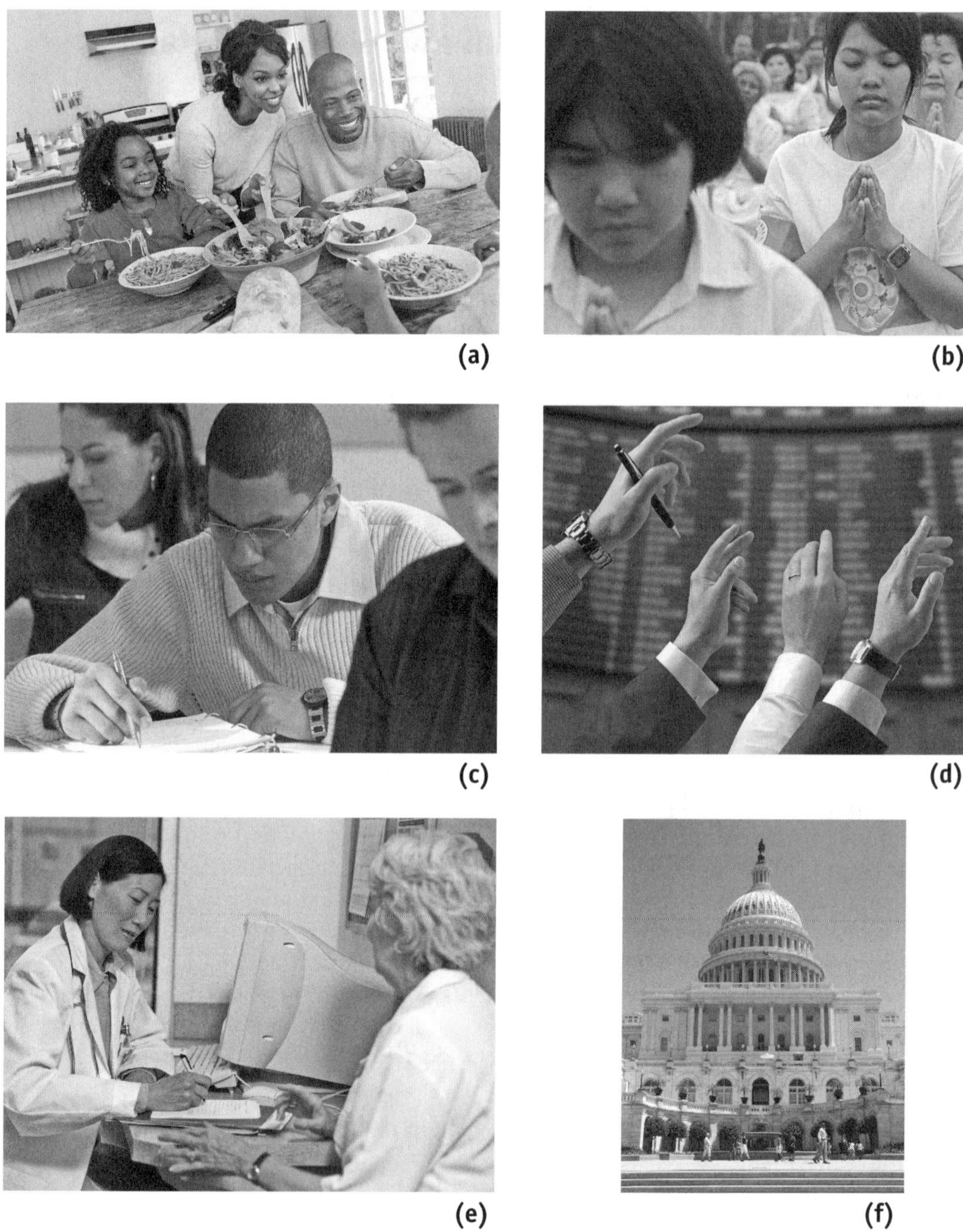

Figure 1–1 Criminal justice is part of the social institution of government, one of a number of institutions in society.
Source: *(a) © BananaStock/Jupiterimages; (b) © Bianda Ahmad Hisham/ShutterStock, Inc.; c) © Photos.com; (d) © ene/ShutterStock, Inc.; (e) © Ryan McVay/Photodisc/Getty Images; (f) © Jonathan Larsen/ShutterStock, Inc.*

Table 1–1 Examples of Criminal Justice Functions by Branches and Levels of Government

	Legislative Branch	Judicial Branch	Executive Branch
Federal Level	• U.S. Penal Code	• U.S. courts	• Federal law enforcement (e.g., FBI) • Federal Bureau of Prisons
State Level	• State penal codes	• State courts	• State police • State adult and juvenile corrections systems
Local Level	• County and city ordinances	• County and city courts (sometimes including juvenile court)	• County and city law enforcement • County and city jails and detention centers

The **legislative branch** is of much importance because it passes laws defining crimes, the corresponding punishments, and certain procedures for processing cases (see Chapter 3). Although this branch does not contain police, court, or correctional agencies per se, legislators also approve budgets and appropriate funding for these agencies. In addition, legislatures hold hearings to investigate allegations of wrongdoing.

The **judicial branch** consists of courts responsible for processing cases to determine issues of guilt or innocence and for sentencing those deemed guilty. This branch also consists of courts responsible for hearing appeals on issues of fairness and constitutionality (see Chapters 5 and 6). In some jurisdictions, the judicial branch is also responsible for certain functions related to corrections, such as diversion or probation (see Chapter 8).

The **executive branch** consists of elected and appointed public officials and the various units subordinate to their authority. Both law enforcement (Chapter 4) and corrections (Chapters 7–13) agencies generally operate from the executive branch. Among other things, law enforcement agencies are responsible for identifying those suspected of crime, apprehending those who appear guilty, and gathering evidence to support prosecution. Corrections agencies are responsible for carrying out the sentences imposed by courts.

Similarly, **juvenile justice**, the agencies and processes established to address crime committed by people under the legal age of adulthood in a jurisdiction (see Chapter 14), also involves all three branches of government. However, few juvenile justice agencies operate at the federal level; most function at the state and local levels.

Criminal Justice as an Academic Field of Study

In colleges and universities across the nation, there are academic fields and departments devoted to the study of all the major social institutions. Examples include sociology, family studies, religious studies, education, economics, and political science, as well as health sciences and medicine. During the past three to four decades, crime and justice have increasingly become objects of study, not only in more general and traditional disciplines such as sociology and political science, but also in more specialized programs with titles like criminology, criminal justice, police science/studies, corrections, and juvenile justice. Today, thousands of students take such courses at numerous college and university programs across the country.[3]

As an academic field, criminal justice involves three interrelated activities: (1) teaching and learning about criminal justice operations and issues, (2) conducting research on topics related to the field, and (3) performing services for the profession, such as consulting and technical assistance. It is the first of these that you will be engaged in as you study this book.

Criminal justice is sometimes thought of as a **multidisciplinary** field, drawing on insights from a number of the disciplines mentioned earlier (e.g., sociology, psychology, political science, geography, and biology). It is also sometimes considered an applied field of study, meaning that there is focus on application of the material being studied to legislative, policing, court, corrections, or juve-

Corrections in the Real World

People who complete a prison sentence in one state sometimes face extradition to another jurisdiction (i.e., either the federal government or another state) to complete a sentence for a criminal conviction in that jurisdiction. This is a result of the fact that criminal justice operates in various units of government, with the units being somewhat independent.

nile justice activities. However, this does not mean that the study of criminal justice must be entirely concrete and unconcerned with explanations or theories about why and how things operate as they do.[4] Along these lines, criminal justice is sometimes distinguished from criminology, which is considered the scientific study of crime—its volume, types, characteristics and distribution, causes, and control. It is in the area of crime control, however, that the study of criminology and criminal justice closely merge. Over time, the two fields have grown more alike.

A distinction worth noting is that between criminal justice *academic education* and *training*. There are certainly similarities between the two with respect to topics covered and sometimes even in the approaches taken to coverage. However, the two differ in relative emphasis. The emphasis of criminal justice education is broader, more academic, and grounded in social science and liberal arts orientations. Criminal justice training is typically more specialized, vocational, and professional in orientation. It is focused around teaching people what they need to know to work effectively in and improve police, court, correctional, and juvenile justice agencies where they are employed. Both training and academic education are important. This is why persons who hold college degrees often complete training academies before starting work in agencies. It is also why some people who have attended numerous trainings and worked in agencies for years enroll in criminal justice courses to pursue degrees. The two can complement one another.

Overview of the Evolution of Criminal Justice

Later chapters in this book trace some of the main historical developments particular to the areas of law, policing, courts, corrections, and juvenile justice. This section examines broader trends in the evolution of criminal justice as a whole. Keep in mind that, as an institution, criminal justice changes slowly across time.

Historically, in both Europe and America, crime was regarded as sin against God and as a wrong against other people. As such, traditional means of dealing with crime often involved a fragmented, unsystematic combination of responses from private individuals and private groups (especially tribes and kinship groups), as well as from clergy and government officials. However, changes gradually took place over the four to five centuries of the Renaissance (latter 1300s through 1600s) and culminating in the **Enlightenment**—a cultural and intellectual movement spanning the latter 1600s through the 1700s that emphasized human reason, science, freedom, and democratic government as an alternative to religion as a world view. Governing officials in England, the country from which much American jurisprudence directly descends, pursued increased centralization and coordination of criminal justice. This trend was at times coupled with efforts to treat accused persons more fairly.

In Colonial America, there was little alternative but to rely heavily on private citizens, and indeed vigilante groups, to address problems of crime. Such minimal government as existed in many areas was subservient to British authority. Accordingly, the colonists implemented a number of English legal customs in the New World. Though never replacing private justice during colonial times, these customs (as well as those imported from other nations such as France and Holland) were expanded and institutionalized. This transpired as population growth and urbanization rendered the informal system of justice increasingly impractical.[5]

It was against this backdrop that Enlightenment thinkers with interests in law and justice, especially the Italian Cesare Beccaria (1738–1794) and Englishman Jeremy Bentham (1748–1832), built upon the core ideas of Enlightenment figures such as Locke, Voltaire, and Montesquieu.[6] Beccaria and others challenged the existing system of justice (which frequently relied on torture and capital punishment) as arbitrary, unfair, barbaric, and ineffective at controlling crime. In a famous 1764 book, *Dei delitti e delle pene* (*On Crimes and Punishments*), Beccaria argued for a system of written laws and punishments that would make people clear about which actions were illegal and the consequences of these actions. Furthermore, laws need to be applied in a uniform and equal manner according to the harm resulting from the crime, rather than according to the status of the offender. Punishments, Beccaria suggested, could deter crime if imposed with certainty, swiftness, and sufficient severity to outweigh rewards gleaned from crime.

Following the American Revolution, the new nation was free to pursue its own system of justice without interference from England. The U.S. Constitution, which divided government into the three branches described earlier, was developed at the apex of the Enlightenment and put into effect in 1789. The Bill of Rights was added two years later to help protect citizens from the power of government (see Chapter 3). At the dawn of the nineteenth century, then, the stage was set for the development of the criminal justice system as we know it today. Especially in the northeast, the nation witnessed the rapid rise of urban police forces (see Chapter 4) and the penitentiary system (see Chapter 10). By the close of the century, a separate justice system was being developed for juveniles (see Chapter 14). Nevertheless, private informal justice was still common on the frontier as the nation steadily pushed westward.

The Civil War took place in the first half of the 1860s and had catastrophic effects on the nation.[7] The decades both before and after the war included heightened crime and disorder. Vigilante violence was common in the South and elsewhere. Gangs developed in northeastern and other cities. Bank and train robberies became legendary in the west. Public corruption was increasingly common and publicized. The situation had not improved drastically by the turn of the twentieth century. The Great Depression (which began in 1929 and did not really end until World War II), along with prohibition (1920–1933), only worsened matters. Throughout this era, outlaws such as Jesse James and Billy the Kid, and then later John Dillinger, Bonnie and Clyde, and Al Capone, became American folk icons.

Professionalism of Criminal Justice

Amid all the problems just highlighted, the Hoover administration established the Wickersham Commission (officially known as the National Commission on Law Observance and Enforcement) in 1929. This was an early national criminal justice study group charged to examine the causes of crime and disorder (especially surrounding Prohibition) and make recommendations for control. This commission, though short-lived, is illustrative of early efforts to professionalize criminal justice. Over the next three decades, the trend toward **professionalism** in criminal justice continued. The movement was spearheaded by a host of agencies and organizations such as the Federal Bureau of Investigation (FBI), the International Association of Chiefs of Police, the American Bar Association, and the National Prison Association (which became the American Correctional Association in 1954).[8]

A defining point in the movement to professionalize criminal justice came during the widespread civil unrest of the 1960s that surrounded the civil rights movement and the war in Southeast Asia. In 1965, President Lyndon Johnson's administration created the **Commission on Law Enforcement and Administration of Justice** to study and improve criminal justice policy. The commission released its report in 1967, which encouraged that criminal justice be treated as a system and as a profession.[9] In 1968, the **Law Enforcement Assistance Administration** (LEAA) was established (as part of the Omnibus Crime Control and Safe Streets Act). Over the next 14 years, LEAA put considerable federal funding toward crime control programs as well as research, education, and training initiatives at the state and local levels of government. LEAA, in combination with the expansion of vocationally oriented community college programs across the nation, provided the initial thrust for the later development of criminal justice as a field of academic study (discussed previously).[10]

The professionalism movement in criminal justice has gained momentum since the 1960s. The movement has involved efforts to accumulate a body of knowledge about effective criminal justice policies and practices. It has also involved efforts to staff police, court, and correctional agencies with people trained and educated in such knowledge so that agencies can operate according to standards reflecting the knowledge base.

Growth in the Criminal Justice System

The most recent data available indicate that during fiscal year 2004, the financial cost of operating criminal justice agencies at all levels of government was $193 billion, representing a 4 percent increase over 2003.[11] **Figure 1-2** shows

Controversies

As criminal justice moves toward greater professionalism, controversy is likely to continue over the need for people to have formal college credentials to staff, and especially to administer, police and correctional agencies. Some suggest that because professions (such as accounting, medicine, and law) are founded on an established body of knowledge and skills, it is essential for people to complete relevant postsecondary degree programs prior to entering the profession. However, others suggest that in-service training and experience are really all that is necessary for police and corrections personnel to acquire the skills to perform their jobs. Additionally, the argument is often made that many jurisdictions do not pay police and correctional staff enough to insist on them having college degrees, especially for entry-level positions. As a reflection of this controversy, there is considerable variation across agencies and jurisdictions in required educational credentials.

increases in criminal justice expenditures by function (policing, courts, and corrections). As shown, spending increased dramatically between 1982 and 2004 for every function.

In March 2003, there were 2.3 million law enforcement, court, and correctional employees.[12] Of these employees, the vast majority worked in law enforcement, and most of these staff members were on the payroll of local government agencies. Although courts employed fewer staff than police or corrections, they still

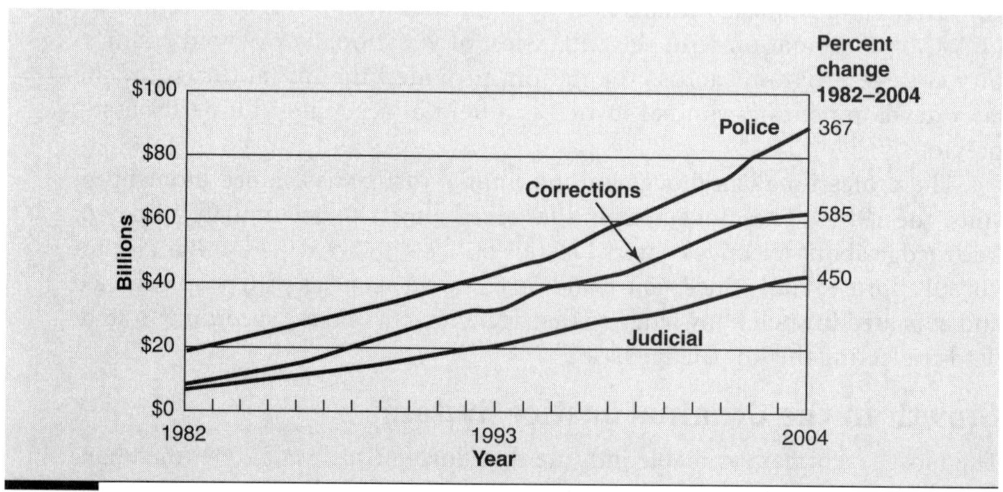

Figure 1–2 Expenditures by criminal justice function, 1982–2004.
Source: *Reproduced from U.S. Department of Justice, Bureau of Justice Statistics, www.ojp.usdoj.gov/bis/eande.htm.*

accounted for almost a half million employees. Of this total, about 10,000 judges sit in more than 3,000 courts and interpret the laws of the federal government, 50 states, and local jurisdictions.

The demands on staff in criminal justice systems are high. Each year, there may be as many as 200 million 9-1-1 calls.[13] The FBI reported that there were some 14.1 million arrests in 2005; most of these arrestees passed through a juvenile detention center or jail.[14] Although many cases were dismissed or did not proceed (either because of lack of evidence or some other factor), millions of the remaining defendants had their day in court.

Given the large scope of the criminal justice system in the United States, it can be difficult to understand its operations and processes. The complexity is compounded by the variation in structure and practice that can be found across jurisdictions; there are 3,139 counties in the United States, and each has a distinct criminal justice system. The next section of this chapter describes the basic structure and processes that are common to most jurisdictions.

The Contemporary Criminal Justice System

Despite variation that exists across the country, there are three common, interacting, and interdependent components in any **criminal justice system**: police, courts, and corrections. Each component plays a unique role that supports society's need for justice and safety, and each operates by exercising **discretion** within the parameters set by law. (The concept of discretion is discussed more thoroughly in Chapter 3.) Although introduced only briefly here, the remaining chapters of this book provide comprehensive coverage of each component.

The Police

There are a variety of law enforcement agencies at the federal, state, and local levels. Each is responsible for upholding the law in its respective jurisdiction (see Chapter 4). These policing agencies enforce the law, try to prevent crimes from occurring, investigate reported crimes, and provide social services in an effort to ensure domestic peace and tranquility. The police are usually the first to respond to a crime, either after observing the act being committed or in response to a report from a victim or other citizen. As such, the police are often referred to as the "gatekeepers" of the criminal justice system. They use their discretion to decide what action should be taken in each case. In less serious cases, a suspect may be warned and released. In more serious cases, a suspect may be arrested and booked into the local jail. Police officers' jobs do not end with arrest. They are responsible for talking to victims and other witnesses and for gathering evidence that can be used in court proceedings. They work closely with the prosecutor to prepare a case for trial.

The Courts

The United States has a dual system of state and federal courts that usually operate independently of one another; however, there is overlap at times (see Chapter 5). State court systems generally include courts of limited jurisdiction that hear misdemeanor and other less serious cases, and courts of general jurisdiction that hear more serious misdemeanor and felony cases. It is these courts, and U.S. District Courts in the federal system, that are most commonly associated with popular conceptions of the criminal justice system. Appellate courts at both the state and federal levels do not hear original cases; instead, they review cases and render decisions regarding outcomes from lower courts, with a focus on the defendant's constitutional rights to due process.

The judge plays a prominent role in court. It is the role of judge, as a "neutral" party, to interpret the law, determine matters of fact in criminal cases, preside over courtroom hearings and trials, and impose sentences on those determined to be guilty. Under the adversarial system found in our state and federal courts, the prosecutor and defense counsel present opposing cases to the court. The prosecutor's role is to prepare and present the government's case that proves the guilt of criminal defendants *beyond a reasonable doubt* so that a conviction results. The defense counsel's role is to ensure that the defendant's constitutional rights are protected and assist the defendant in building a case that will ensure that the defendant is not wrongly convicted for a crime he or she did not commit. Despite the adversarial court system in which they operate, defense attorneys and prosecutors generally become part of a cooperative courtroom work group that develops norms about how to handle cases cooperatively and efficiently.

It should be noted that in the United States we have separate courts for juvenile offenders (see Chapter 14). Although the courtroom work group and processes are similar, the juvenile court is less formal and adversarial than adult courts. Additionally, the overriding purpose of the juvenile court is "to serve the best interests of the child," whereas adult courts are more neutral toward the parties involved.

Corrections

Corrections is composed of several key agencies, including probation, jails, prisons, and parole, each of which is designed to carry out the punishments that the court has imposed on convicted defendants. Probation is a front-end community-based sanction; that is, a judge has determined that rather than being sentenced to jail or prison, the offender deserves a chance to be placed on probation in the community where he or she is subject to certain conditions of release, including supervision by a probation officer and participation in work, school, and treatment (see Chapter 8).

Jails and prisons, although both institutional sanctions for convicted offenders, are very different institutions. Jails are usually operated at the local level and serve a multitude of purposes (see Chapter 7). They house defendants in criminal cases who are awaiting trial as well as sentenced offenders awaiting transportation to prison. Jails also house offenders sentenced for serious traffic violations and misdemeanor offenses, and because of prison crowding in most states, they house felony offenders who have been sentenced to one year or less in prison.

Most prisons operate at the federal or state level, but there are many private prisons operating across the nation that provide correctional services to governments on a contractual basis. Whether public or private, prisons are responsible for the care and custody of felony offenders sentenced by the court for a period of incarceration that exceeds one year (see Chapters 10–12). Although a primary purpose of prisons is the incapacitation of offenders, they also provide mental health, recreation, and educational services designed to help rehabilitate offenders. With the *war on drugs* and *tough on crime* movements, prison populations soared, and many prisons are operating beyond their capacity.

Parole is a back-end sanction designed to protect the public and ease offenders' transition back into the community (see Chapter 13). Selected offenders are released early based on good behavior in prison and the perceived likelihood of their reoffending in the community. As with probation, offenders on parole are subject to supervision by a parole officer and are often required to work, attend school, and/or participate in some type of counseling or other social services. Offenders who violate the conditions of their parole can be returned to prison for the remainder of their sentence.

Although juvenile services are provided through different entities, the correctional subcomponents are very similar (see Chapter 14). Different terminology is used to reflect the more rehabilitative approach taken with juveniles. For instance, jails are called "detention centers," prisons are referred to as "youth development centers" (or some variation thereof), and parole is typically called "aftercare."

The Components as a System

Although presented earlier as separate entities, the police, courts, and corrections comprise a *loose system* of criminal justice with each component interacting with and being influenced by the others. As the gatekeeper of the system, the police determine, to a large degree, the number and types of cases to be prosecuted. Their investigative work directly affects a prosecutor's ability to secure a conviction in a case. The prosecutor's decisions about prosecution and plea bargaining have a direct impact on the court's workload and the number of cases that go to trial. A prosecutor who is less willing to negotiate a plea for a lesser charge or more lenient sentencing recommendation will contribute to higher rates of trials.

The courts' decisions determine the number and type of offenders to be placed in various correctional agencies. Judges and lawmakers who have adopted tough on crime strategies are largely responsible for the prison crowding experienced throughout the United States. In turn, crowded prisons have contributed to the early release of offenders into the community, where in a type of feedback cycle, many of them are once again a risk to the community and a target for police surveillance and intervention.

Caveats About the System Analogy

The components of the criminal justice system have some overlapping functions, must regularly interact, and are influenced by one another. To some extent, criminal justice personnel must work cooperatively to achieve their agency goals and the overriding goals of justice and safety. At the same time, it is misleading to think of criminal justice as a *single, tightly coordinated system* in which components operate and communicate smoothly and in unison toward achieving well-defined and agreed-upon objectives. As should be apparent, there is no single criminal justice system in the nation; instead, there are multiple systems. In addition, there are frequent breakdowns in communication, mutual suspicion and competition among components, as well as diverse and even conflicting interests and objectives. These things can promote inefficiency and a lack of coordination both within and between agencies.

This is why John Hagan describes criminal justice as a **loosely coupled system** in which components, though influenced by one another, still maintain some independence. The components "are only loosely linked to one another and to activities, rules are often violated, decisions often go unimplemented, or if implemented have uncertain consequences, and techniques are often subverted or rendered so vague as to provide little coordination."[15]

The system analogy is useful for highlighting the overlap of functions and the interaction and interdependence that characterize criminal justice. But like any analogy, this one should not be taken too far. Criminal justice is not a system in the same sense as an automotive, computer, or home heating/cooling system. The criminal justice processes described later provide another way of understanding the nature of criminal justice.

Criminal Justice Processes

The **criminal justice process** refers to a series of stages or phases that a criminal case goes through, during which justice system agents make various discretionary decisions and perform various actions within the boundaries of law.[16] The process can be complex and varied. **Figure 1–3** depicts a flowchart that encompasses the major processes through which a criminal case proceeds. Each process described here will be more fully discussed in later chapters.

Arrest—System Entry

The first step in the process is the criminal act. However, the criminal justice process is only triggered when such an act is reported to or observed by the po-

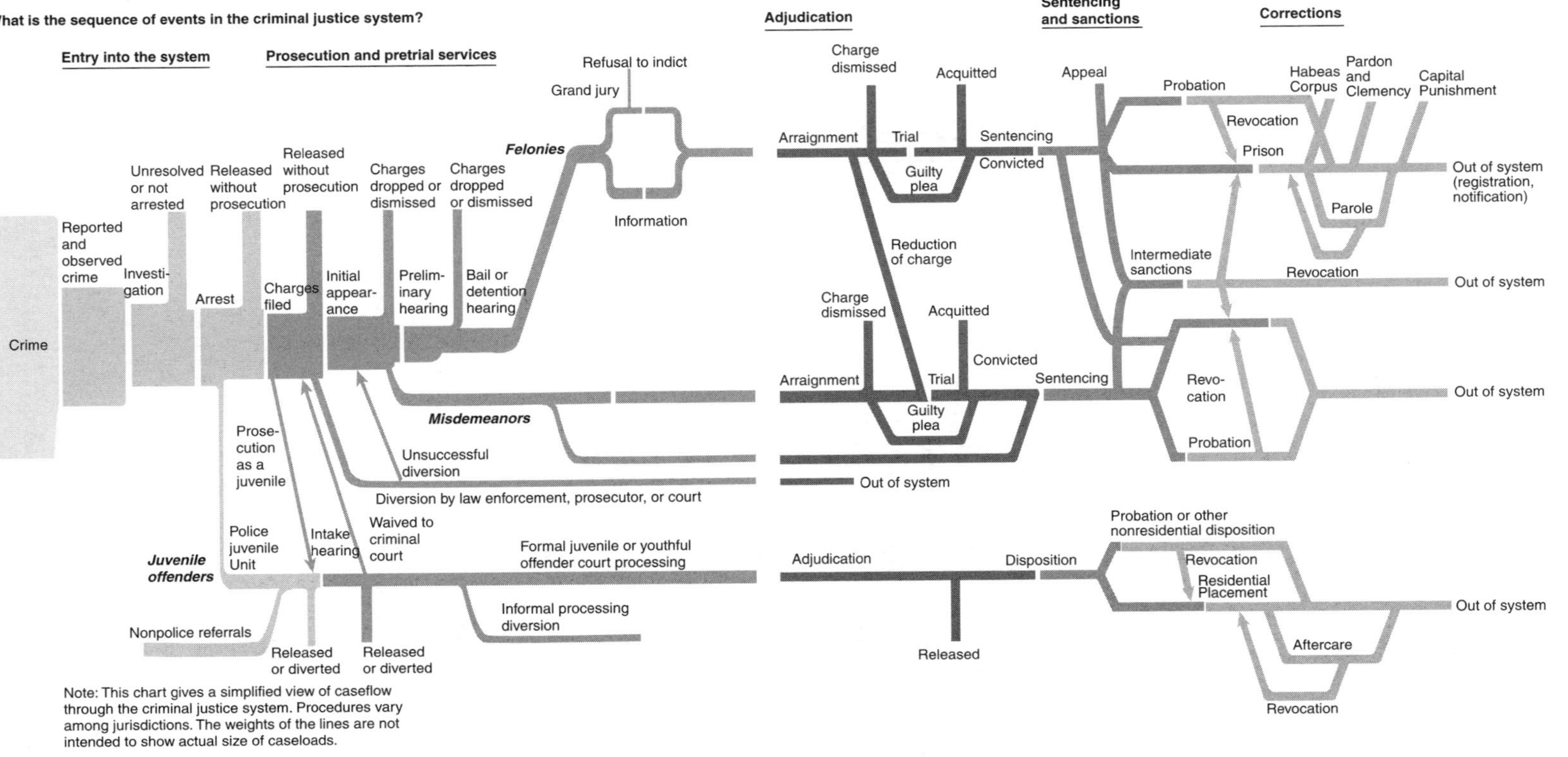

Figure 1–3 The Criminal Justice Process.
Source: *Reproduced from U.S. Bureau of Justice Statistics, www.ojp.gov/bis/flowchart.htm.*

lice, and many crimes go undetected or unsolved. After a crime becomes known to police, an investigation ensues. If, through the investigation, the police have **probable cause** to believe that a crime in fact occurred and a certain suspect committed it, they can initiate an arrest. At this stage, police use their discretion to determine whether to arrest or warn and release. This decision is based on many factors, including the seriousness of the offense, the suspect's past criminal history, the suspect's demeanor during the investigation, the victim's wishes, the police agency's priorities, and the police officers' personal philosophies.[17] If the suspect is arrested, he or she is booked and may be placed in the local jail or detention center. Charges are filed with the prosecutor if the police believe the evidence so warrants.

Pretrial Stage

Two primary processes occur during the pretrial stage. The first pertains to the prosecutor's role in determining how to proceed with the case. The prosecutor reviews the police report and investigation, conducts further investigation if needed, and determines if there is sufficient evidence to pursue prosecution. If not, the charges are dismissed. If there is sufficient evidence, formal charges are filed, and the adjudication process is set in motion.

The second important process involves a pretrial investigation and hearing for the purposes of setting bail. **Bail**, which is fully discussed in Chapter 5, is a sum of money or property that is given to the court in exchange for the defendant's release from jail to ensure his or her appearance in court. The amount of bail is set by the court based upon several factors that indicate the defendant's likelihood of appearing in court (e.g., residency, employment status, family ties). Some defendants who are deemed a good risk to appear are released on their signature, without having to post bail. Other defendants, who are charged with a serious crime and believed to be a threat to the public safety, are detained until trial.

Adjudication

The **adjudication** stage is the fact-finding stage during which several court hearings are conducted for the purpose of determining an appropriate method of case resolution. During an **arraignment**, the formal charges are read by the prosecutor, and the defendant is asked to enter a plea. In serious cases, the defendant typically enters a "not guilty" plea and a date is set for trial. During the arraignment, the presiding judge may hear arguments from the prosecutor and defense counsel regarding unresolved issues of bail and pretrial release.

During the adjudication stage, a pretrial hearing is conducted whereby the judge, defense counsel, and prosecuting attorney meet to discuss the case and determine the best way to proceed. The judge serves as a neutral third party who is present to hear evidence and legal arguments, interpret the law, and determine the rights and obligations of both parties. Outcomes of the pretrial hearing vary greatly, but dismissals of charges and plea agreements are common. Plea agreements generally entail the defendant agreeing to plead guilty in exchange for a reduction in charges or a recommendation for a lenient sentence. A small proportion of cases cannot be resolved through plea negotiations and are set for trial.

During a trial, prosecutors present evidence to prove beyond a reasonable doubt that the defendant is the person who committed the crime in question. After the prosecuting attorney has presented the government's case, the defense counsel is given the opportunity to present evidence supportive of their client's case. During the trial, the judge serves as a neutral third party whose purpose is to interpret and apply the law in a manner that protects the rights of both parties. In the case of a bench trial, the judge makes a determination regarding the defendant's guilt or innocence. In the case of the jury trial, the judge instructs the jury as to what should be considered during their deliberations, and the jury determines guilt or innocence.

Sentencing

A plea of guilty (or a finding of guilt in the event of a trial) concludes the adjudication stage of the criminal justice process and sets the stage for sentencing. In most felony cases, the judge will refer cases to the probation department for a presentence investigation and report to assist them in determining the most appropriate sanction for the offender. The judge considers the seriousness of the offense and the offender's past criminal history and social background when making sentencing decisions. Sanctioning options available often include fines, probation, some type of intermediate sanction (e.g., electronic monitoring), or incarceration.

Corrections

It is the responsibility of the corrections agencies to carry out the punishment imposed by the court. As previously indicated, correction's subcomponents include probation, jails, prisons, and parole. Each of these sentencing options involves measures that result in varying degrees of incapacitation and rehabilitation. On probation, offenders are incapacitated (i.e., their freedom of movement is restricted) to a small degree through certain conditions and ongoing supervision by their probation officer. Most offenders on probation are required to participate in some type of rehabilitative service (e.g., school, work, substance abuse treatment) designed to lower their risk of reoffending. Offenders sentenced to jail or prison are confined for varying lengths of time. During that time, they may choose to participate in some type of rehabilitative service provided in the institution. Release on parole is frequently determined by the offender's behavior in the institution and his or her efforts at self-improvement. After offenders have completed the terms of their sentence, their cases are terminated.

The Criminal Justice Process as a Funnel

When trying to understand how cases that enter the criminal justice process flow through the process toward prison as an ultimate stopping point, it is helpful to use the analogy of a funnel. Envision a large funnel containing many holes where cases can exit the track anywhere before getting to prison at the bottom. Relatively few cases that enter the process (and only a tiny minority of all cases of crime committed) make it to prison (see **Figure 1–4**).

Many crimes never result in arrest; nationally, 45.5 percent of violent crimes and 16.3 percent of property crimes known to the police are cleared by arrest.[18] Either suspects are not located, or the police decide not to make an arrest. Even if a suspect is arrested, the case will eventually be scrutinized by a prosecuting attorney who will decide whether to dismiss the case or proceed to court. Court statistics reveal that of the 56,146 felony cases filed in the state courts of the nation's 75 largest counties during May 2002, 48,847 (87%) were adjudicated. These decisions are shaped by the strength of the case and by priorities and interests of the federal or district attorneys. In most jurisdictions, district attorneys are elected, and the reality of being an elected official contributes to a decision to bring a case to court. Only half of the adjudicated cases resulted in a felony conviction (the remaining charges were either reduced or dismissed), and even fewer (10,580 or 19% of the 56,146 felony cases filed) resulted in prison sentences.[19]

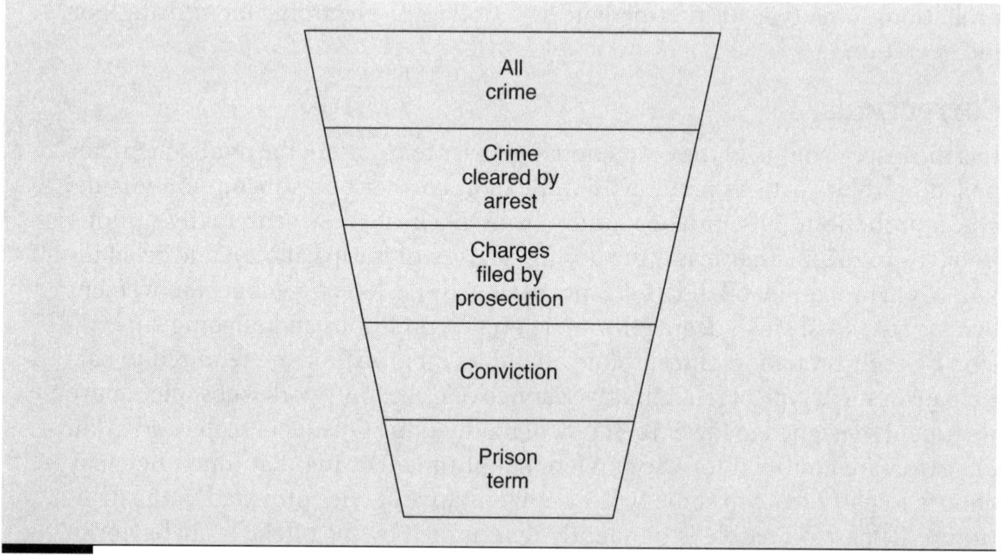

Figure 1–4 The criminal justice funnel.

Race and Gender in Corrections

The president of the National Association of Criminal Defense Lawyers criticized Michael Nifong, the district attorney in the Duke lacrosse case, stating that the criminal justice system was used for "personal political gain."[20] Despite DNA evidence that would cast doubt on the guilt of the suspects in the Duke lacrosse case, Michael Nifong pursued the case against three Duke lacrosse players to show to Durham's African American community that he was tough on crimes of the privileged. Nifong eventually resigned as district attorney and was disbarred (lost his license to practice law).

Models and Images of Crime and Justice

The Media, Crime, and Justice

Gamson and colleagues observe that "we walk around with media-generated images of the world, using them to construct meaning about political and social issues."[21] Much of our understanding about crime, criminals, and the operations of criminal justice systems comes from media images, most of which are received from viewing television (although many people are now using the Internet as their primary source of news). There are a number of advantages to this easy access to information. It can be retrieved or viewed quickly and inexpensively, and the visual images increase the public's interest and aid in their understanding of an issue.

On the other hand, much of the content reported about crime and justice is highly selective, meaning that we are most likely to hear about the cases that television producers or news directors believe will shock the public or otherwise capture their interest. As a result, usually only the most outrageous cases, such as vicious assaults, rapes, robberies, or murder, are reported or featured. In fact, in a book titled *If It Bleeds, It Leads*, Matthew Kerbel shows how the goal of many television executives is to report only the most sensational stories.[22] In addition, because most Internet or television reports are fairly brief, they often convey only minimal information about an issue. As a result, we do not learn many background factors that might have contributed to the response of the justice system—just what was "caught on tape" or the end result.

The public is fascinated by crime, and this shapes the types of programs that are shown on television. A 1997 study by the Center for Media and Public Affairs found that

> *Since 1993 crime has been the most heavily covered topic on the network evening news with 7448 stories, or 1 out of every 7 stories on all topics. . . . News about murders rose even more sharply. During 1990–92 murder coverage averaged 99 stories per year. Since 1993 the coverage increased to 714 stories per year, a jump of 721% at a time when the real world homicide rate was dropping.*[23]

One of the problems of watching television programming that features reports on crime so prominently is that heavy viewers of local news are likely to be more fearful of crime.[24] Thus, watching news accounts of crime not only distorts our thinking about crime and justice but can also make us more afraid of being victimized.

Some cases draw an inordinate amount of media attention because of the characteristics of the victim (e.g., persons who are famous, or those who are especially vulnerable—such as infants or the elderly), the nature of the crime (e.g., a multiple murder or an act of unusual cruelty), or the offender (who may be famous). Recent examples include the Lacey Peterson murder, the O. J. Simpson trial, or the incarceration of Martha Stewart and Paris Hilton. In his "wedding cake" model of criminal justice, Samuel Walker calls these **celebrated cases** and locates them as the small top layer of the cake. Although they represent only a tiny fraction of all criminal cases, the amount of media attention, combined with the fact that the system sometimes handles such cases in an atypical manner, can distort the public's ideas about the nature of crime and how the justice system operates.[25]

For instance, the O. J. Simpson trial lasted more than eight months, from January 24 to October 3, 1995, and involved a "dream team" of defense attorneys, numerous expert witnesses, and consultants. Yet only a small percentage of criminal cases end in a trial (less than 5% in most jurisdictions) and few trials last longer than a day or two, although a murder trial may take slightly longer. In addition, Simpson also had access to a team of skilled attorneys. Most defendants, however, are indigent (meaning that they have no resources), and they rely upon public defenders to represent them in court. Despite the fact that many of these attorneys are competent and hard-working, their caseloads are large, and they do not always have access to resources to independently test (and verify) the prosecutor's evidence or retain consultants or other expert witnesses, making it difficult to mount an effective defense.[26]

In addition to television news, fictional television programs also influence the public's notions of crime and justice. An increased number of television programs about forensic science, such as the three *CSI* series, *Bones*, and *NCIS* all depict the use of science to capture criminals. One problem is that offenders do not often leave samples of usable biological evidence at crime scenes. A second challenge for investigators occurs when the techniques and technology portrayed in these television programs do not exist, and they have to explain to grieving families the differences between television and what actually exists (or is feasible). Moreover, investigators rarely have the ability to resolve a case in a day or two and may wait months before reports on forensic evidence are available.[27] Last, it has been speculated that the expectations of jurors are influenced by their experiences watching crime programs that focus on forensic science (e.g., they may have unrealistic expectations about evidence that might favor the defendant or the prosecutor), and some scholars have labeled this the "CSI effect."[28]

According to Walker's wedding cake model, by far the most numerous and typical cases the system deals with are misdemeanors and less-serious felonies (Chapter 3 distinguishes misdemeanors from felonies).[29] Misdemeanors are the largest base layer of the cake, and less-serious felonies (e.g., theft) are the large layer immediately above. These cases are typically dealt with in routine, assembly-line fashion by criminal justice agents who work together to get the cases processed quickly and efficiently. Many cases are dismissed, diverted from the system, and plea bargained (as opposed to being tried). These cases usually receive little if any mention in the media. In short, if a show like *Entertainment Tonight* is symbolic of celebrated cases (publicized *ad nauseam* and handled meticulously by the system), a fitting cultural symbol for the lower-level cases is Larry the Cable Guy's "Git-R-Done" approach.

Packer's Due Process and Crime Control Models

There are a number of ways of understanding the operations of the police, courts, and corrections. In 1968, Herbert Packer published a book titled *The Limits of the Criminal Sanction*, and this work is still considered to be one of the best ways of conceptualizing the operations of the justice system. Packer outlines two competing viewpoints, the **crime control model** and **due process model**, that can be used to analyze criminal justice system philosophies, operations, and outcomes.[30] First, the crime control approach places a priority on the crime-fighting operations of the police and prosecutors. Individuals who support this approach believe that ensuring public safety through the suppression of crime should be the primary goal of the criminal justice system. Supporters of this position believe that justice should be applied in a quick and efficient manner—what some scholars have termed **assembly-line justice** to describe the rapid and routine handling of cases.[31] Efforts must be made to ensure that technicalities and loopholes do not restrict the authority of the police and prosecutors to fight crime.

Supporters of the due process model, by contrast, believe that upholding the rights of the individual arrestee or defendant is the most important consideration when somebody comes into contact with the justice system. In opposition to the crime control approach, advocates of the due process model believe that the search for the truth is more important than speed and efficiency; criminal justice should be more like an obstacle course than an assembly line. Moreover, supporters of this position believe in fairness and that defendants should receive procedural protections to ensure that the state does not trample the rights of the innocent. A popular saying for supporters of this model is, "It is better for 10 guilty men to go free rather than to punish an innocent man."

Most readers will identify more with one model than with the other, as do most politicians and policymakers. Many of the laws politicians have enacted to respond to crime fall in with one or the other model, but in recent times, such laws most typically support the crime control model. For instance, sentencing laws such as "three strikes" that call for lengthy prison terms for repeat felony offenders (a topic we explore in Chapter 6) are consistent with the crime control approach. Supporters of the due process model, by contrast, support Constitutional protections that provide citizens with freedom from unwarranted searches or the right not to incriminate ourselves (e.g., be silent) when confronted by a law enforcement officer. Although these supporters would include some politicians, it is probably more likely that criminal defense attorneys, social critics, philosophers, and legal scholars would support the due process approach.

Samuel Walker has relabeled Packer's approaches as the conservative and liberal crime control models, with the conservatives supporting the crime control approach and liberals advocating for due process protections.[32] Although both Packer and Walker use these labels, it is important to acknowledge that few people will identify with the extreme ends of the continuum. Most people fall somewhere in the middle, meaning that most of the public want to see society being protected and criminals being held accountable, while also treating offenders fairly. The hazard of rushing to arrest or judgment, for instance, is that some persons will be wrongfully accused and sentenced. By midyear 2007, more than 200 persons had been freed from prison because of wrongful convictions, usually proven innocent as a result of DNA evidence. Most experts believe that this number of offenders represents only the "tip of the iceberg." Ronald Huff and colleagues used a survey of legal officials to estimate that approximately 0.5 percent of all Americans convicted each year are thought to be innocent.[33] Given that there are an approximately 1.1 million persons sentenced on felonies each

Corrections and Policy

The PATRIOT Act was passed rapidly by Congress in the aftermath of September 11, 2001, and is discussed in more detail in Chapter 3. Many people consider this legislation a good illustration of the demands of the crime control model taking precedence over those of the due process model.

year,[34] if Huff and colleagues are correct, there would be more than 55,000 persons mistakenly or wrongfully convicted of felonies alone each year.

Wrongful convictions pose a significant problem to advocates of either position. Supporters of the crime control model, for instance, are worried that if we wrongfully convict an individual, the guilty person will remain in the community and continue to commit crimes. Advocates for the due process model, by contrast, argue that wrongful convictions erode the public's confidence in the justice system. Tom Tyler, for instance, argues that when the justice system is seen as just and fair, citizens are more likely to follow the law.[35] Yet, these problems are not unique to the United States. Justice systems throughout the world have confronted the problem of wrongfully convicting and punishing innocent persons as well as other tensions between the crime control and due process models.

In many respects, then, the tension between Packer's two models captures the fundamental dilemma of American jurisprudence (see **Figure 1–5**). How can the need to protect citizens from criminal victimization and hold offenders accountable for their actions be balanced against the need to provide the rights, freedoms, and fairness promised by a democratic form of government? In short, how does one balance justice and safety? This dilemma is at the basis of the issues covered in the following pages.

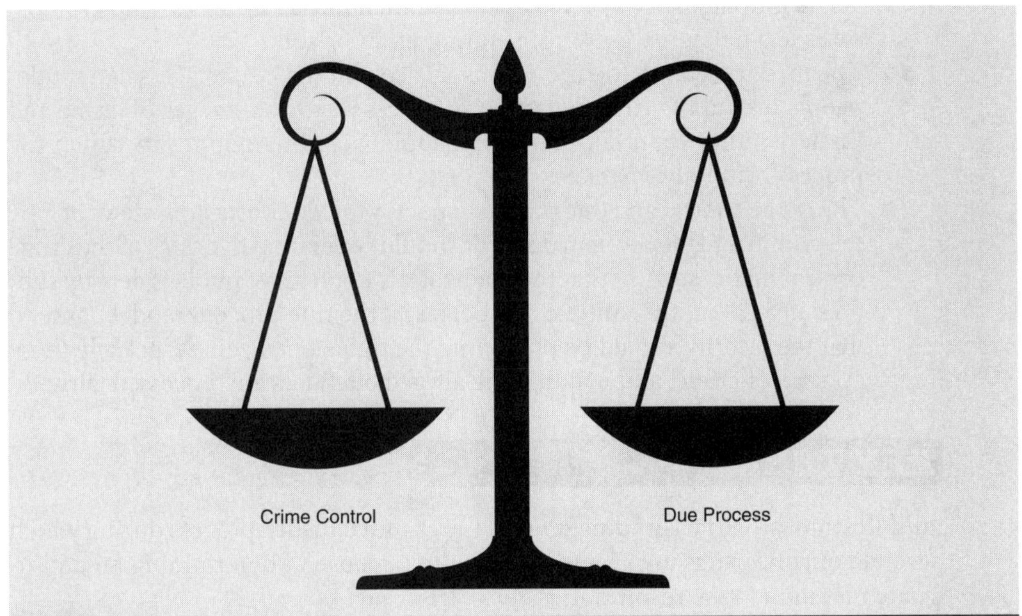

Figure 1–5 The fundamental dilemma of American jurisprudence.
Source: *© Travis Klein/ShutterStock, Inc.*

READY FOR REVIEW

- Criminal justice is a social institution of government meant to promote justice and safety.
- Criminal justice operations are performed at the federal, state, and local levels of government and involve the legislative, judicial, and executive branches of government.
- Criminal justice is an academic field of study that has expanded rapidly in recent decades.
- Over many centuries, justice and safety have evolved from being handled by private citizens, as well as by church and governing authorities, to matters involving a high degree of government control and professionalism.
- In recent decades, criminal justice systems across the nation have experienced much growth in personnel and resource expenditures.
- Three common and interdependent components of any criminal justice system include the police (who enforce the law), courts (which adjudicate legal issues), and corrections (which carries out sentences imposed by the courts). However, criminal justice is a "system" in only a loose sense of the term; it is not always well coordinated and efficient.
- The criminal justice process is a series of stages during which agencies perform various discretionary actions. The major phases include arrest, pretrial processing, adjudication, sentencing, and carrying out of the sentence.
- The criminal justice process operates much like a funnel with holes. Most cases leave the process before ending up in prison.
- The media tends to focus on unusual "celebrated" cases. This can render people fearful of crime and often produces distorted images of crime and justice, rather than familiarizing people with the manner in which the process routinely operates.
- The crime control and due process models provide contrasting views of how the criminal justice system does (or should) operate. Advocates of the crime control model suggest that the main task is to preserve public safety by suppressing crime. By contrast, advocates of the due process model contend that the priority should be preserving the rights of citizens, especially those accused of crime, and making sure alleged offenders are processed fairly.

KEY TERMS

adjudication The fact-finding stage of the criminal justice process during which several court hearings are conducted for the purpose of determining an appropriate method of case resolution

arraignment Hearing during which the defendant is formally charged and asked to enter a plea

assembly-line justice The rapid processing of cases (especially misdemeanors) in court

bail A sum of money or property that is given to the court in exchange for the defendant's release from jail

celebrated cases Criminal cases, such as the 45-day Paris Hilton jail incarceration, that draw an inordinate amount of attention from the media, the public, and criminal justice officials either because of the people involved or the unusual nature of the crime

Commission on Law Enforcement and Administration of Justice A national body created by the Johnson administration in 1965 to study and improve criminal justice policy

crime control model The philosophy that the primary goal of the justice system should be the suppression of crime by the police and prosecutors

criminal justice process A series of stages that a criminal case goes through during which justice agencies perform various discretionary actions

criminal justice system Interdependent collection of agencies consisting of police, court, and correctional agencies

discretion Authority to make decisions within certain legal bounds

due process model The philosophy that upholding the rights of the individual arrestee or defendant should be the most important consideration when somebody comes into contact with the justice system

Enlightenment A cultural and intellectual movement spanning the latter 1600s through the 1700s that emphasized human reason, science, freedom, and democratic government as an alternative to religion as a worldview

executive branch Branch of government out of which most law enforcement and correctional agencies operate

judicial branch Branch of government composed of courts responsible for determining issues of fact, sentencing those found guilty, and hearing appeals on issues of fairness and constitutionality; in some jurisdictions, this branch also performs certain correctional functions

juvenile justice The police, court, and correctional agencies and processes established to address crime committed by people under the legal age of adulthood in a jurisdiction

Law Enforcement Assistance Administration Established in 1968 to provide funding for crime control programs as well as research, education, and training initiatives at the state and local levels

legislative branch Branch of government that passes laws defining crimes, punishments, and certain procedures; also approves budgets and appropriates funding for criminal justice agencies and holds hearings to investigate allegations of wrongdoing

loosely coupled system A system in which components, though influenced by one another, still maintain some independence from one another

multidisciplinary Field that draws together insights from various disciplines in an effort to understand the object of study

probable cause The set of circumstances and facts surrounding the case that would cause a prudent and reasonably intelligent person to believe that a particular person has committed a specific crime in question

professionalism A movement in criminal justice to accumulate a body of knowledge about effective criminal justice policies and practices as well as to staff police, court, and correctional agencies with people trained and educated in such knowledge

social institutions Stable and interdependent groupings of people, beliefs, and practices established to meet societal needs

YOU ARE THE CORRECTIONS PROFESSIONAL SUMMARY

1. Because you have only 20 minutes allotted for the presentation, what are the most important topics that you should cover to provide the audience the widest breadth of knowledge regarding corrections and juvenile justice? Begin with an introduction to the criminal justice system and its main components, noting that there are separate court and correctional agencies for juveniles. Then, briefly discuss the justice process in the order in which an alleged offender would encounter the various stages. Mention the key actors and activities that typically take place. Draw a few key contrasts with celebrated cases.
2. Most of your audience probably has at least a cursory knowledge of policing and the courts, two of the "three Cs" of the criminal justice system. Nevertheless, few probably have much accurate knowledge of corrections or juvenile justice. How will you use that existing knowledge to inform them about corrections and juvenile justice and how the various aspects of the criminal justice system are interrelated and dependent upon one another? When discussing the criminal justice system and process, emphasize the fact that each decision made by the police and the courts has implications for all subsequent components and stages. Additionally, when discussing corrections and juvenile justice topics, refer back to the decisions made earlier in the criminal justice system that so often radically affect corrections and juvenile justice.

NOTES

1. This is consistent with Garland's argument that punishment should be viewed as a social institution. See D. Garland, *Punishment and Modern Society: A Study in Social Theory* (Chicago: University of Chicago Press, 1990).
2. I. Robertson. *Sociology*, 3rd ed. (New York: Worth, 1987).
3. A useful listing of such programs by program type and state can be located by visiting www.allcriminaljusticeschools.com/find/. On the growth and development of criminal justice education, see R. Langworthy and E. Latessa, "Criminal Justice Education: A National Assessment," *Justice Professional* 4(1989): 172–188. Also: F. Morn, *Academic Politics and the History of Criminal Justice Education* (Westport, CT: Greenwood, 1995).
4. See P. B. Kraska, "Criminal Justice Theory: Toward Legitimacy and an Infrastructure," *Justice Quarterly* 23(2006): 167–185.
5. Information in this paragraph and the preceding one draws on H. A. Johnson, *History of Criminal Justice* (Cincinnati, OH: Anderson, 1988).
6. G. Newman and P. Marongiu, "Penological Reform and the Myth of Beccaria," *Criminology* 28(1990): 325–346. See also: G. Newman, *The Punishment Response* (Albany, NY: Harrow and Heston, 1985).
7. Johnson, *History of Criminal Justice.*
8. Johnson, *History of Criminal Justice.* Also see: J. P. Crank, *Imagining Justice* (Cincinnati, OH: Anderson, 2003); L. J. Siegel and J. J. Senna, *Introduction to Criminal Justice*, 11th ed. (Belmont, CA: Thompson/Wadsworth, 2008).
9. President's Commission on Law Enforcement and the Administration of Justice, *The Challenge of Crime in a Free Society* (Washington, DC: U.S. Department of Justice, 1967).
10. Good treatment of the role of community colleges can be found in Crank, *Imagining Justice.*
11. U.S. Department of Justice, Bureau of Justice Statistics, *Expenditure and Employment Statistics*, www.ojp.usdoj.gov/bjs/eande.htm (accessed June 15, 2007).
12. K. Hughes, *Justice Expenditure and Employment in the United States, 2003* (Washington, DC: Bureau of Justice Statistics, 2006).
13. National Emergency Number Association, *9-1-1-Fast Facts* www.nena.org/pages/Content.asp?CID=144&CTID=22 (accessed October 16, 2007).

14. Federal Bureau of Investigation, *Crime in the United States, 2005* (Washington, DC: FBI, 2006).
15. J. Hagan, "Why Is There So Little Criminal Justice Theory? Neglected Macro- and Micro Level Links Between Organization and Power," *Journal of Research in Crime and Delinquency* 26(1989): 119.
16. M. R. Gottfredson and D. M. Gottfredson, *Decisionmaking in Criminal Justice: Toward the Rational Exercise of Discretion* (Cambridge, MA: Ballinger, 1980).
17. F. Smallenger, *Criminal Justice Today*, 9th ed. (Upper Saddle River, NJ: Prentice Hall, 2007).
18. Federal Bureau of Investigation, *Crime in the United States, 2005* (Washington, DC: U.S. Department of Justice), www.fbi.gov/ucr/05cius/offenses/clearances/index.html (accessed June 19, 2007).
19. Bureau of Justice Statistics, *State Court Processing Statistics: Felony Defendants in Large Urban Counties, 2002* (Washington, DC: U.S. Department of Justice, 2006).
20. National Association of Criminal Defense Lawyers, "News Release: NACDL President Martin S. Pinales Blasts Durham District Attorney Michael Nifong," www.nacdl.org/public.nsf/newsreleases/2006mn025?OpenDocument (accessed June 16, 2007).
21. W. A. Gamson, D. Croteau, W. Hoynes, and T. Sasson, "Media Images and the Social Construction of Reality," *Annual Review of Sociology*, 18(1992): 373.
22. M. R. Kerbel, *If It Bleeds, It Leads: An Anatomy of Television News* (Boulder, CO: Westview Press, 2001).
23. Center for Media and Public Affairs, *In the 1990s TV News Turns to Violence and Show Biz* (Washington, DC: Center for Media and Public Affairs, 1997).
24. S. Eschholz, T. Chiricos, and M. Gertz, "Television and Fear of Crime: Program Types, Audience Traits, and the Mediating Effect of Perceived Neighborhood Racial Composition," *Social Problems* 50(2003): 395–415.
25. S. Walker, *Sense and Nonsense About Crime and Drugs*, 6th ed. (Belmont, CA: Wadsworth, 2005).
26. American Bar Association, *Gideon's Broken Promise: America's Continuing Quest for Equal Justice* (Chicago: ABA, 2004).
27. G. L. Mays and R. Ruddell. *Making Sense of Criminal Justice* (New York: Oxford University Press, 2008).
28. T. Tyler, "Viewing CSI and the Threshold of Guilt: Managing Truth and Justice in Reality and Fiction," *Yale Law Journal* 115(2006): 1050–1085.

29. Walker, *Sense and Nonsense.*
30. H. Packer, *The Limits of the Criminal Sanction* (Stanford, CA: Stanford University Press, 1968).
31. Mays and Ruddell, *Making Sense*, 22.
32. Walker, *Sense and Nonsense.*
33. C. R. Huff, A. Rattner, and E. Sagarin, *Convicted but Innocent: Wrongful Conviction and Public Policy* (Thousand Oaks, CA: Sage, 1996).
34. M. R. Durose and P. A. Langan, *Felony Sentences in State Courts, 2002* (Washington, DC: Bureau of Justice Statistics, 2004).
35. T. Tyler, *Why People Obey the Law* (Princeton, NJ: Princeton University Press, 2006).

2

The Picture of Crime and Punishment in the United States

Chapter Objectives

- Understand the various methods of collecting criminal justice and corrections data in the United States among adults.
- Understand the various methods of collecting data about involvement in crime and delinquency in the United States.
- Understand the benefits and limitations of each of those methods.
- Understand demographic predictors of both criminal offending and criminal victimization.
- Understand demographic patterns of involvement in corrections.

CASE STUDY

You are a juvenile probation officer for the state of Mississippi. Because of the quality education that you received at State University, your supervisor has asked for your assistance in responding to a grant proposal. This grant proposal requires the agency applying for funding to: (1) identify a problem in corrections in the local community that affects both juveniles and adults; and (2) demonstrate a need for funding to help alleviate that problem by convincing the funding agency that the problem existing in your community is worse than that experienced by other communities throughout the United States. Your supervisor suggests that the presence of guns among both juveniles and adults in your community is a potential project for funding.

1. Where would you obtain data to help you with your project?
2. Are there limitations of the data that you will be using? If so, what are they?
3. Will the data you obtained help you in predicting what will happen with your organization? Why or why not?

Introduction

To fully understand corrections and criminal justice in the United States, it is helpful to examine the nature and state of crime. By doing so, we can have a better understanding of the incidence, causes, and types of crime.

Figure 2–1 Every day, hundreds of arrests are made in jurisdictions throughout the United States.
Source: *© Corbis.*

Sources of Crime Data in the United States

A **crime** is "an intentional act or omission of an act that violates criminal statutory or case law and for which the state provides a punishment."[1] In the United States, many people argue that crime is society's greatest problem. Although official statistics indicate that crime rates have decreased in recent years, members of the public do not appear to find comfort in these statistics. In fact, public opinion polls indicate that, despite the decreasing or stabilizing crime rate, many citizens remain concerned and/or fearful of crime, and a small percentage believe that crime is the "most important problem facing this country today."[2]

This concern about crime, and the subsequent lack of confidence that many have in the institutions that deal with crime in society, has had a dramatic effect in the lives of many people. In fact, virtually every adult in the United States has an opinion about the prevalence, the causes, and the cure for crime. An individual's opinion and attitude toward crime are informed by several factors, including published governmental statistics, informal conversations with others, the media, and their own personal experience. Data about the crime picture in the United States come from three main sources: the Uniform Crime Reports (UCR), the National Crime Victimization Survey (NCVS), and self-report surveys.

Uniform Crime Reports

The most comprehensive and frequently cited data source about crime in the United States is the Uniform Crime Reports (UCR). The **Uniform Crime Reports** have been compiled by the Federal Bureau of Investigation (FBI) since 1930. The UCR consists of reported crime and arrests submitted annually to the FBI by local law enforcement agencies throughout the country. Although the FBI is responsible for the data collection, the organization does not have the authority to require reporting by all state and local jurisdictions. Even though participation with the UCR is voluntary, the national UCR Program encompasses 94 percent of the U.S. population.[3]

The UCR presents information about two sets of crime and categorizes these crimes into Part I and Part II offenses. There are eight **Part I crimes** covered in the UCR, each of which is defined in **Tables 2–1** and **2–2**. These offenses include both serious violent and serious property crimes. The serious violent crimes include criminal homicide (including both murder and nonnegligent manslaughter), forcible rape, aggravated assault, and robbery. The serious property crimes include burglary, larceny-theft, motor vehicle theft, and arson.

Each month, law enforcement agencies throughout the country tabulate the number of crimes known to the police (reported crime). Most crimes known to the police are reported by victims, although these reported crimes may also include crimes discovered by police in the execution of their everyday duties as well

Table 2-1 Uniform Crime Reports Part I Violent Crimes Definitions and Incidence

Crime	Definition	2005 Numbers (Rate per 100,000)	% Change from 2004 Rate
Part I Violent Crimes	Violent crimes involve force or threat of force.	1,390,695 (469.2)	−1.4
Murder and nonnegligent manslaughter	The willful (nonnegligent) killing of one human being by another.	16,692 (5.6)	+2.4
Forcible rape	The carnal knowledge of a female forcibly and against her will.	93,934 (31.7)	−2.1
Aggravated assault	An unlawful attack by one person upon another for the purpose of inflicting severe or aggravated bodily injury	862,947 (291.1)	+0.9
Robbery	The taking or attempted taking of anything of value from the care, custody, or control of a person or persons by force or threat of force or violence and/or putting the victim in fear.	417,122 (140.7)	+2.9

Source: Table adapted from *Crime in the United States, 2005*, Tables 1 and 1A.

Table 2–2 Uniform Crime Reports Part I Property Crimes Definitions and Incidence

Property Crime	Definition	2005 Numbers from (Rate per 100,000)	% Change 2004 Rate
Part I Property Crimes	The object of the theft-type offenses is the taking of money or property, but there is no force or threat of force against the victims.	10,166,159 (3,429.8)	−2.4
Burglary	The unlawful entry of a structure to commit a felony or theft.	2,154,126 (726.7)	−0.5
Larceny-theft	The unlawful taking, carrying, leading, or riding away of property from the possession or constructive possession of another. It includes crimes such as shoplifting, pocket-picking, purse snatching, thefts from motor vehicles and accessories, and bicycle thefts.	6,776,807 (2,286.3)	−3.2
Motor vehicle theft	The theft or attempted theft of a motor vehicle. This offense includes the stealing of automobiles, trucks, buses, motorcycles, motor scooters, snowmobiles, etc.	1,235,226 (416.7)	−1.1
Arson	Any willful or malicious burning or attempt to burn, with or without intent to defraud, a dwelling house, public building, motor vehicle or aircraft, personal property of another.	64,062 (26.9)	−2.7

Source: Table adapted from *Crime in the United States, 2005*, Tables 1 and 1A.

as crimes reported by witnesses. The UCR provides both the number of Part I offenses and a crime rate. The crime rate is calculated by dividing the number of crimes reported to the police by the number of people in the country. The result is often expressed as a rate of crime per 100,000 people (Number of crimes × 100,000/Population under study).[4]

The UCR also records and publishes arrest data for additional crimes, which are identified as Part II offenses. **Part II crimes** include offenses such as simple assault, prostitution, drug offenses, and illegal gambling, among others; each of the 22 offenses is listed in **Table 2–3**. Although the UCR reports all Part I offenses known to the police (reported crime), for Part II offenses only arrest data are available. Because of this, some argue that the UCR provides more accurate data for Part I than for Part II crimes. Because of the number of Part II crimes (22) and the Part II arrests every year (in 2005, there were more than 11 million arrests for Part II offenses compared to more than two million arrests for Part I offenses), others also believe that it is beyond the scope of the UCR Program to collect information about all the Part II crimes reported to the police. Further, because Part II offenses are viewed as less serious than Part I offenses (i.e., public order violations vs murder), it is possible that the resources necessary to collect data about crimes reported to the police for Part II offenses would be better spent in other areas.[5]

Data based on reported offenses generally do not include the age, gender, race, or ethnic group of the offender, whereas arrest data do. Furthermore, not all crimes reported to the police are "cleared." A crime is cleared when an arrest is made and at least one person is charged with the crime (**clearance by arrest**). Crimes are also cleared when the police know the location and identity of the suspect and have information to support the arrest and subsequent prosecution of the suspect, but are prevented from taking action against the suspect by consequences outside of their control (the suspect is dead, the victim refuses to cooperate, etc.). This is referred to as **clearance by "exceptional means**."[6]

The clearance rate varies by the type of crime. For example, in 2004, only 46.3 percent of the serious violent crimes and 16.5 percent of the serious property crimes were cleared by arrest or exceptional means. Because people are generally much more likely to report serious crimes than less serious crimes (although forcible rape is an exception to this rule), it is possible that the offenses reported in the UCR may not accurately reflect actual offenses that occurred in jurisdictions reporting to the UCR.[7]

Another criticism is that the UCR collects no data on victims of crime (with the exception of murder, where data about the victim are recorded). As such, a variety of important data are unavailable through the UCR. Additionally, both Part I and Part II crimes usually come to the attention of police when they respond to citizens' complaints. Whether or not a citizen reports a crime to the

Table 2-3 Uniform Crime Reports Part II Crimes Definitions and Incidence

Offense	Arrests
Other assaults	1,301,392
Forgery and counterfeiting	118,455
Fraud	321,521
Embezzlement	18,970
Stolen property: buying, receiving possessing	133,856
Vandalism	279,562
Weapons; carrying, possessing, etc.	193,469
Prostitution and commercialized vice	84,891
Sex offenses (except forcible rape and prostitution)	91,625
Drug abuse violations	1,846,351
Gambling	11,180
Offenses against the family and children	129,128
Driving under the influence	1,371,919
Liquor laws	597,838
Drunkenness	556,167
Disorderly conduct	678,231
Vagrancy	33,227
Suspicion	3,764
Curfew and loitering law violations	140,835
Runaways	108,954
All other offenses	3,863,785
Total Arrests for Part II Offenses[1]	**11,885,120**

Source: Adapted from *Crime in the United States*, 2005, Table 29.

[1]Does not include suspicion

police depends on a number of factors. For example, half of all violent crimes (50%) and slightly more than one in three property crimes (39%) are reported to the police. The most important determinant of whether a citizen reports a crime to the police is the seriousness of the crime. Respondents are much more likely to report crimes they consider to be serious (robbery, aggravated assault with injury) than they are to report crimes they do not (larceny-theft of small amounts, etc.). Typically, approximately three in five robberies and aggravated assaults and approximately one in three completed thefts are reported to the police.

The decision to report a crime also depends on the type of crime. Only one in three rapes were reported to police, even though rape is considered the second most serious crime. Factors other than seriousness of the crime may explain the small percentage of people who report rape and sexual assault (relationship to the offender, embarrassment, and so on). On the other hand, motor vehicle theft, which may be considered to be a less serious offense than robbery or aggravated assault with injury, was reported 85 percent of the time (so automobile owners can collect insurance). Other reasons offered by victims of crime as to why they do not report their victimizations also include the following: they feel that nothing could be done about their victimization; the police do not want to be bothered; or their victimization was not important enough to bother the police.[8]

Data reported in the UCR may also be affected by police departments that report those data. Although the UCR covers 94 percent of the population, rural police departments are much less likely to report to the UCR than are larger, more professional departments. This is particularly a problem in rural states such as Mississippi. In any given year, for example, up to half of the counties in Mississippi may not report crime data to the UCR. Additionally, in two states (Illinois and Kentucky), the vast majority of agencies do not report data to the UCR at all.[9]

Furthermore, organizational pressures to get the crime rate up or down as well as differences in state laws and local law enforcement policies affect the validity of UCR data. Political pressures may incite police departments to "crack down" on certain types of crime when the public views this type of crime as a problem. This increased attention given to that type of crime (for example, burglary) will often cause the burglary arrest rate to go up, giving the appearance that burglary is increasing in an area, when actually only burglary *arrests* are increasing. Furthermore, law enforcement agencies may also respond to pressures to lower the crime rate. By classifying the criminal action of entering a building and taking a stereo illegally as a larceny-theft, a police department can reduce the burglary rate while not actually reducing the number of buildings that are burglarized.[10]

These limitations of UCR data have motivated criminologists to attempt to develop more reliable measures of crime and victimization. The additional surveys were originated with the hope of providing a more thorough picture of crime in the United States by uncovering much of the crime not reported to the police and supplying more detailed information on situational factors (location and time of crime, etc.), as well as characteristics of offenders, that might help in determining why people commit crime. One of these alternate sources of crime data is the NCVS.

National Crime Victimization Survey

The **National Crime Victimization Survey** (NCVS)[11] is an official attempt to deal with crime not reported in the United States. The NCVS is compiled by the Bureau of the Census for the Bureau of Justice Statistics, and it measures six of the eight index crimes measured by the UCR (rape, robbery, aggravated assault, burglary, larceny-theft, and motor vehicle theft). The NCVS does not compile data on murder or arson. Data on murder are not collected because the victim is dead, and arson is excluded because of the difficulty in determining if the fire was intentionally set by the property owner or someone else.

The NCVS originated in 1972 and is conducted every six months by the Department of Justice. The NCVS consists of surveys of 60,000 households, involving about 130,000 people. The NCVS tracks households for a three-year period. Each household is surveyed seven times: once upon entering the study and then at six-month intervals over the next three years. Participants are asked to report about victimizations that have occurred in the previous six months. Even if a family moves, the Department of Justice continues to survey that household or the new residents of that household. Of the 60,000 households sampled, about 50,000 complete the interview. A portion of the sample of households changes every six months so that one-third of the households in the sample is replaced in any given year.

The NCVS assesses crime victimization through a two-step process. The first step is to ask a series of "screen questions." If the respondent answers that he or she has been victimized by one or more crimes, the interviewer continues to ask a series of questions designed to provide further detail about the victimization incident. Information collected includes: (1) time and place of victimization; (2) extent of the property damage or physical injury; (3) medical costs; (4) whether the victim engaged in self-defense; (5) whether the assailant was a stranger; (6) estimates of offenders' race, gender, and approximate age; and (7) whether the victim reported the crime to the police.

The NCVS relies on three types of respondents. A "knowledgeable adult" answers questions that pertain to the household. All household members age 12 and older are then asked about their background characteristics and personal victimizations. "Proxy" respondents are used to answer the survey for household members not competent enough to answer on their own. The NCVS reports victimizations (number of people victimized) and incidents (number of criminal acts involving one or more victims). Annually, the NCVS uncovers more victimizations than incidents.

As is the case with the UCR, there are problems with NCVS data. One of the major problems facing the NCVS is "telescoping," or remembering incidents as occurring more recently than they actually occurred (**forward telescoping**) or as occurring in the more distant past than they actually occurred (**backward telescoping**). The hope of the Bureau of Justice Statistics is that the use of six-month intervals will reduce the impact of both of these problems. Results of the NCVS are also subject to interviewer effects; that is, some interviewers are able to uncover a larger number of victimizations from respondents than other interviewers are. Another problem faced by the NCVS is related to the sampling design of the NCVS. Those respondents who have been interviewed before realize that if they answer a screen question about their criminal victimization affirmatively, they will be questioned much more extensively about the victimization incident. There is evidence to indicate that those who have been interviewed before report less victimization, although this effect seems to be relatively small.

A further problem with NCVS data involves the use of a single household respondent to answer questions about household crime. The person being interviewed may not be aware of all the victimizations that occurred to the household during the preceding six months. This may lead to an underreporting of household victimizations, particularly those that involved domestic violence and those that occurred for victims under the age of 12. Finally, there seems to be a positive relationship between education and victimization. This finding may be caused by differential respondent efficiency. In other words, more-educated respondents may be more cooperative and more at ease during the interview, and thus may recall more victimizations.

Despite the shortcomings of the NCVS, it offers two major benefits. First, it provides information about the victims of crime that is not available through the UCR. These victimization data can be used to examine which groups are most at risk of being victimized. Second, the NCVS provides an alternate measure of crime, one that is independent of changes in official responses to crime, and perhaps a more accurate measure of the true level of crime in the United States.

Comparison of the UCR and NCVS

A comparison of the UCR and the NCVS suggests that for all crimes (except motor vehicle theft) covered by both the UCR and the NCVS, more than two times as many crimes are recorded in the NCVS as appear in the UCR. Although the NCVS always uncovers two to three times as much crime as the UCR, the rank order of type of crime is the same in both. In other words, aggravated assault is the most often occurring serious violent crime in both the UCR and the NCVS, while larceny-theft is the most frequently occurring serious property crime. Excluding murder, rape is the least occurring serious violent crime in both the NCVS and the UCR, while motor vehicle theft is the least occurring serious property crime (arson excluded). It is clear that, although the two data sources reveal different amounts of crime, there is similarity in the patterns of crime that each uncovers. It is also clear, however, that no method of collecting crime data will enumerate the absolute amount of crime in the United States.[12]

Although UCR data offer a variety of information about criminal offenders, and the NCVS offers a wide variety of information about crime victims, neither provide insight into the reasons that people commit crime. Because of this, another method of data collection has emerged: the self-report survey.

Self-Report Surveys

Self-report surveys do not examine police records or ask individuals about their victimization experiences. Instead, self-report surveys ask samples of respondents about their own participation in criminal activities.

Self-report surveys became an acceptable criminological research design in the 1950s, largely as a result of the work of James Short and F. Ivan Nye. Short and Nye demonstrated convincingly that people would indeed reliably report their own criminal behavior. In fact, people not only admitted to criminal behavior, but they also admitted much more criminality than was uncovered by official records. Since the studies of Short and Nye, the self-report survey has been the primary source of data for examining the etiology of criminal behavior.[13]

Unlike the UCR and the NCVS, there is no single set of crimes examined or any single design by which self-report data are gathered. Different researchers use vastly different questions and methods to collect data via self-report. Although the type of information gathered varies from one self-report survey to the next, almost all self-report studies include basic demographic information about the respondent's age, gender, race, and family socioeconomic status (SES). Other questions pertain to factors believed to be associated with crime or delinquency such as school performance, substance abuse, religion, and family structure. The types of delinquency that respondents have been questioned about range from trivial offenses (e.g., bought liquor; been loud, rowdy, or unruly in a public place) to more serious offenses (e.g., used a club, knife, or gun to get something from someone).

Although self-report surveys offer insight about crime not found in the UCR or the NCVS, self-report data also have limitations. In the vast majority of cases, self-report data are based on small local or regional samples. Thus, data garnered from self-report surveys offer little insight into the geographic distribution of crime. Furthermore, because the focus of self-report surveys is most often on the offender, very little effort is expended to gather information about crime victims. Self-report surveys are also most often used with adolescent, in-school samples, and thus their coverage is often limited.[14]

Until Short and Nye's groundbreaking work, many researchers believed that self-report surveys would not work because people would not admit to their deviant behavior. The results from self-report surveys indicate that respondents do admit to a sizable number of delinquent and criminal offenses. Although respondents may underreport (or even overreport) their involvement in crime, a large number of respondents are not too embarrassed to admit to delinquent behavior.

Several studies suggest that answers given by respondents on self-report surveys demonstrate striking consistency, and thus can be considered to be reliable. Additionally, there are moderate to strong correlations between: (1) respondents' self-reported number of times picked up by the police; (2) their self-reported number of times referred to the courts by police; and (3) their self-reported official contacts with official measures of these three variables. Nevertheless, potential validity problems have been discovered among blacks in that the agreement between official and self-report measures is lower for blacks than it is for whites.[15]

Convergence and Divergence of Crime Data

As mentioned previously, it is virtually impossible to determine the exact amount of crime in the United States. Each method of data collection has problems that make it impossible to use that method as an absolute measure of the amount of crime in the United States. For example, UCR rates depend on the reporting of crimes to the police or the direct detection of crimes by the police. Furthermore, there are also discrepancies in the amount of crime police departments report to the UCR as a result of factors such as police discretion, reporting practices of police departments, and many others. The NCVS and self-report studies depend on factors such as willingness to report criminal offense and victimization, respondent recall, and so on. It is important to understand, however, that there are several areas of agreement among the three sources of data about crime. The following is a review of the major patterns revealed through these varied methods of crime measurement.

Patterns of Crime

Crime is not experienced evenly across the population. Some groups are disproportionately involved in and affected by crime. In fact, data from the UCR and NCVS indicate that crime is intragroup. In other words, those groups most likely to engage in criminal activity are also the groups most likely to be victimized by crime. The most significant groups in which crime is experienced differentially are gender, race, class, and age.

Gender and Crime

The demographic variable that has the strongest association with crime and criminality is gender. In virtually every society, males are much more likely to commit crime than females are. In 2005, the UCR indicated that males were more than three times as likely as females to be arrested for all crimes. For serious violent crimes, the difference was even greater (males were more than four times as likely as females to be arrested). Males were nine times more likely to be arrested for murder, robbery, and burglary, approximately five times more likely to be arrested for aggravated assault and arson, and six times more likely to be arrested for motor vehicle theft. Even for larceny-theft, the Index crime where gender differences are the smallest, males were almost twice as likely to be arrested as females. In fact, for only two crimes were the majority of those arrested female—prostitution and running away.[16]

Some researchers argue that the UCR may reflect differences in official responses to males and females rather than true differences in their levels of crime. Proponents of the **chivalry hypothesis** argue that criminal justice agents might be more "chivalrous" toward the women that they arrest than they are to men, and thus might deal with the female offender informally, possibly allowing the female to go with only a warning.[17] Nagel and Hagan, however, suggest that women who violate traditional sex-role stereotypes, specifically violent offenders, receive harsher treatment from criminal justice officials. This argument is commonly entitled the **evil woman argument**.[18] Even so, the evil woman argument, if true, is more applicable at the sentencing stage than at the stage of initial arrest.

Even though the chivalry hypothesis and the evil woman argument may have some effect on the differential crime rates of males and females as reported in official statistics, findings from the NCVS and self-report studies suggest that a gender gap truly exists. In fact, for serious violent and property crimes measured by the NCVS (which asks questions about characteristics of the offenders in the victimizations), as well as serious property and violent crimes measured in self-report surveys, the results strongly resemble those of the UCR. For example, the UCR indicates that in 2005, there were 114,616 people arrested for robbery, slightly more than 10 percent of which were female, or a ratio of about nine males arrested for every female.[19] Data from the NCVS, which asks the respondent to identify the gender of the perpetrator in their victimization incident, strongly resemble the UCR data for robbery as well as most other crimes.[20] Self-report studies also highlight gender differences in crime commission, although not quite to the same extent as official data.[21]

Just as males are more likely to commit the vast majority of crimes, males are also more likely to be victimized by crime. Males are three times more likely than females are to be murdered and are almost twice as likely as females are to

Race and Gender in Corrections

Regardless of the source of data, nonwhites and males are more likely to be involved in crime, particularly violent crime, than are their white and female counterparts. Thus, clients of corrections and juvenile justice agencies are primarily male and disproportionately nonwhite.

be victims of robbery or aggravated assault. On the other hand, females were *20 times* more likely to be victimized by rape or sexual assault than were males.[22]

Racial/Ethnic Minority Status and Crime

In virtually every society, there are differences in crime rates among certain racial and ethnic groups. Generally speaking, the most recent immigrant group gains the reputation as most criminal, although there are some notable exceptions. This phenomenon is not limited to the United States. In England, a disproportionate amount of crime is attributed to people of Irish descent, even though the Irish Republic has one of the lowest crime rates among industrialized countries in the world.[23]

Discussing racial differences in crime, however, is a much more difficult matter than discussing gender differences in crime is. It is much easier to determine the gender of an offender or victim than it is to determine his or her race. A person's race, or ethnic identity, is often based on social, cultural, linguistic, or historical affiliations that are much harder to define and often exist largely in the preferences of the person. In fact, it is difficult for social scientists to even come to agreement on the definition of the term *race*. Gordon argues that there are no pure races and that race is only differential collections of gene frequencies responsible for attributes that, so far as we know, are limited to physical displays such as hair form and skin color.[24] As such, social use of the term *race* commonly is based primarily on visible factors such as skin color. Hraba, however, argues that the term *race* has come to have such an ambiguous and inconsistent meaning that it is no longer useful in the study of ethnic groups.[25] He prefers the term *ethnic group* and argues that ethnic groups are self-conscious clusters of people who share a common origin or a separate subculture and perpetuate a contrast between themselves and others.[26]

The United States is commonly considered to have five major ethnic groups, with many smaller subgroups within these major groups. These groups include Caucasian Americans (whites), African Americans (blacks), Hispanic Americans (including people of Puerto Rican, Mexican, or Cuban descent, among others), Asian Americans (including people of Chinese, Japanese, or Korean descent, among others), and American Indians. Many sociologists dichotomize these groups into majority and minority groups. The majority group is usually considered to be all whites, whereas the minority groups include Native Americans, Asian Americans, African Americans, and Hispanic Americans, who share the common experience of being objects of majority group prejudice and discrimination.[27]

According to the 2000 census, of these groups, non-Hispanic whites make up approximately 75 percent of the United States population, while non-Hispanic blacks make up approximately 12 percent of the population. Hispanics also make up approximately 12 percent of the United States population, while Asian Americans make up 3 percent, and Native Americans make up less than 1 percent.[28]

When focusing on the association between minority groups and crime, two of the minority groups most often mentioned are blacks and Hispanics. Although the UCR does not include Hispanic as a racial category, the NCVS does. Official statistics indicate that both these groups are responsible for a disproportionate share of crime, particularly violent crime, in the United States. In 2005, the percentage of blacks arrested for all crimes was more than twice the percentage of blacks in the population. Almost half of those arrested for murder and more than half of those arrested for robbery in 2005 were black.[29] Additionally, blacks are disproportionately arrested for every other crime except driving under the influence and liquor law violations. Black males are also imprisoned at a rate of 3,145 per 100,000 residents, a rate six times higher than the national average.[30]

Critics of official statistics argue that the reason some minority groups, particularly blacks, are disproportionately represented among arrest statistics may be the discrimination and prejudice that pervade society, particularly the criminal justice system. Jeffrey Reiman argues that official statistics show a distorted image of crime in society. Official statistics, Reiman argues, are subject to biases and prejudices of the criminal justice system, and thus may not reflect an accurate picture of crime in society.[31] Critics of official statistics argue that self-report studies and victimization surveys should be used in lieu of arrest statistics to study the breakdown of criminal activity by race in society. The findings from self-report surveys seem to provide some support for this argument with whites and blacks reporting involvement in about the same number of delinquent acts. There is some evidence to suggest, however, that some of the discrepancy among the official statistics of blacks' involvement in crime and self-reported instances of blacks' involvement in crime can be attributed to the fact that blacks, and black males in particular, tend to underreport their involvement in delinquency, particularly for serious offenses.[32]

Victimization studies, on the other hand, allow victims to identify characteristics of the offender who victimized them (race, gender, approximate age, etc.). These studies further indicate that blacks are disproportionately represented among offenders at levels similar to those found in arrest statistics. Although victimization surveys do not include a global measure of all crime, findings from victimization surveys can be generalized to the most serious crimes.[33]

Victimization studies also indicate that blacks and Hispanics are not only more likely to engage in criminal activity, but they are also more likely to be victimized by violent crime than any other racial group. In 2003, there were 29.1 violent crime victimizations per 1,000 black persons, 21.5 per 1,000 whites, and 16 per 1,000 persons in other racial categories. For robbery, there were even larger discrepancies in the rates, with 5.9 robbery victimizations per 1,000 blacks, 1.9 per 1,000 whites, and 3.4 per 1,000 persons in other racial categories.[34]

Regardless of the explanation chosen for the racial differences in criminality in the United States, the fact still remains that some minority groups, particularly blacks, commit a disproportionate amount of crime, particularly violent crime. This holds true when analyzing official statistics, victimization surveys, or self-report surveys.

Social Class and Crime

The association between social class and crime has fascinated criminologists for many years. Researchers commonly argue that poverty causes crime. This assertion has come under sharp attack, and presently this debate is far from being resolved.[35]

Most **aggregate level data** indicate an **inverse relationship** between social class and delinquency; in other words, the lower the SES, the higher the rate of delinquency. For example, Braithwaite reviewed 53 studies of the class and delinquency relationship based on aggregate data analysis. He determined that 44 of the studies indicated that delinquency rates were higher among lower-class juveniles than among middle-class juveniles.[36]

Studies examining the relationship between social class and crime/delinquency in aggregate data most often use official measures of crime/delinquency and social class.[37] Many researchers argue that reliance on official data for measures of class and crime/delinquency may bias the results against members of the lower class, in that official law enforcement agents are more likely to concentrate their efforts in lower-class neighborhoods and thus uncover more crime and delinquency. Furthermore, most official data on crime do not include information about social class of the offenders. For this reason, many researchers suggest using self-report data to measure the relationship between social class and crime.[38]

Individual level data collected in self-report studies often uncover contradictory findings about the relationship between social class and crime. Self-report studies most often indicate that the relationship between class and crime is negligible at best. Tittle and his associates, in one of the first meta-analyses (a statistical technique that analyzes the existing research on a topic to identify patterns within that research) examining the relationship between class and delinquency, reviewed 35 studies that examined this relationship. Their analysis revealed only a slight inverse relationship between social class and delinquency.[39] In an effort to challenge the conclusions of that study, several other researchers attempted to reexamine the relationship between class and crime/delinquency. Thus, 12 years later, Tittle and Meier again reviewed a number of studies specifying the SES/delinquency relationship in various ways. After reviewing numerous studies using both official and self-reported data, they determined that "there is no pervasive relationship between individual SES and delinquency."[40] Braithwaite, on the other hand, also reviewed 47 self-report studies and found that 18 "reported significantly higher levels of delinquency" among lower-class adolescents; 7 studies "provided qualified support," and 22 did not find any significant differences among classes.[41] Thus, the debate about the relationship between class and delinquency is still far from being resolved.

Examining the relationship between class and crime often proves to be an arduous task. Although both the UCR and the NCVS provide data on race and gender of the offender, neither provides data on class of the offender. With the addition of self-report surveys as a method to collect data about crime, self-report surveys became the most often used method of examining the relationship between class and delinquency. Many argue that this often presents a problem, however, in that most self-report surveys use samples of adolescents. Thus, most of the research examining the association between class and crime really measures the association between class and delinquency.

The preceding discussion therefore implies that the social class and delinquency debate has not been resolved. In other words, it is impossible to say, without doubt, that an empirical relationship between social class and delinquency does or does not exist—in fact, this relationship is likely largely dependent on how crime and delinquency are defined (e.g., street crime, white-collar crime, victimless crime) and how class is measured. The relationship between victimization and social class, however, is much more conclusive. People with lower household income are much more likely to be victimized by violent crime than any other income group. Motor vehicle theft and larceny–theft do not conform to this trend; those whose household income is lowest have the lowest victimization rates for motor vehicle theft and larceny–theft. It is plausible to suggest that the reason their rates are lower for these two crimes is that those with lower household income have fewer assets to steal.[42]

Age and Crime

Official statistics indicate that young people are responsible for the largest portion of criminal activity. The violent crime rate peaks at age 18, while the property crime rate peaks at age 16. In 2004, youths between the ages of 13 and 19 years accounted for almost one in four (24.2%) arrests for Part I violent crimes, even though they accounted for one tenth (10%) of the population of the United States (author calculations using FBI and Census data). Annually, this age group accounts for approximately one in four murders and forcible rapes, one in three robberies, and approximately one in five aggravated assaults. The association between age and property crime is even more alarming. Youths age 13 to 19 are responsible for approximately one in three arrests for motor vehicle theft, burglary, and larceny-theft.[43] It is therefore apparent that much of the crime problem, and concern about this problem, can be attributed to young people.

Just as younger people are more likely to commit violent crime, they are also more likely to be victimized by it. Youths age 16 to 19 have the highest victimization rates for *all crime* and for violent crime. Those 65 years of age and older have the lowest victimization rates.[44]

Weapon Availability and Crime

Firearms and other weapons are used in many violent crimes. In 2005, in those incidents where the murder weapon was known, four in five offenders used a gun, a knife, or some other cutting instrument to commit the murder; 68 percent used a firearm. In more than half of these murders, a handgun was used. In approximately half of all robbery arrests, the perpetrator used either a knife or a firearm.[45]

It is evident from the review presented earlier that crime is a problem, particularly among certain groups (e.g., males, blacks, and adolescents). Because of the wide variety of data about crime and victimization now available from the UCR and the NCVS, we have a better understanding of patterns of crime and of factors that contribute to crime. Additional data sources from corrections and juvenile justice enhance our understanding about crime and provide additional information about common responses to crime. These data sources are described in the following section.

Sources of Data for Adult Corrections

A wide variety of sources that provide data on adult corrections is available. Many of these sources are available at no cost electronically at www.ncjrs.org and www.ojp.usdoj.gov/bjs/correct.htm. These data provide a fairly comprehensive picture of the field of corrections.

Data on Adult Prisoners

Each year, researchers with the Bureau of Justice Statistics and the U.S. Census Bureau compile a report summarizing various data about prisoners. The data are collected from each of the 50 states, territories (American Samoa, Guam, and the U.S. Virgin Islands), and commonwealths (Northern Mariana Islands and Puerto Rico), the U.S. military, and the Federal Bureau of Prisons. With the exception of data collected from Alaska, Connecticut, Delaware, Hawaii, Rhode Island, and Vermont, all states that combine their jail and prison systems, the data do not typically include persons confined in facilities operated by local authorities (e.g., jails).[46]

At the end of 2005, there were 1,525,924 prisoners incarcerated in federal and state facilities. This represented a 1.9 percent increase from 2004 (see **Figure 2–2**). The federal prison population totaled 187,618, while the number of prisoners under state custody totaled 1,338,306. The prison incarceration rate (number of individuals incarcerated × 100,000/Population under study) was 491 per 100,000 residents, a 20 percent increase from the 1995 rate of 411 per 100,000. Louisiana (797 per 100,000), Texas (691), Mississippi (660), Oklahoma (652), and Alabama (591) had the highest incarceration rates; Maine (144), Minnesota (180), Rhode Island (189), New Hampshire (192), and North Dakota (208) had the lowest incarceration rates. Seven percent of inmates were housed in privately operated facilities, and approximately 5 percent were housed in local jails. There were also 15,735 persons incarcerated in territorial prisons, 19,562 persons incarcerated in detention facilities operated by the U.S. Bureau of Immigration and Customs, and 2,322 persons incarcerated in military facilities.[47]

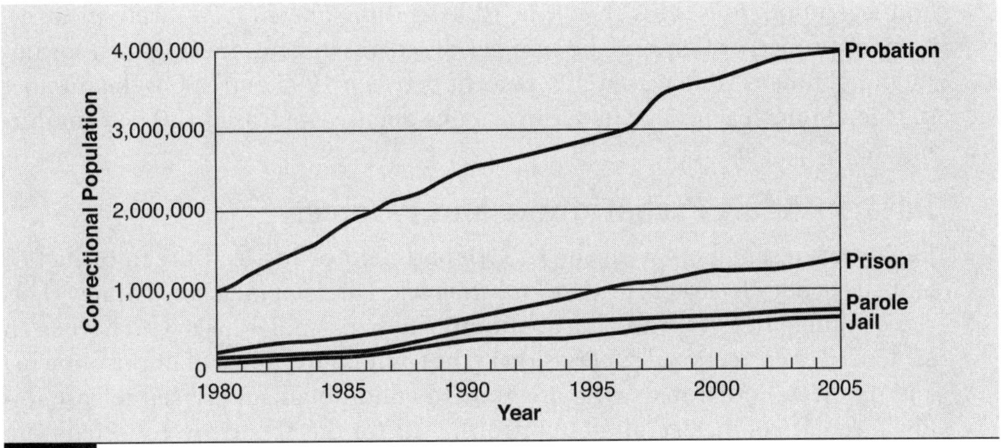

Figure 2–2 Adult correctional populations, 1980–2005.
Source: *Reproduced from Bureau of Justice Statistics Correctional Surveys (the Annual Probation Survey, National Prisoner Statistics, Survey of Jails, and the Annual Parole Survey) as presented in* Correctional Populations in the United States, Annual, Prisoners in 2005, *and* Probation and Parole in the United States, 2005.

At the end of 2005, there were 107,518 (7% of all prisoners) females incarcerated in state or federal prisons. This figure was 57 percent higher than the 68,468 females incarcerated in state or federal prisons at the end of 1995; the rate of annual increase among female prisoners (4.6%) was 53 percent higher than the rate of annual increase for male prisoners (3%) during that time period. Nevertheless, men remained more than 14 times more likely to be incarcerated than women were. One-third of the female inmates were incarcerated in Texas (13,506), the federal system (12,422), and California (11,667). The states with the highest female incarceration rates were Oklahoma (129 per 100,000), Idaho (110), Mississippi (107), Louisiana (99), and Texas (97). The states with the lowest female incarceration rate were Rhode Island (10 per 100,000), Massachusetts (12), Maine (17), and New Hampshire (20).[48]

At the end of 2005, 562,100 (45%) inmates incarcerated in state prisons were non-Hispanic black inmates, 453,400 (36%) were non-Hispanic white inmates, and 219,200 (17%) were Hispanic. At the end of 2003 (the most recent year for which data are available), half (51.8%) of state inmates were incarcerated for violent offenses, while one in five were incarcerated for drug (20.0%) and property (20.9%) offenses. Less than 1 in 10 (6.9%) were incarcerated for public order offenses. White prisoners (26.5%) were more likely to be incarcerated for property offenses than were either blacks (17.9%) or Hispanics (17.0%). Hispanics (22.9%) and blacks (23.7%) were more likely to be incarcerated for drug offenses than were whites (14.3%).[49]

More than half (55%) of all federal inmates were sentenced for drug offenses, while 1 in 10 federal inmates were sentenced for immigration, weapons, and violent offenses. Less than 1 in 10 federal prisoners (7.1%) were incarcerated for property offenses. The number of federal inmates incarcerated for immigration offenses increased 394 percent between 1995 and 2003; the number of federal inmates incarcerated for weapons offenses increased 120 percent during that same period.[50]

Data on Adult Probationers and Parolees

Each year, researchers with the U.S. Department of Justice Bureau of Justice Statistics compile a report titled *Probation and Parole in the United States*. This report summarizes various data about probationers (individuals who have been sentenced to correctional supervision in the community in lieu of imprisonment) and parolees (individuals who are granted conditional supervised release following a term of incarceration in prison). Data for this report are collected from the annual probation and parole surveys sent to correctional authorities in all 50 states and the federal system.

At the end of 2005, there were 4,162,536 individuals under probation supervision (Figure 2–2). This figure represents an increase of 0.5 percent over the 2004 probation population and an increase of 35.2 percent from the 1995 population. At the end of 2005, there were 784,408 individuals under parole supervision (Figure 2–2). This figure was 1.6 percent higher than the 2004 total population on parole and 15.5 percent higher than the 1995 parole population. Probationers and parolees accounted for 70 percent of the correctional population at the end of 2005; probationers alone accounted for 59 percent of the correctional population. The rate of individuals on probation in 2005 (1,858 per 100,000 residents) was nearly four times the rate of individuals incarcerated (491 per 100,000), whereas the rate of adults on parole was 350 per 100,000 residents, slightly less than the incarceration rate presented earlier.

The demographic characteristics of probationers and parolees, along with the states with the highest and lowest rates of probationers and parolees, are presented in **Table 2–4**. Massachusetts had the highest rate of probationers, whereas New Hampshire had the lowest; Pennsylvania had the highest rate of parolees, whereas Maine had the lowest. The vast majority of both probationers (77%) and parolees (88%) were male, and the majority of probationers were white (55%). Approximately equal percentages of parolees were white (41%) and black (40%).

Half of all adult probationers were sentenced to supervision for a misdemeanor offense; almost half were sentenced to probation supervision for a felony. Convictions for drug law violations (28%) accounted for the highest percentage of probationers; 15 percent of adults on probation in 2005 were convicted of driving while intoxicated, and 12 percent of adults on probation were convicted of larceny-theft. Three in five (59%) adults whose probation supervision ended in 2005 had successfully completed their probation sentence.

One in four individuals on parole at the end of 2005 were serving a sentence for a violent or property offense; 37 percent were serving a sentence for a drug

Table 2–4 Characteristics of Probationers and Parolees in the United States, 2005

Total Individuals on Probation	**4,162,536**	**Total Individuals on Parole**	**784,408**
States with the Highest Rates of Probationers	*Rate/100,000*	*States with the Highest Rates of Parolees*	*Rate/100,000*
Massachusetts	3,350	Pennsylvania	787
Rhode Island	3,091	Arkansas	782
Minnesota	2,988	Oregon	766
Delaware	2,828	Louisiana	712
States with the Lowest Rates of Probationers	*Rate/100,000*	*States with the Lowest Rates of Parolees*	*Rate/100,000*
New Hampshire	457	Maine	3
West Virginia	533	Florida	34
Utah	578	Rhode Island	41
Nevada	709	North Carolina	47
Kansas	723	Nebraska	50
Percentage of Probationers by Gender		*Percentage of Parolees by Gender*	
Male	77	Male	88
Female	23	Female	12
Percentage of Probationers by Race		*Percentage of Parolees by Race*	
White	55	White	41
Black, non-Hispanic	30	Black, non-Hispanic	40
Hispanic	13	Hispanic	18

Source: Adapted from *Probation and Parole in the United States, 2005.*

offense. Almost half (45%) of parolees whose parole supervision ended in 2005 had successfully completed their sentence; 38 percent of those whose supervision ended in 2005 were incarcerated as a result of the commission of a new offense or a rule violation.[51]

Data on Capital Punishment

Each year, researchers with the U.S. Department of Justice Bureau of Justice Statistics compile a report titled *Capital Punishment*. This report summarizes various data about: (1) persons under the sentence of death in the United States and (2) information on the status of death penalty statutes in the 50 states, the District of Columbia, and the federal government. Data about persons under the sentence of death are collected annually as part of the National Prisoner Statistics Program, which covers all persons who were held in a federal nonmilitary or state correctional facility under sentence of death at the end of the calendar year. Data collected on the status of death penalty statutes are collected from the Office of the Attorney General from each of the 50 states, the District of Columbia, and the federal government.[52]

At the end of 2005, there were 3,254 prisoners under sentence of death in federal and state facilities. This number represented a 2 percent decrease from 2004, the fifth consecutive year that population had decreased. The results presented in **Table 2–5** reveal that 53 people were executed in 14 states in 2006. Texas alone executed almost half of those prisoners; in fact, two-thirds of all executions occurring between 1977 and 2006 occurred in five states: Texas, Virginia, Oklahoma, Missouri, and Florida.

At the end of 2005, there were 52 females under sentence of death in the United States. Nevertheless, men remained more than 60 times more likely to be on death row than were females. More than half of the females under sentence of death were incarcerated in California (15), Texas (9), and Pennsylvania (5). Of the 38 states authorizing the use of capital punishment, 19 had a female incarcerated under sentence of death at the end of 2005. At the end of 2005, more than half (55%) of inmates under sentence of death were white; 42% were black, and the remaining 2 percent were of some other race. Thirteen percent were of Hispanic origin.

Controversies

Until 2005, individuals who were 16 or 17 years of age who committed a capital offense in those states that used capital punishment could be sentenced to death. The U.S. Supreme Court decision *Roper v. Simmons* forbid imposition of the death penalty on offenders who were under the age of 18 at the time they committed their capital offense.

Table 2–5 Capital Punishment Statistics, 2005/2006

Total Individuals on Death Row, 2005	**3,254**	**Total Persons Executed, 2006**	**53**
States with the Largest Death Row Populations, 2005		*States Performing Executions, 2006*	*Number of Executions*
California	646	Texas	24
Texas	411	Ohio	5
Florida	372	Florida, North Carolina, Oklahoma, Virginia	4
Pennsylvania	218	Alabama, California, Indiana, Mississippi, Montana, Nevada, South Carolina, Tennessee	1
Number (%) of People on Death Row by Gender, 2005		*Executions by Gender, 2006*	
Male	3202	Male	53
Female	52	Female	0
Number of People on Death Row by Race, 2005		*Executions (%) by Race, 2006*	
White	1805	White	32 (60.4)
Black, non-Hispanic	1372	Black, non-Hispanic	21 (39.6)
American Indian	31		
Asian	34		
Unknown race	12		

Source: Adapted from *Capital Punishment, 2005*. The 2006 data were retrieved electronically from the U.S. Bureau of Justice Statistics, "Capital Punishment Statistics," www.ojp.usdoj.gov/bjs/cp.htm.

Conclusion

This chapter has discussed a wide variety of data sources that provide us information on criminal justice and corrections in the United States. Numerous data sources are widely available on the Internet as well. Thus, no matter what the topic, researchers can generally find some type of data to inform them about crime and justice in the United States. Nevertheless, students interested in research need to remember the shortcomings of each of these data sources and temper their conclusions about the data with that knowledge in mind.

All of you have probably heard the following expressions: (1) "There are three types of lies: lies, 'damn lies,' and statistics"; and (2) "You can use statistics to prove anything you want." Although the knowledge you have gained from this chapter about the types and limitations of criminal justice data will not help you discern between "lies" and "damn lies," the knowledge you have gained from this chapter will make your search for "truth" in criminal justice data much more informed.

READY FOR REVIEW

- The three most important sources of crime data in the United States include the UCR, the NCVS, and self-report surveys.
- The UCR is published annually by the FBI and compile data provided to them by police departments from throughout the United States. The UCR provides the best source of data regarding offenders, but these data are limited to crimes that are known to the police.
- The NCVS is conducted annually by the Bureau of Census for the Bureau of Justice Statistics and provides the best source of data regarding victims of crime in the United States. These data are limited in their reliance on reports of victimization among the selected respondents.
- Although the NCVS annually uncovers two to three times as many crimes as does the UCR for the six crimes for which they both collect data (forcible rape, robbery, aggravated assault, burglary, motor vehicle theft, and larceny-theft), the rank order of occurrence is the same for both sources of data and longitudinal trends for crime data generally match one another.
- Self-report surveys provide an additional source of data regarding crime and delinquency and are the primary source of data used to understand theoretical predictors of crime and victimization.
- In general, males, nonwhites, those of lower SES, and young people are disproportionately represented for most crimes for which we have data.
- The U.S. Bureau of Justice Statistics issues annual reports providing data regarding adult corrections in the United States. For 2005 data, these reports were titled *Prisoners in 2005*, *Probation and Parole in the United States, 2005*, and *Capital Punishment, 2005*, among others.

KEY TERMS

aggregate level data Data collected from or about groups of persons or things (e.g., ethnic groups, cities)

backward telescoping Remembering victimization as occurring in the more distant past than it actually occurred

chivalry hypothesis A suggestion that criminal justice agents might be more "chivalrous" toward the women that they arrest than they are toward men, and thus might deal with the female offender informally, possibly allowing the female to go with only a warning

clearance by arrest A crime cleared when an arrest is made and at least one person is charged with the crime

clearance by exceptional means Crimes cleared when the police know the location and identity of the suspect and have information to support the arrest and subsequent prosecution of the suspect, but are prevented from taking action against the suspect by consequences outside of their control (the suspect is dead, the victim refuses to cooperate, etc.)
crime an intentional act or omission of an act that violates criminal statutory or case law and for which the state provides a punishment
evil woman argument A suggestion that sex-role attitudes of criminal justice agents may cause women who violate traditional sex-role stereotypes, specifically violent offenders, to receive harsher treatment from criminal justice officials because of this sex-role violation
forward telescoping Remembering victimization incidents as occurring more recently than they actually occurred
individual level data Data collected from or about individual persons or things
inverse relationship A pattern of change between two variables, or factors, in which one variable increases as the other decreases (e.g., as social economic status increases, involvement in delinquency decreases)
National Crime Victimization Survey A report published annually by the Bureau of Justice Statistics that uses household victimization surveys to estimate the extent to which the residents of the United States are victimized by crime. This report provides data about characteristics of the victims of crime in the United States
Part I crimes The eight crimes classified as serious crimes by the Federal Bureau of Investigation in the Uniform Crime Reports. These crimes include murder and nonnegligent manslaughter, forcible rape, aggravated assault, robbery, burglary, larceny-theft, motor vehicle theft, and arson
Part II crimes The 22 crimes classified as less serious by the Federal Bureau of Investigation for which arrest data are collected and reported in the Uniform Crime Reports. These crimes include such crimes as prostitution and drug abuse violations
self-report surveys Surveys that ask samples of respondents to report their own participation in criminal activities
Uniform Crime Reports A report published annually by the Federal Bureau of Investigation consisting of reported crime and arrests submitted annually by local law enforcement agencies to the FBI throughout the country

YOU ARE THE CORRECTIONS PROFESSIONAL SUMMARY

1. Where would you obtain data to help you with your project? There are a number of sources through which you may obtain data. At the very least, you could look at the UCR Part II offenses weapons violations

for the United States and compare those numbers to those in your local community.

2. Are there limitations of the data that you will be using? If so, what are they? No matter what data source you use, there will be limitations to the data. A wise data user knows all of the limitations of the data and uses that knowledge to inform decisions based on the data.
3. Will the data you obtained help you in predicting what will happen with your organization? Why or why not? Official data are often used to project trends for the future. As such, given the limitations of the data, you can often forecast what will happen in the very near future based on trends in the crime data over the last few years.

NOTES

1. S. T. Reid, *Crime and Criminology*, 7th ed. (Fort Worth, TX: Harcourt Brace College Publishers, 1994), 686.
2. Federal Bureau of Investigation (FBI), *Crime in the United States, 2005: Uniform Crime Reports* (Washington, DC: U.S. Department of Justice, 2006); K. Maguire and A. L. Pastore, eds., *Sourcebook of Criminal Justice Statistics, 1994* (Washington, DC: U.S. Department of Justice, Bureau of Justice Statistics, USGPO, 1996), 128.
3. R. O'Brien, *Crime and Victimization Data* (Beverly Hills, CA: Sage Publications, 1985).
4. O'Brien, *Crime and Victimization*; FBI, *Crime in the United States, 2005.*
5. O'Brien, *Crime and Victimization*; FBI, *Crime in the United States, 2005.*
6. O'Brien, *Crime and Victimization*, 22.
7. FBI, *Crime in the United States, 2005.*
8. O'Brien, *Crime and Victimization*; S. M. Catalano, *Criminal Victimization, 2004*, NCJ 210674 (Washington, DC: Bureau of Justice Statistics, National Crime Victimization Survey, 2005).
9. FBI, *Crime in the United States, 2005.*
10. O'Brien, *Crime and Victimization.*
11. The discussion of the NCVS is drawn largely from O'Brien, *Crime and Victimization.*
12. FBI, *Crime in the United States, 2005*; U.S. Department of Justice, *The Nation's Two Crime Measures*, NCJ 122705 (Washington, DC: U.S. Department of Justice, 2004); NCVS findings are available at www.ojp.usdoj.gov/bjs/cvictgen.htm.

13. O'Brien, *Crime and Victimization.*
14. O'Brien, *Crime and Victimization.*
15. O'Brien, *Crime and Victimization*; M. J. Hindelang, M. R. Gottfredson, and J. Garofalo, *Victims of Personal Crime: An Empirical Foundation for a Theory of Personal Victimization* (Cambridge, MA: Ballinger, 1978).
16. FBI, *Crime in the United States, 2005.*
17. M. Chesney-Lind, "Judicial Paternalism and the Female Status Offender," *Crime and Delinquency* 23(1977): 121–130.
18. I. H. Nagel and J. Hagan, "Gender and Crime: Offense Patterns and Criminal Court Sanctions," in M. Norval, and N. Tonry, (eds.) *Crime and Justice, Volume IV*, (Chicago: University of Chicago Press, 1982), 91–144.
19. FBI, *Crime in the United States, 2005.*
20. Catalano, *Criminal Victimization, 2004.*
21. M. J. Hindelang, T. Hirschi, and J. Weis, *Measuring Delinquency* (Beverly Hills, CA: Sage Publications, 1981).
22. S. M. Catalano, *Criminal Victimization, 2003*, NCJ 205455 (Washington, DC: Bureau of Justice Statistics, National Crime Victimization Survey, September 2004).
23. J. Q. Wilson and R. J. Hernstein, *Crime and Human Nature* (New York: Simon & Schuster, 1985).
24. M. M. Gordon, *Assimilation in American Life: The Role of Race, Religion, and Natural Origin* (New York: Oxford University Press, 1964).
25. J. Hraba, *American Ethnicity*, 2nd ed. (Itasca, IL: F. E. Peacock Publishers, 1994).
26. Wilson and Hernstein, *Crime and Human Nature*; Gordon, *Assimilation in American Life* (New York: Oxford University Press, 1964).
27. Hraba, *American Ethnicity.*
28. U.S. Bureau of the Census. (2000). Decennial Census data, 2000 [Data file], http://www.census.gov/prod/cen2000/dp1/2kh00.pdf (accessed October 16, 2007).
29. FBI, *Crime in the United States, 2005.*
30. P. M. Harrison and A. J. Beck, *Prisoners in 2005*, NCJ 215092 (Washington, DC: U.S. Department of Justice, Bureau of Justice Statistics Bulletin, November 2006).
31. J. Reiman, *The Rich Get Richer and the Poor Get Prison*, 4th ed. (Boston: Allyn & Bacon, 1990).

32. Hindelang, Hirschi, and Weis, *Measuring Delinquency*.
33. R. Chilton and J. Galvin, "Race, Crime, and Criminal Justice," *Crime and Delinquency* 31, no. 1 (1985): 3–14.
34. Catalano, *Criminal Victimization, 2003*.
35. J. Hagan, "The Poverty of Classless Criminology: The American Society of Criminology 1991 Presidential Address," *Criminology* 30 (1992): 1–19.
36. J. Braithwaite, "The Myth of Social Class and Criminology Reconsidered," *American Sociological Review* 46 (1981): 36–57.
37. Braithwaite, "Myth of Social Class."
38. C. R. Tittle and R. F. Meier, "Specifying the SES/Delinquency Relationship," *Criminology* 28, no. 2 (1990): 271–299.
39. C. R. Tittle, W. J. Villemez, and D. A. Smith, "The Myth of Social Class and Criminality: An Empirical Assessment of the Empirical Evidence," *American Sociological Review* 43 (1978): 643–656.
40. Tittle & Meier, "Specifying."
41. Braithwaite, "Myth of Social Class."
42. Catalano, *Criminal Victimization, 2004*.
43. FBI, *Crime in the United States, 2005*.
44. Catalano, *Criminal Victimization, 2004*.
45. FBI, *Crime in the United States, 2005*.
46. Harrison and Beck, *Prisoners in 2005*.
47. Harrison and Beck, *Prisoners in 2005*.
48. Harrison and Beck, *Prisoners in 2005*.
49. Harrison and Beck, *Prisoners in 2005*.
50. Harrison and Beck, *Prisoners in 2005*.
51. L. E. Glaze and T. Bonczar, *Probation and Parole in the United States, 2005*, NCJ 215091 (Washington, DC: U.S. Department of Justice, Bureau of Justice Statistics Bulletin, November, 2006). All statistics included in this section are from this report.
52. T. L. Snell, *Capital Punishment, 2005*, NCJ 215083 (Washington, DC: U.S. Department of Justice, Bureau of Justice Statistics Bulletin, December 2006). Unless otherwise noted, all statistics included in this section are from this report.

Crime and the Law

3

Chapter Objectives

- Display knowledge of how law fits into the structure of government in the United States.
- Define discretion, and explain what it means to say that criminal justice agencies exercise discretion within the parameters of law.
- Summarize how law evolved historically, with particular reference to the English common law.
- Differentiate the civil and criminal law as well as substantive and procedural law.
- Identify the various sources of law.
- Describe the three components of the *corpus delicti* as well as the key principles of substantive law revolving around this concept; show how these principles apply to criminal acts.
- Discuss the various principles of procedural law in terms of the distinction between the ideal and the real; identify the main due process rights granted under the U.S. Constitution, and state how discretion and court rulings can shape these rights.
- Summarize the following approaches to categorizing crimes: (a) *mala in se* versus *mala prohibitum*; (b) felonies versus misdemeanors; (c) classification by offender characteristics; and (d) academic approaches, including distinctions between person offenses, property offenses, public order offenses, syndicated crime, white-collar crime, and political crime.
- Give illustrations of how the field of law applies to the field of corrections.

CASE STUDY

Your supervisor in your state's central office has called your attention to a jail facility located in a medium-size city. In recent weeks, the facility has gotten some negative publicity from the local media because of reports of inmates being assaulted by other inmates. Some of the reports involve sexual assaults, but others involve violence that appears unrelated to sex. The state has already determined that the reports are founded. As such, your supervisor asks you to examine the procedure being used to classify inmates upon admission to the facility, specifically the procedure used to determine each inmate's living unit and the restrictions placed on the inmate inside the facility so as to help contain threats of violence. Your supervisor tells you that after you have examined the classification procedure, you are to recommend improvements that are in compliance with law and policy.

1. What are some sources of law and policy you would consult to learn about how the jail is supposed to be performing classification?
2. What, if anything, else will you need to do before making recommendations to your boss?

Introduction

As pointed out in Chapter 1, in the United States criminal justice and corrections are primarily the domain of government. Although the private sector plays an important role, as evidenced, for instance, by the growth of privately operated prisons, private agencies are subject to governmental regulations. A prerogative of government is the creation and maintenance of law, which is the main subject of this chapter. Just as criminal justice agencies function as part of government, so too these agencies function within the boundaries or parameters of law. American government is based upon the **rule of law**—the principle that behaviors and interactions within society are regulated by a written system of rules and penalties that are to be applied in ways that are as fair and uniform as possible. Law, then, is the very foundation for policing, courts, corrections, and juvenile justice.

Law and Government Structure

Recall from Chapter 1 that all three levels of government (federal, state, and local) have their own sets of laws. Depending on the level, these laws may be referred to as codes, regulations, ordinances, or various other terms. Units of government at the local level (counties, cities, towns, etc.) must ensure that their laws conform to relevant state and federal guidelines. Similarly, the laws of each state have to meet certain federal requirements. For example, states that permit capital punishment must be sure their death penalty legislation is consistent with standards established by federal lawmakers and the federal courts.

Criminal law is primarily the domain of the legislative and judicial branches of government. The legislative branch defines what behavior is criminal (including the associated penalties) through its constitutionally based authority to write and pass laws. Legislators also define procedures related to the processing of offenders, such as procedures related to arrest or sentencing. The judicial branch contains the appellate courts, which are responsible for interpreting laws and procedures to determine if these meet legal standards.

Law and Discretion

Saying that criminal justice agencies operate within the parameters of law raises the issue of **discretion**. Dictionaries tend to define *discretion* as the freedom to act or exercise judgment as one sees fit. But a definition like this is really too broad and implies too much decision-making leeway to accurately capture discretion in criminal justice. Although the decisions of criminal justice officials are necessarily subjective to some extent, some consistency and objectivity are introduced by the law as well as by the agency policies and training to which officials are subjected. In other words, these officials do not have leeway to take whatever action they desire. Permitting them to do so would be counter to the rule of law. Instead, they can select between various courses of action, as defined by law and policy. Sometimes these courses of action are very narrowly defined, and other times they are vaguer and broader. It is in this sense that laws and policies structure discretion. However, laws and policies cannot tell officials exactly what action to perform (or even the ones from which to select) in every situation. Laws and policies must be interpreted in view of the situation, and this is where officials have latitude or flexibility.[1]

This matter is ultimately a question of balance. Realistically, laws and policies cannot be so fixed and rigid that they can be applied mechanically the same way to every case. Instead, they need to be adaptable enough to apply to the many similar, yet different, contexts and circumstances that officials encounter, meaning that interpretation is necessary. At the same time, laws and policies that invite excessive latitude for interpretation set the stage for inconsistency, unfairness, and abuse of power. In short, although discretion must be structured and controlled, it cannot be eliminated. Such a balance is not easily achieved.

The Evolution of Law

Laws are never isolated from history; they evolve and develop over time. The main (but certainly not the only) historical root of laws and legal processes in the United States is the **common law** that developed in England over the years following the eleventh century.[2] Before there were constitutions and legislative bodies, as we know them today, English monarchs charged judges to begin implementing a consistent system of law and procedure that was common across the entire country, as a replacement for customary ways of addressing problems or disputes that varied from location to location. Traditionally, disputes and wrongs had been treated as private matters to be addressed by private parties, often kinship groups.[3] But gradually, English judges began issuing case decisions and disseminating their rulings as precedents to guide future decision making. Over time, these decisions evolved into a body of law.[4] As common law steadily evolved, more and more harmful behaviors came to be regarded as wrongs against the central government, rather than against individuals. Hence, the government gradually assumed responsibility for establishing laws and dealing with lawbreakers.

English **jurisprudence**, or legal philosophy, was applied in the American colonies and, with modifications, was used until independence from Great Britain in 1776.[5] As will be apparent in this chapter and others, following independence the development of jurisprudence in the United States continued to reflect many common law principles. However, in response to the perceived excess power of the central English government, the founders of the American political system were very concerned that the federal government not be allowed too much authority. The founders therefore opted for a decentralized legal system, one that granted much of the authority for legal matters to individual states and local jurisdictions. This is why today the majority of criminal law enactment and handling of offenders are done at the level of the individual states, instead of at the federal level.

As will become apparent in this chapter, the common law is not a source or type of law today. Instead, it is a heritage established by English judges and other officials upon which modern legal practices are founded. But this heritage is important because it has guided legal developments in the United States for many years.

Civil and Criminal Law

In the field of law, a distinction is typically drawn between two primary, yet overlapping areas: civil law and criminal law. Although some cases are subject to prosecution under both civil and criminal law in separate court proceedings, other cases are subject to only one or the other of these forms of prosecution. The main concern throughout this book is with criminal law, but a useful means of better understanding criminal law is by contrasting it with civil law.[6]

Civil Law

Whereas the criminal law focuses on public interests, the **civil law** is more geared (but not exclusively so—see accompanying Controversies box) toward regulating private interests and private relations. The civil law is concerned with such things as private wrongs or torts, property disputes, and contract disputes. Instead of violations being viewed as offenses by a defendant against a unit of government, violations are conceptualized in terms of one party versus another. The party initiating legal action is known as the plaintiff, while the one that is the target of the action is the defendant. Although a government entity can be a party in civil litigation, the emphasis is on determining the extent to which each party is blameworthy, culpable, or liable (as opposed to guilty or not guilty). Intentions may be considered, but consideration is also given to other mental states, such as recklessness and negligence. The standard for winning a civil action is preponderance of the evidence, a less stringent standard than proof beyond a reasonable doubt. The civil standard is less stringent because the penalties are considered less severe. The penalties are geared toward achieving just compensation (usually in monetary terms) for the party suffering a loss. Rather than punishing the party deemed to be guilty, the idea is to collect damages from that party. Although in certain civil cases punitive damages may be ordered, these damages are generally monetary and do not involve a sentence to a period of community supervision or incarceration.

Criminal Law

There are some important distinguishing characteristics of **criminal law**. Its ultimate goal is to protect public interests and safety. As such, violations of the criminal law are viewed as crimes against the government and are said to be prosecuted by the government on behalf of the people. This is why the titles of criminal cases contain the name of the particular unit of government (e.g., the United States, the state of Kentucky) and the name of the individual defendant. Under criminal law, emphasis is placed on establishing what the defendant mentally *intended* to accomplish through his or her actions. The standard for determining the defendant guilty is proof beyond a reasonable doubt, the most stringent standard under law. The standard is stringent because the penalties under criminal law are generally considered to be more severe. The criminal law emphasizes punishing those found guilty to achieve retribution for the wrong imposed on society and to protect members of society from future harm. Typical punishments include fines, supervised sentences in the community, or incarceration.

Controversies

An excellent illustration of controversy over the use of civil law to help regulate public interests is the case of Seung-Hui Cho, the deranged gunman who, before committing suicide, murdered 32 people and wounded many others on the campus of Virginia Tech University in Blacksburg, Virginia, on the morning of April 16, 2007. Prior to the incident, Cho had shown numerous signs of bizarre and even dangerous behavior. But because he had not been caught committing illegal acts that could have resulted in arrest and incarceration in jail or prison, the only way Cho could have legally been removed from the community was for him to have been civilly committed for mental health care by the mental health professionals who interacted with him. After the tragedy, the mental health and legal systems were strongly criticized for not making greater use of civil commitment to protect the public.

Substantive and Procedural Criminal Law

Just like it is conventional to distinguish between civil and criminal law, within criminal law, a distinction is usually drawn between substance and procedure. As in the case of the civil–criminal distinction, these categories are not completely separate. But the distinction is nevertheless useful because it helps promote a better understanding of the different aspects of criminal law.

Substantive Law

The substantive aspect of the criminal law specifies: (1) what conduct is considered crime and (2) the range of sentences for such conduct. **Substantive criminal law** is designed to protect members of society from perpetrators and would-be perpetrators of criminal offenses. It seeks to regulate the activities of all members of society, and as will be seen in the next section, its primary source is written statutes or penal codes. The laws defining and forbidding murder, and laws specifying the possible sentences for a murder conviction, represent instances of substantive law.

Procedural Law

The **procedural criminal law** spells out the rules according to which the substantive law can be formulated by legislative bodies and then implemented or carried out by police, court, and correctional agencies. That is, procedural law is concerned with defining the processes and procedures for applying the substantive law in practice. Criminal justice processing consists of a series of steps or phases, ranging from the authorities becoming aware of a crime, to final completion of an offender's sentence. During this process, many discretionary decisions have to be made, and many actions have to be taken by the various government agencies involved. The overarching goal of procedural law is to protect suspected, accused, and convicted individuals from these government agencies violating legal rights.

Unlike substantive law, where the focus is on crime control, the procedural law is intended to promote due process. As such, procedural law exists to regulate the criminal justice process and the actions of officials who carry out that process. Like substantive law, it comes from statutes, but it also derives from case decisions and constitutions (see **Figure 3–1**). Rules that state that persons accused of murder (and other criminal offenses) have a right against self-incrimination, and then stipulate how this right is to be protected, are examples of procedural law. So are laws that prohibit a convicted person from being subjected to cruel and unusual punishment while they are under the supervision of a correctional agency.

Sources of Law

The distinction between civil and criminal law, and the one between substantive and procedural law, do not refer to sources of law. These are simply categories that students of law use to aid comprehension and understanding. What sources, then, would a person consult to locate actual laws? Four main sources are covered in the following subsections.

Statutory Law

Statutory law refers to both criminal and civil laws passed by legislators at all levels of government. It is found in the federal penal code, in the penal codes of the various states, as well as in documents at different local units of government, such as county and city ordinances. Statutory criminal law is substantive law in that it defines what actions (or lack thereof) are criminal and the corresponding penalties. However, statutes also establish procedures that need to be followed in processing cases through the system, so statutory law is also procedural in nature.

Many of the major offenses found in modern statutes (e.g., murder, rape, robbery, theft) originated historically, not from statutes, but from the collection of judicial decisions and practices that made up the English common law.[7] As stated earlier, this took place as English judges began to document and disseminate decisions in specific cases so that subsequent cases could be decided across varying locations in a way consistent with earlier rulings. It was not until later that legislative bodies, as we know them today, gradually codified these offenses into statutory law. Today offenses of this nature are sometimes called common law crimes to denote their origin in the common law.

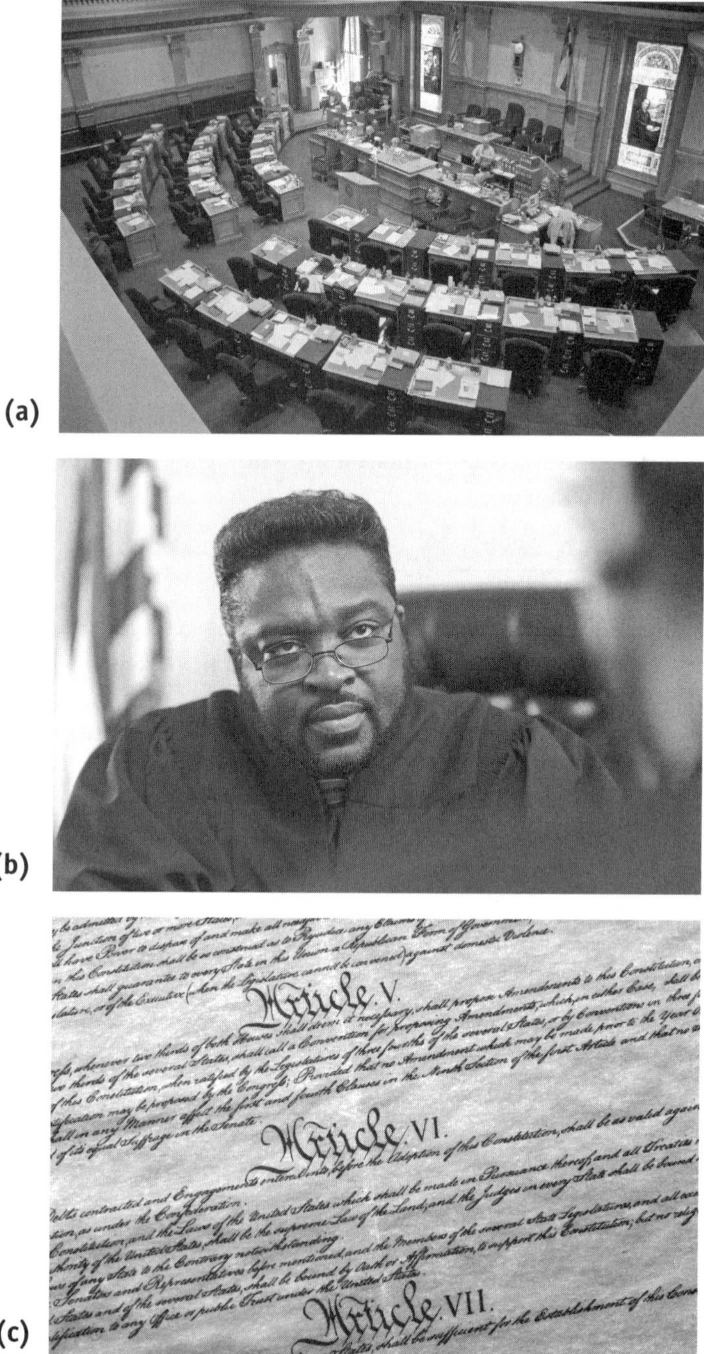

Figure 3–1 Procedural law comes from statutes, court decisions, and constitutions.
Source: *(a) © Andre Nantel/ShutterStock, Inc.; (b) © Corbis/age fotostock; (c) © Rich Koele/ShutterStock, Inc.*

Corrections and Policy

A good example of statutory law affecting policies and procedures in corrections is the Prison Litigation Reform Act (PLRA). The PLRA was enacted by the federal government in 1995 to curb the ability of prisoners to file federal lawsuits by making them first exhaust administrative channels available at the prison, such as institutional grievance procedures, before filing federal claims pertaining to prison conditions.

Constitutional Law

Constitutional law is law contained in the federal and state constitutions. Constitutions grant legislative bodies the legal authority to pass statutory law, but these documents also set forth laws addressing both substance and procedure. In fact, the domains of substantive and procedural law blend together in the area of constitutional law. Constitutions set guidelines on how statutes can be written, something known as substantive due process. For instance, a constitution may prohibit *ex post facto* laws, laws meant to punish people for behaviors committed before the law was ever enacted. Constitutions also contain laws establishing procedural due process. These are guidelines governing the procedures used to process suspected and accused offenders through the system. An example is found in the Fourth Amendment to the U.S. Constitution, which prohibits unreasonable searches and seizures of evidence. As such, constitutional law is designed both to regulate the creation of statutory law and to regulate the process by which a person suspected or accused of crime is brought to justice and punished.

The U.S. Constitution is the ultimate standard to which most laws must conform, irrespective of their source and the level of government involved. Indeed, a main task of the U.S. Supreme Court is to review the substance of criminal and civil laws, as well as the procedures used for applying these laws, to interpret compliance with the U.S. Constitution. The role of the Court in doing so was established early in the history of this nation in 1803 when the Court decided *Marbury v. Madison* and asserted for itself the authority to review the actions of legislatures and lower courts.[8] This is knows as the Court's power of judicial review.

Case Law

Case law (also known as precedent law) is law that evolves out of decisions rendered by judges. In modern times, this mainly refers to decisions rendered by judges deciding cases on appeal from lower courts (see Chapter 5). Recall that under English common law, judges' decisions were recorded and circulated so that subsequent decisions could be made in view of earlier decisions to promote consistency. This practice was supported by the common law principle of ***stare decisis***, meaning to let the decision stand as precedent to guide future case decisions. The goal was to establish a stable and consistent system of law.

Today lawyers and judges routinely refer to previous decisions, especially those of the U.S. Supreme Court and other appeals courts, when considering cases. Although court officials can argue that precedent somehow does not apply to a particular case at hand, or even overrule aspects of precedent to establish it anew, they need to articulate rationales for doing so.

The case of *Marbury v. Madison* mentioned earlier is an example of case law. One of the most well known pieces of precedent law is *Roe v. Wade* (1973), in which the Supreme Court legalized abortion across the United States.[9] Important, but less well known, cases are cited throughout this book where they apply to the material being considered. Most of these cases are considered to be *landmarks*, a term used to describe court decisions that have enormous impact on future decisions.

Regulatory Law

Regulatory law (also known as administrative law) refers to laws enacted by legislatures working in collaboration with administrative agencies at various levels of government. At the federal level, for example, the Occupational Safety and Health Administration, the Internal Revenue Service, and the Environmental Protection Agency control specific areas of activity related to worker safety, taxation, and pollution, respectively. Much regulatory law is civil in nature, but infractions of regulatory law can be prosecuted under criminal law and carry criminal penalties.

Principles of Substantive Law

Crimes can refer either to acts of commission, when a person does something such as robbery that is forbidden by law, or acts of omission, when a person fails to do something required by law such as paying required amounts of income tax on time. In either instance, the substantive law will specify what conduct is forbidden or required as well as the penalties for lawbreaking.

A central principle of substantive law is the common law principle of ***corpus delicti***. This refers to the body or elements of a crime. It is the *corpus delicti* that, when taken as a whole, defines a crime, such that each crime has its own unique set of elements. Successful prosecution or conviction on a particular charge requires proof of the various elements of the crime by the evidence presented to the court.

The *corpus delicti* of any crime consists of three basic components. The first is the forbidden (or required) behavior itself; this is defined later as the ***actus reus***. The second component is the state of mind that a person must be in to commit the crime; this is defined later as the ***mens rea***. Third are the conditions or circumstances that have to surround the act and state of mind for the crime to occur. An example is the definition of murder found in the U.S. Code (Title 18: Section 1111).

> *Murder is the unlawful killing of a human being with malice aforethought. Every murder perpetrated by poison, lying in wait, or any other kind of willful, deliberate, malicious, and premeditated killing; or committed in the perpetration of, or attempt to perpetrate, any arson, escape, murder, kidnapping, treason, espionage, sabotage, aggravated sexual abuse or sexual abuse, burglary, or robbery; or perpetrated from a premeditated design unlawfully and maliciously to effect the death of any human being other than him who is killed, is murder in the first degree.*

In this definition, the forbidden behavior is "the unlawful killing of a human being," meaning simply that the killing cannot have been somehow justified or excused under law (e.g., an accident or self-defense). This is what distinguishes murder from the more general term *homicide*, which is the killing of one human being by another. Criminal homicide is homicide that is not excused or justified under law. As will be shown, the perpetrator's state of mind defines the type of criminal homicide, with murder being only one type.

In the preceding example, the state of mind is "with malice aforethought," meaning that the offender harbored an evil intention to kill (malice) and gave the act at least some advance thought or planning (aforethought). Crimes like this one are often divided into various degrees of seriousness, and when this is done, it is aforethought that distinguishes first- from second-degree murder; the latter involves only malice. If neither malice nor aforethought is present, a criminal homicide may be termed manslaughter. Then, a distinction is typically drawn between voluntary and involuntary manslaughter. Voluntary manslaughter takes place when a person intentionally kills another in the heat of passion arising from provocation. With involuntary manslaughter, the homicide is not intentional but results from criminal recklessness or negligence.

The remainder of the definition from the U.S. Code refers to circumstances or conditions that can distinguish first-degree murder. First-degree murder is said to occur when the circumstances show that the person acted in a premeditated fashion, or when the person killed someone while simultaneously committing or attempting to commit another of the crimes listed (arson, escape, kidnapping, etc.).

Another means of conceptualizing the *corpus delicti* of a crime follows. This discussion considers the act, state of mind, and circumstances in further detail and brings additional factors into consideration. It can be thought of as a discussion of the ingredients for a crime.[10]

Legality and Punishment

A person cannot be prosecuted for a crime without the prior existence of a law that: (1) defines the act (or the omission of a required act) as criminal and (2) stipulates punishment for the crime. As stated at the outset of this chapter, U.S. government is based on the rule of law. A person may do something that proves harmful to himself or herself or someone else (such as smoking as an adult), or something that is highly objectionable to another party, but the harmful or objectionable action is not a crime in the absence of a law defining it as such and setting forth the possible penalties. Furthermore, as stated earlier, a law may not be *ex post facto* or retroactive in nature; it cannot punish a behavior committed prior to the passage of the law.

The *Actus Reus*

Actus reus is a common law term referring to the guilty act. It means the commission of prohibited action or omission of required action. In general, it is not sufficient for a person to have thoughts of committing a crime or to have a certain status (e.g., drug addict) often associated with criminal behavior. However, there are certain latitudes surrounding the *actus reus* requirement. For instance, a person could conspire or plot with someone else to commit a crime, never actually commit the crime, but still be found guilty of conspiring; in this case, the plotting is considered the illegal act. Likewise, one could encourage another person to commit a crime and be deemed guilty of solicitation or be convicted as an accessory for helping conceal the crime of another person from the authorities. In addition, it is not necessary that crimes be fully completed; a person can be convicted of attempting an illegal act.

Harm

As a general rule, offenses must produce some type of demonstrated harm or damage to an entity protected under law. The protected entity can vary (e.g., a person, a property, or even a reputation). In addition, the harm can be directed toward the offender's self, as in the case of so-called victimless crimes like drug use.

According to Hall, "Harm is the focal point between criminal conduct on the one side, and the punitive sanction, on the other."[11] In other words, harm is used to justify punishment, and in general, efforts are made to balance the magnitude of harm with the severity of the punishment.

Causation

Causation simply means it needs to be demonstrated that the *actus reus* produced the harm. For example, it would be difficult to convict an offender of murder if the person the offender assaulted later died of a condition completely unrelated to the assault. For a murder conviction, it would be necessary to link damage from the assault to the cause of death.

Concurrence

The point was made earlier that state of mind is one of the three basic components of the *corpus delecti* of a crime. Concurrence means that the *actus reus* and the state of mind or *mens rea* (see later) have to be connected. Burglary is usually defined as breaking and entering a structure with the intent to commit a felony therein. Theft or larceny, on the other hand, is defined as taking the property of someone else with the intent to deprive the owner permanently of his or her property. Suppose that during a life-threatening winter storm, an individual traveling on foot broke into a house without anyone at home to seek shelter and, while there, stole one of the homeowner's vehicles. Upon apprehension, the individual could be convicted of theft but not necessarily of burglary. To be convicted of the latter, it would need to be shown that the intent to steal the vehicle existed at the time of the forced entry into the house; otherwise, the act and requisite mental state would not concur.

The *Mens Rea*

Just as the *actus reus* refers to the guilty act under common law, the *mens rea* refers to the guilty mind. In other words, to be held criminally responsible for a certain act, the person must have committed it while in the particular state of mind specified by law. This does not necessarily mean that the person acted intentionally, because *mens rea* is not synonymous with intent. Under law, *mens rea* can be defined in relation to a number of mental states, as defined in the following subsections.[12]

Intentional Behavior

When a person behaves intentionally, he or she has a purpose in mind for the behavior. The purpose can be to cause harm to a specific target (such as when a firearm is pointed toward a particular person and discharged with the intent of harming that person), or the purpose can involve harm in general (such as when an individual intentionally discharges random shots in a crowded room). The former is sometimes referred to as specific intent, while the latter is called general intent. Another possibility is transferred intent, where, for example, a perpetrator might intend to shoot another person, but instead miss that person and hit a third party; here the intent to harm could transfer to the third party. The logic behind transferred intent is similar to the logic of so-called felony–murder laws. These laws hold that if an offender intentionally commits a felony (say, arson of a home), and if a person dies in the process, the offender can also be held accountable for murder even though the person's death was not specifically intended.

Knowing, Reckless, and Negligent Behavior

Under law, knowing action means that the offender knew of the wrongful nature of his or her conduct and ignored the fact that harm was very likely to result from the action. For example, a person who allows another person into his or her vehicle, knowing the other person to be in possession of stolen jewelry, could be considered guilty of transporting stolen goods. In contrast to knowing action, with reckless action, the likelihood of harm need not have been extremely high; the risk was simply unacceptable under law. Reckless action is that which demonstrates unjustified disregard for the risk of harmful consequences. Reckless driving is perhaps the most well-known example. A person who is criminally negligent is one who fails to take reasonable precautions to avert harm. His or her conduct is so careless that, in the eyes of the law, a reasonable person would have behaved differently. An example of negligent action is a parent or guardian leaving a baby alone in a car with the windows up on a very hot day while the adult shops at a store.

Exceptions, Justifications, and Excuses

One category of crimes, known as strict liability offenses, represents an exception to the *mens rea* requirement in that the state of mind is irrelevant. With strict liability crimes, the *actus reus* alone is sufficient for an offense to take place. An example to which many people can relate is speeding in a vehicle. Another example is found under regulatory law and involves the violation of some pollution and environmental protection laws (**Figure 3–2**). The *corpus delecti* of such offenses does not specify a *mens rea*.

In addition, the law recognizes a number of justifications and excuses that revolve around the *mens rea* concept and can be used to defend against criminal liability. *Justification* means the person responsible for the illegal act has done the right (or at least the permissible) thing under the circumstances, while *excuse* means the person cannot be considered criminally responsible for his or her action even though the action was unjustified.[13]

Insanity Although rare, perhaps the most controversial excuse defense is "not guilty by reason of insanity." When this defense is used, the argument is that, at the time of committing the crime, the defendant could not form the necessary *mens rea* as a result of a problematic mental state and, therefore, cannot be considered guilty. Insanity at the time of the crime is not the same as being judged mentally incompetent to stand trial. Competency to be tried means the defendant understands the charges against him or her and is capable of participating in a defense. A person could have been sane at the time of the crime, but then later have been deemed incompetent to stand trial. Alternatively, a person might have been insane at the point of the crime and competent by the trial.[14]

There is lack of agreement on how the concept of insanity should be defined for legal purposes; various standards or tests have been used. Although mental health professionals are frequently involved in insanity cases, it is important to recognize

Figure 3–2 Some environmental protection law violations do not specifically require *mens rea*.
Source: *© AbleStock.*

that the standards or tests used to define insanity are legal in nature, not psychiatric or psychological. *Insanity* is a legal term, not a psychiatric or psychological one.

An early legal test for insanity derives from British law and is known as the **M'Naughten Rule**, after Daniel M'Naughten who claimed that delusions led him to kill someone who he believed to be the British prime minister. In effect, this rule asks whether the defendant could distinguish right from wrong behavior at the time of the crime. The defendant is considered insane if he or she could not draw this distinction because of mental disease or defect. As a supplement to this right–wrong test, the **irresistible impulse test** asks whether the defendant was compelled by mental illness to commit the illegal act even though he or she knew the action was wrong; the assumption is the mental illness led to a lack of control over behavior. An alternative test, the **Durham Rule**, contends that the defendant is excused from criminal responsibility if it can be demonstrated that he or she committed the unlawful act as a result of mental disease or defect; ultimately, this issue is left to the judge or jury to ascertain.[15] Finally, the **substantial capacity test** inquires as to whether the defendant possessed sufficient mental capacity to appreciate the wrongful nature of the behavior or the capacity to control the action.

States that allow for a not guilty by reason of insanity plea use one of the preceding rules or a variation thereof. After John Hinckley's attorneys successfully used the insanity defense in 1982 to acquit Hinckley of shooting President Ronald Reagan, controversies over this defense grew heated and, in response, many states revised their legislation to make successful insanity defenses more difficult; a few states abolished the defense completely. Some states have adopted a plea of guilty but insane (or mentally ill) to either replace or supplement the not guilty by reason of insanity plea. A defendant who successfully uses this plea can anticipate being transported to prison once his or her mental condition has been addressed at a treatment facility or may be sent to prison with an order that mental problems be addressed there. By contrast, a person judged not guilty by reason of insanity will be treated in a mental health facility and then released upon making the improvements deemed necessary.

Age The excuse of age (also called infancy or immunity) holds that the perpetrator of an illegal act could not form the necessary *mens rea* because of young age. Under common law, individuals under age 7 were presumed incapable of forming *mens rea*, and today, states tend to follow this convention.[16] Under common law, those between 7 and 14 years were presumed incapable of *mens rea* unless convincing evidence could be offered to the contrary, and those older than 14 were considered capable. With establishment of the juvenile court at the turn of the nineteenth century, a separate system was adopted for dealing with juvenile offenders. Today, offenders falling between a jurisdiction's minimum age limit for legal responsibility (e.g., 7) and the jurisdiction's age of adult majority (generally 16, 17, or 18) are usually processed through the juvenile justice system; the exception is when they are processed by the adult court system (see Chapter 14).

Intoxication Some defendants have argued that they should be excused from criminal responsibility because of their intoxication at the time of the crime. In certain cases, voluntary intoxication can be considered a mitigating factor and thus used to lessen blameworthiness (e.g., a reduction from first- to second-degree murder). But to be an effective defense against guilt, the intoxication must be shown to have taken place involuntarily, such as when a drug is secretly added to someone's soft drink.

Mistake of Fact It has often been said that ignorance or mistake of the law is no excuse, and this is true insofar that people cannot excuse illegal behavior by saying they were unaware of its illegal nature. On the other hand, mistake of fact can be considered an acceptable defense. For instance, an individual leaving a gathering could take something belonging to another person under the genuine impression that the item is his or her own if the individual owns a very similar item.

Entrapment An excuse that sometimes arises out of undercover police work is the defense of entrapment. **Entrapment** means that the offender was not mentally predisposed to commit the crime without police encouragement and would not have done so without excessive inducement from the police. The question is which party instigated the offense or initiated the *mens rea*. Police officers wishing to avoid entrapment must avoid overtly encouraging people to undertake criminal activity. The intention, knowledge, recklessness, negligence, or similar state of mind must originate with the offender, not the police.

Compulsion The excuse of compulsion is also known as duress or coercion. It involves a contention by the defendant that he or she was forced to commit the crime. In general, the defendant has to show that commission of the crime was the only way to avoid death or serious injury to self or others.[17] A person ordered to commit a theft lest his or her friend, who is being held at gunpoint, be killed could claim this defense to the *mens rea* requirement.

Self-Defense Just as the previous excuses can be used to argue that an individual should not be held criminally responsible for illegal action, certain justifications can be employed to argue that the behavior was permissible in view of the circumstances surrounding it. Self-defense is one of the most well-known of these justifications.

Generally speaking, for a defendant to make a credible case of self-defense, it needs to be shown that: (1) the defendant believed he or she was under the immediate threat of death or great physical harm; (2) there were no reasonable or readily available alternatives to exercising self-defense, such as escape from the situation; and (3) the defendant used only that degree of force required to avert physical harm. Defensive behaviors need not be directed only at protecting the self. The law generally recognizes that these behaviors can be directed toward protecting other people and even property. However, because the law considers human life more valuable than property, the use of deadly force (or force intended to produce extreme injury) is usually not justified to protect property alone.[18] The level of force used must be reasonable in relation to what the defender is trying to protect.

Victim Consent As part of their *corpus delecti*, some offenses (e.g., forcible rape and larceny) require lack of consent on the part of the victim. For other offenses (e.g., homicide), consent is less, if at all, relevant. As such, the crime in question is an important consideration in determining whether consent is a viable justification. Rape provides a good illustration of this point. In the case of forcible rape, the law generally requires that the act be committed against the victim's will. But in a case of statutory rape, the victim is considered too young to be capable of granting consent, so consent could not be used to justify sex with a person legally considered a minor. Another important consideration is whether the consent was granted knowingly and voluntarily. Voluntary consent cannot be obtained by placing the victim under a state of duress or through the use of deception. Nor can it be obtained from someone who is mentally incapable of granting it (e.g., a person who has passed out from intoxication).[19]

Necessity In some respects, this justification is similar to the excuse of compulsion (see the description earlier), although it does not require that the perpetrator be coerced by another to commit the illegal act. With necessity, it is the circumstances (rather than coercion from another person) that justify the act. At the same time, it needs to be shown that the act was required to avert harm greater than that caused by the act itself. A motorist who is endangered on the highway by the vehicle of a second driver who has fallen ill behind the wheel could justify reckless driving that causes a third vehicle to wreck if the motorist could show that reckless driving was necessary to avoid death or serious injury to himself or herself.

Battered Woman/Spouse Syndrome Some jurisdictions have recently revised their laws to allow women who have suffered patterns of physical abuse from a spouse (or significant other) to ask the court to consider such abuse as justification for illegal action (e.g., killing the abuser). Some states enacted actual battered woman/spouse defenses, but others incorporated this defense into extant laws defining self-defense, compulsion, or necessity.[20]

Other Defenses

In addition to the preceding defenses that revolve around *mens rea*, defendants sometimes claim other defenses to avoid conviction. These usually revolve around the law or around the actions of law enforcement officials described elsewhere in this book. One example of the former is **double jeopardy**, in which defendants claim that they cannot be subjected to criminal prosecution more than once in the same jurisdiction for the same crime. Although a person acquitted of a criminal charge cannot be retried on the same charge in the same jurisdiction, the person could be prosecuted in civil court for the action that resulted in the filing of criminal charges; this happened with O. J. Simpson following his acquittal for murder. Another example is the claim that the statute of limitations has expired. Most jurisdictions have laws for most crimes (except murder and possibly other crimes considered very serious) stating that prosecution must begin within a certain time period; this is done to ensure that the evidence is not stale and that the person receives a speedy trial. A final example involves a claim that the statute under which the person has been charged is excessively vague and open to overly general interpretations of the defendant's conduct.[21]

Principles of Procedural Law: The Ideal and the Real

Just as basic principles underlie substantive law, there are also certain principles upon which procedural law is founded. One such principle, the rule of law, was identified at the outset of this chapter. This idealistic principle states that the handling of supposed offenders is to be ruled by the objective letter of the law (versus the subjective whims of individuals), but as shown early in the chapter, there is usually room for legal officials to exercise discretion. Again, the law establishes the parameters in which officials exercise discretion. This point is evident in the discussion that follows of some additional principles underpinning procedural law. In the case of each principle, the tension between idealism and the reality of discretion will be apparent.[22]

Fundamental Fairness

The doctrine of **fundamental fairness** is a premise of procedural justice in the United States. The doctrine holds that all steps taken to process cases through the criminal justice system, from creation of laws initially through release of individuals from correctional supervision, should be fair in balancing the interests and freedoms of the offender with the collective interests of the public in being protected from crime. This ideal is reflected in rules governing the way police collect evidence, the way courts try cases, and the way correctional agencies handle sentenced persons.

The reality, of course, is that not all parties with an interest in a given case always perceive things as fair. Victims, their families, and the public may sense unfairness when what they see as a "legal loophole" allows offenders to escape the punishments seen as deserved. Likewise, accused persons may sense unfairness when they do not believe a government-appointed attorney has represented them well, or prisoners may sense it when institutional staff do not protect them from being assaulted by other prisoners. The point is that fundamental fairness is an ideal to be striven toward. As a result of discretion and subjective perceptions, it is not something that can be accomplished in every case to the satisfaction of everyone involved.

Equal Protection

Under ideal circumstances, the law is supposed to be applied impartially without regard for so-called extralegal considerations, such as a person's power, wealth, race, gender, and so forth. Here again, however, the **equal protection** principle is more of an ideal to strive toward than something accomplished in all (or even most) cases. There is little doubt that, in many instances, persons and groups possessing considerable power and wealth receive more lenient treatment than do people lacking these, even when the actions of the former are more, or just as, socially harmful.[23] Similarly, there is a long and well-documented history of ethnic minority groups receiving less than equal protection at various stages of the criminal justice process. The quotation in the accompanying Race and Gender in Corrections box captures one of the most controversial manifestations of this point. In addition to racial bias in the area of lawmaking, Cole describes racial and class biases in the areas of policing, court processing, and sentencing/corrections. His work illustrates the clash between the ideal of equal protection and the reality of discrimination.

Race and Gender in Corrections

Quotation from law professor David Cole:

> *A predominantly white Congress has mandated prison sentences for the possession and distribution of crack cocaine one hundred times more severe than the penalties for powder cocaine. African Americans comprise more than 90 percent of those found guilty of crack cocaine crimes, but only 20 percent of those found guilty of powder cocaine crimes.*[24]

Propriety

The principle of **propriety** holds that, ideally, the quality of the evidence against an accused person should be given as much, or more, weight than the quantity. This is because, ultimately, it is considered more important to protect the innocent than to convict the guilty. For example, there can be much evidence linking a person to a crime, but if important aspects of that evidence were obtained illegally, then in theory, the case should be much more difficult to prosecute. In reality, though, prosecutors routinely "stack" charges against arrested persons, filing charges for as many different crimes as possible. One effect of this practice is to help ensure that if one charge cannot be substantiated because of a lack of evidence, others will remain. A lengthy list of charges can also be useful during plea negotiations because defendants can be encouraged to plead guilty to a particular charge or set of charges in exchange for others being dropped. In reality, then, quantity can (and frequently does) take precedence over quality, and concern with convicting the guilty can take priority over protecting the innocent.

Presumption of Innocence

Citizens of the United States are taught to believe that suspects are presumed innocent until proven guilty. The burden of proving guilt lies with the prosecution, and the task of the defense is to rebut the prosecution's case. Indeed, the **presumption of innocence** principle is at the very heart of the adversarial system of justice. If innocence is not presumed, there is little basis for the prosecution and defense to assume the role of adversaries, a process from which the truth, supported by facts, is supposed to emerge.

Like most ideals, there is some truth to this one. It is, in fact, incumbent on the prosecutor, as the representative for the government, to build a case against the defendant that is supported by evidence. This is clearly different from the practice in some nations of making the defendant take the initiative in proving his or her innocence. But in reality, innocence becomes less tenable in the minds of criminal justice officials the more evidence mounts against the defendant.[25] Obviously, most police officers do not make it a practice to arrest suspects who they assume are innocent, and prosecutors do not make it a practice to accept such cases. Cases with weak evidence, or serious evidence problems, tend to get screened out very early in the process, well before a trial. As evidence mounts, defense attorneys usually become more willing to engage in plea negotiations with the prosecution to strike a "better deal" for their clients. Judges oversee the process and are well aware of what is happening. In short, the presumption of innocence has often waned or vanished before proof of guilt has been firmly established.

Due Process of Law

The Fourteenth Amendment to the U.S. Constitution forbids any state from depriving "any person of life, liberty, or property, without due process of law. . . . " The assumption behind this **due process** principle is that all citizens have certain rights, protections, or safeguards due to them by virtue of American citizenship. The ideal is that during the course of criminal justice processing, an accused person will be accorded the protections he or she has coming. As such, governmental actions cannot be arbitrary and disregardful of one's rights under law.

The main due process rights applicable to criminal justice procedure are presented in the next section. As these are considered, bear in mind the reality of discretion. Although it may be tempting to think of due process as an unchanging absolute, the reality is that the protections accorded an accused individual vary according to the specific case at hand, depending on the interpretations of the judge, prosecutor, and other officials handling the case. These protections also vary with the time era (e.g., the 1930s versus the present) and the location (i.e., the particular jurisdiction in which the case is being processed), depending primarily on the interpretations of lawmakers and appellate courts. Consequently, conceptions of due process are to some extent relative to person, place, and time.[26]

Due Process and the U.S. Constitution

The U.S. Constitution, which took effect in 1789, is the single most important source of procedural due process affecting criminal justice operations across the United States. As mentioned earlier in this chapter, procedural law is also set forth in state constitutions, in federal and state statutes, as well as in case law at the federal and state levels. However, these other sources of procedure are subservient to the U.S. Constitution because, ultimately, their legality is a question of how well they conform to constitutional requirements and guidelines, as interpreted by the U.S. Supreme Court (**Figure 3–3**).

Figure 3–3 The U.S. Supreme Court.
Source: *© Jason Maehl/ShutterStock, Inc.*

Certain aspects of the Constitution are more relevant than are others in defining criminal procedure. The first 10 amendments to the Constitution are known as the **Bill of Rights**. The Bill of Rights was added to the Constitution in 1791 to protect the freedoms of citizens from the power of government. It is the Fourth, Fifth, Sixth, and Eighth of these 10 amendments, along with the Fourteenth Amendment (which was ratified in 1868), that are most pertinent to criminal justice processing (see **Table 3–1**).

Fourth Amendment

As Table 3–1 reveals, the Fourth Amendment grants the right to privacy by prohibiting unreasonable searches for and seizures of evidence. It also establishes that warrants must be based on probable cause and that warrants have to be specific. As will become apparent in Chapter 4, however, police and court officials must define the terms *unreasonable* and *probable cause*. It is at this point that discretion enters into play. Although lawmakers and courts have established guidelines for interpreting such terms, the actual interpretations are made by police and court officials who apply the guidelines to real cases on a daily basis.

Table 3–1 Amendments to the U.S. Constitution Most Pertinent to Criminal Procedure

Fourth Amendment	The right of the people to be secure in their persons, houses, papers, and effects, against unreasonable searches and seizures, shall not be violated, and no Warrants shall issue, but upon probable cause, supported by Oath or affirmation, and particularly describing the place to be searched, and the persons or things to be seized.
Fifth Amendment	No person shall be held to answer for a capital, or otherwise infamous crime, unless on a presentment or indictment of a Grand Jury, except in cases arising in the land or naval forces, or in the Militia, when in actual service in time of War or public danger; nor shall any person be subject for the same offence to be twice put in jeopardy of life or limb; nor shall be compelled in any criminal case to be a witness against himself, nor be deprived of life, liberty, or property, without due process of law; nor shall private property be taken for public use, without just compensation.
Sixth Amendment	In all criminal prosecutions, the accused shall enjoy the right to a speedy and public trial, by an impartial jury of the State and district wherein the crime shall have been committed, which district shall have been previously ascertained by law, and to be informed of the nature and cause of the accusation; to be confronted with the witness against him; to have compulsory process for obtaining witness in his favor, and to have the Assistance of Counsel for his defense.
Eighth Amendment	Excessive bail shall not be required, nor excessive fines imposed, nor cruel and unusual punishment inflicted.
Fourteenth Amendment (Section 1)	All persons born or naturalized in the United States, and subject to the jurisdiction thereof, are citizens of the United States and of the State wherein they reside. No State shall make or enforce any law which shall abridge the privileges or immunities of citizens of the United States; nor shall any State deprive any person of life, liberty, or property, without due process of law; nor deny to any person within its jurisdiction the equal protection of the law.

Fifth Amendment

The Fifth Amendment requires three main things in relation to criminal justice procedure. First, it requires that indictments or formal charges by a court generally must precede criminal prosecutions. Second, it prohibits double jeopardy.

Third, it protects against **self-incrimination**, or compelling people to say things that might result in their being judged guilty. These topics are covered in Chapters 4 and 5. The most controversial of these has been the right against self-incrimination. As will be shown, considerable discretion is involved in deciding what constitutes self-incrimination and also in determining what is involved in compelling a person to be a witness against himself or herself.

Sixth Amendment

This amendment grants a number of important rights, including the rights to: (1) a speedy, public, and impartial jury trial; (2) advance notice of the charges; (3) call and confront witnesses; and (4) the assistance of counsel. These rights are taken up in more detail in Chapter 5, but it is obvious that each entails considerable discretion on the part of officials. For example, terms such as *speedy* and *impartial* have to be interpreted and translated into action. In addition, much controversy has surrounded the right to counsel. At precisely what point in the process does this right apply? What constitutes sufficient counsel? What about people who cannot afford to pay for the services of counsel? These and other questions reveal the discretion involved in applying the Sixth Amendment.

Eighth Amendment

The shortest of the amendments under consideration here outlaws excessive bail as well as cruel and unusual punishment; the bail issue is addressed in Chapter 5, while cruel and unusual punishment is addressed at various points throughout this book. Here the important point is to appreciate the discretion involved in determining how much bail is "excessive" or what type or amount of punishment is considered "cruel and unusual." These are subjective terms that court and correctional officials have worked to define for many years. They are terms with enormous ramifications for persons detained in jail awaiting further court processing, for persons serving time in the nation's prisons, and for persons sentenced to death.

Fourteenth Amendment

The language of Section 1 of the Fourteenth Amendment (see Table 3–1) is deceptively simple. The idea, or so it would seem, is to require that the various states, during the course of making and enforcing their laws, adhere to the same due process and equal protection requirements as those applying to the federal government through the Bill of Rights. When the Bill of Rights was added to the Constitution in 1791, the rights contained therein applied only to the federal government. People facing prosecution for crimes under laws of the various states (which is exactly how most criminal behavior was and is prosecuted) were not entitled to these rights. In fact, in the 1833 case of *Barron v. City of Baltimore*, the U.S. Supreme Court ruled explicitly that only federal officials were bound to follow the Bill of Rights. Chief Justice John Marshall summarized the Court's logic:

> *The constitution was ordained and established by the people of the United States for themselves, for their own government, and not for the government of the individual states. Each state established a constitution for itself, and in that constitution, provided such limitations and restrictions on the powers of its particular government, as its judgment dictated. The people of the United States framed such a government for the United States as they supposed best adapted to their situation and best calculated to promote their interests. The powers they conferred on this government were to be exercised by itself; and the limitations on power, if expressed in general terms, are naturally, and, we think, necessarily, applicable to the government created by the instrument. They are limitations of power granted in the instrument itself; not of distinct governments, framed by different persons and for different purposes.*[27]

When the Fourteenth Amendment was ratified a few years after the Civil War, the states did not automatically begin adhering to the Bill of Rights. By that time, the states had a long history of resisting a strong centralized federal government, one that could mandate how states should process criminal cases; the states wanted autonomy to decide such matters. The U.S. Supreme Court was reluctant to assume responsibility for overseeing states' criminal procedures, and additionally, the Court did not have a well-developed body of case law to use in setting up guidelines for states to follow. So, the Court made little systematic effort to require state adherence to the Fourteenth Amendment. In general, the Court overruled states only when the justices perceived gross violations of fundamental fairness.

This situation began to change after Earl Warren, a former prosecutor, became Chief Justice of the Supreme Court in 1953. For years before Warren's appointment, there had been debate about the extent to which the Fourteenth Amendment was intended to include or incorporate the Bill of Rights because the amendment itself does not mention the matter. The issue is important because to the extent that the amendment is viewed as incorporating the Bill of Rights, the states are bound to observe those rights.

The Warren Court spearheaded a policy of **selective incorporation**. This means that instead of rendering a wholesale decision that all provisions in the Bill of Rights apply to the states, the Court determined that certain rights would become binding on the states by virtue of its decisions rendered on a case-by-case basis. As such, the Warren Court gradually applied the Bill of Rights to the states.

A major U.S. Supreme Court decision in this regard was *Mapp v. Ohio*, decided in 1961.[28] Years earlier in 1914, the Court had established a precedent known as the **exclusionary rule** in the federal case of *Weeks v. U.S.*[29] This rule held that evidence obtained by law enforcement in violation of a person's constitutional rights (under the Bill of Rights) cannot be used to prosecute the person in federal proceedings; even though such evidence might be quite relevant, it must be excluded. In *Mapp v. Ohio*, the Court made the exclusionary rule, which initially had applied only to the federal government, binding on the states.

> *Our holding that the exclusionary rule is an essential part of both the Fourth and Fourteenth Amendments is not only the logical dictate of prior cases, but it also makes very good sense. There is no war between the Constitution and common sense. Presently, a federal prosecutor may make no use of evidence illegally seized, but a State's attorney across the street may, although he supposedly is operating under the enforceable prohibitions of the same Amendment. Thus the State, by admitting evidence unlawfully seized, serves to encourage disobedience to the Federal Constitution which it is bound to uphold.*

Later in the decision, the Court noted:

> *The ignoble shortcut to conviction left open to the State tends to destroy the entire system of constitutional restraints on which the liberties of the people rest. Having once recognized that the right to privacy embodied in the Fourth Amendment is enforceable against the States, and that the right to be secure against rude invasions of privacy by state officers is, therefore, constitutional in origin, we can no longer permit that right to remain an empty promise. Because it is enforceable in the same manner and to like effect as other basic rights secured by the Due Process Clause, we can no longer permit it to be revocable at the whim of any police officer who, in the name of law enforcement itself, chooses to suspend its enjoyment. Our decision, founded on reason and truth, gives to the individual no more than that which the Constitution guarantees him, to the police officer no less than that to which honest law enforcement is entitled, and, to the courts, that judicial integrity so necessary in the true administration of justice* Mapp v. Ohio, *367 U.S. 643 (1961).*

The important thing about this decision is the willingness of the Court to spell out rather detailed restrictions for how states can process cases in an effort to hold states accountable to the Fourteenth Amendment and, hence, the Bill of Rights. The string of landmark cases that followed the *Mapp* decision further extended this trend. For example, two years after *Mapp* in *Gideon v. Wainwright* (1963), the Court required state courts to furnish defense counsel in felony cases when defendants were indigent and could not afford counsel; this was intended to protect the Sixth Amendment right to counsel.[30] The next year, the Court established important precedent in *Cooper v. Pate* for granting certain constitutional freedoms to prison inmates.[31] Then, in what some consider the most important landmark case in criminal justice (and undoubtedly the most important Fifth Amendment case), the Court ruled in *Miranda v. Arizona* (1966) that, upon being arrested and before being questioned, suspects must be read their rights concerning self-incrimination and defense counsel (the famous Miranda Warnings).[32]

Collectively, these decisions, and others like them handed down by the Warren Court, are known as the **due process revolution** in criminal justice. These decisions were controversial. Some politicians and criminal justice officials took a conservative stance and argued that the Supreme Court was hindering the ability of lawmakers, police, courts, and correctional officials to control crime and preserve law and order. The liberal position, on the other hand, saw the decisions as advancing civil rights.

The composition of the Supreme Court gradually changed to become more conservative. Warren Burger became Chief Justice in 1969, and in 1986, William Rehnquist became Chief. To some extent, the Burger Court continued the due process trends of the Warren Court; for example, in 1974 certain due process protections were extended to prisoners facing disciplinary hearings in institutions.[33] In fact, as a general rule today, the Bill of Rights is considered binding on state criminal justice proceedings.

Although it remains to be seen how the Court headed by Chief Justice John Roberts (nominated and confirmed in 2005) will affect matters, it was becoming increasingly clear even during the 1970s that the due process revolution had reached a peak. This is evidenced by decisions that have recognized exceptions to some of the protections mandated by the Warren Court. For example, in 1984 the Court recognized two exceptions to the exclusionary rule, known as the "good faith exception" and the "inevitable discovery exception."[34] The first holds that evidence obtained using a search warrant that is ultimately determined to be invalid may be allowed in court if the police were acting in good faith on what they believed to be a valid warrant. The second holds that evidence obtained in violation of a defendant's rights may be allowed if it can be shown that the evidence would have been discovered via legal means at a future point in time.

The turn toward a more conservative era in criminal justice is evident not only in the actions of the Supreme Court, but also in those of lawmakers. A good illustration is the controversial **PATRIOT Act** passed by Congress in the aftermath of September 11, 2001. This complex and multifaceted legislation added to and revised existing federal law to improve efforts to control terrorism. Among other things, the act loosens certain legal restrictions on such federal agencies as the Federal Bureau of Investigation (FBI) and the Central Intelligence Agency to render it easier to gather evidence against suspected terrorists, especially regarding information from electronic sources.[35] Some claim that if the nation is to be better protected from future attack, it is essential to give those officials who are responsible for public safety more latitude to do their jobs. Others say that provisions in the PATRIOT Act threaten to erode progress in civil rights made during the due process revolution. This debate is a classic example of the potential for tension between the value our society places on having people feel safe and the value it places achieving just and fair treatment for those considered a threat to safety.

Defining and Categorizing Crime

It is a mistake to think that the official, legal definition of crime equates with what all people and groups consider harmful, wrong, or deviant from normative behavior. An act such as abortion or malfeasance by a large corporation might be considered wrong or harmful by some groups and yet not even be defined as illegal under the law. Similarly, an act such as growing marijuana on one's private land might not be considered wrong by some, even though it is illegal.

This illustrates both the subjective and political nature of the way crime is defined.[36] To a very important extent, what is considered wrong and/or illegal is relative to person, place, and time. A key factor in shaping what the law defines as crime, for any given person or group at any give place or point in time, is the success of various interest groups (e.g., members of the religious right, civil libertarians) at achieving the power, money, and other resources needed to codify their views into law through the political process.[37] Law is inherently political; it is the outcome of such political realities as negotiation, compromise, and partisanship.

Not only is there a lack of consensus on what the law should define as crime, but there is also no single agreed upon way to classify crimes into categories. Recall from Chapter 2 that the FBI categorizes crime into more serious Part I offenses and less serious Part II offenses. Some other categorizations systems are examined in the following subsections.

Mala In Se versus *Mala Prohibitum*

Under English common law, a distinction evolved between crimes that are ***mala in se***, or wrong or evil by their nature, and ones that are ***mala prohibitum***, or wrong because they are defined as such by law. *Mala in se* offenses, such as murder or rape, came to be considered illegal as the decisions of judges were established as precedent under the doctrine of *stare decisis*. By contrast, *mala prohibitum* offenses, such as drug use or not paying taxes, came to be considered wrong after being defined as crime under statutes passed by the government.

David Skelton ties this distinction to the religious origins of ideas about crime.[38] Some acts (*mala in se*) are widely viewed as inherently immoral or sinful. For other acts (*mala prohibitum*), there is less social consensus about their wrongful nature, and religious prohibition is less intense. In addition to revealing the relationship of religion to law, the *mala in se* versus *mala prohibitum* distinction also illustrates the subjective and political nature of crime described earlier. What one person or group sees as *mala in se* another might view as *mala prohibitum*.

Felonies versus Misdemeanors

Another basic distinction that emerged under common law is that between felonies and misdemeanors, and unlike the *mala in se* versus *mala prohibitum* differentiation, this distinction is used widely today. A **felony** is considered a more serious crime and is usually punishable by death or incarceration for a year or more in a state or federal prison. By contrast, a **misdemeanor** is considered less serious and is typically punishable by fine and/or less than a year incarceration in jail (as opposed to prison). Some jurisdictions' codes recognize a third category considered less serious than misdemeanors; these offenses are called infractions or violations.

As pointed out previously in this chapter, felonies and misdemeanors are often divided into degrees, such as the distinction between first- and second-degree murder discussed earlier. In addition, collections of felony and misdemeanor crimes that are considered to be of roughly similar seriousness are frequently categorized into classes of seriousness (e.g., a Class A versus a Class B felony). For example, under Indiana law manufacturing methamphetamine is considered a Class B felony, unless the manufacturing takes place within 1,000 feet of school property, in which event the act is a Class A felony and punished more severely. Likewise, Indiana law considers involuntary manslaughter (the intentional killing of another person in the heat of passion) a Class B felony, but if the offense is committed with a deadly weapon, it is a Class A felony.

Classification by Characteristic of the Offender

The FBI system, the *mala in se* versus *mala prohibitum* distinction, and the felony versus misdemeanor distinction are all means of classifying crimes according to perceived wrongfulness or seriousness. Another way to classify crimes is according to some characteristic of the offender.

In the United States, the legal systems for dealing with adults and juveniles have, to some extent, been separate for more than a century (see Chapter 14). There are separate sets of laws, separate police procedures, separate courts, and separate correctional agencies. Although terminology varies across jurisdictions, behaviors on the part of people under the age of legal majority (usually 16, 17, or 18 years) that violate the criminal law are often called criminal, public, or delinquent offenses. Behaviors such as robbery, assault, or theft are considered criminal regardless of age (as long as the offender is old enough to be considered capable of forming the *mens rea*). But juveniles can also be convicted of **status offenses**, behaviors that are illegal only when committed by a person who is legally a juvenile or minor. Examples include running away from home, being incorrigible or truant at school, smoking, and drinking alcohol. These behaviors are not considered illegal for an adult.

Another illustration is found in laws that focus on the criminal history of the offender. Many jurisdictions have laws that provide stiffer punishments for people who have been convicted of lawbreaking multiple times. Such persons may be called habitual, repeat, or career criminals. For example, the Indiana Code (35-50-2-8) provides that "except as otherwise provided in this section, the state may seek to have a person sentenced as a habitual offender for any felony by alleging . . . that the person has accumulated two (2) prior unrelated felony convictions." In some jurisdictions, if repeat crimes are sex-related, offenders may be processed as sexual predators. The point is that the same illegal behavior can take on a different meaning under the law depending on whether it is committed by a first-time offender or by someone who has a record of prior convictions.

Academic Classifications

Criminologists, legal scholars, sociologists, political scientists, and other academics have offered numerous categorization schemes. Because it is impossible to cover all these, an integration of some representative sources is summarized in the following subsections.[39]

Person Offenses

In addition to the FBI's four violent Part I crimes that were covered in Chapter 2 (murder and nonnegligent manslaughter, forcible rape, aggravated assault, and robbery), the laws of many jurisdictions recognize additional person offenses. Many of these are regarded as Part II offenses by the FBI. One example is simple assault, which is distinguished from aggravated assault in that simple assault usually involves less serious injury and/or the absence of a deadly weapon. Some more illustrations include the following:

- Kidnapping—unlawfully detaining a person against his or her will
- Mayhem—unlawfully dismembering or disabling another
- Hate or bias crimes—directing criminal behavior at another person or group because of hostility toward that person or group's race, religion, sexual preference, or other characteristic
- Stalking—repeated following and harassment of a person
- Abuse/neglect of a child

There are sex-related crimes in addition to forcible rape. Examples include the following:

- Statutory rape of a minor incapable of consent
- Rape of a spouse
- Sexual assault—subjecting a person to nonconsensual sexual contact other than rape
- Engaging in sexual contact when knowingly infected by HIV/AIDS
- Sexual harassment—such as making unwanted advances or requests for sexual favors in the workplace

Property Offenses

The FBI recognizes four Part I property crimes (burglary, larceny-theft, motor vehicle theft, and arson) as well as a number of Part II offenses related to property (see Chapter 2). These Part II offenses partially overlap and include forgery (falsely writing or altering a document so as to defraud the recipient), counterfeiting, fraud (intentionally deceiving someone for personal gain), embezzlement (which occurs when a person who has been entrusted with someone else's property illegally converts or misappropriates that property for his or her own gain), handling stolen property, and vandalism or property destruction.

In many places, there are additional property offenses not explicitly listed as Part II crimes by the FBI. Examples are as follows:

- Criminal mischief—similar to vandalism or property destruction but with malicious intent to injure the owner
- Criminal trespass—illegal intrusion onto the property of another
- Criminal littering
- Extortion or blackmail—gaining property from someone by threatening harm or coercion in the future
- Identity theft—illegally using the identity of another for personal gain
- Joyriding—taking a vehicle without the owner's consent, but without the intention to deprive the owner permanently

The proliferation of computer technology has led to a category of property offenses known as computer or cybercrime. These are distinct from some of the traditional offenses that can be committed using a computer (e.g., pornography offenses, blackmail, theft). As Frank Schmalleger points out, "Only crimes that employ computer technology . . . and . . . could not be committed without it" qualify as computer crime.[40] Schmalleger identifies five somewhat overlapping categories of computer offenses: (1) use of a computer to fraudulently gain or convert property, (2) trespass or illegal manipulation of stored information, (3) stealing computerized services without authority, (4) illegal use of a computer to cause physical harm, and (5) tampering or disseminating viruses.

Controversies

Many of the person and property offenses just covered, as well as some of the other offenses described later, occur with considerable regularity inside the nation's prisons and jails, with inmates victimizing inmates, staff victimizing inmates, and inmates victimizing staff. A recent controversial acknowledgment of this fact is the Prison Rape Elimination Act (PREA) that was signed into law by President George W. Bush on September 4, 2003, as the first federal law to address sexual assault in prisons and jails. Although some state and local officials minimize or deny the existence of rape behind bars and others resent federal intrusion into their affairs, advocates hail PREA as a milestone in gaining better control of sexual victimization in correctional facilities.[41]

Offenses Against Public Order, Justice Administration, and Morality

This category includes offenses that are considered either threatening to public order, threatening to the administration of justice or government, or in violation of public decency. Public order crimes include disorderly conduct, public intoxication or drunkenness, vagrancy/loitering, illegal carrying or possession of weapons, unlawful assembly, and rioting. John Klotter also places certain acts related to terrorism in this category.[42] Obviously, people who perpetrate terrorist acts can be convicted of other offenses (e.g., murder, property destruction, kidnapping). But even before September 11, 2001, there were specific counterterrorism laws in the United States. The PATRIOT Act, discussed earlier in the chapter, was enacted thereafter to broaden existing legislation. For example, this act makes it illegal to provide material support or resources to foreign terrorist organizations.

Offenses against the administration of justice or government include such acts as resisting arrest, perjury (lying under oath), acting with contempt or disrespect toward a court, paying or accepting bribes for special favors, obstructing justice, failing to report crimes, tampering with witnesses or evidence, misconduct by an official, and treason. Escape from the custody of criminal justice officials also falls into this category.

Morality offenses are also called decency, vice, or victimless crimes. These offenses are quite controversial because different people and groups do not agree on the extent to which the activities should be regulated by government, if at all. Some argue that the government is responsible for upholding the morality of citizens and that no crime is truly victimless—even when it involves consenting parties, as in the case of prostitution. Others counter that because the issue is whose version of morality the government is going to uphold, bias and unfairness are inherent in any such effort. Morals offenses can be conveniently grouped into three categories, as discussed in the following subsections.

Sex-Related Offenses Sex-related morality offenses include prostitution, pimping, violations of regulations concerning pornography and obscenity, indecent exposure (sometimes called lewd or lascivious conduct), and voyeurism. Other such offenses involve consensual contact. Depending on the jurisdiction, examples can include fornication, adultery, polygamy, incest, and forms of sexual intercourse considered to be deviant. The latter could include bestiality (sex with an animal) and sodomy, a term used to describe anal and/or oral sex. Some jurisdictions rely on sodomy statutes to regulate homosexuality, whereas others have laws explicitly prohibiting homosexual relations.

Obviously, many of these laws, especially those related to consensual contact, are very difficult, if not impossible, to enforce. Opponents to criminalizing such behaviors argue that this is yet another reason for the government to avoid regulation of morality. But proponents counter that enforcement difficulties, though real, do not absolve the government of its responsibility to uphold public decency.

Controlled Substance Offenses The law distinguishes two basic uses of drugs, including medical and recreational purposes, and both are regulated by law. Although the distinction can become blurred, as in the case of athletes using strength-enhancing drugs, recreational use is typically what people have in mind when they think of drug violations as offenses against morality. So, this subcategory of morals offenses involves governmental efforts to regulate the manufacture, distribution, and consumption of alcohol and other drugs for recreational purposes.

Some of these drugs, such as alcohol and nicotine, are generally considered legal as long as they are made, sold, purchased, and used under conditions set by law. Others, such as heroin, crack/cocaine, and marijuana, are generally considered illegal unless the law specifies conditions for legal use.

Drugs are controlled under federal legislation, states' laws, and local legislation. Some of the first federal laws were the Harrison Narcotics Act of 1914 that sought to regulate opiate drugs and the Marijuana Tax Act of 1937. Furthermore, alcohol was prohibited during the 1920s. A wave of federal legislation followed over subsequent decades.[43] Initially, much federal drug control legislation sought to raise tax revenue (as was true of the Marijuana Tax Act of 1937), but gradually, drugs came to be defined as a crime problem—a prerogative of the U.S. Justice Department rather than the Treasury Department.[44] Today, the federal government and most states group drugs into categories or "schedules" based primarily upon (1) the extent of known medical use and (2) the potential for abuse and addiction. Stronger penalties are attached to those drugs considered by government officials to be the most harmful.

Gambling Offenses Like drug use, gambling is another activity that has legal and illegal dimensions. What is prohibited depends largely on the jurisdiction, the particular location within a jurisdiction (as with casinos located on riverboats or reservations), and time era in question. For example, some jurisdictions allow lotteries, bingo, betting on horse or dog races, and/or casino games. Others forbid virtually all types of gambling. Federal law regulates gambling on federal property and the interstate transmission or transportation of gambling information and devices.[45] Some jurisdictions presently allow forms of gambling that they previously prohibited, and some now prohibit what they previously allowed.

Syndicated (Organized) Crime

Unlike the categories of crime discussed earlier, **syndicated crime** (as well as white-collar and political crime—both discussed later) is distinguished in important ways by certain characteristics of those who commit it. Those who commit syndicated, white-collar, and/or political crimes certainly can, and sometimes do, commit the offenses described previously. But because of the roles and criminal opportunities available to them by virtue of their culture or jobs, they are also in positions to carry out various other illegal activities.

Crime syndicates consist of people who group together to perform illegal activities, mostly for monetary gain. Such syndicates have varying structures and levels of organization, depending largely on the nature of the illegal activity.[46] Some are highly structured and organized, whereas others have less complex, looser organizational structures. Likewise, some are quite large (and even international in scope), while others are very small and geographically confined. Crime syndicates even exist in some jail and prison systems in the form of prison gangs, and these groups often have ties to syndicates outside.

The basic motive for syndicated crime is easy to understand. Any time the law prohibits certain goods and services that members of a society (or sectors of a society) demand, there is an incentive for crime syndicates to form and provide those goods and services illegally. The structure and organization of a given syndicate is a function of how the illegal goods and services in which the syndicate deals can be provided most efficiently and profitably.

The stereotype of syndicated crime in the United States is the infamous Sicilian or Italian mafia. However, like most stereotypes, this one is misleading. Crime syndicates have involved, and do involve, numerous other groups of people. These groups frequently, but not always, share a common ethnic heritage (e.g., Russian, Latino, Asian) and often have close working ties with corrupt officials in the "legitimate" world of business and government. An overarching, national Italian mafia does not have a monopoly on syndicated crime in the United States.

The illegal activities of crime syndicates include many of the crimes already covered. For instance, two property crimes that are common among crime syndicates include extortion (defined previously) and loan sharking—illegally providing financial loans at excessive interest rates, typically to people having difficulty obtaining loans through legal means. In addition, vice offenses are especially attractive to syndicates because of opportunities for sizable profits through provision of illegal goods and services that are in high demand. Major examples include drug trade activities, prostitution, pornography, gambling, and illegal dealing in weapons. Such offenses are sometimes called "black market" crimes.[47]

But as Beirne and Messerschmidt point out, syndicated crime extends beyond these crimes to include illegal activities related to legitimate business and government.[48] For example, money obtained illegally through a syndicated drug trade can be invested in various legitimate businesses to create added profits and help provide a cover for criminal activities. In addition, because having cash money secured from black market activities can increase the chances of being detected, syndicates sometimes engage in money laundering; this means they use various means of concealing black market profit and disguise it to create the appearance that it came from legitimate operations. Some syndicates are also involved in racketeering. This means that their members infiltrate legitimate businesses or organizations such as labor unions and then use extortion or fraud to obtain profits. Racketeering can be done in various ways, such as by looting the organization's accounts or by diverting the organization's funds to illegal enterprises. Finally, many syndicate figures have connections with corrupt government officials. These figures (or persons associated with them) interact with select criminal justice and political officials to perform mutual favors and make profits. In fact, a solid argument can be made that it is difficult for a syndicate to sustain illegal operations successfully without ultimately having connections in the legitimate world of business, government, or both. For instance, prison syndicates frequently develop cooperative relations with certain prison employees as a means of getting drugs inside the institution.

White-Collar Crime

The term **white-collar crime** was coined in 1949 by the sociologist Edwin Sutherland to refer to offenses committed by respectable individuals occupying higher-level social positions during the course of their jobs.[49] Two basic types of white-collar crime include occupational crime and corporate crime.[50]

Occupational Crime As the name implies, **occupational crime** refers to crimes committed by people for personal gain during the course of their jobs. An example, previously mentioned as a property crime, is embezzlement. Embezzlement takes place when an employee misappropriates money that has been legally entrusted to the employee's control (e.g., a stockbroker diverting investor funds for personal use). By contrast, employee theft involves outright larceny from the employer or the customers, such as stealing a computer from the workplace or intentionally overcharging or shortchanging customers for personal profit. Additionally, there are various forms of occupational fraud. One example, illustrative of what some scholars call "professional crime," is a physician who defrauds Medicare by performing unneeded tests.[51] Another example is insider trading, which takes place when an employee uses information that is not available to the general public to gain an upper hand in the trading of stocks.

Corporate Crime The main difference between corporate crime and occupational crime is that **corporate crime** is committed to further the objectives of the company (rather than for the profit of a particular employee) with support from other members of the company. Corporate crime differs from syndicated crime in that the corporation is not in existence primarily for illegal purposes.

Because some corporations have powerful ties to lawmakers and other policy officials (through lobby efforts, campaign contributions, and the like), corporate officials can be in a position to affect the way their activities are legally defined. Thus, although some corporate wrongdoing is defined as crime under criminal law, a good deal of it is defined as civil infractions under administrative or regulatory law (where penalties are generally milder).

One fairly well-known type of corporate offending includes violation of antitrust laws, which exist to keep the market competitive. For example, these laws prohibit corporations from conspiring to fix prices at a particular level (so that consumers have no choice but to pay a certain rate) and also from illegally monopolizing markets through corporate buyouts, mergers, and acquisitions. Another fairly well known type of corporate crime is false advertising, wherein deceptive practices are used to market products or services. Corporations can also perpetrate fraud against either the public or the government, such as when "costume" products are sold to the public at authentic prices or government defense officials end up paying unreasonably high prices for items or services under their contracts with corporations. Similarly, corporations can bribe either private or government organizations to gain profit or other advantages.[52]

Corporate crime is not usually thought of as being violent. Yet there is good evidence that some corporate offenses can be very damaging to people's health and safety.[53] One example takes place when companies knowingly violate codes regulating worker safety, leading to accidents or diseases that culminate in disability or death. Another example is found in the knowing violation of consumer protection laws that regulate the testing and marketing of products such as food, medical supplies, and vehicles. A final example involves the knowing violation of pollution laws resulting in the contamination of air, water, and even entire towns. On occasion, the response of some corporations to stronger efforts to control these activities in the United States has been to illegally "dump" products not approved for marketing in the United States on other countries that have lower legal standards or to locate operations in nations that have less-restrictive pollution or worker safety codes.[54] Although it may be true that most "violent" corporate offenses are ultimately committed to further profit objectives, the end results can be the same as the results of violent street crime (i.e., death, disease, and serious injury).

Political Crime

In much the same way that there are laws to regulate the activities of persons performing their occupations, or the activities of corporations pursuing their objectives, there are also laws meant to regulate the behavior of those who hold or aspire to hold public offices or various government posts. Such persons attain their positions either through being elected or appointed, or they are hired by elected or appointed officials. In general, these laws address the way political power is acquired, maintained, and exercised as well as the way public monies are handled.[55] **Political crime** refers to violation of these laws.

Of course, government officials can and do commit some of the crimes described previously. But because of their positions as public servants, these officials can be presented with motives and opportunities for additional offenses. For example, there may be opportunities to accept (or offer) bribes or kickbacks of money in return for political favors; such behavior sometimes involves collusion with syndicated crime figures or corporations. Likewise, there are often ample opportunities for government officials to make personal use of government funds or services, such as transportation or hotel accommodations.

In addition, the environment in which most government officials operate is highly competitive and political; there is considerable maneuvering for power and position. This environment produces both motives and opportunities for offenses such as violation of election laws, illegal campaign financing, and illegal cover-up schemes to contain potentially damaging information. Related illustrations include various illegal dealings with the officials or citizens of foreign nations to advance certain U.S. interests abroad.[56]

Political crime can assume at least two other forms. One involves government officials violating such civil liberties as free speech, free assembly, and privacy to repress dissent. Examples of these repressive activities are illegal monitoring of hard-copy and electronic correspondence, illegal entries of homes or places of business to conduct searches, illegal surveillances, and violations of various treaties and agreements with various domestic and foreign groups.

The other form that political crime can take is politically motivated illegal actions perpetrated against government. Treason and terrorism have already been mentioned in this chapter, and both are good illustrations of this form of political crime. As part of trying to advance their causes, some historically oppressed groups, such as ethnic minorities, feminists, and organized laborers, have engaged in violence directed against the government that falls far short of constituting treason or terrorism. Other offenses against government are nonviolent and include illegal demonstrations, acts of civil disobedience, and espionage or spying against the government.[57]

Application to Corrections

Although some of the material presented in this chapter may, at first glance, seem rather far removed from adult and juvenile corrections, it is all connected. A case in point involves this chapter's coverage of discretion. Clearly, it is the discretion of police and court officials that determines the cases that enter correctional agencies. People do not serve time on probation or in prison without first having been arrested, prosecuted, convicted, and sentenced. And as with police and court officials, the discretion of correctional officials is exercised within the boundaries of law and policy. Because the work of correctional officials is not as high profile and often more removed from the public, correctional discretion has traditionally been less constrained by public/media scrutiny. But the same tensions between idealistic principles and the reality of discretion that pervade police and court processing also pervade correctional processing.

The criminal justice system's common law heritage is reflected no less in corrections than in other areas. Substantive law defines the behaviors (or lack thereof) for which people are placed under correctional supervision. The vast majority of these people are under correctional supervision for FBI Part I common law crimes, FBI Part II crimes, public order crimes, and morality offenses (especially controlled substance crimes). Although less frequently, correctional agencies also supervise those convicted of behaviors discussed in this chapter under the heading of syndicated crime. Less frequently still, these agencies supervise white-collar and political offenders. Indeed, the common law distinction between felonies and misdemeanors often has profound impact on the kind of correctional supervision an offender receives. The same applies for legal distinctions drawn according to characteristics of the offender (e.g., juvenile versus adult and criminal history).

Procedural law specifies certain procedures according to which these persons must be handled by correctional agencies. References to the operation of correctional organizations are found in the various sources of law—statutes, constitutions, cases, and administrative regulations. Perhaps most noteworthy in this respect is constitutional law reflected through case decisions. The due process movement launched by the Warren Court had far-reaching impact on the procedures used by correctional officials because these procedures relate to the Bill of Rights. Today, correctional officials can be subjected to both civil and criminal prosecution for failing to protect the rights of those under correctional supervision.

Increasingly today, a challenge for the law is how best to encourage cooperation among police, court, and correctional officials in promoting public safety, while also protecting civil rights. One area where this challenge is evident is the growing need to monitor and control interactions and communications between street gangs, crime syndicates, and terrorist organizations operating outside correctional facilities, on the one hand, and persons serving sentences in

Corrections in the Real World

A probation officer responsible for supervising a felon who has been assigned to intensively supervised probation could have five face-to-face contacts with the offender in one week. An officer responsible for supervising a person placed on probation for a misdemeanor might not have five face-to-face contacts with the offender during the entire duration of the sentence.

these facilities on the other hand. Although the number of prisoners with ties to gangs, syndicates, and/or terrorist organizations operating on the streets is probably small (relative to the number lacking such ties), the threat is potentially serious. Controlling the threat in a constitutionally fair manner will necessitate collaboration between law enforcement, the judiciary, and corrections. Another area of this challenge involves the need to better control white-collar and political forms of crime. Although much of this behavior is hidden from scrutiny and handled lightly (in some instances condoned) by law, such activities exert great physical, financial, and social costs on society—costs that have traditionally far exceeded the effort and resources devoted to control. Achieving better control will require close collaboration among the three main elements of criminal justice as well as with lawmakers.

READY FOR REVIEW

- Law, as passed by legislative bodies and interpreted by the judiciary at all levels of government (federal, state, and local), is the foundation for the operation of both the adult criminal justice and the juvenile justice systems.
- Criminal and juvenile justice officials exercise discretion within the guidelines spelled out by the rule of law. These guidelines range from very broad to very narrow. Although law can structure and control discretion, law cannot eliminate it entirely.
- The modern system of jurisprudence in the United States has its main roots in the common law of England.
- The criminal law is meant to regulate public interests, and violations are prosecuted by the government. The civil law regulates private interests, and actions are often litigated between private parties.
- The substantive aspect of law defines what behavior is criminal and the corresponding sanctions, while the procedural aspect lays out the rules according to which the government can formulate and implement the substantive law.
- The major sources of substantive and procedural law include statutes, constitutions, court cases, and administrative regulations.
- As a central principle of substantive law, the *corpus delicti* (or body of the crime) includes the illegal act (*actus reus*), the appropriate mental state (*mens rea*), and the appropriate accompanying circumstances. There can be no crime without a written law and corresponding punishment for the act in question. Additionally, in general, offenses must cause some harm or damage, and the *actus reus* and *mens rea* have to be connected. A number of justifications and excuses based on *mens rea*, such as insanity, can lead to exoneration of an otherwise guilty person.
- Principles of procedural law illustrate the contrast between what is ideally supposed to occur and what often really does.
- The concept of fundamental fairness holds that criminal justice processing should fairly balance the interests of the offender with the interests of the public. Realistically, not all interested parties always (or even usually) perceive things as fair.
  - The law is supposed to protect and be applied to all persons equally. Realistically, groups with greater power and wealth often derive more benefits from the system.
  - The principle of propriety holds that the quality of evidence is more important than the quantity, but in reality, quantity is emphasized just as heavily, if not more.
  - Citizens are supposed to be presumed innocent until proven guilty. Although it is incumbent on the government to build a case against the

defendant, the presumption of innocence is often untenable long before full proof is offered.
  - Ideally, all citizens are entitled to due process of law, particularly as specified in the U.S. Constitution's Fourth, Fifth, Sixth, Eighth, and Fourteenth Amendments. In reality, the due process actually received depends on the particular case—the discretion exercised by officials, the time era, and the location. As an illustration, the U.S. Supreme Court did not begin requiring states to observe the federal Bill of Rights until the 1960s, and since that time, the Court has recognized important exceptions to the requirement.
- The manner in which crime is defined by law is both subjective and political, reflecting the success of various lobby and interest groups at mobilizing the power and resources to achieve their agendas.
- Crimes can be categorized in various ways, as follows.
  - *Mala in se* (wrong by nature) versus *mala prohibitum* (wrong due to prohibitive legislation)
  - Felony (more serious and punishable by imprisonment) versus misdemeanor (less serious and punishable by fine or jail)
  - By offender characteristics (e.g., juvenile versus adult, criminal history of the offender)
  - Academically (person offenses, property offenses, public order offenses, syndicated crime, white-collar crime, and political crime)
- The law is tightly bound up with the operation of adult and juvenile corrections.

KEY TERMS

actus reus The illegal act element of a crime. It refers to the commission of prohibited action or, conversely, omission of required action

Bill of Rights The first 10 amendments to the U.S. Constitution added in 1791 to protect citizens' freedoms from government

case (precedent) law Law that evolves out of judicial decisions, based on the principle of *stare decisis*, or let the decision stand

civil law Law geared mainly toward regulating private interests and relations. Violations are defined in terms of one private party (plaintiff) versus another (defendant). Emphasis is placed on ascertaining blameworthiness, culpability, or liability rather than guilt versus nonguilt. The standard of proof is preponderance of the evidence

common law A body and heritage of law that gradually evolved in England following the eleventh century from judicial decisions and that served as a foundation for development of the U.S. legal system

constitutional law Law found in federal and state constitutions that regulates both statutes and procedures

corporate crime Crime committed to further the objectives of a company (rather than for the profit of a particular employee) with support from other members of the company

corpus delicti A common law term referring to the body or elements of a crime

criminal law Law designed to protect public interests and safety. Violations are defined as violations against the public and prosecuted by the government. Emphasis is placed on the intent driving defendant actions, and the standard of proof is beyond a reasonable doubt

discretion The latitude to exercise judgment and select between various courses of action, as defined by law and policy

double jeopardy A prohibition against prosecuting a defendant more than once for the same offense in the same jurisdiction

due process The idealistic principle of procedural law that states that, during the course of criminal processing, people have particular rights and protections coming to them by virtue of U.S. citizenship

due process revolution An era of expanding rights for persons suspected, accused, and convicted of crime, spurred on by certain landmark decisions of the U.S. Supreme Court under Chief Justice Earl Warren

Durham Rule Contends that a defendant is excused from responsibility for a crime if a judge or jury determines that the crime resulted from mental disease or defect

entrapment An excuse to criminal liability where it is argued that the culpable state of mind originated with the authorities rather than with the defendant

equal protection An ideal principle of procedural law stating that the law is to be applied impartially across persons and groups

Exclusionary Rule A ruling established by the U.S. Supreme Court that evidence obtained in violation of an individual's constitutional rights cannot be used to prosecute the person

felony A crime generally considered to be more serious and usually punishable by death or incarceration for a year or more in a prison setting

fundamental fairness An ideal principle of procedural law stating that all steps taken to process a case through the system should fairly balance the interests of the accused with the interests of the public

irresistible impulse test A supplement to the M'Naughten Rule (see later definition) that asks whether the defendant was compelled by mental illness to commit a crime even though he or she knew the behavior was wrong

jurisprudence The legal philosophy underpinning a system of law and justice.

mala in se Offenses considered wrong by their nature

mala prohibitum Offenses considered wrong because they are defined as such by law

mens rea The element of a crime that describes the state of mind a person must be in to commit a given criminal act

misdemeanor Offenses generally considered to be less serious than felonies (see earlier definition) and usually punishable by fine and/or less than a year in a jail setting
M'Naughten Rule A test for insanity based on whether the defendant could distinguish right from wrong at the time of his or her crime
occupational crime Crimes committed by people for personal gain during the course of their jobs
PATRIOT Act Legislation passed by Congress in the aftermath of the September 11, 2001, terrorist attacks to improve efforts to control terrorism by, among other things, loosening certain legal restrictions of key federal agencies
political crime Crime involving the violation of laws regulating the manner in which political power is acquired, maintained, and exercised as well as the manner in which public monies are handled by politicians. Such crime also includes politically motivated illegal actions perpetrated against the government, such as treason
presumption of innocence The idealistic principle of procedural law that states that people accused of crime are presumed innocent until proven guilty by the government
procedural criminal law Those aspects of the criminal law that spell out the rules according to which the substantive criminal law (see later definition) may be formulated and implemented
propriety An idealistic principle of procedural law that holds that the quality of evidence against a person is more important than the quantity
regulatory (administrative) law Law enacted by legislatures working with administrative agencies at various levels of government
rule of law The principle that behaviors and interactions within society are regulated by a written system of rules and penalties to be applied in ways that are as fair and uniform as possible
selective incorporation The doctrine that certain rights in the federal Bill of Rights are binding on the states, as decided by courts on a case-by-case basis
self-incrimination Compelling persons suspected or accused of crime to say things that might result in a judgment of guilt; prohibited by the Fifth Amendment
stare decisis A principle of common law (see earlier definition) meaning to let the judicial decision stand
status offenses Behaviors that are illegal only when committed by a person who is legally a juvenile or minor (e.g., runaway)
statutory law Law passed by legislators at all levels of government
substantial capacity test A test for insanity based on whether the defendant had enough mental capacity to appreciate the wrongful nature of a behavior or to control his or her behavior
substantive criminal law Those aspects of the criminal law that specify what action is illegal as well as the sanctions for such action

syndicated crime Offenses committed by people who group together for the purpose of carrying out illegal activities, typically meant to achieve monetary gain by providing outlawed goods and services

white-collar crime Offenses committed by respectable individuals occupying higher-level social positions during the course of their jobs; basic types of white-collar crime include occupational crime and corporate crime (see earlier definitions)

YOU ARE THE CORRECTIONS PROFESSIONAL SUMMARY

1. What are some sources of law and policy you would consult to learn about how the jail is supposed to be performing classification? You would want to review the local jail's internal policy documents because these pertain to classification procedures (assuming such a policy exists). These should be compared with any relevant state-level regulations. To the extent that the state has authority over the local jurisdiction, local policies should be in compliance with state regulations. You should also review state and local statutes, appellate court decisions at both the state and federal levels, and perhaps the state constitution because these apply to conditions, practices, and violence in local jails. As part of these reviews, you may also be led to review standards or information provided by various professional organizations (e.g., the American Jail Association, the American Correctional Association, or the National Institute of Corrections).
2. What, if anything, else will you need to do before making recommendations to your boss? Careful observations and interviews will need to be conducted at the jail. Even if the jail has a sound classification policy in place (something unlikely in many jails), it is possible the policy is not being carried out as intended. The reality of discretion is that practice can be a far cry from policy, especially when policy is vague. Ask yourself what is actually being done in the jail to differentiate inmates according to such factors as offense seriousness, current legal status, criminal history, propensity toward violence, and propensity toward becoming a victim.

NOTES

1. M. R. Gottfredson and D. M. Gottfredson. *Decision Making in Criminal Justice: Toward the Rational Exercise of Discretion* (Cambridge, MA: Ballinger, 1980). See also: K. C. Davis, *Discretionary Justice: A Preliminary Inquiry* (Baton Rouge: Louisiana State University Press, 1969).

2. For its part, English law grew out of Roman law, Scandinavian law, and church law. Other roots of American law include French and Spanish law. See C. A. Cripe and M. G. Pearlman, *Legal Aspects of Corrections Management*, 2nd ed. (Boston: Jones and Bartlett, 2005).
3. R. Elias, *The Politics of Victimization: Victims, Victimology and Human Rights* (New York: Oxford University Press, 1986).
4. D. T. Skelton, *Contemporary Criminal Law* (Boston: Butterworth-Heinemann, 1998).
5. H. A. Johnson and N. T. Wolfe, *History of Criminal Justice* (Cincinnati, OH: Anderson, 1996).
6. The distinction below draws on a useful discussion found in L. J. Siegel and J. J. Senna, *Introduction to Criminal Justice*, 11th ed. (Belmont, CA: Thompson Wadsworth, 2008).
7. J. Samaha, *Criminal Law*, 6th ed. (Belmont, CA: West/Wadsworth, 1999).
8. *Marbury v. Madison*, 5 U.S. 137 (1803).
9. *Roe v. Wade*, 410 U.S. 113 (1973).
10. The discussion is based on the classic work of Jerome Hall. See J. Hall, *General Principles of Criminal Law*, 2nd ed. (Indianapolis, IN: Bobbs-Merrill, 1960).
11. Hall, *General Principles*, 213.
12. See J. C. Klotter, *Criminal Law*, 7th ed. (Cincinnati, OH: Anderson, 2004). See also: F. Schmalleger, *Criminal Law Today*, 2nd ed. (Upper Saddle River, NJ: Prentice Hall, 2002).
13. L. Katz, M. S. Moore, and S. J. Morse, eds., *Foundations of Criminal Law* (New York: Oxford University Press, 1999).
14. Schmalleger, *Criminal Law Today*.
15. Skelton, *Contemporary Criminal Law*.
16. Schmalleger, *Criminal Law Today*.
17. Siegel and Senna, *Introduction*.
18. Schmalleger, *Criminal Law Today*.
19. Schmalleger, *Criminal Law Today*.
20. Klotter, *Criminal Law*.
21. Klotter, *Criminal Law*.
22. The discussion is patterned after the ideas of Donald Newman as summarized in P. R. Anderson and D. J. Newman, *Introduction to Criminal Justice*, 5th ed. (New York: McGraw-Hill, 1993).
23. J. Reiman, *The Rich Get Richer and the Poor Get Prison* (Boston: Allyn & Bacon, 2004).

24. D. Cole, *No Equal Justice: Race and Class in the American Criminal Justice System* (New York: New Press, 1999), 8.
25. H. L. Packer, *The Limits of the Criminal Sanction* (Palo Alto, CA: Stanford University Press, 1968).
26. A discussion of the changing nature of justice and due process can be found in D. Garland, *Punishment and Modern Society* (Chicago: University of Chicago Press, 1990).
27. *Barron v. City of Baltimore*, 32 U.S. 243 (1833).
28. *Mapp v. Ohio*, 367 U.S. 643 (1961).
29. *Weeks v. U.S.*, 232 U.S. 383 (1914).
30. *Gideon v. Wainwright*, 372 U.S. 335 (1963).
31. *Cooper v. Pate*, 378 U.S. 546 (1964).
32. *Miranda v. Arizona*, 384 U.S. 436 (1966).
33. *Wolff v. McDonnell*, 418 U.S. 539 (1974).
34. See *U.S. v. Leon*, 468 U.S. 897 (1984) and *Nix v. Williams*, 467 U.S. 431 (1984).
35. The PATRIOT Act can be viewed at www.epic.org/privacy/terrorism/hr3162.html.
36. See H. S. Becker, *Outsiders: Studies in the Sociology of Deviance* (New York: Free Press, 1963). See also: R. Quinney, *The Social Reality of Crime* (Boston: Little, Brown, 1970).
37. G. B. Vold, *Theoretical Criminology* (New York: Oxford University Press, 1958).
38. Skelton, *Contemporary Criminal Law*.
39. The summary draws on the following: P. Beirne and J. Messerschmidt, *Criminology*, 4th ed. (Los Angeles: Roxbury, 2006); Klotter, *Criminal Law*; Schmalleger, *Criminal Law Today*, and Skelton, *Contemporary Criminal Law*.
40. Schmalleger, *Criminal Law Today*, 395. See also: R. Moore, *Cybercrime: Investigating High-Technology Computer Crime* (Cincinnati, OH: LexisNexis/Anderson, 2005).
41. To learn more about PREA, visit www.nicic.org/WebGateway_54.htm.
42. Klotter, *Criminal Law*.
43. Schmalleger, *Criminal Law Today*.
44. Beirne and Messerschmidt, *Criminology*.
45. Schmalleger, *Criminal Law Today*.
46. Beirne and Messerschmidt, *Criminology*.

47. R. J. Michalowski, *Order, Law, and Crime* (New York: Random House, 1985).
48. Beirne and Messerschmidt, *Criminology*.
49. E. H. Sutherland, *White Collar Crime* (New Haven, CT: Yale University Press, 1949).
50. Beirne and Messerschmidt, *Criminology*.
51. Michalowski, *Order*.
52. Michalowski, *Order*.
53. Reiman, *Rich Get Richer*.
54. Beirne and Messerschmidt, *Criminology*.
55. Michalowski, *Order*.
56. Beirne and Messerschmidt, *Criminology*.
57. Beirne and Messerschmidt, *Criminology*.

4

Policing and Corrections

Chapter Objectives

- Understand the history of American policing and the important influences in its evolution to twenty-first-century policing.
- Understand the structure and organization of policing in the United States.
- Understand the roles, functions, and activities that police perform as part of their daily activities
- Understand the various legal decisions that have affected policing.
- Understand the importance of policing in determining who becomes involved in the correctional filter and how the fields of policing and corrections need to be symbiotic to ensure the most effective criminal justice system.
- Understand how corrections and law enforcement work together to ensure community safety.

CASE STUDY

As a juvenile probation officer, a large percentage of the cases that you supervise are youth who are involved in gang-related behavior. You realize that one of the most crucial aspects of your supervision is to ensure that your clients avoid their friends who are currently involved in gangs and the opportunities for illegal behavior those gang members present to your clients. Unfortunately, this is often a difficult job, and you suspect that one of your clients (David) is associating with his buddies who are in a gang.

Because you have a good working relationship with the local police department, you contact one of the officers with whom you have worked closely before. You inform the officer of David's suspected involvement in gang activities and ask for her assistance. She suggests that you ride with her on her patrol of David's neighborhood that evening (**Figure 4–1**). After arriving in David's neighborhood, you notice that David is standing on the street corner talking to Kevin and Rick, both of whom you suspect are gang members.

1. What legal options do you have at this point?
2. Does the presence of the police officer affect your decision on how to handle this situation? Explain.
3. What action would you take? Why?

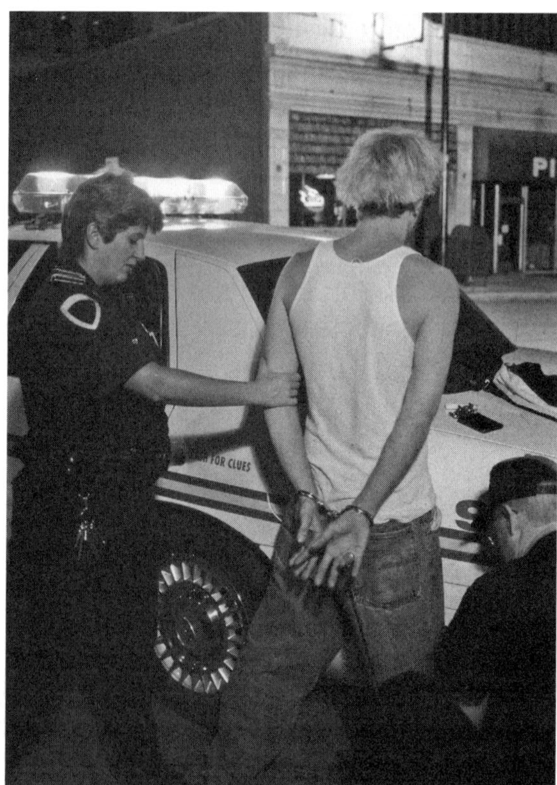

Figure 4–1 The police officer is one of the most visible actors in the criminal justice system.
Source: *© Creatas.*

Introduction

The first exposure most people have to the criminal justice system is the police. Unlike many other jobs in criminal justice and corrections that operate behind the closed door of the courtroom or the prison, police are constantly in the public eye. Of all the jobs in the criminal justice area, policing is perhaps the most prestigious, yet also the most dangerous. A whole genre of television shows glamorize the role of the police in society, including *Cops*, *48 Hours*, *The Shield*, *Law and Order* (and all its spin-offs), and *CSI*.

Despite this media exposure and constant surveillance by the public, many individuals aspire to be police officers from a very early age. Ask many children what they want to be "when they grow up" and "a police officer" is always an answer many children give. They want to chase "bad guys," "fight crime," or just help society in general. Nevertheless, these aspirations are often based on inaccurate ideas about what the role of the police in society entails. In this chapter, we discuss: (1) the history and evolution of policing; (2) its structure and organization in the United States; (3) the roles, functions, and activities of police officers; (4) legal issues in policing; and (5) the implications of policing for corrections and juvenile justice.

History and Evolution of Policing

Police operations in the twenty-first century differ significantly from those of earlier times. Police departments in the twenty-first century use computerized mapping to uncover trends in the geographic location of criminal activity, Special Weapons and Tactics or Emergency Services teams to defuse highly volatile situations, and DNA evidence to help solve crimes. Despite the dramatic advances in technology and weaponry that emerged during the latter part of the twentieth century, the primary functions of police today remain the same: (1) prevent and investigate crimes; (2) apprehend offenders; (3) help maintain domestic peace and tranquility; and (4) enforce and support the laws of society in which they are a part.[1]

Policing in Europe Prior to the Nineteenth Century

One of the earliest recorded types of policing is often referred to as **kin policing**. Kin policing was a policing strategy in which the family, clan, or tribe enforced informal rules and customs of conduct.[2] Under this strategy, each member of the family group had at least some authority (and even responsibility) to enforce the informal rules of the family group. This type of behavior control predominated in most societies until the arrival of the Greek and Roman empires, where more formal police/military officials began to enforce the laws. After the fall of those empires, policing returned to the responsibility of the family clans.

An important development in the history of policing in Europe began in the twelfth and thirteenth centuries, when kings began to assume responsibility for administration of the law. One aspect of that administration included strengthening the **nightwatch**, a group of citizens whose responsibility was to patrol their local area at night, constantly searching for fires, fights, and other problems. These positions were typically volunteer positions.[3]

In eleventh-century England, the **shire reeve** (from which the modern word *sheriff* evolved) appeared on the scene. The sheriff was appointed by the king to levy fines and make sure the **frankpledge** system worked. The frankpledge system was based on an ordering of tithings (groups of 10 households, headed by tithing-man), hundreds (groups of 100 households, headed by a reeve), and shires (groups of several hundreds). One role of the sheriff was to maintain law and order in the tithings. The sheriff served as a deputy for the king and regulated military, judicial, and fiscal matters. This office was seen as a pawn of the king, and officeholders were never popular among the citizens they regulated. By the late 1200s, many of the powers of the sheriff had been reduced and given to the **coroner**, a locally elected official whose job it was to monitor the sheriffs and make rulings on many civil matters. Just before his death, the king of England Edward I removed the position of sheriff from under the control of the king and made the sheriff a locally elected position as well. This trend has continued in England until today.[4]

Much like the sheriff, the **constable** can be traced to historical Europe. King Edward I established the office of constable in every parish during his reign. Because of the concern with defense that took up much of the time of the citizenry, it became the job of the constable to pursue lawbreakers. The constable also held a variety of other duties, including serving as magistrate and collecting taxes.[5]

By the seventeenth century, the vast majority of citizens were unwilling to serve as constable and the office of constable became one primarily used for financial gain, usually through corrupt enterprise. By the end of the eighteenth century, the collapse of the offices of sheriff and constable, along with the concurrent growth of large cities, crime, and civil disobedience, forced changes to be made.[6]

Policing in Nineteenth-Century England

By the late eighteenth century, largely as a result of the criticisms of the frankpledge system discussed earlier, several civic associations were created to enforce the law in specific areas in London. In 1746, Henry Fielding was appointed magistrate of the Bow Street Court in London. He established a police office as part of his court and paid his officers from the fines he collected from individuals violating the laws in his court. He published an essay on police reform in 1751 titled "Enquiry into the Causes of the Late Increase of Robbers," in which he argued that there were a number of social conditions prevalent in London that were associated with increases in robbery and other types of crime. His efforts at reform, which were continued by his brother John Fielding who succeeded him as magistrate in the Bow Street Court, were instrumental in the changes in policing that occurred in London at the beginning of the nineteenth century.[7]

In 1822, Robert Peel, the British home secretary, introduced the Constabulary Act in Ireland, and the Constabulary Police of Ireland was formed. In 1829, Peel was instrumental in the passage of the Metropolitan Police Act in Britain. This act created the first organized British metropolitan police force. Peel, assisted by commissioners Charles Rowan and Richard Mayne, was diligent in recruiting the most competent personnel possible, with a goal of establishing a military-style, professional police force. Together, they collaborated to produce the *First Instructions* to the new police force. In these instructions, they set forth a number of guiding principles that have continued (along with other decisions made by this triumvirate) to guide British and American policing into the twenty-first century. Some of these principles and decisions include the following:

1. The basic mission of the police should be crime prevention.
2. Police officers must be easily identified to ensure public confidence in the police. As such, police should be uniformed.
3. Police headquarters should be centrally located and easily accessible to citizens.
4. Officers would carry no weapons.
5. The police force would function on a 24-hour basis.[8]

Although this new police force did not replace the existing system of peace officers, by 1839, most of the other peace officers in London had been merged with the new police force. Although the new force was not initially popular with London citizens (because of their fear of one centralized police authority), the professionalism and impartial nature of the officers eventually won wide public support and the London police force became a model for municipal police forces worldwide.[9]

Policing in Colonial America

Policing in the colonial America time period (mid- to late eighteenth century) was similar to law enforcement practices in England during that time period. Colonial America policing originally consisted of volunteer constables and night watchmen, who were appointed or elected in cities and villages; their duties typically mirrored those of their counterparts in England. Nevertheless, the military presence of Britain, and the resistance to that military presence on the part of the colonists, caused numerous riots and disturbances, so much so that the duties of public safety were eventually given to military forces. After the Revolutionary War, law enforcement authority was returned to local governments.[10]

Policing in the rural areas of the South often differed from practices used in the colonial cities of the North. Southern governments used slave patrols whose job was to search for runaway slaves and return them to their owners. Because of this variation, it is difficult to neatly classify colonial policing into one or two historical categories. Nevertheless, at least in the vast majority of towns and cities in colonial America after the Revolutionary War, policing in America looked remarkably similar to policing in England.[11]

Policing in the Early to Mid-1800s

Following the Revolutionary War, many larger cities established paid police forces. By 1850, police forces in Philadelphia, Boston, and New York looked remarkably similar to one another and to the police force in London. Outside of the colonial cities, however, policing took on a different appearance. In areas where organized law enforcement did not exist (such as the western frontier), citizens often took control and formed **vigilance committees** to protect the life and property of community members. These committees often were the primary method of law enforcement until more formal policing agencies and methods emerged in those areas. Popular figures in the nineteenth century such as Judge Roy Bean, Wyatt Earp, Pat Garrett, and Bat Masterson often took it upon themselves to organize these vigilance committees (sometimes referred to as posses) to police the frontier.[12]

Along with the vigilance committees, private industry and the railroads formed their own protection agencies to protect their assets from criminals and other threats. One example of these private agencies was the Pinkerton Agency, whose original task was to investigate criminal activity on the railroads. Henry Wells and William Fargo formed a freight service about that same time (early 1850s) and provided their own protection for the goods and valuables that they transported (these agencies are discussed in detail later in this chapter).[13]

Policing in the Civil War Era

After the Civil War, southern states enacted laws that were designed to regulate the actions of the newly emancipated slaves. During this time period, white police officers routinely victimized blacks with little worry of punishment or regulation. The fourteenth and fifteenth Amendments (which gave equal protection and the rights of citizenship to blacks) reduced these victimizations, and blacks began to be hired in police departments as early as 1867 (in Selma, Alabama). Nevertheless, these black police officers were not considered equal to the white officers in those departments and were restricted in making arrests, often being assigned only to black areas and being prohibited from arresting whites.[14] When the U.S. Supreme Court issued the ***Plessy v. Ferguson*** **(1896)**[15] decision, which upheld the constitutionality of a Louisiana law that allowed segregation, many cities interpreted this decision as giving them the authority to fire most of their black police officers and return to police forces that were composed of all (or almost all) white officers. Consequently, at the turn of the twentieth century in the United States, most urban police forces consisted largely of white, male officers.

Policing from 1890 to 1930

At the turn of the twentieth century, policing in urban areas was characterized by political patronage. Uniformed officers were often appointed on an annual basis by city council members and mayors and were heavily influenced by the political forces in power in their city. As such, political and economic corruption was commonplace in most city police departments. Because police work during this period was often neighborhood oriented and decentralized, individual officers had a tremendous amount of discretion when dealing with the violations they witnessed. This combination of political patronage and individual officer discretion often meant that individuals of certain families, incomes, and political connections were treated differently from other citizens for engaging in the same activities.

By the 1890s, this system of policing came under increased criticism and efforts began to reform the police. These efforts coincided with the **Progressive Movement**. The Progressive Movement played a large role in reforming police departments, mainly through increasing the professionalism of policing. Increased professionalism was spurred primarily by two factors: the birth of professional associations and the Wickersham Commission.

In 1893, police administrators from throughout the United States created the National Chiefs of Police Union; the union later changed its name to the International Association of Chiefs of Police in 1902. The purpose of this association was to encourage cooperation and mutual assistance between city police departments, primarily in the East. Other issues addressed at early meetings included civil service standards and use of the telegraph to assist communication between police departments. This organization, along with reformers such as Theodore Roosevelt (police commissioner in New York prior to his presidency) and **August Vollmer** (chief of police in Berkeley, California), was a leading advocate of increased professionalism in policing. Their efforts led to decreased interference from political officials, increased standards for hiring of police officers, motorized police forces that used radio communication to increase their effectiveness, and increased numbers of women in policing, among other reforms.[16]

In 1929, President Herbert Hoover created the National Commission on Law Observance and Enforcement and appointed Attorney General George Wickersham to chair the committee. This committee was commonly referred to as the **Wickersham Commission** and was mentioned briefly in Chapter 1. The

Wickersham Commission conducted the first national study of the criminal justice system in the United States. August Vollmer was charged with surveying the police community and reporting the results to the commission. The commission's report was released in 1931. The report made a number of recommendations, including (1) the police department should be led by an individual who had been selected for competence and could be removed from office only after charges had been brought against him and a public hearing had been held; (2) salaries should permit decent living standards, with both health benefits and life insurance; (3) complete records should be maintained; and (4) state bureaus of criminal investigation and information should be established in every state. The report laid the groundwork for reform efforts throughout the United States that continued until World War II.[17]

Post–World War II Developments in Policing

After World War II ended, policing in the United States fully embraced the professional approach that had been recommended by the Progressive reformers. Post–World War II reformers were highly educated, held positions of public influence, and were responsible for leading the movement toward increased and improved training, and toward creating criminology and police science programs at colleges and universities throughout the United States. These reformers were also able to establish a systematic body of knowledge regarding policing in the United States. Some of these reformers included **O. W. Wilson**, **V. A. Leonard**, and **J. Edgar Hoover**.

Beginning in the late 1950s and continuing through the 1960s, police throughout the United States (but particularly in large cities and the South) were tasked with controlling tensions brought about by racial unrest, the Vietnam War, and the civil rights movement. In a number of situations, police attempting to control these situations were criticized for use of excessive force; these situations deepened a racial divide between police and citizens that dated back to the Civil War. As a result of these tensions, President Lyndon B. Johnson declared a "war on crime" in 1965 and formed the President's Commission on Law Enforcement and Administration of Justice (see Chapter 1).[18]

The commission issued its summary report, *The Challenge of Crime in a Free Society*, in 1967 and followed that publication with nine supplemental reports regarding a wide variety of criminal justice issues. Each of these reports offered a number of recommendations for improvement of policing in the United States. After these reports were published, Congress created the Law Enforcement Assistance Administration (LEAA), which was mentioned in Chapter 1, to assist government agencies at all levels in implementing recommendations offered by the Commission. This legislation was referred to as the Omnibus Crime Control and Safe Streets Act of 1968 and provided billions of dollars of assistance to law enforcement organizations and universities and other agencies assisting with law enforcement training throughout the 1970s.[19]

Since the end of LEAA in 1980, law enforcement has continued to undergo tremendous change into the twenty-first century. There are higher percentages of nonwhite and female officers than ever before, and the percentage of officers with a college degree continues to grow. Despite these positive trends, police agencies face many challenges. These challenges include a fluctuating crime rate that peaked in the early 1990s then rapidly declined afterward; an upsurge in drug-related violence throughout the 1980s; and various forms of domestic and international terrorism, particularly prevalent after the September 11, 2001, terrorist activities in New York, Washington, D.C., and Shanksville, Pennsylvania. All of these issues continue to challenge law enforcement agencies. Strategies to deal with these challenges include a return to "community policing" (discussed later in this chapter), neighborhood crime prevention programs, computerized crime mapping, and problem-oriented policing. Each of these policing strategies is discussed in more detail in the pages that follow.

Structure and Organization of Policing in the United States

Number of Police in the United States

There are approximately 800,000 sworn law enforcement officers now serving in the United States. Ten percent of these officers are female and almost one in four (23%) are nonwhite. There are approximately 12,666 local police agencies, 3,070 sheriff's departments, 1,376 local agencies with special jurisdictions, 49 state police agencies, and 42 federal law enforcement agencies.[20] The types of and characteristics of these agencies and officers are discussed in the following subsections.

Federal Law Enforcement Agencies in the United States

As of June 2002, there were more than 93,000 full-time law enforcement officers employed in federal law enforcement in the United States (**Table 4–1**). These officers were employed in numerous law enforcement agencies distributed among 11 U.S. governmental services. Table 4–1 lists federal agencies employing full-time law enforcement officers and provides each agency's Web site for readers who want more information about individual agencies. The largest employers of law enforcement officers at the federal level include the Department of Homeland Security (which includes the U.S. Secret Service, U.S. Immigration and Customs Enforcement, and U.S. Customs and Border Protection, among others) and the

Table 4–1 Federal Agencies Employing Full-Time Personnel with Authority to Make Arrest and Carry Firearms, June 2003

Agency	Web Site
Department of Agriculture	
USDA Forest Service, Law Enforcement and Investigations	www.fs.fed.us/lei
Department of Commerce	
Bureau of Industry and Security, Office of Export Enforcement	www.bxa.doc.gov/enforcement
Technology Administration, National Institute of Standards and Technology, NOAA, National Marine Fisheries Service, Office of Law Enforcement	www.nist.gov www.nmfs.noaa.gov/ole
Department of Defense	
Pentagon Force Protection Agency	www.dtic.mil/dps
Department of Energy	
Office of Transportation Safeguards, Transportation Safeguards Division	www.doeal.gov
Department of Health and Human Services	
Food and Drug Administration, Office of Regulatory Affairs	www.fda.gov/ora
Office of Criminal Investigations, National Institutes of Health	www.nih.gov/od/ors/dps/police
Office of Research Services, Division of Public Safety, Police Branch	
Department of Homeland Security	
Bureau of Customs and Border Protection	www.cbp.gov
Bureau of Immigration and Customs Enforcement	www.bice.immigration.gov
Federal Emergency Management Agency	www.fema.gov
Transportation Security Administration	www.tsa.gov
U.S. Coast Guard	www.uscg.mil
U.S. Secret Service	www.secretservice.gov
Department of the Interior	
Bureau of Land Management, National Law Enforcement Office	www.blm.gov/nhp/pubs/brochures/law
Bureau of Indian Affairs, Office of Law Enforcement Services	www.doi.gov/bureau-indian-affairs.html
Bureau of Reclamation, Hoover Dam Police	www.lc.usbr.gov
National Park Service, Division of Ranger Activities	www.nps.gov
National Park Service, U.S. Park Police	www.nps.gov/uspp/
U.S. Fish and Wildlife Service, Division of Law Enforcement	www.le.fws.gov
Department of Justice	
Drug Enforcement Administration	www.usdoj.gov/dea
Federal Bureau of Investigation	www.fbi.gov
Federal Bureau of Prisons	www.bop.gov
Bureau of Alcohol, Tobacco, Firearms and Explosives	www.atf.gov
U.S. Marshals Service	www.usdoj.gov/marshals

Table 4–1 Federal Agencies Employing Full-Time Personnel with Authority to Make Arrest and Carry Firearms, June 2003, continued

Agency	Web Site
Department of State	
Bureau of Diplomatic Security, Diplomatic Security Service	www.ds.state.gov
Department of Transportation	
Federal Aviation Administration, Federal Air Marshals	www.tsa.gov
Department of the Treasury	
Bureau of Engraving and Printing, Police	www.moneyfactory.com
Internal Revenue Service, Criminal Investigation	www.ustreas.gov/irs/ci
U.S. Mint Police	www.usmint.gov/about_the_mint/mint_police
Department of Veterans Affairs	
Veterans Health Administration, Office of Security and Law Enforcement	www.va.gov/osle
Other	
Administrative Office of the U.S. Courts, Federal Corrections and Supervision Division	www.uscourts.gov/misc/propretrial.htm
Amtrak, Police	www.amtrak.com
Central Intelligence Agency, Security Protective Services	www.cia.gov
Library of Congress, Police	www.loc.gov
Smithsonian National Zoological Park, Police	natzoo.si.edu
Tennessee Valley Authority, Police	www.tva.gov/abouttva/tvap
U.S. Environmental Protection Agency, Office of Criminal Enforcement	www.epa.gov/compliance/criminal
U.S. Government Printing Office, Police	www.gpo.gov
U.S. Capitol Police	www.uscapitolpolice.gov
U.S. Postal Service, U.S. Postal Inspection Service	ww.usps.com/postalinspectors
U.S. Supreme Court, Police	www.supremecourtus.gov

Note: Table excludes federal military officers in the Army, Navy, Air Force, Marines, and Coast Guard.

Source: Adapted from B. Reaves and L. Bauer, *Federal Law Enforcement Officers, 2002*, NCJ 199995 (Washington, DC: U.S. Department of Justice, Bureau of Justice Statistics Bulletin, 2003).

Department of Justice [which includes the Federal Bureau of Investigation (FBI); the Bureau of Alcohol, Tobacco, Firearms and Explosives; the Drug Enforcement Administration; the U.S. Marshals Service; and the Bureau of Prisons]. Along with the 40 agencies distributed among the 11 U.S. governmental services, practically every cabinet-level federal agency has its own **Office of the Inspector General**, an officer who has investigative powers within an organization. The latest available figures suggest that 14.8 percent of all federal agents were women, while 1 in 6 (16.8%) federal law enforcement officers were Hispanic or Latinos, and just over 1 in 10 (11.7%) were black.[21]

Federal law enforcement agencies draw their enforcement authority from the United States Constitution. In Article I, Section 8 of the U.S. Constitution, Congress is given the authority to make laws necessary to execute the powers of the government. Some of these powers include the power to coin money and punish counterfeiting, the power to regulate commerce between states and the United States and other countries, and the power to lay and collect taxes and duties. These constitutional powers provide the federal authority under which federal law enforcement agencies operate. Generally, when Congress enacts a law, it designates which federal agency is responsible for enforcement of that law.[22]

Federal Bureau of Investigation (FBI)

The most well known federal law enforcement agency is probably the FBI. The FBI began as the Bureau of Investigation in 1908 as the investigative arm of the Bureau of Justice. In 1924, J. Edgar Hoover was appointed to direct the bureau, eventually making it the professional organization it is today and hiring only college graduates. Prohibition made the bureau famous, as its battles with well-armed organized criminals became well known, contributing to the popular conception of the "G-men." In 1935, the Bureau of Investigation changed its name to the FBI.

More than 28,000 people work for the FBI in 56 field offices and 400 satellite offices throughout the United States, more than 11,000 of whom are special agents. These special agents investigate a wide variety of federal crimes, including kidnapping, bank robbery, and drug distribution, among others. The FBI also maintains the National Crime Information Center, a computerized database of criminal justice information, and operates the Combined DNA Index system, a database of DNA profiles of various known and unknown offenders. The FBI also maintains one of the largest and most comprehensive crime laboratories in the world, providing lab services to law enforcement agencies throughout the United States. The FBI also annually publishes the Uniform Crime Reports (discussed in Chapter 2).[23]

After the tragic events of September 11, 2001, the FBI has made modernization of information technology systems a top priority. Under the guidance of FBI director Robert Mueller, the agency has taken a variety of steps to make information services for law enforcement agencies more secure and efficient.[24]

State Police Agencies

Although the Texas Rangers, the first state police agency, was created in 1835, most other state police agencies were created in the late nineteenth or early twentieth century. There are currently 49 state police departments operating in every state except Hawaii. These agencies employ more than 56,000 full-time sworn law enforcement officers. More than 2 in 3 of these officers (69%) were patrol officers while 1 in 10 (11%) were investigators. California has the largest state police agency, with more than 6,600 full-time sworn officers; Texas, at about half the size (3,119 sworn officers), is the second largest. Three states (North Dakota, South Dakota, and Wyoming) have less than 200 sworn state police officers.[25]

State police agencies typically use one of two models: a **centralized model** or a **decentralized model**. In a centralized model, criminal investigation tasks are combined with the patrol of the state highways. In other words, little distinction is typically made between traffic enforcement and investigation. Pennsylvania, Michigan, and New York are a few states that use this model. In the decentralized model, a clear distinction is drawn between traffic enforcement on the state highways and other state-level enforcement functions. Typically, these states have two or more agencies designed to enforce the law in that state. North Carolina, South Carolina, Kentucky, and Georgia all use this model. In this model, states typically have a "highway patrol" and a state bureau of investigation.[26]

Local Agencies

Local police agencies include municipal (city) police, sheriff's departments, and a number of other specialized law enforcement agencies (e.g., campus police, transit police). The most recent estimates suggest that there are 12,666 local police agencies, 3,070 sheriff's departments, and 1,376 agencies with special jurisdiction. Local police agencies employ approximately 440,000 sworn officers, while sheriff's departments employ approximately 165,000 officers, and special jurisdiction agencies employ approximately 43,000. Nine in 10 (89.4%) local law enforcement officers are male, and 3 in 4 (77.3%) are white. Approximately 12 percent are black and 8 percent are Hispanic. The vast majority of local agencies (11,277 out of 12,666, or 89%) in the United States employ 50 officers or fewer. About 46.5 percent of all departments have less than 10 officers, and only 47 have 1,000 or more officers.[27]

City/Municipal Police

A **municipality** is any political unit incorporated for local self-government and management.[28] As such, there are literally thousands of municipalities throughout the United States with less than 1,000 residents; in many cases, these municipalities have their own small police agency, sometimes consisting of only one full-time sworn officer. Most municipal police agencies are led by a police chief or commissioner. These chiefs are typically selected by the city council or appointed by the city mayor. The jurisdiction of these agencies is typically limited to the geographic boundaries of their community.[29]

New York City has the largest local police department in the United States, with more than 40,000 sworn officers. Four other cities have more than 5,000 sworn officers: Chicago (13,466), Los Angeles (9,341), Philadelphia (7,024), and

Houston (5,343). Nevertheless, the vast majority of municipal police departments are far smaller in size, employing fewer than 10 full-time officers.

Sheriff's Departments

Outside of municipal areas, the **sheriff** and the sheriff's deputies are normally responsible for law enforcement. Sheriff's departments are generally also responsible for maintaining security in courtrooms, serving a variety of court papers (e.g., warrants and subpoenas), and operating the county jail. Although the largest sheriff's department (Los Angeles County) has more than 8,400 full-time sworn officers, nearly two-thirds (60%) of all sheriff's departments have less than 25 sworn officers.

A total of 844 sheriff's offices employ fewer than 10 full-time sworn personnel. About one in eight (12.5%) sworn officers in sheriff's deputies are female, and about one in six (17.1%) are black or Hispanic.[30]

Private Protective Services

Private protective services officers typically work for corporate employers and private entities. These officers typically provide personal or property protection to private individuals or organizations for a fee. Although the exact number of private protective officers is unknown, the Bureau of Labor Statistics estimates that there are more than 1 million officers in the United States alone,[31] and private police probably outnumber "public" police by a ratio of at least two to one. The number and scope of these agencies have grown rapidly over the last three decades; in fact, some estimates put the number of private agencies at more than 13,000.

As discussed earlier, private policing has a long history; in fact, private policing predates public policing. In the United States, the emergence of private policing is usually traced to William Pinkerton, who formed the first private security contract operation, the Pinkerton National Detective Agency, in 1851. Pinkerton, along with Wells and Fargo, was largely responsible for protection of resources going into the West until the early twentieth century. Other important private security pioneers include William Burns, who founded the Burns agency in 1909 (now the second largest contract security agency in the United States) and George Wackenhut, who, along with three other former FBI agents, formed the Wackenhut Corporation in 1954. In 2000, Securitas Security Services completed its acquisition of both Pinkerton and Burns. Securitas and Wackenhut now rank as the largest private security firms in the United States.[32]

A report released by the National Institute of Justice (NIJ) in 2001 suggests that private protective services are transforming law enforcement in the United States and serve as an essential supplement to existing public law enforcement operations. This report suggests that the number of private police outnumber public police in most industrialized countries and the majority of crime prevention efforts are undertaken by private sector agencies.[33]

Roles, Functions, and Activities of Police

As mentioned at the beginning of this chapter, the general public has a number of different (and often incorrect) ideas about what police officers do during their daily activities. These ideas are structured largely by movies and television, media that often exaggerate, embellish, or, in some cases, deceive the viewers about the activities in which police officers regularly engage (see **Figure 4–2**). As such, it is important to objectively examine the roles, functions, and activities in which police officers engage as part of their daily activities.

Most scholars agree that police perform four basic functions: (1) enforce the law; (2) prevent crimes from occurring; (3) investigate crimes to determine the identification of the perpetrator(s) and apprehend offenders in an effort to ensure safety of the larger society; and (4) provide social services and ensure domestic peace and tranquility. Each of these functions is discussed in detail in the following subsections.

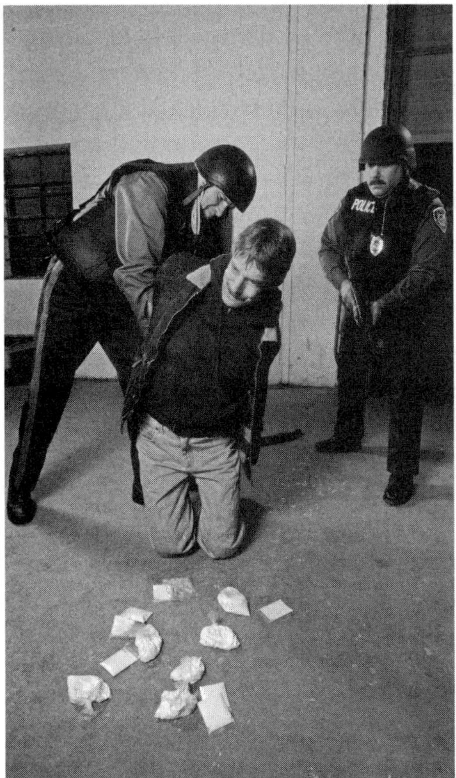

Figure 4–2 Despite media portrayals of the role of police as largely crimefighters, the majority of a police officer's time is spent in roles other than that of the crimefighter.
Source: *© Corbis.*

Enforce the Law

Police often view themselves as "crimefighters," a view the media regularly reinforces. Despite this perception, officers spend the majority of their time in non-crime-fighting enterprises. In fact, research demonstrates that less than 20 percent of an officer's time is spent enforcing the law and fighting crime.[34] Furthermore, even when police are busy enforcing the laws, constraints on time, personnel, and financial resources prevent them from enforcing *all* the laws. Police typically structure their enforcement efforts to coincide with those of the citizens that they serve. For example, if a community is upset about prostitutes actively soliciting customers in a neighborhood, the local police department may "crack down" on enforcement of laws against prostitution; in other communities, officers may arrest only the most visible and egregious prostitutes. As such, when police enforce the law, they do so through use of discretion, discussed earlier in Chapter 3 and later in this chapter.

Prevent Crimes from Occurring

To prevent crime, police have to act before a crime happens, thus stopping potential crimes from occurring. Police do this through a variety of strategies and programs. Crime prevention strategies regularly used by police include targeted surveillance (keeping a watchful eye on areas where crime is likely to occur) and active patrol. **Crime prevention programs** are organized efforts by police designed to promote citizen involvement and responsibility in averting crime in their neighborhoods and homes. These programs typically revolve around the police and the community working together in some capacity to prevent or reduce the likelihood that crime will occur (discussed in greater detail later).[35]

Investigate Crime and Apprehend Offenders

Although some offenders are apprehended during the act of committing a crime (commonly referred to as an **on-view arrest**), many arrests occur after an investigation into the criminal incident is conducted by police. Despite the popular portrayal of "crime-solving" detectives, in many cases, the police officer who arrives first on scene becomes the arresting officer, even if the officer did not witness the actual criminal event occurring. Technological advances such as the growing use and availability of DNA evidence and better communication and databases for all law enforcement agencies have the potential to increase the likelihood that criminal investigations will end in the arrest of the perpetrator who committed the criminal activity.[36]

Provide Social Services and Ensure Domestic Peace and Tranquility

Despite the fact that police agencies are primarily designed to enforce the law and prevent and investigate crimes, much of a police officer's time each day is spent answering nonemergency calls and providing social services to citizens. These services often include providing assistance to stranded motorists, answering questions about police practices and other related (or unrelated, for that matter) topics, dispersing troublesome groups, and helping citizens with other problems that they encounter. Many officers also work closely with the community, particularly schools, to promote a positive image of police and increase comfort levels that children have with police in their neighborhood. Thus, police officers are not only crimefighters, but peacemakers and citizen assistants as well.

Police Management

Police management involves directing law enforcement officers to achieve organizational goals in an effective and efficient manner.[37] Police managers perform the administrative activities of controlling, directing, and coordinating police personnel, resources, and activities to assist in reducing crime. Police managers include all sworn law enforcement personnel who have administrative authority, from the rank of sergeant to chief or sheriff.

Generally, police organizations are arranged in a hierarchical structure known as the chain of command. In this chain of command, the higher the position, the greater the power and influence of that individual occupying the position. Typically, this chain of command for police contains three levels of police managers. The lowest level of managers is often called lower managers (sergeants, field training officers); these lower managers supervise and direct the activities of the lowest-level employees in a police agency (patrol officers, detectives, and civilian personnel). The second level of managers is often called middle managers (lieutenants and captains); their responsibilities typically include developing plans for implementing decisions from above. Finally, the third (and highest) level of police management is often called top managers. Top managers (police chief, assistant police chief) make policy decisions and develop the overall goals of the organization and are also typically responsible for making policy and resource allocation decisions.[38]

Policing Styles

As discussed earlier, the police organization has evolved significantly over time. With that evolution, a number of different styles of policing have been developed. A **policing style** is how a particular police agency sees its purpose and chooses the strategies through which it accomplishes that purpose. James Q. Wilson, a noted policing author, first described these styles in 1968, but these descriptions continue to be as applicable today as they were when they were introduced.[39]

Watchman Policing Style

The **watchman policing style** is concerned with order maintenance. In this style, officers control both disruptive and illegal behavior. Officers using this style use discretion liberally and maintain order as much through informal actions as formal activities. This style of policing is characteristic of policing in lower-class communities. An example of the watchman policing style would be when police in a community continually harass young men or women whom they suspect of engaging in prostitution until those alleged prostitutes leave the jurisdiction, even though they may never have received payment for sexual services in that jurisdiction.

Legalistic Policing Style

The **legalistic policing style** is concerned with enforcing the strict letter of the law. An officer using this style of policing will ticket an individual who failed to signal when changing lanes in traffic. Police officers working in legalistic departments typically avoid involvement with activities that may be immoral, but not illegal. This style may go overboard in terms of enforcing the law, but often takes a completely hands-off approach to some activities that, although not illegal, may lead to crime (disruptive youth congregating in the streets, teenagers and adults cruising looking for drugs or prostitutes). Police adopting the legalistic policing style arrest only those prostitutes and their customers that they "catch in action" while ignoring cruising cars and scantily clad women walking alone or in small groups.

Service Style of Policing

The **service style of policing** is concerned with helping the public rather than strict enforcement of the law. Officers in these departments work hand in hand with social service agencies and other criminal justice agents (e.g., judges, prosecutors, probation officers) to improve the community by using strategies to address both criminal and noncriminal behaviors. An example of the service style of policing would be when police in a community that has a number of alleged prostitutes bring literature on drug treatment and social service workers to these prostitutes, with the idea that "fixing" the factors leading to prostitution is better than reacting to the problem with arrests.

Police–Community Relations

As mentioned earlier, the turbulence of the 1960s brought about by the civil rights movement, the Vietnam War, and other social movements dramatically reinforced the need for the police and the community to work more closely to address crime and the problems that both precede and result from crime. In an effort to manage these problems, many police departments moved from a primarily legalistic style of policing to the newer service-oriented policing style. For the first time in the history of many departments, police became concerned about **police–community relations**.

Proponents of stronger police–community relations stressed the need for the community and the police to work together. As a result, many agencies went to models of policing that held police to higher levels of accountability to the citizenry than ever before. A number of projects originated from these efforts, including: (1) Neighborhood Watch Programs (programs designed to empower citizens to increase vigilance and awareness of crime in their own community); (2) Project ID (using police equipment to mark items of value to help identify them in case of theft); and (3) Victim's Assistance Programs (programs designed to provide services to crime victims, including counseling, financial assistance, and familiarization with the criminal justice system). Because of these and other efforts to improve policing, policing today only slightly resembles 1960s policing.[40]

Scientific Police Management

One of the key contributions of LEAA discussed earlier was the funding it provided for research in the area of policing. Between 1969 and 1982, a number of research efforts were conducted that yielded findings applicable to police administration and management. This tradition of conducting research on policing strategies to determine their effectiveness and using the results to guide police management and administrative strategies has become known as **scientific police management**. Purposes of scientific police management include: (1) increasing the effectiveness of police; (2) reducing the frequency of citizen complaints; and (3) enhancing the efficient use of available resources.[41]

Today, federal support for research in policing and criminal justice continues under the Bureau of Justice Statistics and the NIJ, both part of the Office of Justice Programs created by Congress in 1984 to provide federal assistance in preventing and controlling crime. With increased demands on federal monies brought about by the efforts to combat terrorism (both domestically and internationally) and the continued U.S. involvement in the Middle East, funds for research in policing have become more limited.[42] As such, the vast majority of influential studies regarding policing effectiveness were conducted during the 1970s and 1980s; four of those studies are discussed here.

The **Kansas City Preventive Patrol Experiment** is by far the most famous example of research guided by the principles of scientific police management. Published in 1974, this study was funded by the Police Foundation and was a year-long experiment in which the southern part of Kansas City was divided into three areas; in each of those areas were five "beats." In one area, officers conducted patrols in a normal manner; in a second area, twice the normal number of officers patrolled, thereby doubling the amount of patrol activities in that area. The final area had no patrols assigned to it, and no uniformed officers traveled into that part of the city unless they were answering a call. The program was kept secret so that residents of the city were unaware the research was being conducted.[43]

The results of that study were somewhat surprising. "Preventable crimes" (classified by the authors as burglary, robbery, auto theft, larceny, and vandalism) showed no significant differences in the rate of occurrence from one group of beats to the other. Similarly, citizens didn't seem to notice the change, and fear of crime did not increase. Thus, this report contradicted popular perceptions regarding the value of increased patrol; its publication caused police managers throughout the United States to reconsider their use of patrols and, perhaps more important, cemented the importance of scientific research in the area of policing.

Because the Kansas City Preventive Patrol Experiment, along with other subsequent experiments, indicated that traditional patrol was probably not the most effective manner of policing, police managers reconsidered the use of traditional patrol in their jurisdictions. One of the results of these experiments was the creation of **directed patrol**. Directed patrol is a police management strategy designed to increase the productivity of patrol officers through the application of scientific analysis and evaluation of patrol techniques. A police manager using a directed patrol strategy varies the number of officers assigned to an area based on the time of day or the frequency of reported crimes in an area; this maximizes the effectiveness of a group of patrol officers by putting them in the areas they are most needed at the times they are most needed. This idea represents a significant improvement over traditional patrol strategies.[44]

Another important example of scientific police management is the **Kansas City Gun Experiment**. In this experiment, a special gun detection unit was assigned to Kansas City using stop-and-frisk procedures, traffic stops, and arrests for other non-gun-related crimes in "hot spots" of the city, or locations identified by computer analysis as areas having the highest rates of gun-related crimes. While the program was in operation, gun crimes declined by 49 percent in the targeted area, while increasing slightly in another area in the city used as a comparison. Additionally, drive-by shootings dropped from seven in the previous six months to one in the six months while this study was in effect.[45]

A third example of the impact of scientific police management is the 1984 **Minneapolis Domestic Violence Experiment**. This experiment compared the impact of arrest with other forms of treatment (e.g., conflict resolution at the scene of the domestic violence, ordering one partner to leave the premises) in domestic violence situations. In this experiment, researchers determined that offenders who were arrested were less likely to reoffend than those who were handled in some other fashion.[46] This study influenced many jurisdictions to implement mandatory arrest policies for persons involved in domestic violence incidents. A follow-up study conducted in Florida replicated those findings, but only for employed offenders. Other follow-up studies have had a number of divergent results.

A final example of the impact of scientific police management is the 2001 **Boston Operation Ceasefire** evaluation. This research examined the impact of collaborative efforts where police, probation officers, and prosecutors work together to reduce the incidence of gun violence, especially gang-related homicides. In this project, criminal justice agencies agreed to collaborate with one another to increase the enforcement of gun laws and trafficking laws and adopted a zero tolerance policy for both gun-related and gang-related activity. The findings from this study suggest that these collaborative efforts were successful; subsequent replications and extensions in Indianapolis and St. Louis have upheld this finding.[47]

Based on the findings of these studies (and several others), police managers began reconsidering traditional random patrol policing in the mid-1970s. Several alternative policing strategies have since been developed; each of these is discussed in detail in the following subsections.

Strategies Guiding American Policing

Because of the civil unrest of the 1960s, and the increased concern with police–community relations that resulted from those disturbances, police departments throughout the United States began to seek creative management philosophies to increase both the effectiveness of the police and their interaction with the community. Four of those strategies are discussed here.

Team Policing

The idea of team policing began in the mid-1970s. **Team policing** differs from traditional patrol in a number of ways. Under team policing, officers are divided into small teams that are assigned permanently to neighborhoods or small geographic areas. These officers are given the responsibility of being largely accountable for all aspects of law enforcement in that area, including patrol, crime investigation, problem solving, and police–community relations. Although this concept was later abandoned by most departments because of the strain it placed on departmental resources, it served as the precursor for the community policing movement that began in the early 1980s.[48]

Corrections and Policy

A good example of policing affecting policies and procedures in corrections is the Operation Ceasefire efforts discussed earlier. These collaborative efforts (and similar task forces designed to achieve collaborative goals like this one) not only serve to reduce violent and gun-related crime, but, perhaps more important, serve as conduits for communication among corrections, courts, policing, juvenile justice, and other social service agencies. This communication between agencies makes these agencies more effective in accomplishing their tasks and makes the community safer in general.

Community Policing

In the early 1980s, the concept of **community policing** emerged as one of the dominant policing strategies in the United States. This strategy focuses on exerting police influence on order maintenance and law enforcement in the context of the community, with citizen involvement as its most essential component. Although definitions of this concept do vary, the definition provided by the late Robert Trojanowicz and Bonnie Bucqueroux describes the major elements of this philosophy:

Community policing is a philosophy and organizational strategy that promotes a new partnership between the people and the police. Community policing is based on the idea that the police and the community need to work together as equal partners to identify, prioritize, and solve community problems such as crime, fear of crime, disorder (both social and physical), and overall neighborhood decay, with the goal of improving the overall quality of life in the area.[49]

The tenets of community policing evolved out of the work of Robert Trojanowicz and his colleagues and their analysis of foot patrol programs in Newark, New Jersey, and Flint, Michigan. The neighborhood foot patrol program in Flint provided one of the earliest examinations of a community policing effort in the United States. This experimental policing effort operated from 1979 through 1981 and involved 27 police officers working in 14 selected neighborhoods. The developers of the foot patrol program established 10 goals and measures of success for the final evaluation report. Using arrest data for automobile theft, assault, vandalism, robbery, criminal sexual assault, and larceny from a home, person, or vehicle, the authors of the evaluation of the program reported decreases in the actual and perceived crime in the foot beat patrol areas. At the conclusion of the study period, these crimes had decreased 8.7 percent in the experimental area, while crime in the city of Flint had generally increased.[50]

Effective community policing typically involves four elements: (1) community-based crime prevention, partnering the police with the community members in efforts to identify potential causes of crime; (2) reorientation of patrol activities to emphasize the importance of nonemergency services and to challenge officers' mindset that their only responsibility is law violations; (3) increased accountability of police to the public; and (4) decentralization of command, empowering individual officers to make decisions and solve problems without waiting on decisions by higher levels of management.

The philosophy of community policing promotes the idea of a partnership between the police and citizens, an idea diametrically opposed to the policing strategies that existed between 1930 and 1970 in the United States. Community policing continues to be a popular policing strategy today. In fact, recent estimates suggest that almost all (87%) local police officers were employed by a department that provided community policing training for some or all new recruits.[51]

Problem-Oriented Policing

Another core component of community policing is problem solving. Community policing was focused on getting to the root causes of issues necessitating repeated responses by the police. Thus, the tenets of community policing call for a dramatic ideological reorganization, utilizing police–citizen cooperation to solve community problems. The philosophy of community policing therefore encompasses more than crime; it expands the role of community police officers to problem solvers and citizen advocates, with the hope that this new role would make for better police–citizen relationships.

The philosophy of community policing brought forth a complementary perspective, that of **problem-oriented policing**. Herman Goldstein first used this term out of frustration with the overemphasis on response time that he felt was pervasive in policing at that time. Problem-oriented policing suggests that, rather than just respond to crimes after they occur, police should attempt to identify the underlying causes of recurring crimes and causes of disorder. Using this method, police identify problems, analyze them closely, develop reasonable response strategies, and then evaluate the effectiveness of those strategies.[52] Although problem solving is not new in policing, regular application of problem-solving techniques such as Scanning, Analysis, Response, and Assessment throughout the police culture is new and continues to evolve as a strategy in policing.[53]

Intelligence-Led Policing

Since September 11, 2001, policing strategies in the United States have taken a dramatic turn. Although the four basic functions of police discussed earlier have not changed as a result of the terrorist activities of that day, law enforcement agencies at all levels now devote a tremendous amount of time and other resources to gathering intelligence to both help prevent terrorist attacks and help prepare for those attacks in their community. Some of the steps that local police departments have taken to prevent further attacks include the following:

1. Strengthening relationships with federal, state, and local emergency management agencies, including fire departments and other police departments
2. Refining emergency preparedness and response training to include terrorist threats such as attacks by weapons of mass destruction
3. Providing additional security at public events and certain public spaces (nuclear plants, landmarks, places of worship, etc.)
4. Using intelligence sources and sophisticated technological innovations (such as X-ray scanners to detect chemical, biological, or radiological attacks)[54]

In 2005, the U.S. Department of Justice introduced the concept of **intelligence-led policing**. Under this policing strategy, intelligence derived from a number of sources (surveillance, newspapers, interviews, and interrogations) is used to structure police decision making to respond effectively to threat information (both criminal and terroristic) that agencies receive. Many law enforcement agencies have combined the strategies discussed earlier with this strategy to combat both criminal and terrorist activity more effectively in their communities.

Working Personality of a Police Officer

Despite the similarities between police departments in their management strategies and policing styles, it is difficult to make broad generalizations about police officers based on solely those criteria. Each officer brings his or her own unique experiences and styles to the police academy when the officer is first hired. Nevertheless, over time, many of these officers, often from diverse backgrounds, develop similar personality characteristics. A number of researchers have discussed these characteristics;[55] the most commonly mentioned aspects of the **police working personality** are listed in **Table 4–2**.

These aspects of the police personality have both positive and negative implications for the daily life of a police officer. For example, because police regularly deal with people who lie to them, it is important that they be suspicious and not accept everything they hear as factual. Additionally, because police officers often make decisions that have the potential to alter individuals' lives dramatically (e.g., during testimony in court, when deciding to make an arrest or issue a verbal warning), it is also important that they be honorable. Although incidents such as the videotaped Rodney King incident that occurred in Los Angeles in 1991 (where King was kicked in the stomach, face, and back and left with missing teeth, a crushed cheekbone, and numerous skull fractures) lead many to question the honor of police officers, even the most severe critics of police officers

Table 4–2 Characteristics of the Police Personality

Authoritarian	Individualistic
Conservative	Insecure
Cynical	Loyal
Dogmatic	Prejudiced
Efficient	Secret
Honorable	Suspicious
Hostile	

Source: Adapted from Frank Schmalleger, *Criminal Justice Today*, 9th ed. (Upper Saddle River, NJ: Prentice Hall, 2007), 291.

would admit that a sense of honor is a characteristic found in the personality of most police officers.

Some aspects of the police officer personality may be detrimental to the execution of the police officer's daily activities. For example, because police regularly deal with emotionally disturbed individuals, an officer who is overly hostile, dogmatic, and/or authoritarian with someone experiencing emotional distress will probably be less effective than an officer who is more accepting and flexible in dealings with that individual. Likewise, these personality traits may be sources of problems when dealing with victims of crime, particularly serious violent crimes.

Because of the types of individuals attracted to police work (e.g., former military members and former athletes), recruits entering the field of policing often bring with them some of the characteristics of the police personality (e.g., conservative, honorable, authoritarian). Some aspects of the personality, however, probably develop once the recruit is fully socialized into the policing subculture.

Police Subculture

A *subculture* can be defined as "a set of norms and beliefs that are separate from the main culture and represent the beliefs of a comparatively small number of people."[56] As such, a **police subculture** could be defined as norms and beliefs among police officers that are separate from the main culture and represent the beliefs of police officers throughout the profession. From this subculture, then, police learn a set of informal values that characterize the police force as a distinct community with a common identity.

A number of characteristics of the police subculture have been identified. Some of these characteristics include a sense of mission, a combination of suspicion/paranoia, a view of the police as a separate community, a resistance to change, gender-based chauvinism, a prejudiced view toward women and minorities, and an emphasis on realism and pragmatism.[57]

Legal Aspects of Policing

The Amadou Diallo and Abner Louima incidents in New York in the late 1990s graphically reinforce the fine line between legal and illegal police behavior. In the Diallo incident, four New York City police officers who were searching for a rape suspect knocked on Diallo's door. When he came to the door, he reached inside his jacket (for what later was found to be his wallet); at that point, the officers shot 41 times, hitting him with 19 of the bullets, and killing him. This incident occurred only two years after the 1997 Abner Louima incident (where Louima was sodomized with a toilet plunger handle inside a precinct office) in the same city.[58]

These incidents, along with the Rodney King incident mentioned earlier, are graphic portrayals of what can occur when officers engage in illegal behaviors. Yet, even these incidents are not without controversy. In the Diallo case, all four officers in question were acquitted of any criminal charges. In the Louima case, one officer pled guilty to assault and was sentenced to 30 years in prison, one officer was convicted of assault for helping to hold Louima down while he was assaulted, and four officers were convicted of lying to authorities.[59] In both cases, there was tremendous public outcry for greater accountability among police officers.

This outcry for an increased protection of the individual rights of citizens (and limitations on the powers of the police) was not the first time citizens expressed concern about police abuse of power. In fact, as early as the mid-nineteenth century in London, citizens were concerned about abuse of power among police. As mentioned earlier, one of Peel's primary concerns with the introduction of professional police was that citizens' rights not be violated. This discussion of restraint of police powers has continued until today.

Until the mid-twentieth century, police powers were virtually unchecked by the judicial or legislative branches of the U.S. government. As discussed in Chapter 3, this situation began to change after Earl Warren, a former prosecutor, became Chief Justice of the Supreme Court in 1953. Under his leadership, the U.S. Supreme Court made numerous decisions that protected individual rights.

In Chapter 3, we discussed the three amendments to the U.S. Constitution that are most pertinent to any discussion of the legal aspects of policing (see Table 3–1 for the language of these amendments). As we learned in that chapter, these

Race and Gender in Corrections

Although the proportion of female and nonwhite officers has continued to grow over the past decade, a special report by the U.S. Department of Justice titled *Police Departments in Large Cities, 1990–2000* indicates that, in cities with populations higher than 250,000 residents, females still represent less than one in four sworn police officers. This same report suggests that slightly more than one in three police officers in a large city are nonwhite (black, Hispanic, Asian/Pacific Islander, or American Indian), an increase of almost 33 percent from 1990 to 2000. Although we have no recent accurate data about the race and gender of officers in departments that are rural or in small towns, most evidence suggests the proportion of police who are female and/or nonwhite is lower in those areas. As such, the number of females in policing will probably continue to grow throughout the twenty-first century.

amendments (and the U.S. Supreme Court decisions based on these amendments) have defined the idea of due process in the United States and have identified what is fair and legal conduct on the part of the various actors in the criminal justice system. These requirements of the Fourth, Fifth, and Sixth Amendments mandate that justice system officials should respect the individual rights of accused citizens throughout the criminal justice process, but are particularly applicable to police.

Although Chapter 3 provides a broad overview of the idea of due process and the amendments to the Constitution that established the guidelines for due process, in the following sections, we discuss important cases that have clarified due process specifically for the police. Most due process requirements pertain to the police in the areas of search and seizure, arrest, and interrogation. As such, we present a brief discussion of the landmark court cases that have defined each of the aforementioned amendments' contributions to due process requirements for police. A more detailed discussion of many of these cases is available at www.landmarkcases.org.

Fourth Amendment

As reviewed in Chapter 3, the Fourth Amendment to the U.S. Constitution ensures protection against unreasonable searches and seizures and arrests without probable cause. The first landmark case concerning search and seizure was *Weeks v. U.S.* (1914) that we reviewed in Chapter 3.[60] In this case, the U.S. Supreme Court (hereafter referred to as the Court) established the exclusionary rule. As you may recall, the exclusionary rule prohibited the use of illegally gained evidence in federal prosecutions; when the Warren Court ruled on *Mapp v. Ohio*[61] in 1961, they applied the exclusionary rule (and all other previous federal court decisions pertaining to police conduct) to the states.

The Court continued to lay the foundation for evidentiary requirements under due process in *Silverthorne Lumber Co. v. U.S.* (1920).[62] In the Silverthorne case, the Silverthornes were accused of avoiding federal tax payment. Federal agents raided their lumber company and seized their books without a warrant. The Silverthornes' lawyer asked that the books be returned, citing the *Weeks* decision, and the judge ruled that the books had to be returned. At trial, however, the prosecution produced copies of the incriminating evidence from the books and the Silverthornes were subsequently convicted. Their lawyer appealed, and the appeal reached the U.S. Supreme Court. The Court overturned the Silverthornes' conviction and ruled that just as illegally seized evidence cannot be used in federal prosecutions, neither can evidence that is derived from an illegal search. This decision established a principle now known as the **fruit of the poisoned tree doctrine**.

Another important Warren Court case regarding the Fourth Amendment was *Chimel v. California* (1969).[63] Ted Chimel was arrested without a **search warrant** in his home for burglary of a coin shop. During a search conducted while executing the arrest warrant, officers uncovered evidence that led to his eventual conviction. Upon appeal, the Court ruled that the search of Chimel's residence was illegal because it went beyond the area subject to his immediate control. This decision set a precedent that, unless police had a search warrant, the only area they could search upon arrest was the alleged offender and any area physically within reach of the offender (known as the wingspan). A number of subsequent court decisions have clearly suggested that people can expect privacy in their own homes, and that any search conducted in the home must be conducted with a search warrant.

As we learned in Chapter 3, subsequent court decisions have recognized exceptions to some of the Court decisions of the Warren era. Two of these exceptions directly affect Fourth Amendment protections: the "good faith exception" (which states that evidence obtained in violation of the Fourth Amendment may be allowed in court if the police were acting in good faith on erroneous information), and the "inevitable discovery exception" (which states that evidence obtained in violation of a defendant's rights may be allowed if it can be shown that the evidence would have been discovered via legal means at a future point in time).[64] There is also the **computer errors exception to the exclusionary rule.** This exception resulted from *Arizona v. Evans* (1995),[65] a Court decision that held that a traffic stop in which marijuana was seized was legal even though officers conducted the stop because of an arrest warrant that should have been deleted from the computer database prior to that stop.

Another important decision concerning due process protections afforded by the Fourth Amendment in the area of policing is the **plain view doctrine**. The plain view doctrine was established by the Court in *Harris v. U.S.* (1968).[66] In this case, a police officer inventorying a vehicle after arrest discovered evidence of a robbery. Harris was convicted of robbery, and upon appeal, the Court held that objects found in "plain view" of an officer who had a right to be where he was are subject to seizure without a warrant. Later Court rulings clarified the plain view doctrine further. In *Arizona v. Hicks* (1987),[67] the Court ruled that officers could not move or dislodge objects to make the evidence in "plain view." Furthermore, the Court has also ruled in *Horton v. California* (1990)[68] that even though most evidence allowed under the plain view doctrine is discovered inadvertently, inadvertence is not a necessary condition of plain view. In this case, the officer who requested the search warrant for Terry Brice Horton's home alluded to weapons used in the robbery of a jewelry store, but the warrant was issued for stolen jewelry. When the officers searched the home, they found weapons used in the robbery but no stolen jewelry. The Court rejected Horton's appeal that the weapons should not be allowed to be introduced as evidence in the trial because the officers suspected the weapons were in his home and thus their discovery was not inadvertent.

The Court has also ruled that, in some cases, emergency searches are needed and officers do not need a warrant to conduct those searches. An **emergency search** is a search conducted by the police without a warrant, the use of which is justified by clear danger to life, escape, and/or the removal or destruction of evidence.[69] The Court has authorized emergency searches of residences when alleged offenders enter those residences while fleeing from police officers[70] or when officers are serving an arrest warrant in a residence.[71]

Searches and Arrests

According to the Supreme Court, an arrest occurs when a law enforcement officer restricts a person's freedom to leave. A Miranda warning (see following) need not be issued in an arrest until just prior to questioning. Nevertheless, for an arrest to occur, a police officer must have **probable cause** to believe that the individual they are questioning is responsible for committing the crime in question. Probable cause is the set of circumstances and facts surrounding the case that would cause a prudent and reasonably intelligent person to believe that a particular person has committed a specific crime in question.[72] Arrests may also occur when a police officer witnesses a crime in progress. Most of these arrests are for misdemeanors; in fact, many states do not allow officers to make an arrest for a misdemeanor unless they actually witnessed the crime in progress.

As part of an arrest, the Court has ruled that officers have the right to search a person being arrested (regardless of the officer's or alleged offender's gender) and the area under the alleged offender's immediate control (discussed earlier). One of the landmark cases that cemented this right was *U.S. v. Robinson* (1973).[73] In this case, the Court ruled that officers could search without a warrant for their personal protection, and the evidence found during that search could be used against the defendant for additional charges, if necessary. This decision reinforced *Terry v. Ohio* (1968),[74] in which the Court ruled that law enforcement officers can search citizens without probable cause if they have **reasonable suspicion** to believe that there exists a need to protect themselves and other prospective victims of violence. According to the Supreme Court, *reasonable suspicion* is a belief, based upon consideration of facts at hand and inferences drawn from those facts, that would induce an ordinarily prudent and cautious person under the same circumstances to generally conclude that criminal activity is taking place or has recently occurred.[75]

The *Terry* case has become the basis for the field interrogation or the "stop and frisk" procedures used today. In a "stop and frisk" procedure, an officer, without probable cause, can stop and briefly detain someone and/or conduct a "pat-down" of an individual's outer garments if the "totality of circumstances" of the situation provides the basis for an officer to be legitimately suspicious that the individual is currently engaged in an illegal activity.[76]

Vehicle Searches

Along with protection against unreasonable searches and seizures of individuals, the Court has also ruled that the Fourth Amendment protects individuals from unreasonable searches and seizures of their vehicles. The first significant Court case involving a vehicle search was *Carroll v. U.S.*, in which the Court determined that a warrantless search of an automobile or other vehicle is allowed if it is reasonably believed that contraband is present. Other Court decisions have set forth the standard that, generally speaking, the Fourth Amendment allows investigatory stops of a vehicle if the officer has reasonable suspicion to believe a crime has been committed; officers can search a stopped car without a warrant if that search is based on probable cause. Furthermore, warrantless vehicle searches can extend to any area of the vehicle if the vehicle occupant has given the officer permission to search the vehicle or if the officer has probable cause to believe that the vehicle contains illegal substances or other illegal articles.[77]

Suspicionless Searches

Although each of the aforementioned searches was based on either probable cause or reasonable suspicion, the Court has also ruled that, in some cases, **suspicionless searches** are also allowed. A suspicionless search occurs when an officer conducts a search of an individual without either a warrant or any suspicion that the individual in question has committed an illegal activity. The Court has ruled that, in some cases, these searches are permissible because of an overriding concern for public safety that requires the search. The landmark case in this area is *Florida v. Bostick* (1991).[78] In this case, the Court ruled that officers can conduct suspicionless searches if they: (1) ask for permission; (2) do not force participation with the search; (3) do not convey the message that compliance is mandatory; and (4) receive voluntary permission from the subject they are searching. In recent years, these suspicionless searches have become a routine part of most individuals' lives, as terrorist activity has increased searches at airports, athletic events, and many other public gatherings. Despite controversy about the violation of individual rights caused by these searches, the Court has not reversed its position, and thus, these searches have become a regular, if unwanted, part of many people's lives.

Use of Deadly Force

Although not regularly considered as a search or seizure, the Supreme Court ruled in *Tennessee v. Garner* (1985)[79] that the use of deadly force is the most intrusive type of seizure because it deprives an individual of his or her life. In this case, the Court held that when an officer is chasing a fleeing suspect, the officer may use deadly force only to prevent escape when the officer has probable cause to believe that the suspect poses a threat of serious physical injury or death to the officer or others. In this case, two Memphis police officers gave chase to a fleeing burglary suspect, Edward Garner. One of the officers ordered Garner to halt as he climbed a six-foot fence; when Garner continued climbing, the officer shot and killed him. A purse from the burglarized house, but no weapon of any type, was found on Garner.

Fifth and Sixth Amendments

As we learned in Chapter 3, the Fifth Amendment protects individuals from self-incrimination in that individuals cannot be forced to be a witness against themselves and cannot be deprived of liberty without due process of law. The Sixth Amendment gives individuals the right to be informed of the nature and type of crimes for which they are accused and the right to an attorney to aid them with their defense. Thus, the Fifth and Sixth Amendments are also directly applicable to the area of policing. Most Fifth and Sixth Amendment protections for citizens in their interactions with police are concerned with two major areas: arrest and interrogation. The landmark cases dealing with these topics are discussed in detail in the following subsections.

Interrogation

Interrogation, as defined by the Court, is "any words or actions on the part of the police (other than those normally attendant to arrest and custody) that the police should know are reasonably likely to elicit an incriminating response from the suspect."[80] As such, a police officer can talk with a suspect and may even ask questions of the suspect without the conversation being considered an interrogation. The moment the police officer begins questioning the suspect about the crime, however, interrogation has begun.

The Court has ruled on a number of different aspects of interrogation. Generally, the Court has ruled that although officers can use deception in interrogations, physical coercion, psychological manipulation, or threats of violence may not be used. Additionally, in *Escobedo v. Illinois* (1964),[81] the Court ruled that individuals have the right to have legal counsel present during their interrogation, and in *Edwards v. Arizona* (1981),[82] the Court stipulated that once an individual requests counsel, all questioning must *cease* until an attorney is present.

Undoubtedly, the most well known Court case that regulates police interrogations is *Miranda v. Arizona* (1966),[83] the case that established the **Miranda warning**. In this case, Ernesto Miranda was arrested in Phoenix, Arizona, and accused of kidnapping and raping a young female. Upon his arrival at the police station, Miranda was identified by the victim and, after two hours of interrogation, signed a confession that served as a key piece of evidence for his later conviction. On appeal, the Court ruled that his conviction was unconstitutional because Miranda was not advised of his Fifth and Sixth Amendment rights that protected him against self-incrimination and allowed him to have an attorney present for his defense.

The Miranda warning is used to advise individuals suspected of involvement in illegal activity of their rights prior to questioning by the police. Although the Court did not mandate the language of these warnings, the typical language of the Miranda warning is as follows:

> *You have the right to remain silent. Anything you say can and will be used against you in a court of law. You have the right to speak to an attorney and to have an attorney present during any questioning. If you cannot afford a lawyer, one will be provided for you at government expense.*[84]

Thus, to ensure that suspects are aware of their rights, the Miranda warning is now read to a suspect before any questioning begins, and in many cases, the suspect is required to initial or sign by each of those rights to indicate that the individual is aware of his or her constitutional rights prior to questioning. Interrogation can then begin, but only if the suspect waives the right against self-incrimination and the right to an attorney; if the suspect refuses to sign the Miranda warning, the police officer must wait for an attorney before any questioning about the crime for which the individual is being interrogated begins.

When the *Miranda* decision was first handed down in the mid-1960s, police officers and other critics protested that the Miranda warning would allow many individuals who had committed the crimes for which they were being questioned to escape without punishment as a result of the presence of an attorney. Nevertheless, 40 years later, most critics would concur that the Miranda warning has not hampered the police interrogation process and, in fact, has assured individuals who deal with police of protections that the Constitution requires.

Exceptions to *Miranda*

As mentioned previously, the Warren Court issued a large number of landmark cases; the cornerstone of those cases (at least in the area of policing) may have been the *Miranda* decision. Nevertheless, subsequent Court decisions have eroded some of the individual rights protections afforded by the *Miranda* decision. Two of the most notable decisions have become known as "Miranda exceptions." The case of Robert Williams is a good contrast of the philosophical differences in the Supreme Court under the leadership of Earl Warren and subsequent courts. In *Nix v. Williams* (1984),[85] the Court established what has come to be known as the **inevitable discovery exception *to Miranda*** when it ruled that evidence gathered in violation of the *Miranda* decision can be used in a criminal trial if that evidence would have been discovered anyway. In that same year, the Court also established the **public safety exception *to Miranda*** in *New York v. Quarles* (1984),[86] when it ruled that police can question suspects directly about their involvement in criminal activity without issuing the Miranda warning when consideration for public safety overrides the need for rights advisement prior to limited questioning.

Discretion

In this chapter, we have discussed the evolution of policing from its infancy to its twenty-first century state of existence. One factor has remained relatively constant throughout the growth and development of policing: the ability of police officers to make discretionary decisions that affect individuals' lives.

In Chapter 3, we learned that all criminal justice officials exercise discretion in their daily activities. One could suggest, however, that exercise of discretion is most important among police officers. The discretionary choices of police officers influence the discretionary abilities of all other criminal justice actors that make up the criminal justice system. If a police officer does not make an arrest, or collects evidence in a way the precludes successful prosecution, no other official at any other stage in the criminal justice system will have opportunity to use discretion because the case will never come into contact with other officials. Prosecutors, judges, and all correctional officials depend on the police officer to determine which individuals will be further processed through the criminal justice filter; only those cases officers choose to arrest enter that filter and are exposed to the discretionary choices of other criminal justice actors. As such, it is important to understand what factors influence the discretionary choices of police officers.

Factors Influencing Discretion among Police Officers

A number of studies have reviewed the factors that influence discretionary decisions of individual officers.[87] Some of these factors include the following:

1. *The background of the officer*—Every law enforcement officer brings his or her own life experience, attitudes, and values to the job. Thus, the decisions an individual officer makes are influenced by that background. For example, if an officer lost a sibling in an alcohol-related traffic fatality, that officer may be far more likely to vigorously enforce laws prohibiting drinking and driving than would an officer who enjoys "tying one on" every night after his shift ends.
2. *Characteristics of the suspect*—Suspect characteristics that are most influential on the decisions of the police officer include gender, grooming and style of dress, and demeanor. Belligerent, disrespectful suspects are more likely to be treated harshly than are those who exhibit respectful attitudes toward police; officers will afford well-dressed and well-kempt individuals more respect (and typically more leniency) than they will suspects who dress poorly or are poorly groomed. Finally, some officers treat women more leniently, while others treat them more harshly.

3. *Seriousness of the crime*—Generally, the less serious the criminal activity in question, the more likely the officer is to use his or her discretionary powers to handle the matter informally. In the most serious crimes (murder and forcible rape), officers have little discretionary power and typically handle those situations within the letter of the law.
4. *Departmental policy*—Official departmental policy rarely considers discretion; nevertheless, each department has its own culture and norms of behavior. For example, departments where supervisors expect line officers to adhere strictly to the law, and supervise them closely to ensure they do so, have less opportunity and likelihood for individual use of discretion by the officers.
5. *Community interests*—The attitudes of residents in the community regarding crimes and how the police department should handle those crimes also influence the discretionary decisions of the officers in that community. Contemporary attitudes toward crimes involving children and sexual assault crimes in general have led to strict enforcement of laws governing these behaviors. In areas where public order crimes (marijuana use, alcohol use, prostitution, etc.) are tolerated or ignored by the residents, police rarely enforce the laws prohibiting those behaviors; in areas where the residents voice their concerns about these activities, police often respond with increased enforcement and a large increase in arrests.
6. *Pressure from victims*—Some victims of crime refuse to file a complaint (spousal abuse, robbery of customers by prostitutes); in those cases, there is little that police can do. In other cases, victims are adamant that the police make an arrest, regardless of the severity of the offense. In the latter cases, police officers generally support the victim's wishes and formally process the alleged offenders.
7. *Personal practices of the officer*—Some officers view the violation of certain laws more (or less) seriously than do other officers who witness the same law violation. An officer whose father is disabled may be far more likely to ticket vehicles illegally parked in parking areas designated for handicapped users; that same officer may also use marijuana recreationally and may be more likely to ignore a marijuana user than is a fellow officer on the force.

Implications for Corrections

To this point, this chapter has provided a brief overview of the history of policing; the roles, functions, and activities of police officers; and the court decisions that have structured police officers' interactions with citizens in their everyday activities. The role of the police officer also has considerable implications for the field of corrections.[88] For an offender to be sentenced to jail, prison, or any other form of alternative sanction, the offender must first be arrested by a police officer. That arrest must also be constitutional, and the evidence used at subsequent stages of the criminal justice process to convict the offender must be gained legally by police officers. As much as any other criminal justice entity, then, the police officer affects the field of corrections.

Moreover, in recent years, police and correctional agencies have formed a number of partnerships where staff from both areas collaborate and share information in ways that are beneficial to both organizations. In some cases, these partnerships are formalized; other partnerships are informal and have been created because one or two staff members in one or two agencies identified common needs and problems for their agencies and began working together on an informal basis to mutually benefit their agencies.

One of the best examples of these collaborations is the concept of community probation (see Chapter 8). Community probation evolved from the precepts of community policing (discussed earlier). Community probation officers have redefined their roles to include community involvement, partnerships with other agencies (including police agencies), and problem solving. These officers, much like their community police colleagues, enlist community aid in identifying problems that lead probationers and parolees to violate their supervision directives and use problem-solving techniques originally devised by police officers to attempt to increase opportunities for successful supervision among their clients. Thus, the role of the community police officer is no longer limited to community policing; community probation uses those ideas as well.

Additional examples of police–corrections partnerships also exist. These partnerships can be divided into five broad categories, including: (1) enhanced supervision partnerships; (2) fugitive apprehension units; (3) information-sharing partnerships; (4) specialized enforcement partnerships; and (5) interagency problem-solving partnerships. Each of these partnership categories is discussed in this section.

Enhanced supervision partnerships are partnerships where police and probation and parole officers perform joint supervision and other functions related to offenders in the community. The goal of these partnerships is to reduce the number of crimes committed by persons on probation and parole by increasing the likelihood that violations of conditions of their supervision will be detected. These partnerships also allow the agencies to intervene more quickly and effectively when these violations occur. An example of this type of partnership is Operation Night Light in Boston, Massachusetts, where probation officers and Boston police officers patrol together to check for curfew and other technical violations of probationers who were thought to be criminally active in the community.

Fugitive apprehension units are units in which correctional agencies collaborate with police to locate and apprehend offenders who have absconded (i.e., left without permission) from their probation or parole supervision. An example of this type of unit is the Parolee At-Large (PAL) Apprehension Team, a partnership between the California Department of Corrections and local, state, and federal law enforcement agencies. At eight locations in California, teams of two to six agents work in partnership with local police in their area, as well as state and federal police, to share information on where absconders were last seen, their friends and relatives in the area, and other information that will help lead to their capture. PAL teams also regularly conduct sweeps with local police in neighborhoods where absconders are believed to reside. In their first year of operation, PAL teams arrested 2,125 parolees at large.

Information-sharing partnerships are partnerships where correctional agencies develop formal procedures to share offender information with law enforcement agencies. These partnerships often include programs where law enforcement agencies are notified when sex offenders are released and programs where correctional agencies share gang intelligence information with law enforcement agencies to reduce gang-related activity in the community and in prisons. A number of states use sex offender release notification partnerships. The Connecticut Department of Corrections (CDOC) Anti-Gang Program is an example of an information-sharing partnership designed to reduce gang-related activity. In this partnership, the CDOC uses information supplied from local law enforcement agencies throughout the state to identify gang-involved inmates and make placement decisions. The CDOC then notifies local law enforcement agencies about the pending release of identified gang members back to their community.

Specialized enforcement partnerships are partnerships in which correctional and police agencies work in collaboration with other community agencies to identify and attempt to eradicate particular crime problems. An example of this partnership is Operation Revitalization in Vallejo, California. This program uses a partnership consisting of the Parole and Community Services Division, the Vallejo Police Department, the Solano County Probation Department, and 10 other agencies representing a wide variety of criminal justice and community agencies. This partnership uses suppression strategies (adjudication, prosecution, and supervision) in conjunction with community-oriented prevention, intervention, and treatment services to attempt to reduce both the causes and incidence of criminal activity in the city of Vallejo.

Interagency problem-solving partnerships are partnerships in which correctional agencies and law enforcement agencies work together to identify problems that affect both agencies and implement solutions to those problems. An example of this type of agency is the Parole and Community Services Law Enforcement Consortium (PCSLEC) in California. This consortium consists of 26 members representing eight state-level correctional (e.g., Chief Probation

Officers of California) and law enforcement (California Police Chiefs Association) agencies. The goals of this consortium are: (1) to provide a forum for a discussion of concerns that affect all the agencies in the consortium; (2) to strengthen relationships among these agencies; and (3) to promote the exchange of more accurate information among the members. PCSLEC has developed a training video on parole and law enforcement partnerships and has increased the number of PALs that have been apprehended. Additionally, PCSLEC has also created a parole-law enforcement data system to share parolee information maintained by the California Department of Corrections with law enforcement agencies throughout the state.

The numbers and types of these partnerships continue to grow; in most cases, the collaborative efforts have increased interagency communication and cooperation. Nevertheless, despite the success of some of these partnerships, their long-term effectiveness is currently unknown. As such, future efforts need to evaluate their effectiveness carefully and identify areas that need improvement.

Conclusion

In this chapter, we have learned about the history of policing; the structure, organization, and types of law enforcement agencies in the United States; the roles, functions, and activities of police; and the legal environment that structures police activities. We closed the chapter by introducing you to a number of police–corrections partnerships that are becoming increasingly popular throughout the United States. In the future, these partnerships will continue to flourish and, if present trends continue, will expand to encompass other areas of criminal justice as well. In the decades that follow, corrections, police, the courts, and the prosecutors may work more closely and collaboratively; if this happens, the criminal justice system may move toward an actual "system," rather than the series of disjointed agencies that has characterized criminal justice for the last century in the United States.

READY FOR REVIEW

- Police operations in the twenty-first century have evolved significantly since the first organized police force in the early nineteenth century. Technological innovations such as computerized mapping, radio communications, and vehicles have made the twenty-first century police force a more effective crime-fighting organization.
- Some of the duties of police have not changed that drastically. Police still have a substantial amount of discretion to use in their daily activities, and the most effective police officers utilize citizens and community resources in the performance of their daily duties.
- There are approximately 800,000 sworn local, state, and federal law enforcement officers in the United States. The vast majority of these officers at every level are white males.
- Federal law enforcement agencies employ almost 100,000 sworn law enforcement personnel in a wide variety of agencies.
- Forty-nine state law enforcement agencies employ more than 56,000 sworn law enforcement personnel using one of two models: a centralized model, where all law enforcement tasks are combined under one agency, and a decentralized model, where investigation and traffic enforcement are performed by two separate state police agencies.
- There are more than 12,000 local law enforcement agencies scattered throughout the United States. These agencies employ more than 600,000 sworn law enforcement officers in municipal police agencies and sheriff's departments.
- There are more than one million private protective services agents working for private employers and individuals to supplement the public law enforcement agencies throughout the United States.
- Most scholars agree that police perform four basic functions: (1) enforce the law; (2) prevent crimes from occurring; (3) investigate crimes to determine the identification of the perpetrator(s) and apprehend offenders in an effort to ensure safety of the larger society; and (4) provide social services and ensure domestic peace and tranquility.
- Generally, police agencies are organized in a hierarchical chain of command structure, with at least two levels of management supervising daily activities.
- Police agencies generally use one of three styles of policing: the watchman style, the legalistic style, and the service style.
- Police agencies use scientific police management to increase their effectiveness, reduce the frequency of citizen complaints, and enhance the efficient use of available resources.

- Knowledge gained from a number of research efforts has been used to utilize police resources more effectively. Some of the most well-known research efforts in policing include the Kansas City Preventive Patrol Experiment, the Kansas City Gun Experiment, the Minneapolis Domestic Violence Experiment, and the Boston Operation Ceasefire evaluation.
- In the 1970s, some police departments began using team policing, where officers were divided into small teams that were assigned permanently to neighborhoods or small geographic areas and were largely responsible for all aspects of law enforcement in that area.
- In the early 1980s, team policing was replaced by community policing, a collaborative effort where police departments and community residents work together to decrease criminal activities and identify the causes of those activities.
- Community policing has been supplemented with problem-oriented policing, where police identify problems, analyze them closely, develop reasonable response strategies, and then evaluate the effectiveness of those strategies.
- After the terrorist activities of September 11, 2001, many police departments have moved toward intelligence-led policing, where law enforcement agencies at all levels work collaboratively to gather intelligence both to help prevent terrorist attacks and help prepare for those attacks in their community.
- Police officers have their own unique working personalities. These personalities have both positive and negative impacts on their daily interactions with citizens.
- Within policing, a subculture exists that helps to shape the officer's working personality. Some of the characteristics of the subculture include a sense of mission, a combination of suspicion/paranoia, a view of the police as a separate community, a resistance to change, gender-based chauvinism, a prejudiced view toward women and minorities, and an emphasis on realism and pragmatism.
- A number of landmark Supreme Court cases have had a substantial impact on policing activities. Landmark cases clarifying the Fourth Amendment for police officers include *Weeks v. U.S.*, *Silverthorne Lumber Company v. U.S.*, *Chimel v. California*, *Harris v. U.S.*, *U.S. v. Robinson*, *Terry v. Ohio*, *Carroll v. U.S*, and *Florida v. Bostick*. Landmark cases clarifying the Fifth and Sixth Amendments for police officers include *Escobedo v. Illinois*, *Edwards v. Arizona*, *Miranda v. Arizona*, *Nix v. Williams*, and *New York v. Quarles*.
- Police officers have a large amount of discretion that structures their daily decisions. Their use of discretion is influenced by a number of offense characteristics and offender characteristics and behaviors, as well as the officer's personal background and experiences.

- In recent years, police and correctional agencies have formed a number of partnerships where staff from both areas collaborate and share information in ways that are beneficial to both organizations. Examples of these partnerships include enhanced supervision partnerships, fugitive apprehension units, information-sharing partnerships, specialized enforcement partnerships, and interagency problem-solving partnerships.

KEY TERMS

Boston Operation Ceasefire A project in Boston in which police, probation officers, and prosecutors collaborated to reduce the incidence of gun violence, especially gang-related homicides, by increasing the enforcement of gun laws and trafficking laws and adopting a zero tolerance policy for both gun-related and gang-related activity

centralized model of policing A model of state police organization in which both investigative and traffic enforcement activities are combined

community policing A policing strategy based on the premise that both the police and the community must work together as equal partners to identify, prioritize, and solve contemporary community problems such as crime, fear of crime, social and physical disorder, and overall neighborhood decay, with the goal of improving the overall quality of life in the area

computer errors exception to the exclusionary rule A U.S. Supreme Court decision that stated evidence gained as the result of a search or seizure based on a clerical error is not protected under the exclusionary rule

constable An English official appointed by the king whose job was to maintain order among the hundreds in the frankpledge system

coroner Locally elected official whose position was created in thirteenth-century England and whose job it was to monitor the sheriffs and make rulings on many civil matters

crime prevention programs Organized efforts by police designed to promote citizen involvement and responsibility in averting crime in their neighborhoods and homes

decentralized model of policing An organizational model of state policing in which a clear distinction is drawn between traffic enforcement and investigative activities

directed patrol A police management strategy designed to increase the productivity of patrol officers through the application of scientific analysis and evaluation of patrol techniques

emergency search A search conducted by the police without a warrant, whose use is justified by clear danger to life, escape, and/or the removal or destruction of evidence

enhanced supervision partnerships Partnerships where police and probation and parole officers perform joint supervision and other functions related to offenders in the community

frankpledge A system of government used in early England based on an ordering of tithings (groups of 10 households), hundreds (groups of 100 households), and shires (groups of several hundreds)

fruit of the poisoned tree doctrine A legal principle that prohibits use of any evidence at trial that developed as a result of an illegal search or seizure

fugitive apprehension units Units in which correctional agencies collaborate with police to locate and apprehend offenders who have absconded from (left) their probation or parole supervision

J. Edgar Hoover Served as the director of the Federal Bureau of Investigation from 1924 until his death in 1972. Despite criticisms of the constitutionality of some of his methods, particularly during the 1960s, the FBI became one of the most effective and respected law enforcement agencies in the world under his leadership

inevitable discovery exception to Miranda Evidence gathered in violation of the *Miranda* decision can be used in a criminal trial if that evidence would have been discovered anyway

information-sharing partnerships Partnerships where correctional agencies develop formal procedures to share offender information with law enforcement agencies

intelligence-led policing Collection and analysis of information to provide intelligence to law enforcement officers to assist both strategic and tactical police decision-making efforts

interagency problem-solving partnerships Partnerships in which correctional agencies and law enforcement agencies work together to identify problems that affect both agencies and implement solutions to those problems

interrogation Any words or actions on the part of the police (other than those normally attendant to arrest and custody) that the police should know are reasonably likely to elicit an incriminating response from the suspect

Kansas City Gun Experiment An experiment in which a special gun detection unit was assigned to Kansas City using stop-and-frisk procedures, traffic stops, and arrests for other non-gun-related crimes in hot spots of the city, or locations identified by computer analysis as areas having the highest rates of gun-related crimes

Kansas City Preventive Patrol Experiment An experiment in which the southern part of the city of Kansas City was divided into 3 areas; in each of those areas were five "beats." In one area, officers conducted patrols in a normal manner; in a second area, twice the normal number of officers patrolled, thereby doubling the amount of patrol activities in that area. The final area had no patrols assigned to it, and no uniformed officers traveled into that part of the city unless they were answering a call

kin policing A policing strategy in which the family, clan, or tribe enforced informal rules and customs of conduct

legalistic policing style A style of policing in which officers are concerned only with enforcing the strict letter of the law

V.A. Leonard A post–World War II reformer who authored numerous textbooks about policing and founded the Academy of Criminal Justice Sciences and Alpha Phi Sigma (the national criminal justice honor society).

Minneapolis Domestic Violence Experiment An experiment that compared the impact of arrest with other forms of treatment in domestic violence situations

Miranda warning A series of warnings used to advise individuals suspected of involvement in illegal activity of their rights prior to questioning by the police

municipality Any political unit incorporated for local self-government and management

nightwatch A group of citizens whose responsibility was to patrol their local area at night, constantly searching for fires, fights, and other problems

Office of the Inspector General An officer who has investigative powers within an organization

on-view arrest An arrest that occurs when a police officer actually catches an offender in the act of committing a crime

plain view doctrine A legal term that states that objects found in plain view of an officer who had a right to be where he or she was are subject to seizure without a warrant

***Plessy v. Ferguson* (1896)** A U.S. Supreme Court decision that states that the doctrine of "separate but equal" was constitutional and that states could enact laws that allowed segregation in public areas, in this case on a public train

police–community relations The association between the police department and the community. Good police–community relations involve regular communication and shared priorities between the police department and the community it serves

police subculture Norms and beliefs among police officers that are separate from the main culture and represent the beliefs of police officers throughout the profession

police working personality The personality found among most police officers that consists of a group of personality characteristics whose combination is unique to the profession of policing

policing style How a particular police agency sees its purpose and chooses the strategies through which it accomplishes that purpose

probable cause The set of circumstances and facts surrounding the case that would cause a prudent and reasonably intelligent person to believe that a particular person has committed a specific crime in question

problem-oriented policing A policing strategy whereby police attempt to identify the underlying causes of recurring crimes and causes of disorder by identifying problems, analyzing them closely, developing reasonable response strategies, and then evaluating the effectiveness of those strategies

Progressive Movement A movement in the United States at the beginning of the twentieth century whose motives were to promote honesty and efficiency in government, reduce the power of politicians, and use experts to assist in finding solutions to problems

public safety exception to Miranda Police can question suspects directly about their involvement in criminal activity without issuing the Miranda warning when consideration for public safety overrides the need for rights advisement prior to limited questioning

reasonable suspicion A belief, based upon consideration of facts at hand and inferences drawn from those facts, that would induce an ordinarily prudent and cautious person under the same circumstances to generally conclude that criminal activity is taking place or has recently occurred

scientific police management The tradition of conducting research on policing strategies to determine their effectiveness and using those results to guide police management and administrative strategies

search warrant A written document issued by a judge or magistrate that authorizes police to conduct a search of a person or location for evidence of a criminal offense, and, if evidence is found, seize the evidence

service style of policing A style of policing in which officers are concerned with helping the public to improve the community in which they live

sheriff The elected chief of a county law enforcement agency. The sheriff is usually responsible for operation of the county jail, executing warrants, and law enforcement in the unincorporated areas in a county

shire reeve An official who served as a deputy for the king in early England whose job was to maintain law and order among the people. The modern word *sheriff* evolves from this office

specialized enforcement partnerships Partnerships in which correctional and police agencies work in collaboration with other community agencies to identify and attempt to eradicate particular crime problems

suspicionless search When an officer conducts a search of an individual without either a warrant or any suspicion that the individual in question has committed an illegal activity

team policing A policing strategy where conventional police patrol was reorganized into an integrated and versatile police team of two or more officers assigned to a fixed district and charged with all aspects of law enforcement in that district

vigilance committees Committees of private citizen volunteers formed in areas where organized law enforcement agencies did not exist whose purpose was to protect life and property of the citizens in that community

August Vollmer A Progressive reformer who served as police chief at Berkeley, California, and assisted the Wickersham Commission in preparing its report, among other efforts

watchman policing style A policing style in which the officer is concerned primarily with order maintenance. In this style, officers control both disruptive and illegal behaviors

Wickersham Commission A commission chaired by John Wickersham, attorney general under Herbert Hoover, whose task was to survey the criminal justice system and causes of crime in the United States

O. W. Wilson A post–World War II reformer who published several major textbooks that became the training materials for police managers and students. Wilson served as both a police chief of several agencies and a professor of police administration at the University of California Berkeley

YOU ARE THE CORRECTIONS PROFESSIONAL SUMMARY

1. What legal options do you have at this point? As a juvenile probation officer, you want to continue a good working relationship with the local police department. When you encounter David talking with Kevin and Rick, you and the police officer have a number of options. Because you are supervising David's probation, you have a substantial amount of discretion in this matter. You can ask David to let you search him; your police officer colleague can do the same to both Kevin and Rick if she has reasonable suspicion that they are engaged in illegal activity. You could also order David to return to his home and cease his association with Kevin and Rick, or you could place David under arrest for violating the conditions of his probation. Finally, you could tell the police officer to continue her patrol and ignore the whole matter.
2. Does the presence of the police officer affect your decision on how to handle this situation? Explain. No matter which choice you make, the presence of the police officer should not affect your decision. Your relationship with law enforcement should be a professional one; in other words, you should both act professionally at all times and the presence of one another should be used to mutually benefit both of your agencies, not to encourage or discourage you from performing your duties.
3. What action would you take? Why? The action you would take depends on the actions that you have taken prior to this incident. If this is the first time David has violated his order not to associate with known gang members, you could pull David off to the side, talk to him about

the serious ramifications of his action, and order him to go home. Legally, you could also arrest David for violation of his probation conditions. Ignoring the situation, however, is not an option.

NOTES

1. F. Schmalleger, *Criminal Justice Today*, 6th ed. (Upper Saddle River, NJ: Prentice Hall, 2001).
2. B. L. Berg, *Law Enforcement: An Introduction to Police in Society* (Boston: Allyn & Bacon, 1992).
3. R. Roberg, J. Crank, and J. Kuykendall, *Policing and Society*, 2nd ed. (Los Angeles: Roxbury, 2000).
4. B. Smith, *Rural Crime Control* (New York: Columbia University, 1933); Roberg et al., *Policing and Society*.
5. K. J. Peak, *Policing in America: Methods, Issues, and Challenges*, 2nd ed. (Upper Saddle River, NJ: Prentice Hall, 1997).
6. Peak, *Policing in America*.
7. J. J. Tobias, *Crime and Police in England, 1700–1900* (New York: St. Martin's Press, 1979).
8. J. A. Conser and G. D. Russell, *Law Enforcement in the United States* (Gaithersburg, MD: Aspen Publishers, 2000).
9. Conser and Russell, *Law Enforcement*.
10. Conser and Russell, *Law Enforcement*.
11. Conser and Russell, *Law Enforcement*.
12. Conser and Russell, *Law Enforcement*.
13. Conser and Russell, *Law Enforcement*.
14. Conser and Russell, *Law Enforcement*.
15. *Plessy v. Ferguson*, May 18, 1896. 163 U.S. 537, 16 S. Ct. 1138, 41 L.Ed. 256; U.S. 1896.
16. Conser and Russell, *Law Enforcement*; Peak, *Policing in America*.
17. National Commission on Law Observance and Enforcement, *Report on Police* (Washington, DC: Government Printing Office, 1931).
18. Peak, *Policing in America*.
19. Peak, *Policing in America*.
20. M. J. Hickman and B. A. Reaves, *Local Police Departments*, NCJ 196002 (Washington, DC: U.S. Department of Justice, Bureau of Justice Statistics,

2003); B. A. Reaves and M. J. Hickman, *Census of State and Local Law Enforcement Agencies, 2000*, NCJ 194066 (Washington, DC: U.S. Department of Justice, Bureau of Justice Statistics Bulletin, 2002).

21. B. A. Reaves and L. M. Bauer, *Federal Law Enforcement Officers, 2002*, NCJ 199995 (Washington, DC: U.S. Department of Justice, Bureau of Justice Statistics Bulletin, 2003); F. Schmalleger, *Criminal Justice Today*, 9th ed. (Upper Saddle River, NJ: Prentice Hall, 2007).
22. Conser and Russell, *Law Enforcement*.
23. Schmalleger, *Criminal Justice Today, 2007*.
24. Schmalleger, *Criminal Justice Today, 2007*.
25. Hickman and Reaves, *Census*.
26. Schmalleger, *Criminal Justice Today, 2007*.
27. Hickman and Reaves, *Census*.
28. Definition from Dictionary.com, available at http://dictionary.reference.com/browse/municipality.
29. Schmallager, *Criminal Justice Today, 2007*.
30. M. J. Hickman and B. A. Reaves, *Sheriffs' Offices 2000*, NCJ 196534 (Washington, DC: U.S. Department of Justice, Bureau of Justice Statistics, 2003).
31. U.S. Bureau of Labor Statistics. Estimates derived by author from www.bls.gov/.
32. Peak, *Policing in America*.
33. D. H. Bayley and C. D. Shearing, *The New Structure of Policing: Description, Conceptualization, and Research Agenda* (Washington, DC: National Institute of Justice, 2001).
34. V. E. Kappeler and G. W. Potter, *The Mythology of Crime and Criminal Justice*, 4th ed. (Long Grove, IL: Waveland Press, 2005).
35. Schmalleger, *Criminal Justice Today, 2007*.
36. Schmalleger, *Criminal Justice Today, 2007*.
37. Roberg et al., *Policing and Society*.
38. Roberg et al., *Policing and Society*.
39. J. Q. Wilson, *Varieties of Police Behavior: The Management of Law and Order in Eight Communities* (Cambridge, MA: Harvard University Press, 1968).
40. Schmalleger, *Criminal Justice Today, 2007*.
41. Schmalleger, *Criminal Justice Today, 2007*.
42. Schmalleger, *Criminal Justice Today, 2007*.

43. G. Kelling, T. Pate, D. Dieckman, and C. E. Brown, *The Kansas City Patrol Experiment: A Summary Report* (Washington, DC: The Police Foundation, 1974). Also available online at www.policefoundation.org/pdf/kcppe.pdf.
44. Schmalleger, *Criminal Justice Today, 2007.*
45. L. W. Sherman, D. P. Rogan, and J. W. Shaw, *The Kansas City Gun Experiment*, NCJ 150855 (Washington, DC: U.S. Department of Justice, National Institute of Justice Research in Brief, 1995).
46. L. W. Sherman and R. A. Berk, *Minneapolis Domestic Violence Experiment* (Washington, DC: Police Foundation, 1954).
47. D. M. Kennedy, A. A. Braga, A. M. Piehl, and E. J. Waring, *Reducing Gun Violence: The Boston Gun Project's Operation Ceasefire*, NCJ 188741 (Washington, DC: U.S. Department of Justice Research Report, 2001).
48. Peak, *Policing in America.*
49. R. Trojanowicz and B. Bucqueroux, *Community Policing: How to Get Started*, 2nd ed. (Cincinnati: Anderson Publishing Company, 1998), 6.
50. R. Trojanowicz and B. Bucqueroux, *Community Policing: A Contemporary Perspective* (Cincinnati: Anderson Publishing Company, 1990).
51. M. J. Hickman and B. A. Reaves, *Community Policing in Local Police Departments, 1997 and 1999*, NCJ 184794 (Washington, DC: U.S. Department of Justice, Bureau of Justice Statistics Bulletin, 2001).
52. H. Goldstein, *Problem-Oriented Policing* (New York: McGraw-Hill, 1990).
53. Peak, *Policing in America.*
54. Police Executive Research Forum, *Local Law Enforcement's Role in Preventing and Responding to Terrorism* (Washington, DC: PERF, October 2, 2001). www.policeforum.org/upload/terrorismfinal%5B1%5D_715866088_12302005135139.pdf.
55. J. H. Skolnick, *Justice Without Trial; Law Enforcement in a Democratic Society* (New York: John Wiley, 1966); M. Brown, *Working the Street: Police Discretion and Dilemmas of Reform* (New York: Sage, 1981).
56. Conser and Russell, *Law Enforcement*, 145.
57. S. Herbert, "Police Subculture reconsidered," *Criminology 36*, no. 2 (1998): 343–369.
58. "Amadou Diallo Case." *Washington Post*, 2000, March 5, 2000 www.washingtonpost.com/wp-dyn/nation/specials/aroundthenation/nypd/.
59. Wikepedia, "Abner Louima," http://en.wikipedia.org/wiki/Abner_Louima.
60. *Weeks v. U.S.*, 232 U.S. 383 (1914).
61. *Mapp v. Ohio*, 367 U.S. 643 (1961).

62. *Silverthorne Lumber Co. v. U.S.*, 251 U.S. 385 (1920).
63. *Chimel v. California*, 395 U.S. 752 (1969).
64. See *U.S. v. Leon*, 468 U.S. 897 (1984) and *Nix v. Williams*, 467 U.S. 431 (1984).
65. *Arizona v. Evans*, 514 U.S. 1 (1995).
66. *Harris v. U.S.*, 390 U.S. 234 (1968).
67. *Arizona v. Hicks*, 108 S. Ct. 1149 (1987).
68. *Horton v. California*, 496 U.S. 128 (1990).
69. Schmalleger, *Criminal Justice Today, 2007*.
70. *Warden v. Hayden*, 387 U.S. 204 (1967).
71. *Maryland v. Buie*, 110 S. Ct. 1093 (1990).
72. Schmalleger, *Criminal Justice Today, 2007*.
73. *U.S. v. Robinson*, 414 U.S. 218 (1973).
74. *Terry v. Ohio*, 392 U.S. 1 (1968).
75. Schmalleger, *Criminal Justice Today, 2007*.
76. *U.S. v. Sokolow*, 109 S. Ct. 1581 (1989).
77. *Terry v. Ohio*, 392 U.S. 1 (1968); *California v. Acevedo*, 500 U.S. 565 (1991); *Florida v. Jimeno*, 111 S. Ct. 1801 (1991).
78. *Florida v. Bostick*, 111 S. Ct. 2382 (1991).
79. *Tennessee v. Garner*, 471 U.S. 1 (1985).
80. *Rhode Island v. Innis*, 446 U.S. 291 (1980).
81. *Escobedo v. Illinois*, 378 U.S. 478 (1964). In this decision, the Court ruled that the right to counsel is invoked when the line of questions shifts from one of general investigation to one of accusation directed at a particular suspect.
82. *Edwards v. Arizona*, 451 U.S. 477 (1981).
83. *Miranda v. Arizona*, 384 U.S. 436 (1966).
84. U.S. Constitution Online "*The Miranda warning*," www.usconstitution .net/miranda.html.
85. *Nix v. Williams*, 467 U.S. 431 (1984).
86. *New York v. Quarles*, 467 U.S. 649 (1984).
87. This section is based largely on Schmalleger, *Criminal Justice Today, 2007*, 222–223.
88. The following discussion is derived largely from D. Parent and B. Snyder, *Police-Corrections Partnerships*, NCJ 175047 (Washington, DC: U.S. Department of Justice, National Institute of Justice Issues and Practices, 1999).

5

Courts and Corrections

Chapter Objectives

- Understand the history and evolution of the U.S. court system.
- Understand the structure and organization of the court system in the United States.
- Identify similarities and differences between the federal and state court systems in the United States.
- Identify and describe the various roles of the professional members of the courtroom work group.
- Identify and describe the roles of the nonprofessional courtroom participants.
- Describe the various steps typically taken during pretrial activities.
- Describe the various stages of a criminal trial.
- Describe the impact of the Bill of Rights on criminal trial processes as they pertain to corrections.
- Discuss the implications of both current and emergent court systems for corrections.

CASE STUDY

You are an adult probation officer in the state of Kentucky. You have been assigned supervision of a defendant involved in a case where the incumbent county sheriff was allegedly assassinated by his opposition (your client) during a heated bid for reelection. Although the media coverage around the event portrays the case as a "slam dunk" for the prosecutor, you are not certain that your client is actually the person who was responsible for the death of the sheriff.

1. What kind of factors would need to be considered and excluded to ensure that the alleged defendant receives a fair trial?
2. What kind of trial would most likely take place? Would a jury be involved? If so, what role would they play?
3. Will the alleged defendant be able to receive a quality defense in this trial? If not, what steps could you take to ensure that this trial represents an impartial hearing of the facts of the case rather than becomes an emotional contest?
4. If the defendant is found guilty, what implications will his sentence have for both local and state corrections?

Introduction

On November 5, 2006, Saddam Hussein was found guilty of crimes against humanity for the killing of 148 people in the Tigris River city of Dujail in 1982. He was sentenced to death by hanging. The verdict brought a conclusion to one of the most infamous examples of the importance of the courts in the criminal justice system. The year-long trial that began in October 2005 was a good example of the numerous facets of the criminal trial and presents a number of examples of the difficulty of conducting a criminal trial. Difficulties experienced in this trial include: (1) three of Saddam's defense attorneys were killed; (2) defense attorneys, defendants, and co-defendants were regularly removed from the court because of angry outbursts; (3) a number of witnesses for the defense were arrested for perjury; and (4) the original judge in the trial, Rizgar Amin, resigned three months into the trial.[1]

Although the vast majority of criminal trials flow far more smoothly than the trial of Saddam Hussein in Iraq during the middle of a war, this trial clearly demonstrates the need for an ordered courtroom environment and structure in which the courtroom actors can operate to ensure a fair, impartial verdict for the alleged defendant. Without the courts to decide the guilt or innocence of individuals who have allegedly committed crimes, the activities of law enforcement officers discussed in previous chapters would be meaningless. In this chapter, we discuss how the U.S. court system has evolved and how it currently operates at both the state and federal levels. We then discuss the various actors and their roles in the "courtroom work group" and close with a discussion of the implications of the activities of the courts for corrections.

Philosophy of the U.S. Court System

The legal system in the United States is often referred to as an **adversarial system**, reflecting the idea that justice can be realized when attorneys representing opposing legal teams (typically referred to in criminal cases as the prosecution and the defense) compete in court to determine the facts of the case and determine the most just resolution. The adversarial system is founded on the principle of "innocent until proven guilty."

This system directly contrasts with the **inquisitorial system** used in most countries in Latin America and Europe, where the accused is presumed guilty and must prove innocence to the court. In these courts, an independent judge hears suspects and witnesses and orders searches and subpoenas for other investigations. The judge's goal in the inquisitorial system is not the prosecution of a certain alleged offender, but the finding of truth.[2] Under this system, the judge typically decides if there is a valid case against a certain suspect, then defers the suspect to a tribunal or court that uses an adversarial system very similar to that used in the United States to determine guilt or innocence.

Despite the widely held view that the adversarial system is a competition between two opponents, in actuality a highly developed system of cooperation between defense attorneys, prosecutors, and judges often exists. In this system, members of the courtroom work group develop norms about how certain types of cases should be handled and what sort of pleas are acceptable for certain types of offenses.[3]

In Chapter 1, you learned that the government of the United States is divided into three branches; you also learned in Chapter 3 that criminal law is primarily under the domain of the legislative and judicial branches of government. Furthermore, you learned that the role of the judicial branch of government is to determine matters of fact in criminal cases and to impose sentences on those determined to be guilty. It is through the court system that this role is carried out. Before we can discuss courts in the twenty-first century, however, we need to examine the evolution of that court system.

History of the U.S. Court System

As with many other facets of government in the United States, the U.S. court system has evolved primarily from that used in England. The British court system evolved from the Anglo-Saxon courts that developed through a number of centuries prior to the signing of the Magna Carta by King John in 1215. Anglo-Saxon "justice" was dominated by powerful lords who defined and enforced the law as they deemed necessary on their lands. Public dissatisfaction with this system spurned the efforts that eventually evolved into the Magna Carta; prior to that time, three types of trials were regularly used. These included trial by ordeal, trial by battle, and trial by compurgation (or oath).[4]

A **trial by ordeal** was a trial in which defendants had to prove their innocence by experiencing some test that, without divine intervention, was almost impossible to pass. Common trials included trials by fire or hot water (where all defendants who showed evidence of burns three days after they walked across hot coals or placed their hand in boiling water were found guilty) or trials where the accused was bound hand and foot and then thrown into a body of water and determined to be guilty if he or she floated and innocent if he or she failed to emerge. A **trial by battle** was a trial in which the accused (or a champion fighting on his or her behalf) fought his accusers. If the accused won the battle, he or she was declared innocent; if the accused lost, he or she was executed and the estate went to the king.[5] A **trial by compurgation** required the accused to locate a required number of "compurgators" (typically 12 people who would testify to the truthfulness of the accused's oath that he or she was not guilty of the crime for which he or she was accused). The job of the compurgators was to convince a judge (typically in a civil case) that the defendant's case was more believable than the accuser's. Compurgation was allowed in civil testimonies until 1752; the jury system used today (and its peculiar number of 12 members still used in the twenty-first century) appears to have evolved from this trial of compurgation.[6]

After the Norman invasion of England in the eleventh century, English courts began to use juries. The first juries were used primarily to settle civil disputes. Although these juries were similar to twenty-first-century juries because they were composed of common citizens, these juries were often pressured by the king to find defendants guilty. When they did not, jurors were often punished because the Crown could not seize an innocent person's property. Another key difference was that early jurors were selected for service because they knew the parties involved and could inform the court on their knowledge of the accused and his or her charges. In fact, these early juries typically heard no witnesses and delivered a verdict based on the knowledge of those members of the jury.[7]

By the end of the Middle Ages (around the fifteenth century), population growth made it difficult to find 12 individuals who were familiar with a given case, so judges began to allow citizens who knew about a case to testify as witnesses while the other jurors heard the evidence and issued the verdict. By the sixteenth and seventeenth centuries, juries used outside witnesses to supplement their own personal knowledge of the case, and by the eighteenth century, juries no longer included anyone who knew anything about the case. Over six centuries, then, juries were transformed from bodies in which each member knew the accused and the details of the case to a system where 12 jurors decided on which of two scenarios (one offered by the defense and one provided by the prosecution) was most credible.[8]

Because America first existed as a colony of England, it is important to keep in mind that the first courts in the United States were actually English courts. As such, many of the elements of the court system in the twenty-first-century United States evolved directly from the English court system. Rulings by colonial courts were expected to conform to English law; however, given that the economic and social situation in the colonies was drastically different from that in England in the seventeenth and eighteenth centuries, colonial courts often adapted by including the legal procedures and laws that seemed most appropriate for the colonists' unique situation and then adding their own substantive laws to deal with situations that were uniquely colonial (e.g., slavery and dealing with Native Americans). As such, although the English courts were clearly visible in colonial courts, the colonists also had their own innovations.[9]

In colonial America prior to 1776, judicial functions were typically delivered by a panel consisting of the governor and his appointees who were appointed by the English monarchy. In some cases, accused criminals were even returned to England for trial.[10] These courts were typically small, simplistic, served both civil and criminal functions, and often used the Bible as their source of law.[11] As time progressed and the colonies further developed, jurisdictions began documenting laws and rights to which criminal defendants were entitled and eventually began extending the right to trial by jury to all criminal cases.[12] Eventually, the office of the Justice of the Peace was established to handle less-serious crimes and disputes, and by the end of the seventeenth century, most jurisdictions had developed criminal courts to handle both civil and criminal cases and many of the rights afforded defendants in the Bill of Rights were guaranteed, including the right to post bail, the right to counsel in civil cases, and the right to challenge jurors during the selection phase.[13] Traveling judges and lawyers often rode from town to town (the "circuit") to conduct trials; the structure of these eighteenth-century courts still remains today in many jurisdictions. As such, the colonial courts in the mid-eighteenth century looked more similar to twenty-first-century courts in the United States than did their British predecessors.

In the twenty-first century, the United States uses a dual system of federal and state courts that usually operate independently of one another, although state cases are frequently appealed to the federal courts. This dual court system evolved from the ideas of the nation's founding fathers that individual states need to retain significant legislative authority and judicial autonomy from the federal system. Whereas state court systems evolved from colonial courts, the federal court system was developed after the United States became independent from England to complement the existing state court systems. The current federal court system developed as a compromise between the founding fathers who favored a strong federal government (the Federalists) and those who advocated for states' rights with a limited federal government (the Anti-Federalists).

In the United States in the twenty-first century, there are a number of separate court systems. Although both the military and a number of American Indian tribes have their own court systems, the court systems that most Americans will experience are the federal court system and the state court systems. These court systems are described in detail in the following sections.

Structure and Organization of Twenty-First Century Courts in the United States

The structure of the current federal system is outlined in **Figure 5–1**; the graphic presented in that figure also demonstrates the path through which cases that originate in a state court must travel to reach the federal court system. The structure of the court system of the Commonwealth of Kentucky is diagrammed in **Figure 5–2**; this four-tiered design is the most common design of state court systems. Because most cases originate in the state courts, we begin our discussion with them.

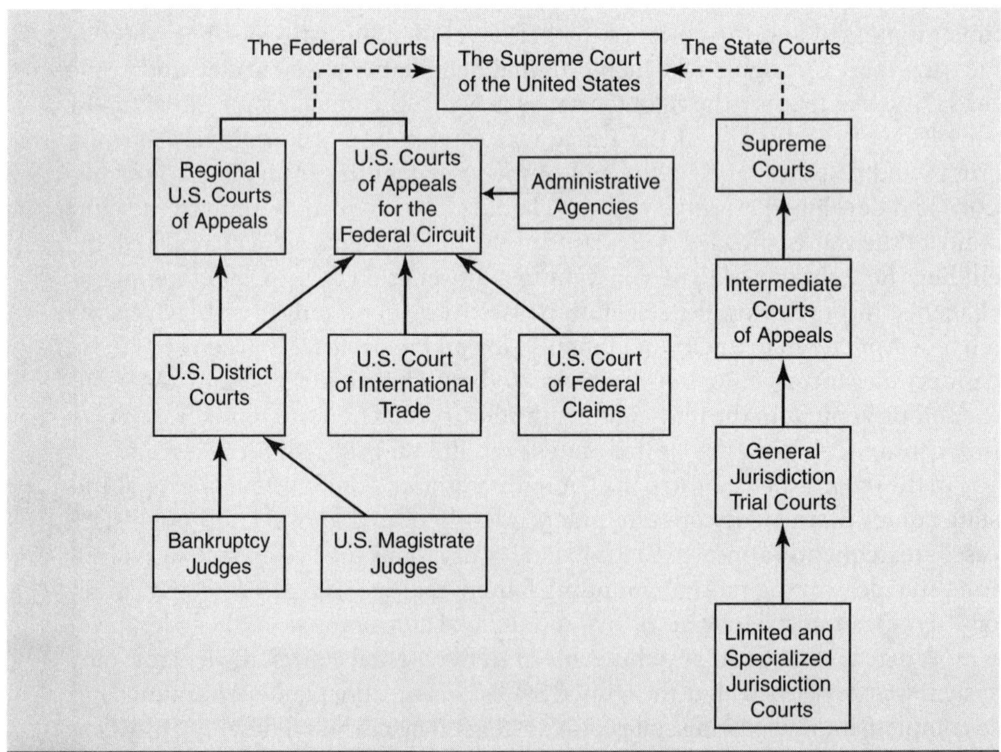

Figure 5–1 Structure of the federal court system.
Source: *Reproduced from United States District Court of Nebraska, www.ned.uscourts.gov/outreach/fedctswhtdo.pdf.*

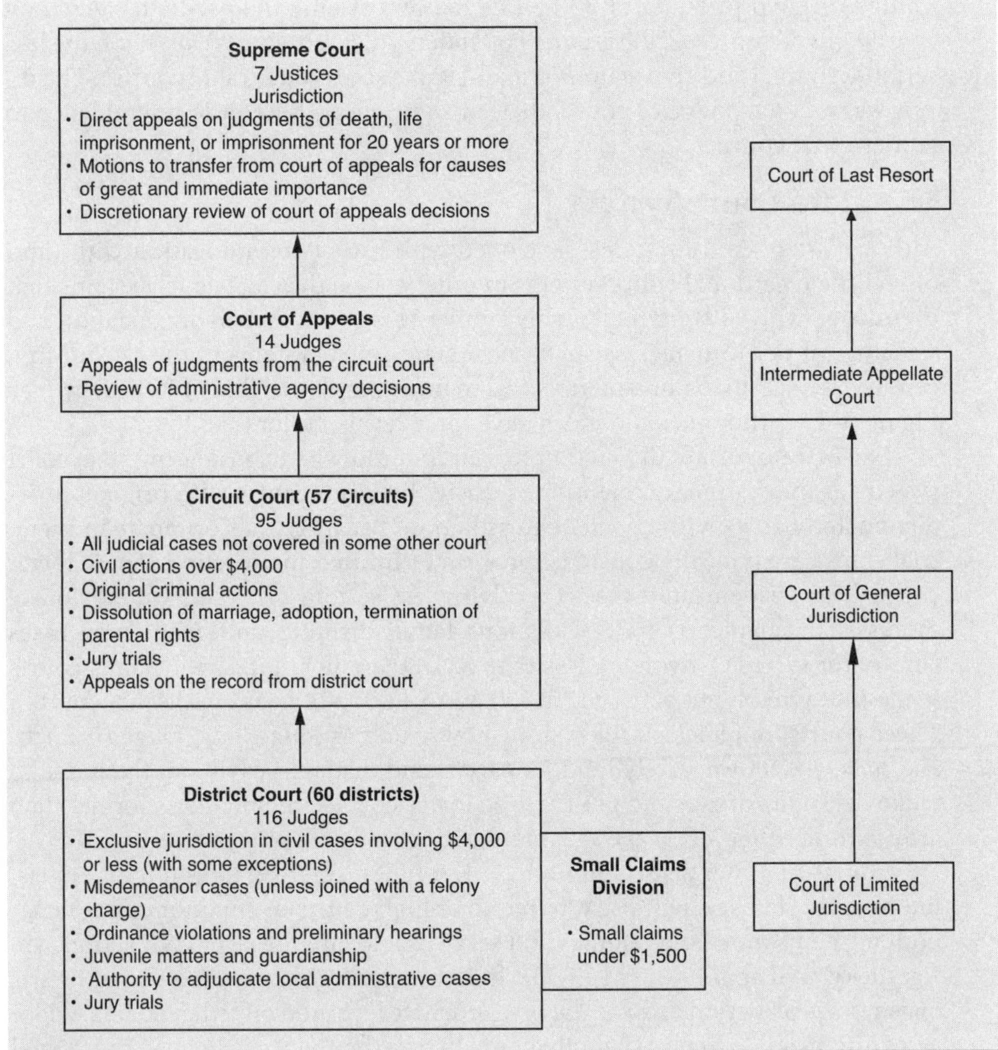

Figure 5–2 Diagram of Kentucky court system.
Source: *Adapted from Kentucky Courts in the Commonwealth Web site (http://courts.ky.gov/courts/) and the Bureau of Justice Statistics* State Court Organization 2004.

As mentioned earlier, each of the colonies had its own court system prior to the American Revolution. After the American Revolution, these courts provided the foundation for the state court systems that developed. Initially, most states did not distinguish between **original jurisdiction** and **appellate jurisdiction**. Many states did not even have provisions whereby a decision could be appealed from one court to another. By the late nineteenth century, however, urbanization, increasing population, and continued expansion into the western part of the United States led to an escalating number of both civil and criminal cases. To manage this increase, state legislatures created an increasing number of courts with limited original jurisdiction to handle matters arising in specific locales (city courts), involving a specific group of offenders (juvenile courts), or involving less serious criminal and civil actions (small-claims courts and traffic courts). The result was a hodgepodge of courts that, in some cases, bore little resemblance to modern trial courts.[14]

The State Court System

Although state court systems developed with little communication with (and often little regard for) court systems in other states, the state court systems that eventually evolved were remarkably similar to one another in organization and structure, if not in name. As such, most state court systems in the twenty-first century are organized in a hierarchical manner similar to that of Kentucky (see Figure 5–2), although some states have three levels rather than four.

The first important distinction to bear in mind when considering state court systems is the distinction mentioned earlier between courts with original jurisdiction and courts with appellate jurisdiction. Because trials originate in them, trial courts are typically considered **courts of limited jurisdiction**. These lower courts hear misdemeanor cases (involving less serious crimes as we previously discussed in Chapter 3), traffic violations, family disputes, and small claims cases (in Kentucky, cases involving less than $1,500 are heard in small claims courts while those involving between $1,500 and $4,000 are heard in district courts). These courts, depicted in television shows such as *Judge Judy*, *Judge Hatchett*, and *Judge Joe Brown*, rarely hold jury trials and depend heavily on the judge to make decisions of fact and guilt. These lower courts are much less formal than are courts at other levels and keep minimal records of court proceedings.[15]

Courts of general jurisdiction (in Kentucky, referred to as circuit courts, but in other states sometimes referred to as high courts or superior courts) hear both misdemeanors and felonies and serve as a forum for lager civil actions (in Kentucky, civil actions over $4,000). In Kentucky and many other states, courts of general jurisdiction provide the first appellate level and offer defendants whose cases originated in a court of limited jurisdiction the chance for a new trial (**a trial *de novo***) or a review of the record of an earlier hearing. These courts use juries, witnesses, defense attorneys, prosecutors, bailiffs, and all the other actors typically associated with American courtrooms. These courts operate far more formally than do courts of limited jurisdiction, use extensive record-keeping practices, and function under the adversarial model discussed earlier.[16]

In the twenty-first century, the next level in 39 state court systems is the **intermediate appellate court level** (in Kentucky, referred to as the court of appeals). The primary function of these courts is to review **appeals** that originate from decisions rendered in the courts of general jurisdiction. The highest court level in all state systems is referred to as the **court of last resort** (titled the Supreme Court in most states). As the name suggests, an appeal can go no further in the state system after this court renders a decision regarding the appeal.[17]

Rather than conducting trials of their own, appellate courts render decisions regarding outcomes from lower courts, with a focus on due process. After an appellate court agrees to review a decision of a lower court, these courts do not conduct a new trial but review the written transcript of the lower-court hearing to verify that the lower-court proceedings were carried out fairly and lawfully. In some cases, these courts allow attorneys from both sides to offer brief oral arguments and generally consider other information or briefs filed by both sides. Most states generally require an automatic review of sentences of death or life imprisonment by the state's court of last resort. Most convictions are upheld in the appellate court; however, if an appellate court believes that an appeal is legally valid, it can reverse and remand the case back to the court from which it originated. Defendants that are not satisfied with a decision by an appellate court may attempt an appeal to the U.S. Supreme Court; however, this appeal must be based on an alleged violation of the defendant's rights as guaranteed by federal law or the U.S. Constitution.[18]

New Directions in State Court Systems

As a response to growing caseloads in the 1980s, a number of innovative approaches were developed to reduce state court caseloads in an attempt to make the criminal justice system operate more efficiently. These new alternative methods of dealing with criminal activity mirror the Native American traditional processes of resolving disputes where they used methods such as peacemaking, elders' councils, or sentencing circles. One of the current and more widely used court alternatives is **dispute resolution centers**, where trained mediators are used to resolve minor civil disputes, divorce and child custody cases, and disputes between tenants and landlords, along with minor criminal cases (e.g., trespassing, shoplifting, passing bad checks). These resolutions take place at an early stage in the criminal justice process to avoid the substantial costs that pretrial preparations involve. Usually, a party that is dissatisfied with the outcome of the mediation may take the case to trial without penalty.[19]

Corrections and Policy

There are a number of dispute resolution centers in various locales throughout the United States. For example, Florida has its own statewide center (The Florida Dispute Resolution Center) that provides education, training, and research in the field of alternative dispute resolution to locales throughout the state. More information about this center is available at www.flcourts.org/gen_public/adr/brochure.shtml.

Community courts are neighborhood-focused courts that use problem solving and partnerships within the context of the justice system to deal with "nuisance" cases (e.g., prostitution, trespassing, disorderly conduct) and cases involving juveniles. Judges in these courts typically sentence offenders to work in partnership with various community agencies to perform community service and receive social services (e.g., drug and alcohol counseling, job training, mental health counseling). The first community court originated in the city of New York in 1993; currently, there are more than 30 active community courts in the United States.[20]

Family courts are designed to serve the multiple and complex needs of troubled families through a unified court model. In this model, a single judge addresses all of the problems facing a family, including child abuse and neglect, divorce and custody issues, delinquency, and domestic violence.[21] The benefits of these specialized courts include more efficient case management, coordination of needed social services, and committed and well-trained judges and staff.[22] According to the National Center for Juvenile Justice, 37 states have implemented some variation of the family court since 1960.[23]

Federal Court System

The federal court system was developed in the U.S. Constitution. Article III, Section 1, established one "supreme court" and as many inferior courts as Congress felt were necessary to establish; Article III, Section 2 dictates that these courts are to have jurisdiction over cases arising under the Constitution, treaties, and federal law. Federal courts are also empowered to settle disputes between states and issue rulings when one party in a dispute is a state. During their formation, federal courts were organized along state lines (even today, there are 11 federal circuit courts organized by region, as depicted in **Figure 5–3**) and, as a testament to the power of states' rights in the judicial system, decisions binding in one federal court circuit are not binding in another unless the U.S. Supreme Court has ruled on the issue in question.[24]

The current federal court system has developed through a number of congressional mandates that have expanded the federal court structure to meet its dynamic needs. The Judiciary Act of 1789, the Judiciary Act of 1925, and the Magistrate's Act of 1968 all played important roles in contributing to the structure of the federal court system in the twenty-first century.

United States District Courts

As depicted in Figure 5–1, the lowest level in the federal court system is the **United States district courts**. These are the federal courts of original jurisdiction, having jurisdiction to hear both civil and criminal federal cases. There are 94 federal judicial districts, including at least one district in each state, the District of Columbia, Puerto Rico, the Virgin Islands, Guam, and the Northern Mariana Islands. These district courts also have exclusive jurisdiction over bankruptcy

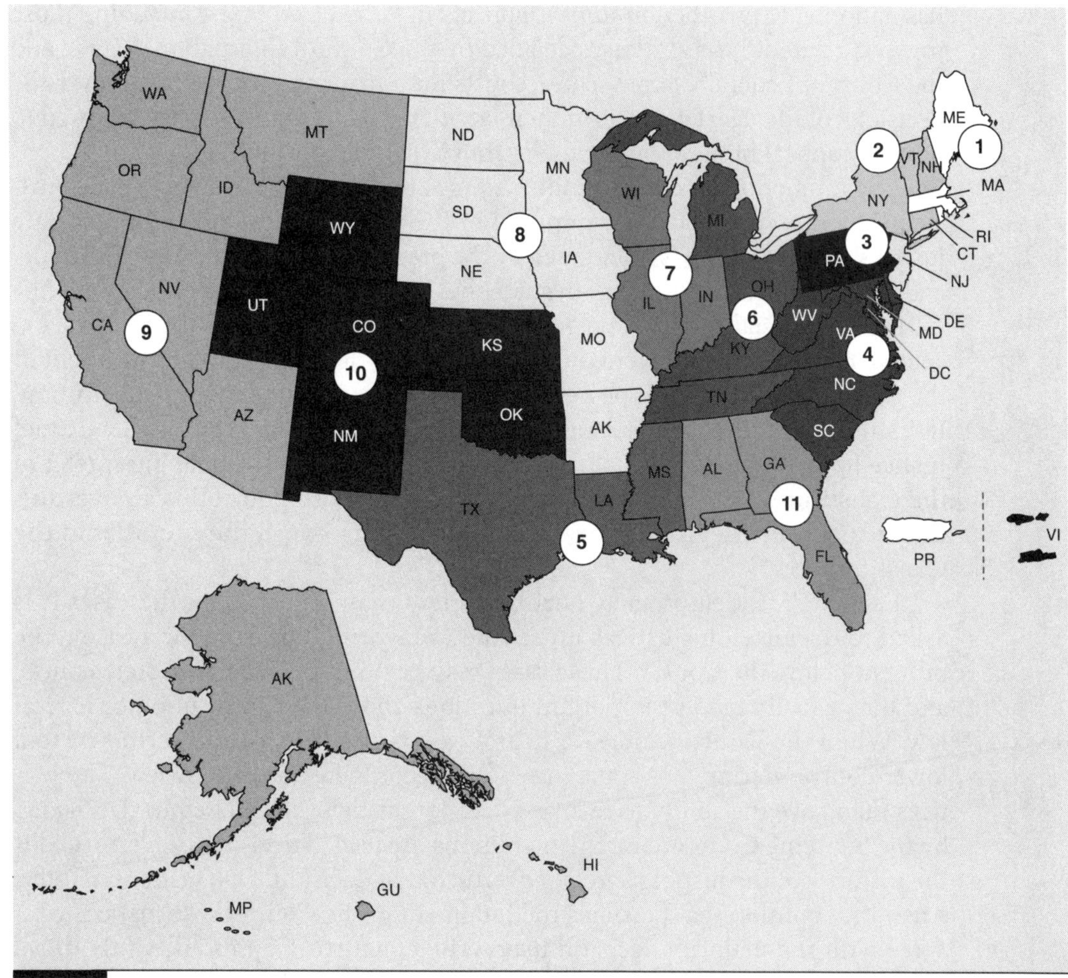

Figure 5–3 Geographic boundaries of the United States courts of appeals and United States district courts.
Source: *Reproduced from the United States Courts, www.uscourts.gov/images/CircuitMap.pdf.*

courts where all bankruptcy cases must be filed. Also, two special trial courts have nationwide jurisdiction over cases involving international trade and customs issues (the Court of International Trade) and over disputes over federal contracts and claims for monetary damages against the federal government (the United States Court of Federal Claims).[25] Judges in these courts are appointed for life by the president with Senate consent.

The next level in the federal court system is the **United States Courts of Appeals.** The 94 U.S. district courts are organized into 12 regional circuits, each of which has a United States court of appeals. These courts hear appeals from the district courts located in their circuit, along with appeals from decisions of federal administrative agencies. Additionally, the U.S. Court of Appeals for the Federal Circuit has nationwide jurisdiction to hear appeals in specialized cases, including those involving patent laws and cases decided by the Court of International Trade and the Court of Federal Claims. These courts have mandatory jurisdiction over the decisions of the district courts in their circuits, which means they are required to hear any appeal brought before them from a district court in their circuit.[26]

The top tier of the federal court system is the **United States Supreme Court**. The U.S. Supreme Court is composed of the Chief Justice and eight associate justices, each of whom is nominated by the president, is confirmed by the Senate, and serves for life.[27] Within certain guidelines established by Congress, the Supreme Court has **judicial review** power that allows its members to invalidate laws, executive decisions, and lower-court decisions that do not keep with the intent of the U.S. Constitution. This power was not explicitly granted in the Constitution. It resulted from the Court's ruling in ***Marbury v. Madison*** (1803) where Chief Justice John Marshall established the Court's authority as the final interpreter of the Constitution when he asserted that the Court's responsibility to overturn unconstitutional legislation was mandated by its sworn duty to uphold the Constitution.

Each year, the Supreme Court hears less than 5 percent of the cases it is asked to decide (four justices must vote in favor of a hearing for a case to be brought before the Court). Those cases may begin in the federal or state courts, and they usually involve important questions about the Constitution or federal law. When the Court decides to hear a case, it issues a *writ of certiorari* to a lower court ordering that court to send its records forward for review. The justices then have the ability to revoke a decision made by a lower court. Decisions by the Supreme Court are rarely unanimous; instead, a decision is overturned if the majority of the justices agree to overturn that decision. One justice compiles a majority opinion that becomes the judgment of the Court; those justices who agree with the majority decision may write concurring opinions, while those justices who do not agree with the decision write dissenting opinions, which often provide new possibilities for successful appeals of future cases.[28]

Race and Gender in Corrections

Only two women have served as justices in the history of the U.S. Supreme Court, Ruth Bader Ginsberg and Sandra Day O'Connor, and both were justices of the Supreme Court as recently as 2005, when O'Connor retired. Thus, for roughly 200 years prior to O'Connor's appointment to the court, the U.S. Supreme Court was an "all-male" institution.

The Courtroom Work Group

In the previous sections, we have reviewed the structure of both the federal and state court systems. Within those court systems, there are a number of individuals whose daily activities in the courts structure the criminal trial process in that jurisdiction. Several of these actors are discussed in this section. Although the term *courtroom work group* is usually reserved for the judge, prosecutor, and defense attorney, we have included a number of other important roles in this discussion as well.

The Judge

Judges are the most easily recognizable members of the courtroom work group, both because of their distinctive courtroom apparel consisting of flowing black robes and because of their unique yet prominent position in the courtroom. The judge's primary duty is to ensure justice is served; judges are responsible for protecting both the rights of the accused and the interests of the public. The criminal trial judge must be a neutral party who is responsible for helping determine whether the prosecution has established the guilt of the accused as required by law.[29] Along with the well-known roles of presider over the trial and sentencing agent, in jury trials judges also instruct the jury at the conclusion of the trial about their duties in determining the guilt of the defendant. During **bench trials**, judges serve as the trial referee and make determinations of guilt or innocence. As part of their referee role, during the trial judges must rule on objections brought forth by either prosecuting or defense attorneys and are tasked to do their best to ensure that each defendant receives a fair and impartial trial.

Judges also are often assigned the task of sentencing a defendant found guilty. In this role, judges must determine appropriate penalties for the crime that has been committed. Although judges often receive input from probation officers through presentence investigation reports (discussed in Chapters 6 and 8), attorneys, and judicial precedent, sentencing often is a difficult task for judges, primarily because of statutory restrictions and limited sentencing options in many cases.[30] Even after sentencing, judges continue to play a role in the court system. In some cases (e.g., capital cases), appeals are mandatory, so appellate judges must review cases tried in lower courts; even in those cases that do not involve mandatory appeal, appellate judges may review lower-court decisions because of courtroom actions deemed unconstitutional by the defendant (e.g., juror or judge mistakes, corrupt prosecutor actions) or because of legal errors accredited to the judge by defense attorneys in the original trial in the lower-court (e.g., allowing prejudicial testimony or illegal evidence, giving incorrect jury instructions). Judges, particularly in smaller districts, are also generally responsible for management of daily courtroom activities. Lower-court judges in larger jurisdictions often have caseloads of hundreds of cases per day and are regularly concerned with "clearing their calendar" in an effort to provide hearings for each and every defendant in a timely manner. Consequently, the role of the judge in the trial is a demanding, stressful, and often misunderstood part of the criminal justice system.

Judicial Selection and Qualifications

As mentioned earlier, federal judges at the appellate levels are appointed by the president and confirmed by the U.S. Senate. These appointments, although often approved routinely by the Senate, sometimes generate a large amount of controversy and discussion. One of the best examples of these controversial appointments involved Justice Clarence Thomas, who was appointed to the Supreme Court in 1991 by President George Bush. The Senate Hearings that surrounded that appointment lasted five months, primarily because of allegations of sexual harassment against Thomas made by Oklahoma law professor Anita Hill. The appointment of Thomas by George Bush was confirmed by the Senate as the result of a vote of 52 to 48, the closest confirmation vote in the twentieth century.[31] Federal judges are appointed for life and are replaced only by impeachment or when they are not able to perform their judicial duties (e.g., physical or mental illness).[32]

Although judicial selection in the federal court system is straightforward (though sometimes controversial), the methods of selecting state-level trial and appellate judges vary widely from state to state. State-level trial court judges are generally elected through either partisan or nonpartisan elections (31 out of 50 states use elections); in these states, potential judges campaign for their positions and are guaranteed to be in office only until the next election. The remaining states use some form of gubernatorial or legislative appointment. Terms of appointment in state courts range from a minimum term in office of one year (in some courts in Alaska, Nebraska, Ohio, Tennessee, and Washington) to a maximum of a lifetime appointment (Rhode Island) or age 70 (Massachusetts and New Hampshire). The vast majority of state court judges are required to have a law degree and to have been admitted to the state bar prior to their election or appointment, although a law degree is not required for a judicial election or appointment in lower courts in several states. [33]

The Prosecuting Attorney

Arguably, the second most important actor in the court system is the prosecuting attorney (commonly referred to as the **prosecutor**). The prosecutor is often the first member of the courtroom work group defendants encounter in their travels through the criminal trial experience. Nevertheless, prior to recent media portrayals of the prosecutor in television courtroom dramas (e.g., the *Law and Order* series and its spin-offs) and the cable Court TV channel, many people had a limited understanding of the role of the prosecutor, and even today, the prosecutor role is still shrouded in secrecy or misunderstanding for most citizens. As such, we provide a fairly detailed discussion of the role of the prosecutor here.

The modern role of the prosecutor did not emerge overnight. Although the role of the prosecutor was found as early as English common law and colonial courts, the role was given new meaning and power when the Judiciary Act of 1789 created the office of the U.S. Attorney General. The U.S. Attorney General is selected by presidential appointment and is the highest-ranking law enforcement officer in the United States. Each federal district also has its own attorney general, who serves as the chief prosecuting attorney of that district. States elect their own attorney general, and county-level equivalents (district attorneys) are usually elected as well, although in some states district attorneys are appointed by the governor.[34]

Because society (commonly referred to as the "state") is viewed as the injured party when individuals violate criminal laws, individuals who violate such statutes are considered to have committed crimes against every member of society. Consequently, the prosecutor doesn't have a "client" in the traditional sense because the prosecutor is representing all members of that jurisdiction by representing the interests of the state. As such, the primary function of the prosecutor is to prepare and present the state's case against defendants in criminal and civil cases. Although many people are aware of this role, few understand that this role, although important, is only one of several roles the prosecutor serves.

Another important role of the prosecutor is to ensure not only that the guilty are prosecuted, but also that the innocent are protected. As such, the prosecutor seeks the most "just" resolution (justice) to each case, not simply the one that provides the most benefit to the state. In some cases, justice may best be served by declining to prosecute a suspect, even when enough evidence might exist to support an **indictment** in the case of a serious criminal offense or an **information** in a less serious offense. In an information, prosecutors initiate charges against an individual accused of a crime based on their own legal opinion that the defendant committed the crime in question. An indictment is issued by a **grand jury**, whose role is to determine whether the prosecutor has presented enough evidence to support the accusations being made against the alleged defendant (discussed in detail later). Consequently, another role of the prosecutor is to determine, based on the resources of the jurisdiction and other practical matters surrounding the case, which cases to pursue and which cases to refuse to pursue so that justice is served in that jurisdiction.[35]

Although the duties and daily activities of prosecutors may vary widely from one jurisdiction to another, prosecutors also perform a number of activities that generally fall into three broad categories: (1) managing the investigation phase of a criminal case; (2) preparing cases for trial; and (3) responding to issues related to appeal. In each task, prosecutors often use their discretion (discussed in Chapter 3) in deciding what to do with a particular case, including whether to dismiss a complaint or file charges, and, if they choose to file charges, which charges to file and how many counts for each charge should be filed.

In fact, some argue that this discretion given to the prosecutor at such an early stage in the criminal justice filter makes the prosecutor the most powerful actor in the criminal justice system. Critics of recent sentencing policies such as "three strikes policies" (discussed in Chapter 6) and other mandatory sentencing policies argue that these policies have dramatically changed the role of the prosecutor. Prior to mandatory sentencing, prosecutors could recommend that the judge provide a less stringent sentence in exchange for a guilty plea; mandatory sentencing laws sharply reduce the judge's discretion, however, and most defendants facing a mandatory sentence now go to trial. Prosecutors can still circumvent the laws, however, by charging the defendant with a less serious charge or dropping some charges altogether. This often gives the prosecutor even greater leverage for plea bargaining (discussed in detail later) when the defendant has committed a series of criminal activities, only one of which may carry a mandatory sentence (for example, using a firearm in a felony). Defendants may be much more likely to accept a plea in these situations to avoid a mandatory sentence than they would have been prior to the enactment of those laws.

Although the discretionary powers of the prosecutor are vital elements of the prosecutor's role in the criminal justice system, prosecutors do not work alone. In fact, the prosecutor not only works closely with all members of the courtroom work group (particularly the judge and the defense attorney) but normally works with law enforcement and community corrections personnel in the jurisdiction as well. Decisions made by prosecutors have significant impact on each of these parties. For example, if a prosecutor decides that he or she will no longer prosecute individuals who possess only small amounts of marijuana, this decision has implications for the police (who may eventually stop arresting individuals in violation of a statute prohibiting possession of small amounts of marijuana) and the judge and those involved in corrections (who may see a reduction in their caseload). As such, the role of the prosecutor continues to be one of the most important (and controversial) roles in the courtroom work group.

Defense Counsel

According to the U.S. Supreme Court, the primary role of the **defense counsel** is to contest the state's case against the alleged defendant, to "put the State's case in the worst possible light" to avoid convicting the innocent.[36] As such, the defense counsel's role is to provide the alleged defendant with legal counsel and assistance to ensure that the defendant's constitutional rights are protected. Additionally, the defense counsel is to assist the defendant in building a case that will ensure that the defendant is not wrongly convicted for a crime the person did not commit. Defense attorneys thus ask the court to consider legal questions such as whether the identity of the alleged defendant matches that of the individual who actually committed the criminal action for which their client is charged, whether enough evidence exists to charge their client with a crime, and whether this evidence is reliable enough to be used in trial; whether mitigating circumstances exist that may have caused their client to violate the law; and in some cases, whether the plea offer given by the prosecutor is a fair punishment for what their client has allegedly done.

The preparation of the defense attorney's argument usually involves conversations between the defendant and his or her attorney. These conversations are protected under the umbrella of attorney–client confidentiality; defense attorneys cannot be forced to disclose information their clients have told them in these conversations. If the defendant is found guilty at trial, the defense attorney will also represent the defendant's interests at the sentencing phase of the trial and may be asked to represent the defendant in an appeal of the court's decision. The defense attorney often counsels the defendant and his or her family about civil matters as well (e.g., debt payment, contractual obligations). As such, the role of the defense attorney is multifaceted and extends far beyond the well-known role of representing the defendant in the criminal trial.[37]

Three Categories of Defense Attorneys

Some defendants in criminal trials are represented by **private attorneys**.[38] These private attorneys (also referred to as retained counsel) have their own legal practices or law firms and charge fees (typically ranging from $100 to $250 per hour) to the defendant in exchange for their representation at trial. The defendants are billed not only for the time the attorney spends representing them at trial but also for the time that it takes for the attorney to build their defense argument prior to trial. Although there are several well-paid, well-known defense attorneys in the United States [e.g., Johnnie Cochran and Robert Shapiro (O.J. Simpson defense attorneys), Pamela Mackey (Kobe Bryant), Mark Geragos (Scott Peterson)], this type of defense attorney is rarely found in a criminal trial. Far more often, defendants are unable to pay for an attorney and thus rely on an attorney appointed by the court to serve as their defense attorney.

Although the "right to counsel" is provided to each citizen of the United States in the Sixth Amendment, until the twentieth century the "right to counsel" was interpreted by the courts to mean that defendants were allowed to hire an attorney to represent their interests in a trial. It wasn't until 1932, in *Powell v. Alabama*, that the Supreme Court ruled that the states had to provide defense attorneys to those individuals who could not afford to hire their own attorneys in capital cases. The right to have an attorney provided for *all* defendants in *federal* courts was provided by the Court in *Johnson v. Zerst* (1938) and was extended to felony defendants in *state* courts in *Gideon v. Wainwright* (1963) and misdemeanants facing possible imprisonment a decade later (*Argersinger v. Hamlin* in 1972).

As such, the Supreme Court has ruled that those defendants who cannot afford to hire their own private attorney still must be able to have an attorney represent them at trial. There are two categories of defense attorneys provided by the state to help those who cannot afford to hire a private attorney in their criminal defense. These attorneys are commonly referred to as **court-appointed counsel** and **public defenders**. In a jurisdiction that uses court-appointed counsel, the court maintains a list of private attorneys from which it appoints attorneys for indigent defendants (those who cannot afford to pay for their own private defense attorney). These attorneys normally have their own private clients and are paid a flat fee for their efforts by the court (which is often much less than the attorneys would charge one of their own clients). Public defenders, on the other hand, are attorneys who are employed by the state to provide defense for indigent defendants. They typically have no private clients.

Ethics of Defense

As we previously discussed, the job of the defense attorney is to provide a client with legal counsel and assistance to ensure that the defendant's constitutional rights are protected and assist in building a case for a client to ensure a client is not wrongfully convicted. Nevertheless, because the defense counsel often knows more about the innocence or guilt of the defendant than does anyone else at trial, it is important that defense attorneys follow strict ethical stan-

Controversies

Both public defenders and court-appointed counsel are often criticized for providing less-adequate legal representation than private attorneys do. The research in this area tends to bear that out, particularly with public defenders, who are more likely than private attorneys to (1) be new attorneys with relatively little experience; (2) have very high caseloads (clients assigned to them); and (3) represent defendants who are indigent and have prior criminal histories. This difficult situation is aggravated by limited financial resources for investigations to assist in the defense preparation and more conservative policy changes that increased the number of defendants as a result of mandatory sentencing policies and sharply increased drug caseloads over the last 20 years. Consequently, some have argued that there are two types of defendants in the United States: the "haves" (those with private attorneys) and the "have-nots" (those with court-appointed counsel or public defenders).

dards to ensure that the justice process works effectively. As such, the American Bar Association has established a number of ethical guidelines for attorneys.[39]

Even with these ethical directives, defense attorneys are not obligated to disclose information obtained from their client without the client's permission. Nevertheless, in all states, defense attorneys are obligated to disclose information obtained from their client if they are aware that their client intends to commit a future crime. Furthermore, the Supreme Court has also ruled that attorneys are prohibited from assisting their client in law violations or offering false evidence.[40] As such, despite the often dubious reputation that defense attorneys may have, their ethical obligation is to uphold the law to ensure that their client is constitutionally protected from false accusation and imprisonment but also to do this in a manner that does not violate the law or the ethical standards established by national and state bar associations.

Bailiff

The **bailiff** is the court officer responsible for maintaining order in the courtroom and is charged with supervising and managing the jurors in a jury trial. The bailiff is usually an armed law enforcement officer (typically a deputy sheriff or deputy marshal) who is assigned to the courtroom to assist with the operation of the court by calling witnesses, announcing the judge's entry into the courtroom, protecting the other courtroom workgroup actors, and preventing the escape of the alleged defendants. The bailiff also supervises and assists the jury when they are sequestered for deliberation. In the federal courts, the bailiffs are typically deputy U.S. marshals.[41]

Although the vast majority of courtroom trials occur without any dangerous incidents, there are notable exceptions. The most recent well-publicized exception occurred in a courthouse in Atlanta, Georgia, in 2005 when Brian Nichols overpowered the bailiff and used the bailiff's firearm to shoot and kill a Fulton County judge, a sheriff's deputy, and a court reporter. Nichols then escaped by carjacking a number of cars; he was eventually recaptured in an apartment complex the next day.[42] Incidents like this one emphasize the critical and often dangerous role that members of the courtroom work group play in the administration of justice.

Court Administrator

Many jurisdictions, particularly larger ones, have created a relatively new role in the courtroom work group, that of the **court administrator** who assists the judge or judges in the operations and management of the court's daily activities. Duties of the court administrator often include administering and managing the court's budget, personnel, and caseloads in a manner so that the court operates efficiently. Despite the authority of the administrator, the ultimate authority for managing the court still rests with the chief judge in that jurisdiction. Court administrators also assist with juror management and track case outcomes to ensure that a defendant is processed through the court in the most efficient manner possible.[43]

Court Reporter

A **court reporter** (also referred to as a court stenographer) transcribes the verbal courtroom proceedings into written form, typically using a stenotype (shorthand machine), audio recorder, or computer, to create the official trial court transcript.[44] Accurate records are essential in criminal trial because appeals are based entirely on the transcript of the court's proceedings. This official trial record typically includes all verbal comments made in the courtroom, including witness testimony, arguments made by the attorneys, the judge's instructions to the jury, and the results of any conference between the attorneys and the judge. In the twenty-first century, court reporters often use computer software that assists in transcribing notes written in shorthand into complete and legible transcripts.

Clerk of Court

The **clerk of court** is a courtroom officer whose responsibilities include maintaining the records of the court and swearing in witnesses, jurors, and grand jurors.[45] The clerk's role also extends outside the courtroom because clerks typically are also responsible for determining when cases are ready for trial, marking and maintaining custody of all exhibits entered as evidence, preparing the jury pool and issuing jury summons, and preparing and issuing **subpoenas**, bench warrants, citations, and other legal documents to assist the daily operations of the court.

Nonprofessional Courtroom Participants

Along with the professional courtroom participants discussed previously, a number of unpaid, "nonprofessional" participants are actors in the courtroom work group. These participants are discussed in detail in the following section.

Witnesses

In general terms, **witnesses** are individuals who have firsthand knowledge about an event and provide evidence that helps to establish the facts in a particular case. There are several different types of witnesses, as discussed in this section.

Lay witnesses are witnesses who appear in the criminal trial because of their firsthand knowledge of (1) the event for which the defendant is accused or (2) an individual involved in the criminal event. Eyewitnesses are individuals who witnessed a criminal event; eyewitnesses are often the most persuasive source of evidence in a criminal trial, particularly a jury trial. Character witnesses provide testimony about the defendant in an attempt to convince the judge or jury that the defendant is not the type of person who would have committed the crime in question. **Victims** are sometimes witnesses as well, providing their version of the alleged criminal activity of which they were victimized as well as testimony about the defendant's involvement in that activity. Lay witnesses are allowed to testify regarding only things about which they have direct knowledge and must do so under oath, swearing or affirming that their testimony is accurate and honest. Witnesses are often mistreated by the criminal justice system by being forced to endure unannounced changes and delays in trial dates; even when they are compensated for time away from work, the compensation is minimal. Despite efforts to improve these processes, these difficulties still remain in some jurisdictions, and the witness in the twenty-first century often testifies out of a sense of duty in an effort to seek justice at his or her own financial expense.

Expert witnesses are persons with specialized knowledge or skills who are recognized by the court to assist in providing context to the judge or jury to assist them in their determination of guilt. Expert witnesses typically are asked to (1) educate the jury about general issues in an area for which they have special knowledge or (2) provide specific information on a key issue in the case (e.g., the mental state of the defendant at the time the person committed the crime). Expert witnesses can testify for either the state or the defendant, but usually are asked to appear for the side (the state or the defendant) their testimony is most likely to assist. Unlike lay witnesses, however, they are usually allowed to express their opinions or draw conclusions based on their particular area of expertise. Because each side often brings its own expert witness to trial, expert witness testimony can be contradictory and thus confusing. Despite this difficulty, most researchers agree that expert witnesses provide more reliable testimony than lay witnesses do.[46]

The Victim

Although not all crimes have an identifiable victim (and in the case of murder, the victim cannot testify), when there is a victim of a crime, he or she is often neglected or mistreated in the criminal justice process. In some cases, the victim may not even be allowed to participate in his or her own justice process (particularly in the vast majority of cases when plea bargaining is used). Although a number of efforts have attempted to make the victim more involved in the criminal justice process, victims are still often unaware of the status of the proceedings or outcome of the process through which the defendant allegedly responsible for their victimization is punished. Victims, like witnesses described earlier, often experience trial delays or cancellations but, when testifying in a criminal trial, also experience the emotional and psychological distress that comes with facing the defendant allegedly responsible for their victimization.

Additionally, testifying and experiencing cross-examination by the defense attorney is often traumatic as well. Some suggest that the victim's experience with the criminal justice process can be considered a second victimization at the hands of the criminal justice system.[47]

Jurors

Jurors are given the authority to participate in trials in Article III of the United States Constitution.[48] Even though most cases are plea-bargained, jurors still play an essential role in the courtroom work group when cases do go to trial. In fact, the concept of the jury often influences plea bargain agreements, as both prosecutors and defense attorneys often base their plea decisions on what the jury might decide if the case were to appear before them in a trial. The role of the juror evolved from colonial times when juries were used to prevent government oppression against citizens, and this is often considered the central role of juries even today.

Although states have the authority to constitute juries with as many members as they desire (the Supreme Court has upheld as few as 6), most states use juries of 12 members. Jury duty is often regarded as one of the duties of citizenship and, with limited exceptions (e.g., emergency services workers, doctors, and the military), those who are called for jury duty must serve. Some groups (e.g., convicted felons, citizens who have recently served on jury duty, juveniles) are excluded from jury service in most jurisdictions. Names of prospective jurors are usually chosen by the clerk of court from voter, motor vehicle, or tax registration records; minimum qualifications typically include a basic command of the English language, local residency, and U.S. citizenship. Jurors are also expected to possess their "natural faculties" (be able to move, see, speak, and hear) although some jurisdictions have recently begun to allow individuals with physical disabilities to perform jury service. Although a jury is to be comprised of one's "peers," the Supreme Court has clarified that it is not necessary for every jury to be entirely representative by race, gender, ethnicity, or religion, as long as jurors are not systematically excluded by court officials because of their demographic characteristics.

As with witnesses and victims, jurors are often inconvenienced by the criminal justice process. Depending on the state, jurors receive compensation for their service that ranges from $5 to $40 per day and are not reimbursed for expenses they may incur as part of their jury service (e.g., parking, travel, food). Jurors may be required to sit for weeks or months of service and are usually prohibited from discussing their work with anyone outside of their peers on the jury. Given the nature of the role of the juror, despite its importance, it is one of the least desired roles in the courtroom work group by citizens in that jurisdiction.

Defendant

With limited exceptions, the **defendant**[49] must be present at trial. Most criminal defendants are poor, uneducated, and disproportionately male and nonwhite.

Many defendants are represented by public defenders or court-appointed counsel; nevertheless, defendants may still have tremendous influence over their trial by deciding what plea to enter or, if they choose to go to trial, what information to provide to or withhold from their attorney and whether or not to testify at the trial. Despite these options, most defendants have a number of disadvantages. First, and foremost, although a person is considered "innocent until proven guilty," the fact that the defendant is accused of a crime often stigmatizes the defendant and negatively influences the opinions of the jury or other courtroom workgroup actors. Coupled with the often pronounced social and cultural differences between the defendant (often a young, poor, nonwhite male) and the members of the courtroom work group (typically white, middle-class males and females), defendants often face an uphill battle as soon as they enter the courtroom.[50]

Pretrial Activities

Prior to the criminal trial, a number of events take place that lengthen the process between the time a defendant is arrested and the time a decision is made regarding the defendant's guilt or innocence. The activities that occur prior to the criminal trial are called pretrial activities and are discussed in detail in this section.

Initial Appearance

After an arrest, defendants usually do not come into contact with a member of the courtroom work group until their **initial appearance**. During the initial appearance, three activities typically occur. First, defendants are informed of the charges against them, are advised of their rights, and are given the opportunity to have an attorney appointed if they are unable to afford a private attorney. Second, **bail** is set (if it has not been set already). Finally, typically the time of the preliminary hearing is scheduled, where a judge or magistrate will determine if probable cause exists to indicate that the alleged defendant is responsible for the criminal activity in question.[51]

Pretrial Release

Although the majority of defendants have bail set at the initial appearance, defendants charged with serious crimes or who are considered "flight risks" are often denied bail and held in jail until trial, a practice called pretrial detention. The decision by the judge or magistrate regarding whether to grant **pretrial release** is structured by background information about the defendant. This information typically includes: (1) the seriousness of the current charge; (2) the defendant's criminal record; (3) information about the defendant's ties to the community (e.g., employment status, housing, family ties); and (4) available alternatives for supervision if the defendant is released.[52]

Bail

Bail is the most common method of pretrial release in more serious cases. Bail involves the posting of some form of property that is deposited or pledged to the court to secure the release of a defendant from custody prior to trial. If the defendant does not appear at trial, the property is forfeited. Some defendants may deposit with the court the full bail amount or property. In most cases, however, the defendant can obtain privately secured bail through a bail bondsman, who charges a fee (typically around 10% of the total bond amount) that the defendant has to pay to the bondsman to secure the defendant's release. In most states, bondsmen are allowed to pursue those defendants who do not appear at trial after the bondsmen have posted their bail. These bondsmen (nicknamed "bounty hunters") often have unrestricted power to pursue and forcibly detain those defendants who have "skipped bail" when they locate them in other states (and sometimes in other countries) without regard for constitutional considerations of the defendant,[53] although recent court decisions have begun to limit those powers.[54] In some jurisdictions, rather than secure the services of a bondsman, the courts allow the defendant to post a percentage (generally 10%) of their bail with the court; those who do not appear in court forfeit their 10 percent payment and are billed for the remaining 90 percent. Another type of bail is unsecured bail, where the defendant does not have to deposit any amount with the court unless he or she fails to appear for trial, when the entire amount is forfeited.[55]

Alternatives to Bail

Although the U.S. Constitution does not guarantee any person bail (the Eighth Amendment only prohibits excessive bail), the vast majority of defendants are granted bail in some fashion. Nevertheless, other alternatives to pretrial detention also exist.

The first and most often used alternative to bail for nonserious crimes is **release on recognizance (ROR)**. Defendants released on their own recognizance simply assure the court that they will return for their subsequent court appearances. In these situations, should the defendant not appear in court, no collateral or bond is forfeited, but a warrant can be issued for his or her arrest. ROR is usually granted to individuals with strong ties to the community who have been accused of less serious crimes; rarely are those accused of murder or rape granted ROR, no matter how much standing they have in the community.

Another alternative to bail is **conditional release** where the judge releases the defendant prior to the trial but imposes conditions on the defendant by which the defendant must abide during the release (e.g., maintain employment, avoid contact with the victim or co-defendants, and submit to urine analyses or electronic monitoring). If the defendant fails to abide by these conditions, the defendant is arrested and returned to detention. Finally, **third-party custody** occurs when a third party (e.g., parent, friend, or family member) promises to ensure that the defendant will return for the next scheduled court hearing if the court will release the defendant to that third party's supervision. This form of release is rarely used.

Preliminary Hearing

The next court appearance for most defendants occurs when they appear at the **preliminary hearing** or before the grand jury (for more serious crimes in those jurisdictions that use grand juries). At the preliminary hearing, the prosecutor files an information document that the judge or magistrate reviews to determine if probable cause (discussed in Chapters 1 and 4) exists to suggest that a crime has been committed and the defendant committed it. At the preliminary hearing, the judge summarizes the charges and the defendant's rights; both the prosecutor and defense attorney may present witnesses or evidence, and both the victim and defendant may testify.[56]

At the grand jury, the hearing proceeds in much the same way, except a jury of private citizens (usually larger in number than trial juries) reviews an indictment filed by the prosecutor and determines whether to issue a "true bill" (stating that the majority of the jury agrees that the indictment should move forward) or dismiss the charges against the defendant. The defendant and the defense attorney are generally not present for the grand jury proceedings.

Arraignment

After an indictment has been returned or the information has been filed, the **arraignment** is held in front of a judge or magistrate. At this hearing, the indictment or information is read to the defendant and the defendant enters a plea. According to the Federal Rules of Criminal Procedure,[57] a defendant may enter one of three pleas: guilty, not guilty, or *nolo contendre* (no contest). Although the no contest plea is treated as a guilty plea, a defendant using this plea is not admitting guilt, and this accomplishes two purposes: (1) the plea cannot be used as an admission of guilt at a civil trial revolving around the same incident; and (2) the plea allows the defendant to withdraw the plea if the court's decision about the incident is overturned upon appeal. Defendants may also choose not to enter any plea (or "stand mute"); for procedural purposes, a defendant who does not enter a plea is treated as if he or she entered a plea of not guilty.

Plea Bargaining

After the preliminary hearing, the most frequent resolution to a criminal case is that the defense and the prosecutor engage in a process known as **plea bargaining** (discussed briefly in Chapter 3). Plea bargaining is a process of negotiation between the defense and the prosecutor in which the defendant agrees to plead guilty to some crimes in exchange for (1) dismissal of some charges, (2) reduction in the severity of the charges, (3) reduction of the number of counts on a

Corrections in the Real World

Despite the controversies surrounding plea bargaining, the criminal justice system in the United States is heavily dependent upon the use of pleas to avoid criminal trials. Almost 9 in 10 criminal cases are resolved through plea bargaining.[58] Without plea bargaining, the backlog of criminal court cases would be unmanageable and the "wheels of justice" would turn even more slowly than they presently do.

certain charge, and/or (4) recommendation for a more lenient sentence. The judge must agree with the negotiated result before the plea can be accepted.[59]

Plea bargaining often benefits all parties involved. Defendants and their attorneys may agree to a plea when they are not sure of their ability to avoid a guilty verdict at trial; prosecutors are often willing to plea bargain because (1) the evidence they have against the defendant is not as convincing as they would like it to be, and (2) plea bargaining gives them the opportunity for a quick conviction without committing the time and resources needed for a criminal trial. Plea bargains require judicial consent; judges are also willing to accept pleas because they reduce the court's workload. Although judges are rarely willing to guarantee a sentence before a plea is entered, the previous experience of both the prosecutors and defense attorneys with the judge generally allows the attorneys to know what sentences can be expected from typical pleas.

Although the Supreme Court has ruled that a guilty plea constitutes a conviction and does not violate the right to a trial set forth in the Constitution,[60] plea bargaining has received criticism from both conservatives and liberals. Conservatives argue that plea bargaining results in unjustifiably light sentences; liberals argue that the plea bargaining process may result in defendants being convicted of crimes they did not commit. Plea bargaining also has the potential to be abused by both prosecutors and defense attorneys who are more interested in a speedy end to the case than they are in finding a just resolution to the matter. To safeguard against abuses of plea bargaining, the Federal Rules of Criminal Procedure require judges to address the defendant in open court and ensure that defendant is entering the plea voluntarily and that defendants understand (1) that they have the right to plead not guilty, or to continue their guilty plea, if they so desire; (2) that they have a right to a trial to confront and cross-examine witnesses and that they are waiving those rights if they accept the plea; and (3) the nature of the plea to which they are agreeing.[61]

Steps of a Criminal Trial

Despite the high percentage of cases that end in plea bargaining (often motivated by a desire of one or both parties to avoid trial) a number of cases still result in a criminal trial. The length of the average criminal trial ranges between three days and one week; occasionally, trials last longer because of the large number of witnesses called to testify and the amount of physical evidence presented. The steps of the criminal trial are presented in **Figure 5–4**. Each of the steps is discussed in detail in the following subsections.

Trial Initiation

Despite the language in the Sixth Amendment of the United States Constitution that guarantees a speedy trial to defendants, the 1974 passage of the Speedy Trial Act, and the *Klopfer* decision (that allows for dismissal of charges when the trial does not start within 180 days after the defendant issues a plea), months of-

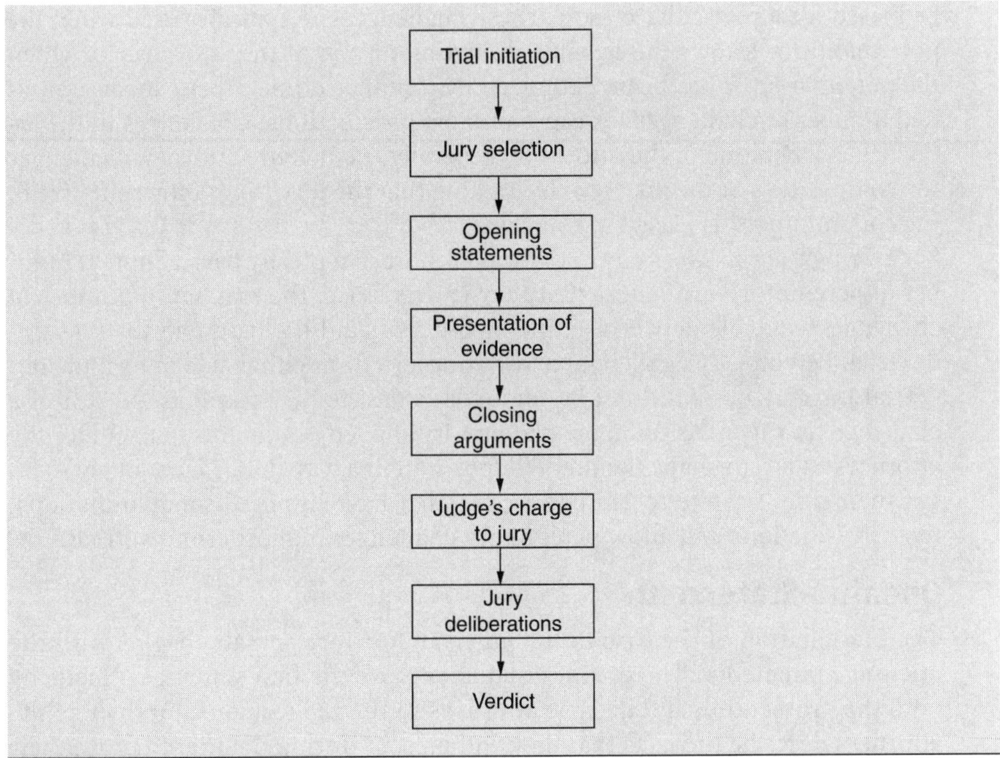

Figure 5–4 Steps in a criminal trial.

ten lapse between the preliminary hearing and the beginning of the criminal trial. This delay may be caused by clogged court calendars, limited judicial resources, delays brought by the defense counsel, or a wide variety of other reasons; nevertheless, most state legislation limits the time between arraignment and trial to 90 or 120 days.[62]

Jury Selection

As reviewed earlier, a number of individuals are prohibited from serving on juries or may be excused from jury duty because of their occupation. Because some people would like to be excused from jury duty and others would like to serve but are not suitable for jury service, a process known as ***voir dire*** is used by attorneys to screen potential jurors in an effort to remove obviously biased jurors and create a jury they believe will help "win" the case for their side. During the *voir dire*, defense and prosecuting attorneys use challenges to ensure the impartiality of the jury's members. Three types of challenges are generally used: **challenges for cause**, **challenges to the array**, and **peremptory challenges.**

Challenges for cause are used when an attorney asks the judge to excuse a potential juror from consideration because the attorney believes the juror would be biased for a particular reason. These challenges are typically used when the potential juror knows the defendant, victim, or one of the attorneys, or when the potential juror has been victimized by a similar crime. There are an unlimited number of challenges for cause in most jurisdictions. Challenges to the array occur when one of the attorneys, generally the defense attorney, challenges the composition of the jury pool by arguing that the jury is not representative of the community or is biased in some way.

During the *voir dire*, attorneys from both sides typically have a limited number of peremptory challenges that they can exercise; the maximum number of challenges available depends on the jurisdiction and the nature of the case under trial. Peremptory challenges allow attorneys from either side to exclude potential jurors who could not be removed from the jury panel as a result of a challenge for cause. As such, the peremptory challenge amounts to a tool for the attorneys to use to shape the jury composition in a way that is most likely to favor their side. Despite recent court cases that have imposed some limitations, most jurisdictions still allow peremptory challenges in most criminal trials.[63]

Opening Statements

The presentation of the case to the judge and/or jury actually begins with the **opening statements**. The opening statements are the first statements made by both the prosecution and defense attorneys to the judge and/or jury where the attorneys describe the facts that they will present during the trial to prove their case. Although the attorneys do not begin the presentation of evidence at this stage, they typically use this statement to outline the key facts and circumstances that they will use during the presentation of evidence. When defense attorneys have little evidence that suggests the defendant client was not involved in the criminal event, their typical strategy is to disprove the case of the prosecution against their client. During the opening statement, attorneys are required to limit the facts they discuss in their statement to evidence they believe will actually be presented during the trial process.[64]

The Presentation of Evidence

The main focus of any criminal trial is the presentation of evidence to the judge and/or jury. In most cases, the prosecuting attorney presents evidence to the court to prove the defendant's guilt. After the prosecuting attorney has presented his or her case, the defense counsel is then given the opportunity to present evidence supportive of the client's case.

Evidence is material that is introduced to the court that is supportive of the facts presented in that trial. Evidence consists of written documents, videotapes, physical objects, witness testimony, and any other matters that may assist the attorneys in convincing the court of the validity of their argument. Evidence can be divided into two general categories: direct evidence and circumstantial evidence. **Direct evidence** is evidence that, if taken at face value, directly indicates a fact. Direct evidence typically speaks for itself; types of direct evidence include videotaped documentation, eyewitness testimony, and weapons that are introduced at the trial. **Circumstantial evidence** is evidence that suggests a fact by inference or implication that requires judgment on the part of the judge or jury regarding the validity of the argument made by the attorney presenting that evidence. Circumstantial evidence often includes the proximity of the defendant to the criminal event, physical evidence that suggests the defendant's involvement in criminal activity, and other facts that may implicate the defendant in the criminal activity. Despite the general perception that direct evidence is preferable to circumstantial evidence, some prosecutors prefer to work entirely with circumstantial evidence because it allows the prosecutor to portray his or her argument in more creative ways than direct evidence does.

One of the important roles the judge plays in the criminal trial is that of "evidence evaluator." When making the decision to allow evidence at trial, judges must consider the relevance of the evidence to the case under trial. If the evidence does not have a relationship with the facts under trial, then the judge is bound to prohibit that evidence from appearing in the court. Judges must also examine the relevance of the evidence, along with the likelihood that the evidence may be so prejudicial or provoking that it biases the judge or jury because of its gory or inflammatory nature. In cases where the evidence introduces bias, judges often preclude that evidence to ensure impartial consideration of evidence. Finally, judges must also examine the situation under which evidence was obtained. In cases where evidence is gained illegally or unconstitutionally (e.g., through an illegal search, through illegal questioning), the defense attorney may file a **motion to suppress evidence** requesting that certain evidence be ruled inadmissible at the trial because the evidence has been obtained by illegal means; if the judge agrees with the motion, then he or she must prohibit that evidence from being introduced at trial.[65]

Along with the type of evidence that will be allowed in the trial, the judge is often required to rule on a wide variety of motions brought forth by both attorneys. A **motion** is a request to the court by an attorney asking for a decision on a legal issue.[66] Other common motions introduced at trial include a **motion for discovery**, a **motion to dismiss charges**, a **motion for continuance**, and a **motion for change of venue**.[67]

Testimony of Witnesses

One of the primary sources of evidence in a trial is **testimony** by witnesses. As reviewed earlier, these witnesses may include the victim, the alleged defendant, or others who have firsthand knowledge of the event in question. Witnesses may be questioned (examined) by both the prosecutor and the defense attorney. During the examination and subsequent cross-examination by the opposing attorney, witnesses are directed to either answer questions about or provide background for the event in question. Witnesses take an oath or affirmation to provide truthful responses to any question and provide truthful statements when asked to testify.

Witnesses who make statements they know are untrue commit **perjury**. Witnesses who provide inaccurate testimony may face criminal sanctions and their testimony may be impeached, or set aside, when it can be demonstrated that they knowingly did not provide truthful testimony. As discussed earlier, lay witnesses are generally required to limit their testimony to their personal knowledge of the situation. Any other statements are considered **hearsay** and are prohibited; the judge will typically instruct the jury to ignore said testimony when it occurs.[68]

Closing Arguments

After both sides have completed the presentation of their evidence and all testimony has concluded, the trial concludes with the closing statement from both the prosecutor and the defense attorney to the jury. These statements summarize the case, reminding the jury of the strength of their argument and the weaknesses of the argument of the opposition, and are designed to persuade the jurors to draw conclusions from the testimony and evidence that support an outcome that is favorable to the speaker's side. The prosecution typically delivers the first closing statement, followed by the defense attorney.[69]

Judge's Instructions to the Jury

After the closing arguments are concluded, the judge typically provides instructions to the jurors regarding their duties and responsibilities during the deliberation process. In these instructions, the judge generally outlines the elements of the crime that must be proven and reminds the jurors that the prosecution must have demonstrated that the defendant in question is **guilty beyond a reasonable doubt**. The judge will instruct jurors regarding their verdict choices (e.g., jurors may have a choice between verdicts of second- and third-degree assault) and remind them that their decision must be based on the evidence that has been presented in the trial. The judge may also provide specific instructions regarding how some information they have learned can be used (e.g., defendant's prior record).

After the charge has been delivered, the jury is removed from the courtroom and the deliberations begin. Deliberations may be brief or may last for several days or weeks, depending on the nature and complexity of the evidence presented at the trial. In those jurisdictions that require a unanimous **verdict**, some juries are never able to agree unanimously. In those cases, the jury is said to be a **hung jury** and no verdict is delivered, meaning that the case may be retried.[70]

Implications for Corrections

To this point, we have provided a brief overview of the history of the courts, the structure and organization of the court systems in the United States, the various courtroom workgroup members, and the various steps in the trial process. In recent years, some jurisdictions have begun to blend courts with corrections to ensure more efficient processing. Two examples of this new idea are discussed in this section.

Drug Courts

In the past two decades, one of the most publicized efforts to combine the court system with corrections has been the **drug court**. The idea of drug courts originated because of the increase in drug cases brought about by the War on Drugs in the 1980s. The first drug court was established in Miami, Florida, in 1989, and since its emergence, the U.S. Office of Justice Programs has awarded millions of dollars to fund drug courts. By the end of 2005, there were more than 1,500 drug courts in operation and almost 400 more being considered.[71]

A wide variety of drug courts exists. Defendants may enter drug court as a diversion program or a pretrial program in some communities; in others, drug court is used as a sentence upon a determination of guilt or as a probation revocation strategy, where a judge agrees to delay the revocation decision if the defendant agrees to participate in the drug court program. Despite these subtle differences, most drug courts have 10 key components. These components[72] are as follows:

1. Substance abuse treatment is integrated with justice system case processing.
2. The prosecution and defense counsel work together as a team to promote public safety while still protecting the due process rights of the accused.
3. Participants eligible for drug court are identified early in the criminal justice process and placed in the drug court as soon as possible.
4. Clients are provided a continuum of alcohol and drug treatment and rehabilitation services.
5. Clients are tested regularly for alcohol and drug use while under drug court supervision to ensure abstinence.
6. The judge, prosecutor, defense attorney, and treatment providers work together in cooperation to ensure that offender compliance is monitored and effectively managed.
7. The judge maintains an active, supervising relationship with the client throughout the duration of the program.
8. Concrete, measurable goals are monitored and evaluated to measure progress and gauge effectiveness of drug court treatment for each client.
9. All participants in the drug court program participate in periodic education and training to promote effective planning, implementation, and operation.
10. Partnerships and enduring relationships are formed with public agencies and community-based organizations to generate local support and enhance drug court effectiveness.

Although there are a wide variety of drug court structures and routes of entry into drug courts, the general goals of drug courts are similar, regardless of the jurisdiction that they serve. These goals include reducing (1) the amount of criminal activity related to drug use and abuse; (2) recidivism among drug offenders; (3) the number of nonviolent drug offenders in jail and prison; (4) the number of drug offenders that fail to appear for trials, sentencing, and probation supervision; and (5) the financial burden on the criminal justice system caused by drug offenses and offenders.[73]

Given these goals, drug courts are designed to manage offenders by focusing on treatment, rather than punishment, for drug abuse. Clients who successfully complete their sentence to drug court usually have the charges against them dismissed or reduced in some way. Finally, the role of the drug court judge is substantially different from that of the trial court judge. Drug court judges are heavily involved in supervision of those sentenced to drug courts because they meet regularly with the drug court clients to ensure clients are avoiding drug use and drug-related crimes. Generally, drug courts are found to reduce crime and recidivism among those who successfully complete their sentences.[74]

Teen Courts

Another innovative blend of the courts and corrections is the **teen court**. Teen courts are generally used for younger juveniles who are first-time property offenders. In these courts, adolescents serve as the actors in the courtroom work group to work together to determine the disposition (from a broad array of sentencing options provided by adults supervising the program) of the case against the juvenile defendant.[75] From their beginning in the 1960s, teen courts are now estimated to be operating in almost 700 communities throughout the United States. These courts were designed as less expensive alternatives to traditional juvenile courts for low-risk, young offenders.

Teen courts are relatively small and are closely affiliated with the juvenile justice system. Young defendants typically volunteer to participate in teen court as an alternative to involvement in the traditional juvenile justice system. Although adults often administer the program and are responsible for planning, budgeting, and personnel issues, in most courts teens serve as bailiffs, court clerks, attorneys, jurors, and, in some cases, judges. These courts are immensely popular with the members of the community that they serve, primarily because of the positive media coverage they receive and the positive experiences of the parents, teachers, and youth involved in the teen court process. Although limited research has evaluated the effectiveness of teen courts, the few studies that exist indicate that teen courts may be linked to lower recidivism rates and increased knowledge of the justice system among the youth involved in the teen court process.[76]

Conclusion

In this chapter, we have discussed the evolution of courts in the United States, the various actors in the courtroom work group, the various steps through which most defendants proceed in a criminal trial, and current collaborative efforts to blend courts and corrections for both adult and juvenile systems. A number of other innovative court efforts have not been discussed, but we encourage you to learn more about the court system in your own jurisdiction, as well as your own state court system. As a responsible corrections professional, you should continue to increase your knowledge of the court system; this knowledge will not only benefit you and your job performance but, more important, it will help you work with your clients to improve both their lives and the lives of members of the community in which they live.

READY FOR REVIEW

- In the United States, an adversarial legal system is used. This system is founded on the principle of "innocent until proven guilty," where attorneys representing opposing legal teams work with a judge to determine the facts of a case that result in a just outcome. This system directly contrasts with the inquisitorial system used in most countries in Latin America and Europe where the accused is presumed guilty and must prove his or her innocence to the court. The American court system has evolved primarily from the court system used in England.
- Courts in the United States consist primarily of a federal court system and 50 state court systems. Both the military and American Indian tribes have their own separate court systems.
- Most state court systems use a four-tiered design. These tiers typically are called (1) lower courts or courts of limited jurisdiction; (2) superior courts or courts of general jurisdiction; (3) intermediate appellate courts; and (4) supreme courts, or courts of last resort.
- The United States federal court system also uses a four-tiered system. These tiers are called (1) district courts (courts of limited jurisdiction); (2) circuit courts (courts of general jurisdiction); (3) courts of appeal (intermediate appellate courts); and (4) the Supreme Court (the federal court of last resort).
- A wide variety of actors comprise the courtroom work group. These actors include the judge, the prosecutor, the defense counsel, the bailiff, the court administrator, the court reporter, the clerk of court, witnesses, the defendant, and jurors.
- There are three categories of defense attorneys. These include private attorneys, court-appointed counsel, and public defenders.
- There are also three categories of witnesses. These categories include victims, expert witnesses, and lay witnesses.
- Activities occurring prior to the trial are called pretrial activities. These activities include the initial appearance, pretrial release, the preliminary hearing, arraignment, and often plea bargaining.
- Bail is the most common method of pretrial release in more serious cases. Alternatives to bail include release on recognizance, conditional release, and third-party custody.
- The steps of a criminal trial include the trial initiation, jury selection, opening statements, presentation of evidence, testimony of witnesses, closing statements by the attorneys, the judge's instructions to the jury, and the determination of the verdict by the jury (in a jury trial) or the judge (in a bench trial).
- Evidence can be divided into two general categories: direct evidence and circumstantial evidence.

- A number of motions are often raised by attorneys during the criminal trial. These motions include motions to suppress evidence, motions for discovery, motions to dismiss charges, motions for continuance, and motions for change of venue.
- One of the primary sources of evidence in a trial is testimony by witnesses. Witnesses who make statements they know are untrue commit perjury; lay witnesses are generally required to limit their testimony to their personal knowledge of the situation, because any other testimony would be hearsay.
- Drug courts are courts designed to be used as alternatives to prison for drug offenders. In these courts, clients are provided a continuum of alcohol and drug treatment, coupled with close supervision by the criminal justice system, in an effort to combat the drug abuse problem that often precipitated their criminal action.
- Teen courts are courts used for younger juveniles who are first-time property offenders in which adolescents serve as the actors in the courtroom work group to work together to determine the disposition (from a broad array of sentencing options provided by adults supervising the program) of the case against the juvenile defendant.

KEY TERMS

adversarial system A legal system in which opposing teams representing the defense and the prosecution compete in court in an attempt to determine the facts of a case. In an adversarial system, the role of the judge is that of a neutral party who oversees the competition in the courtroom. In this model, the accused is presumed innocent until proven guilty

appeal A request that a court with appellate jurisdiction review the decision or judgment rendered from a lower court with the hope of either a reversal or a modification of the decision in favor of the defense

appellate jurisdiction Gives certain courts lawful authority to review a decision made by lower courts. The U.S. Supreme Court has appellate jurisdiction over all courts in the United States

arraignment A brief hearing held before a judge or magistrate at which the indictment or information is read to the defendant and the defendant enters a plea

bail Some form of property that is deposited or pledged to the court to secure the release of a defendant from custody prior to trial. If the defendant does not appear at trial, the bail is forfeited

bailiff The court officer responsible for maintaining order in the courtroom and charged with supervising and managing the jurors in a jury trial

bench trial A trial where the judge makes the determination of guilt without the presence or participation of a jury in the trial

challenges for cause Challenges used when an attorney asks the judge to excuse a potential juror from consideration because the attorney believes the juror would be biased for a particular reason (e.g., juror knows the defendant, victim, or one of the attorneys). There are an unlimited number of challenges for cause in most jurisdictions

challenges to the array Challenges that occur when one of the attorneys, generally the defense attorney, challenges the composition of the jury pool by arguing that the jury is not representative of the community or is biased in some way

circumstantial evidence Evidence that suggests a fact by inference or implication that requires judgment on the part of the judge or jury regarding the validity of the argument made by the attorney presenting that evidence

clerk of court A courtroom officer whose responsibilities include maintaining the records of the court; swearing in witnesses, jurors, and grand jurors; issuing subpoenas and other summons; and performing a variety of other activities essential to the daily operations of the court

community courts Neighborhood-focused courts that use problem solving and partnerships within the context of the justice system to deal with "nuisance" cases (e.g., prostitution, trespassing, disorderly conduct) and cases involving juveniles

conditional release When a court official releases the defendant prior to the trial but imposes conditions on the defendant by which the defendant must abide during the release

court administrator A person whose job is to help the judges manage the daily budget, activities, and schedule of the court

court-appointed counsel An attorney provided to an indigent defendant by the court. These attorneys usually have their own private clients and are paid a flat fee for their efforts by the court and are drawn from a list of private attorneys maintained by the court

court reporter Transcribes the verbal courtroom proceedings into written form, typically using a stenotype (shorthand machine) or audio recorder, to create the official trial court transcript

courts of general jurisdiction Courts that try both misdemeanors and felonies, serve as the venue for larger civil lawsuits, and often serve as the first appellate level for decisions rendered in courts of limited jurisdiction

courts of last resort The highest level of appellate courts in the state court system. After this court renders a decision on an appeal, the appeal can go no higher in the state court system

courts of limited jurisdiction Courts authorized to hear less-serious criminal cases (usually misdemeanors) along with small civil claims cases and other special types of cases (e.g., traffic citations, family disputes)

defendant Any party who has been formally charged or accused of violating a criminal statute or who is required to answer the complaint of a plaintiff in a civil lawsuit

defense counsel The attorney whose role is to represent the individual alleged to have committed the action that brought the charges against them to ensure that innocent individuals are not convicted of crimes they did not commit
direct evidence Evidence that, if taken at face value, directly indicates a fact
dispute resolution centers Informal hearing places where mediators resolve interpersonal disputes without resorting to the traditional arrangements of a criminal or civil court
drug courts Courts where judges closely monitor and supervise clients sentenced for drug-related crimes. Generally, upon successful completion of drug court, the charges against the defendant are dropped or reduced in some way
evidence Material that is introduced to the court that is supportive of the facts presented in that trial
expert witnesses Persons with specialized knowledge or skills that are recognized by the court who assist in providing context to the judge or jury to assist them in their determination of guilt
family courts Courts designed to serve the multiple needs of troubled families in a single court. One judge hears all cases associated with a family, including cases of child abuse and neglect, divorce and custody, delinquency, and domestic violence
grand jury A body of citizens (usually between 12 and 23 in number) assigned to a specific term of service whose primary role is to issue indictments in cases requiring grand jury input. Grand juries may also investigate a wide variety of legal matters arising in the jurisdiction in which they serve as well
guilty beyond a reasonable doubt The highest standard of evidence and the burden of proof the prosecution must meet for conviction in an adult criminal trial
hearsay Testimony based solely on what others have said, not on anything the witness offering the testimony saw, heard, or knew personally
hung jury A jury that cannot reach agreement in a trial requiring a unanimous verdict
indictment A formal statement of charges issued by a grand jury that has determined that sufficient evidence exists to indicate the alleged defendant was responsible for the crime committed.
information A document, typically prepared by the prosecutor, that formally lists the charges against a defendant
initial appearance A hearing before a judge or magistrate in which the defendant is informed of the charges, the terms and conditions of release are set, and a date for the preliminary hearing is scheduled
inquisitorial system A legal system in which the accused is presumed guilty and is required by the court to prove his or her innocence
intermediate appellate court level The level of courts in 39 states between the courts of general jurisdiction and the court of last resort. This level serves as

a filter to reduce the number of appeals that reach the court of last resort in those states

judicial review The power of an appellate court to review decisions made by lower courts and other government agencies

juror A member of a trial or grand jury who is required to decide the facts in a court of law. Jurors are expected to deliver verdicts of "not guilty" or "guilty" with regard to the charges brought against the accused

lay witnesses Witnesses who appear in the criminal trial because of their first-hand knowledge of the event or an individual involved in the criminal event

Marbury v. Madison The Supreme Court case decided in 1803 that established the U.S. Supreme Court's authority as the final interpreter of the Constitution

motion A request to the court by an attorney asking for a decision on a legal issue

motion for change of venue A motion that asks that the trial be moved to another location where prejudice against the defendant is less likely to exist

motion for continuance A motion that seeks to delay the start of a trial.

motion for discovery A motion filed by the defense that asks the judge to allow them to view evidence that the prosecution plans to introduce at trial

motion to dismiss charges A motion entered by either the defense attorney or the prosecutor asking the judge to dismiss the charges against the defendant. A variety of circumstances may bring forth this motion (e.g., plea bargain, successful motion to suppress evidence eliminates the prosecution's case against the defendant)

motion to suppress evidence A motion filed by the defense attorney requesting that certain evidence be ruled inadmissible at the trial because the evidence has been obtained by illegal means

opening statements The first statements made by both the prosecution and defense attorneys to the judge and/or jury where the attorneys describe the facts that they will present during the trial to prove their case

original jurisdiction Gives a court lawful authority to hear cases that originate in a certain geographic area or that involve specific types of law violations (e.g., drug courts have original jurisdiction over drug crimes in an area)

peremptory challenges Challenges that allow attorneys from either side to exclude potential jurors who could not be removed from the jury panel as a result of a challenge for cause

perjury When a witness provides testimony that he or she knows is untrue while under oath or affirmation

plea bargaining A process of negotiation between the defense and the prosecutor in which the defendant agrees to plead guilty to some crimes in exchange for dismissal of some charges or reduction in the severity of the charges. The judge must agree with the negotiated result before the plea can be accepted

preliminary hearing A hearing where the judge or magistrate determines if probable cause exists to suggest that a crime has been committed and the defendant committed that crime

pretrial release The discharge of an accused individual from custody during the time between the arrest and the defendant's appearance at trial on his or her promise to appear for trial

private attorney An attorney hired and paid by the defendant to represent the defendant's legal interests in a trial

prosecutor A government official whose role is to carry out legal proceedings against individuals alleged to have committed a criminal action

public defender An attorney who is employed by the state to provide defense for indigent defendants

release on recognizance (ROR) When the defendant promises the court that he or she will return for subsequent court appearances. Should the defendant not appear in court, no collateral or bond is forfeited, but a warrant can be issued for his or her arrest

subpoena A written order issued by the court requiring an individual to appear in court to give testimony or produce material to be used as evidence. Subpoenas often mandate that papers, books, legal documents, and other items be relinquished to the court

teen courts Courts generally used for younger juveniles who are first-time property offenders where adolescents serve as the actors in the courtroom work group to work together to determine the disposition (from a broad array of sentencing options provided by adults supervising the program) of the case against the juvenile defendant

testimony A statement made by a witness under oath or affirmation

third-party custody A form of pretrial release where a third party (e.g., parent, friend, or family member) promises to ensure that the defendant will return for the next scheduled court hearing if the court will release the defendant to the third party's supervision

trial by battle A trial in which the accused (or a champion fighting on his or her behalf) fought his or her accusers. If the accused won the battle, the person was declared innocent

trial by compurgation A trial in which the accused located 12 people who testified to the truthfulness of the accused's acts of innocence

trial by ordeal An ancient form of justice in which the defendant was forced to experience some ordeal to prove his or her innocence. The basis for this trial was the belief that God would intervene if the defendant was not guilty

trial *de novo* Literally means "new trial." This term is applied to cases that are retried on appeal, rather than simply reviewed on the record

United States courts of appeal The second tier in the federal court system consists of 12 intermediate courts of appeal in the federal system that have mandatory appellate jurisdiction over the district courts in their circuit

United States district courts The lowest level of the federal court system that serve as the courts of original jurisdiction within that system

United States Supreme Court The highest level of the federal court system. It consists of one chief justice and eight associate justices and serves as the most powerful court in the nation because of its power to review the decision of lower courts at both the federal and state levels

verdict The decision of the jury in a jury trial or the judge in a bench trial regarding the guilt of the defendant under trial

victims Individuals who have had a criminal offense committed against them. Victim testimony is often one of the most important types of testimony in a criminal case

voir dire The process through which potential jurors are questioned to determine their suitability for serving on the trial jury

witnesses Individuals who have firsthand knowledge about an event and provide evidence that helps to establish the facts in a particular case

YOU ARE THE CORRECTIONS PROFESSIONAL SUMMARY

1. What kind of factors would need to be considered and excluded to ensure that the alleged defendant receives a fair trial? First and foremost, you need to consider the evidence against the alleged defendant, whether the evidence was legally secured, and whether all of the evidence will be allowed in trial. You will also need to consider what kind of risk the alleged defendant poses to the community; this will affect the type of pretrial supervision you may be responsible for providing to the defendant.
2. What kind of trial would most likely take place? Would a jury be involved? If so, what role would they play? In most jurisdictions, the prosecutor would first present the evidence to the grand jury. If the grand jury issues an indictment, then a jury trial would likely take place. Based on the evidence presented to them, the jury would then decide whether to proclaim the defendant guilty or not guilty.
3. Will the alleged defendant be able to receive a quality defense in this trial? If not, what steps could you take to ensure that this trial represents an impartial hearing of the facts of the case rather than an emotional contest? It would be difficult for the defendant to receive a completely objective (or fair) trial in the jurisdiction where the incumbent sheriff was killed. As such, the defense attorney would likely ask for a change of venue that would allow the criminal trial to take place in another jurisdiction. At best, this would make your job more difficult; in some cases, the supervision of the client may be assigned to the corrections professionals in that jurisdiction where the trial will take place.

4. If the defendant is found guilty, what implications will his sentence have for both local and state corrections? Because of the severity of the charge against the defendant, he will likely be sentenced to a long term in the state prison where you live or, if your state allows it, may be sentenced to death. In either case, the defendant will likely be immediately transferred to a state prison (or reception center) at the completion of the sentencing phase of the criminal trial.

NOTES

1. BBC News. "Timeline: Saddam Hussein Dujail Trial," November 3, 2006, http://news.bbc.co.uk/2/hi/middle_east/4507568.stm (accessed November 18, 2006).
2. C. E. Smith, C. DeJong, and J. D. Burrow, *The Supreme Court, Crime, and the Ideal of Equal Justice* (New York: Peter Lang Publishing, 2003).
3. S. E. Walker, *Sense and Nonsense About Crime and Drugs: A Policy Guide*, 6th ed. (Belmont, CA: Wadsworth, 2005).
4. J. F. Meyer and D. R. Grant, *The Courts in Our Criminal Justice System* (Upper Saddle River, NJ: Prentice Hall, 2003).
5. L. E. Moore, *The Jury: Tool of Kings, Palladium of Liberty* (Cincinnati, OH: W. H. Anderson, 1973).
6. Moore, *The Jury*.
7. Meyer and Grant, *The Courts*.
8. Meyer and Grant, *The Courts*.
9. Meyer and Grant, *The Courts*.
10. G. D. Langdon, *Pilgrim Colony: A History of New Plymouth 1620–1691* (New Haven, CT: Yale University Press, 1966).
11. Meyer and Grant, *The Courts*.
12. L. M. Friedman, *Crime and Punishment in American History* (New York: Basic Books, 1993).
13. Langdon, *Pilgrim Colony*.
14. F. Schmalleger, *Criminal Justice Today*, 9th ed. (Upper Saddle River, NJ: Prentice Hall, 2007).
15. Meyer and Grant, *The Courts*; Schmalleger, *Criminal Justice Today*.
16. D. B. Rottman and S. M. Strickland, *State Court Organization, 2004*, NCJ 212351 (Washington, DC: U.S. Department of Justice, Bureau of Justice Statistics, 2006).

17. Rottman and Strickland, *State Court Organization*; Schmalleger, *Criminal Justice Today*; Meyer and Grant, *The Courts.*

18. Rottman and Strickland, *State Court Organization;* Schmalleger, *Criminal Justice Today*; Meyer and Grant, *The Courts.*

19. R. V. Duizend, "The American Court System: Long Traditions, New Directions," *Issues of Democracy: The Changing Face of U.S. Courts* 8, no. 1 (2003): 6–12.

20. Information retrieved from Center for Court Innovation Web site (www.communityjustice.org/index.cfm?fuseaction=Page.ViewPage&PageID=591¤tTopTier2=true) on December 27, 2006.

21. J. Patton, "Unified Family Courts: Streamlining Justice for Troubled Families," *Children's Voice* 8, no. 5 (1999): 4–6.

22. J. A. Kuhn, "A Seven-Year Lesson on Unified Family Courts: What We Have Learned Since the 1990 National Family Court Symposium," *Family Law Quarterly* 32, no. 1 (1998): 67–94.

23. Information retrieved from National Center for Juvenile Justice Web site (http://ncjj.servehttp.com/NCJJWebsite/faq/familycourt.htm) on June 14, 2007.

24. Meyer and Grant, *The Courts.*

25. Much of the information and some of the wording in this section come from the Administrative Office of the U.S. Courts Web site at www.uscourts.gov/districtcourts.html (accessed December 30, 2006).

26. Much of the information and some of the wording in this section come from the Administrative Office of the U.S. Courts Web site at www.uscourts.gov/courtsofappeals.html (accessed December 30, 2006).

27. A detailed discussion of the history of the Supreme Court and biographies of the current Supreme Court justices is available at www.supremecourtus.gov/about/about.html (accessed December 30, 2006). The chief justice of the U.S. Supreme Court in January 2007 is John G. Roberts. The eight associate justices are John Paul Stevens, Antonin Scalia, Anthony Kennedy, David Souter, Clarence Thomas, Ruth Bader Ginsburg, Stephen Breyer, and Samuel Alito, Jr.

28. Much of the information and some of the wording in this section come from the detailed discussion of the operations of the Supreme Court available at www.supremecourtus.gov/about/about.html (accessed December 30, 2006).

29. Much of the information and some of the wording in this section come from the detailed discussion of the roles of the various actors in the criminal courts published on the American Bar Association's Criminal Justice

Section Standards homepage at www.abanet.org/crimjust/standards/ (accessed December 30, 2006).

30. Meyer and Grant, *The Courts*.
31. The information regarding the confirmation hearings of Justice Clarence Thomas was retrieved from the Wikipedia site at http://en.wikipedia.org/wiki/Clarence_Thomas (accessed January 4, 2007).
32. Meyer and Grant, *The Courts*.
33. Rottman and Strickland, *State Court Organization*.
34. Meyer and Grant, *The Courts*.
35. Meyer and Grant, *The Courts*.
36. *United States v. Wade*, 388 U.S. 218 (1967).
37. Schmalleger, *Criminal Justice Today*.
38. The majority of the discussion regarding the various types of defense attorneys was based on information provided by Meyer and Grant, *The Courts*.
39. American Bar Association, *Criminal Justice Section Standards*, www.abanet.org/crimjust/standards/dfunc_blk.html#1.6 (accessed January 20, 2007).
40. *Nix v. Whiteside*, 475 U.S. 157 (1986).
41. Schmalleger, *Criminal Justice Today*.
42. Channel 2 Action News, "A Very Sad Day for Fulton County," Wsbtv.com, March 11, 2005, www.wsbtv.com/news/4275403/detail.html (accessed January 20, 2007).
43. A number of job announcements that can provide the student a better understanding of the roles of court administrator are available at www.ncsconline.org/D_KIS/jobdeda/Jobs_TrialAdmin(3).htm (accessed January 20, 2007).
44. Wikipedia, "Court Reporter," http://en.wikipedia.org/wiki/Court_reporter (accessed January 20, 2007).
45. Wikipedia, "Court Clerk," http://en.wikipedia.org/wiki/Court_clerk (accessed January 20, 2007).
46. Schmalleger, *Criminal Justice Today*; Meyer and Grant, *The Courts*.
47. Schmalleger, *Criminal Justice Today*.
48. Most of the discussion surrounding the role of jurors is derived from Schmalleger, *Criminal Justice Today*.
49. Wikipedia, "Defendant," http://en.wikipedia.org/wiki/defendant (accessed January 23, 2007).
50. Schmalleger, *Criminal Justice Today*.

51. Much of the section on initial appearance is based on the information provided by the state of Arizona as part of its explanation of criminal proceedings for court-appointed special advocates (CASA) and is available at www.supremecourt.az.gov/casa/prepare/criminalproceedings.pdf (accessed January 26, 2007).
52. Schmalleger, *Criminal Justice Today*.
53. *Taylor v. Taintor*, 83 U.S. 366 (1872).
54. *Herd v. State of Maryland*, 125 Md. App. 77, 90, 724 A.2d 693, 700 (1999).
55. Meyer and Grant, *The Courts*.
56. Much of the section on the preliminary hearing is based on the information provided by the state of Arizona as part of its explanation of criminal proceedings for court-appointed special advocates (CASA) and is available at www.supremecourt.az.gov/casa/prepare/criminalproceedings.pdf (accessed January 26, 2007).
57. Federal Rules of Criminal Procedure are available at www.law.cornell.edu/rules/frcrmp/Rule11.htm.
58. Bureau of Justice Statistics, *The Prosecution of Felony Arrests* (Washington, DC: U.S. Government Printing Office, 1983).
59. Law.com Dictionary, "Plea Bargain," http://dictionary.law.com/default2.asp?selected=1541&bold=bargaining||plea|| (accessed January 28, 2007).
60. *United States v. Jackson*, 390 U.S. 570 (1968); *Brady v. United States*, 397 U.S. 742 (1970); *Santobello v. New York*, 404 U.S. 260 (1971).
61. Federal Rules of Criminal Procedure, No. 11, www.law.cornell.edu/rules/frcrmp/Rule11.htm (accessed May 29, 2007).
62. Schmalleger, *Criminal Justice Today*.
63. Meyer and Grant, *The Courts*.
64. Schmalleger, *Criminal Justice Today*.
65. Meyer and Grant, *The Courts*.
66. Meyer and Grant, *The Courts*.
67. See Schmalleger, *Criminal Justice Today*, 388, for a lengthy discussion of a variety of common motions.
68. Schmalleger, *Criminal Justice Today*.
69. Meyer and Grant, *The Courts*.
70. Meyer and Grant, *The Courts*.
71. National Institute of Justice, *Drug Courts: The Second Decade* (Washington, DC: Government Printing Office, U.S. Department of Justice, 2006).
72. National Association of Drug Court Professionals Drug Court Standards Committee, *Defining Drug Courts: The Key Components*, NCJ 205621

(Washington, DC: Government Printing Office, Bureau of Justice Assistance, 2004), www.ojp.usdoj.gov/BJA/grant/DrugCourts/DefiningDC.pdf.

73. J. R. Brown, "Drug Diversion Courts: Are They Needed and Will They Succeed in Breaking the Cycle of Drug-Related Crime?" in *Criminal Courts for the 21st Century*, ed. L. Stolzenberg and S. J. D'Alessio, 47–76 (Upper Saddle River, NJ: Prentice Hall, 1999).
74. Brown, "Drug Diversion Courts."
75. J. A. Butts and J. Buck, *Teen Courts: A Focus on Research*, NCJ 183472 (Washington, DC: U.S. Government Printing Office, Office of Juvenile Justice and Delinquency Prevention, 2000).
76. Butts and Buck, *Teen Courts*.

Sentencing and Corrections

6

Chapter Objectives

- Describe the different philosophies of punishment (goals of sentencing).
- Understand the sentencing process from plea bargaining to conviction.
- Describe the components of a presentence investigation report.
- Describe changes in the goals of sentencing over time, including the shift from indeterminate to determinate sentences.
- Outline some of the changes in sentencing that made punishments for offenders more severe.
- Describe some of the controversies surrounding use of the death penalty.

CASE STUDY

You are the commissioner for the Department of Corrections in a large state. One of the probation officers in your department failed to file timely paperwork for the arrest of a probationer after the probationer had been charged with felony battery and was held overnight in jail (this man was initially placed on probation for committing assault, which was his most recent conviction). Despite the fact that the offender had violated the conditions of his probation, he was allowed to remain free. He even reported to the probation office but was not detained on a warrantless arrest. One week later, the probationer was involved, with three other men, in the murder of six young people. The probationer was convicted of capital murder and has just been sentenced to receive the death penalty. You are scheduled to participate in a high-profile television interview in an hour to discuss this probationer.

1. How would you reassure the public that the probation officer has been held accountable? Or should the focus be on the actions of the probationer who was involved in these six murders?
2. What is your view on whether first-time offenders convicted of violent crimes should be placed on probation?
3. What are you going to say if you are asked about your views on capital punishment in this case?

Introduction

In the previous chapter, we described the various stages of court processing leading to a finding of not guilty or a conviction. If the defendant is convicted, the **sentencing hearing** or process begins; this is also called the punishment or penalty phase. There are differences between states, but in many cases, the imposition of a sentence is delayed until a **presentence investigation (PSI) report** can be written, especially for more serious offenses. These reports provide the judge with information about the offense and the characteristics of the offender. The probation officers who write PSI reports usually provide the judge recommendations for a sentence. The PSI report is discussed further in this chapter as well as in Chapter 8.

Only a small percentage of offenders are ever convicted of a crime through a trial. In most cases, defendants plead guilty to an offense to receive a less serious sentence after a **plea agreement** (also discussed in Chapters 3 and 5) has been made between the offender's attorney and the prosecuting attorney. As a result, the sentence that an offender has to serve is rarely much of a surprise, because the range of the sentence has generally been negotiated in advance.

Sentencing Options

Judges have a number of options at sentencing, although they are limited by the minimum and maximum penalties outlined in a jurisdiction's penal code. The following are examples of sentencing options available in many jurisdictions.

- Economic sanctions, such as fines (Chapter 9)
- Posttrial diversion (Chapter 8)
- Probation (Chapter 8)
- Home confinement (Chapter 9)
- Boot camp (Chapter 9)
- Jail term (Chapter 7)
- Imprisonment (Chapters 10, 11, 12)
- Death (this chapter)

Though working within the parameters set by law, the judge generally has broad discretion to impose a sentence. In some states, the judge's authority is limited by guidelines that restrict the judge to a number of sentencing options; in some jurisdictions, such guidelines are advisory, but in other jurisdictions judges must provide a written rationale when departing from guidelines.

Judges may grant credit toward the sentence for jail time the offender has served prior to sentencing. Also, a person who has been convicted of more than one crime may be required to serve sentences consecutively (back to back) or concurrently (together). An offender sentenced to two consecutive 2-year terms, for example, would serve four years in prison (assuming no jail time credit, prison good time credit, or early release on parole—see Chapter 13). Alternatively, an individual sentenced to serve two 2-year terms concurrently would serve only two years. The severity of sanction that an offender receives depends on a number of factors, including whether the crime was the person's first offense, the perceived potential for rehabilitation, and whether there were aggravating or mitigating factors.

It is important to understand that there is tremendous variation in the severity of sentences that different offenders might receive for committing the same crime. In 2002, the average sentence for murder and nonnegligent manslaughter in state courts was 217 months, while the average sentence for murder in federal courts was 109 months.[1] Yet not everybody convicted of murder was sent to prison. The same study showed that 4 percent of homicide offenders in state courts were sent to jail for terms of incarceration of less than one year, while 5 percent received a probationary sentence. The following paragraphs outline some of the reasons for variation in sentences for similar crimes, something known as **sentencing disparity**.

There are sometimes significant differences in severity of sentences between jurisdictions. An offender from a rural area, for example, may serve a longer sentence than a city resident does even though both persons have similar criminal histories and committed similar offenses. In rural areas where crime occurs less frequently, there may be higher levels of punishment because these acts may be seen as more threatening. Different sentencing outcomes based on where the case was processed are called **justice by geography**.[2] It has also long been recognized that sentencing outcomes can differ based on factors that have nothing to do with the offense, such as the defendant's race, gender, or social class. Such so-called **extralegal factors** are distinct from legally relevant factors such as offense seriousness and prior convictions, and we examine the issue of differences in sentencing based on these characteristics more closely in the following pages.

Sentences are also based on a number of **goals of sentencing** (some scholars refer to these goals as **philosophies of punishment**, and we use these terms interchangeably in this chapter). The philosophy of punishment a jurisdiction adopts will have a significant impact on the types of correctional programs that are developed. States that have a more rehabilitative emphasis place more inmates in prison-based educational or work training programs (see Chapter 12). States that are more punitive, by contrast, may provide inmates with only basic necessities of shelter, food, and medical care. As a result, inmates in these states may have fewer opportunities to make rehabilitative changes.

Over the past few decades, most states and the federal government have adopted "tough on crime" policies that have increased the severity of sentences for many inmates. Since the early 1990s, punishments such as mandatory minimum sentences, three strike sentencing schemes, and truth-in-sentencing have been introduced or expanded. Although we tend to see these punishments as new, some of these types of sentences have existed for more than a century. However, lengthy prison sentences were imposed less frequently in the past. The reason why we became more punitive is a result of a complex interplay between changing beliefs about the potential to rehabilitate offenders, public opinion (including support for tough on crime policies), the influence of the media on our knowledge about crime, and politicians and policymakers who advocate for harsh punishments.

Philosophies of Punishment

Most scholars suggest that there are four philosophies of punishment that affect sentencing strategies in the United States. We discuss each of those four strategies in the following section.

Retribution, Just Deserts, Deterrence, and Incapacitation

As pointed out in Chapter 10, incarceration in prison is a relatively new means of responding to crime. Prior to the late 1700s and early 1800s, people suspected of being involved in crime were often held in jails until their cases were decided by the courts. Thereafter, rather than being imprisoned, persons found guilty were frequently subjected to some form of corporal or capital punishment. These practices reflected the desire for **retribution**, the leading philosophy of punishment of that era. Retribution (sometimes referred to as retributive justice) can be traced back to the Code of Hammurabi, codified law that goes back to the Sumerian era, almost 4,000 years ago. These early laws stressed revenge, and the principle of retribution is outlined by ***lex talionis*** or "an eye for an eye, and a tooth for a tooth." In other words, society is obliged to administer a payback to the offender for his or her offense.

Retribution rests upon the desire to pay offenders back for wrongs they have inflicted on society. It contains an emotionally charged element of anger, outrage, and vengeance. By contrast, **just deserts** is a more recently articulated philosophy, growing popular during the 1970s.[3] Theoretically, just deserts is distinguishable from retribution because the former does not contain the emotional element of vengeance. Punishment is simply seen as being fair or equitable for criminal behavior. Crime is viewed as creating a state of imbalance in society, and the purpose of punishment is to restore a state of balance. Advocates of just deserts, or the justice model as this philosophy is sometimes called, contend that punishment must be administered and implemented fairly and justly according to principles of law to avoid transforming the existing state of imbalance created by the crime into a new state of imbalance or unfairness. Thus, advocates are very concerned with protecting the rights of offenders during the punishment process.

Retribution and just deserts are similar in that both focus on the offender's past behaviors as the basis for punishment; the issue in determining a suitable punishment is what the offender *has* done. Other philosophies discussed later (deterrence, incapacitation, and rehabilitation in particular) differ in that they are more geared toward the offender's future behaviors; a central issue is what the offender *might* do.

Many of the activities of the police, courts, and corrections are based on the principle of **deterrence**, the belief that future criminal behaviors are prevented when we punish offenders. Deterrence can be broken into two different components, **general deterrence** and **specific deterrence**. Specific deterrence rests on the assumption that the pain of being punished will deter the person from committing future offenses. General deterrence, on the other hand, is the belief that after witnessing or learning about an offender's punishment, other people will be less likely to violate the law. As such, deterrence relies on the threat of future punishment to control future behavior following initial application of punishment.[4]

It is difficult to explain accurately the factors that influence human behavior, and research has generally failed to demonstrate a clear relationship between punishment and deterrence. One fiercely debated issue taken up later in the chapter is whether the death penalty is a general deterrent to murder.

Although the concept of deterrence is attractive in theory, it is harder to apply this approach to the real world. Scholars have suggested that deterrence is more likely to work when punishments are applied with swiftness (also called celerity), certainty, and severity.[5] The trouble, however, is that many crimes are unreported, and many minor offenses tend to go unpunished; the police and other criminal justice officials direct most of their resources to investigating and prosecuting serious offenses. Further, as noted in previous chapters, U.S. criminal justice systems are founded on the principle of providing due process protections to suspects, and by the time a case works its way through the court system it may be months, or even years, before the case is resolved and the offender is sentenced. Case processing times (the time between arrest and the resolution of the case) for defendants charged with more serious offenses are likely to be longer; only half of murder cases are resolved within one year.[6]

Another limitation of the deterrence approach is that punishments for most crimes are not very severe, especially for first-time offenders. A recent federal analysis of probation caseloads, for example, found that almost half of all probationers had pled guilty or were convicted of felony offenses.[7] Even when persons are incarcerated, they might only serve a fraction of their total sentence, especially for nonviolent offenses. In the summer of 2006, for instance, an inmate receiving a one-year sentence in Los Angeles County would serve approximately 10 percent of that time in jail prior to release. The Los Angeles example is not a typical case, and a recent study of rural jails revealed that jail inmates sentenced to a term of one year would serve approximately nine months.[8] Moreover, in some states, prison inmates serve about half of their sentences prior to release. Given these facts, deterrence is unlikely to work effectively.

Another philosophy of punishment is **incapacitation**. Under this philosophy, that purpose of punishment is to make it physically impossible (at least temporarily) for the offender to revictimize members of the free community. If offenders are held apart from society in prison, they are unlikely to commit any further crimes in the community (although some prisoners, such as gang members, have been involved in conspiracies with persons in the community who commit crimes). Jurisdictions that stress incapacitation may provide only the basic necessities of living for some prisoners. Jail inmates who are awaiting trial, for instance, typically receive only the most basic services, such as meals, a bed, and limited access to recreation. Similar to jails, in some state prison systems there are few opportunities for education, counseling, or skill development, and we call this *warehousing*—simply holding the prisoner until release.

Some policy analysts who have examined the offenses of career criminals argue that if we could imprison only the offenders who pose the highest risk to society, we could use incarceration much more effectively. Early studies by the RAND Corporation (a research organization or think tank), found that a relatively small number of repeat offenders are responsible for a disproportionately large number of offenses.[9] As such, if we could identify and incarcerate these high-risk or high-rate offenders, we would all be safer, an idea termed **selective incapacitation**. These beliefs led to the enactment of **three strikes laws** in the early 1990s that enabled legislators and judges to impose lengthy prison sentences for repeat felony offenders. We examine these laws, and how they have been used, in the following pages.

One problem with selective incapacitation is that it is hard to predict whether an offender will be involved in future crimes. As a result, we might sentence a person to a lengthy term of incarceration when that person would not have committed any future offenses. This is the problem of *false positives*. The other problem is *false negatives*; this occurs when we believe that a person will not commit any further crimes, but that person actually goes on to do so. Predicting future criminal behavior is an inexact science, but persons who advocate for strict sentences point out that the needs to hold criminals accountable and protect society outweigh other considerations.

Correctional systems that are guided by the principles of retribution, just deserts, deterrence, or incapacitation are less likely to offer extensive rehabilitative services. Instead, they provide a basic level of care and supervision, along with recreational or work programs that keep prisoners occupied.

Rehabilitation, Restitution, and Restoration

Another set of sentencing goals shifts focus to the offender and his or her capacity to make meaningful life changes that will reduce future crimes and ultimately translate into greater public safety. These philosophies include **rehabilitation**, **restitution**, and **restoration**. Rehabilitation involves positive changes to an offender's attitude and behavior so he or she avoids future crime and contributes positively to society. Restitution, by contrast, involves the offender repaying the victim or community for the damages done; in a sense it is the opposite of retribution (defined earlier) which entails society repaying the offender with punishment. Restoration extends the concept of restitution to include structured and supervised contact between the offender and victim. This philosophy emphasizes providing an opportunity for victims to safely confront the person who has harmed them.

Like retribution, restitution has a lengthy history. Early justice systems in northern Europe, for instance, had the option of forcing an offender to make reparations (called weregild) to the family of a murder victim.[10] In that era, restitution was important for the survival of the victim's family, especially if the victim was no longer able to care for his or her family. Today, however, restitution is used primarily for property offenses and is often a condition of an offender's probation. By repaying a property owner for the damages and losses that occurred in a residential burglary, for example, the victim will recover some economic losses. Yet, we also have to acknowledge that not all of the costs of crime are economic. A burglary victim experiences feelings of violation and victimization and may feel unsafe in his or her own home for years afterward.

The goal of correctional or offender rehabilitation emerged much later in history than restitution did. We discuss in Chapter 10 how the first penitentiaries were founded around the turn of the eighteenth century and into the nineteenth century to help offenders make positive changes that would reduce their future criminality. Rehabilitation programs were usually based on increasing the education of the prisoner, helping prisoners address various life problems, and vocational preparation. Yet, many policymakers were dissatisfied with high rates of recidivism. Many questioned whether there was a more appropriate way of holding offenders accountable while ensuring public safety.

As a philosophy of sentencing, restoration (sometimes called restorative justice) is based on the goal of repairing harm to the victim and community that occurred when the crime was committed. Although victim impact statements (formal reports from victims about the losses they experienced) have become an important component of sentencing hearings in most places, modern justice systems have often taken an impersonal approach to dispensing justice that leaves victims dissatisfied. Moreover, the punishments that are meted out to offenders (especially probationary sanctions) seldom make offenders fully realize the consequences of their actions. Restorative justice, by contrast, includes all of the people who have a stake in an issue, including the victim, the victim's supporters (such as family members), the offender, and representatives from the community, including the police.

Restorative justice interventions focus on harms that were committed and steps the offender must take to repair damage to the victim and community. This approach commonly involves restitution, but it extends beyond giving the victim(s) money. Offenders are often confronted by their victims and hear about the impact of their crimes firsthand. Examples of restorative justice include **family group counseling** and **circle sentencing**, both of which are discussed in the accompanying Corrections in the Real World box.

Corrections in the Real World

Family group counseling is an intervention that is commonly used with juveniles and is intended to acknowledge the harm done to the victim. This approach often involves a structured meeting between offender, victim, community members, and supporters, but it may also include officials from the justice system, including police officers and probation staff. Circle sentencing, by contrast, is oriented for adult offenders, and consequences or sanctions for the offender are decided after consultation with members of the circle. The circle typically includes members of the community, including justice officials and victims. The intent of family group counseling or circle sentencing is to design an intervention that responds to the unique circumstances of the offender, the nature of the offense, and the damage done. These approaches are similar in that they are modeled after interventions used by aboriginal peoples in New Zealand, Australia, and Canada.[11]

Restorative justice acknowledges that criminal acts damage the community and assumes that a solution can be developed to restore the relationships between the offender and the community. Thus, these approaches move away from our traditional and abstract notions of law and justice to concepts that are more easily understood by offenders and victims alike.

Although restorative justice has been used primarily in U.S. juvenile justice systems, it has been used for adult offenders in common law nations such as Australia, Canada, New Zealand, and the United Kingdom. Though many people are skeptical about the effectiveness of this approach, some of the biggest supporters are police officers who believe that the treatment of offenders by traditional justice systems does not have much of a meaningful or long-term impact on the offender.

None of the various philosophies of punishment exist in a pure form, and it is important to realize that local court or correctional systems often blend these goals. As a result, some judges may be more punitive or rehabilitative than their counterparts in neighboring counties are. This sometimes results in disparities. In some cases, disparities are based on geography, race, gender, socioeconomic class, or some other extralegal factor. These differences become troubling when they result in some offenders being punished more harshly than are others who committed similar crimes.

Considerations in Sentencing

The Federal Bureau of Investigation reports that there were about 14.1 million arrests in 2005, and most of these arrestees passed through a juvenile detention

Race and Gender in Corrections

Offenders in federal courts who are sentenced on possession or trafficking of crack cocaine are punished more severely than are offenders with powdered cocaine because possession of one gram of crack cocaine is considered to be the same as 100 grams of powdered cocaine at sentencing. In 1995, 2002, and 2007, the United States Sentencing Commission (USSC), a body that oversees sentencing on federal crimes, found that these policies were discriminatory because crack cocaine was used overwhelmingly by members of minority groups. Despite the fact that the USSC found that sentencing was biased, there have been no changes to federal sentencing laws. When criminal justice policies appear to be discriminatory, they may erode public confidence in the police, courts, and corrections.

center or jail.[12] Eventually, many of these individuals appeared in court, and although many cases were dismissed or did not proceed (either because of lack of evidence or some other factor), millions of the remaining defendants had their day in court. Most of the cases that courts administer are misdemeanors, and these cases are often processed quickly; a judge might spend only a few seconds on each of these matters. These courts are able to process these cases without much controversy because everybody in the courtroom work group applies the going rate to offenses.

The Going Rate

We noted in the previous chapter and in Chapter 3 that the U.S. legal system is said to be adversarial. The truth is supposed to emerge from an oppositional process between the prosecution and defense, with the judge ensuring that legal guidelines are followed. However, a number of scholars argue that most legal systems operate on the basis of a courtroom work group where all the members have a vested interest in getting along over the long term, and relationships are usually not adversarial (refer to Chapter 3).[13]

One of the cornerstones of any courtroom work group is shared understanding of the seriousness of a given offense, what has been termed the **going rate**. The going rate has been defined as the "shared beliefs about appropriate sentence levels for defendants charged with given crimes, who possess similar records, [and] standard operating procedures which regularize sentences and routinize plea bargaining."[14] In other words, the courtroom work group operates with a shared view of the seriousness of an offense and the degree of punishment that should be meted out to the offender.

For example, in Chico, California, a college town of approximately 100,000 residents, underage drinking has been a long-term problem. To respond to this challenge, persons under 21 years of age convicted of being in possession of alcohol normally receive a fine and the loss of their driver's license for one year. This standard punishment is the going rate, and local prosecutors and law enforcement support this sanction to deter underage drinking. Because the prosecutor, defense attorneys, and judge all understand that the loss of the driver's license is part of the going rate, defense attorneys do not take an adversarial position in court at sentencing. The ability to process these cases quickly and harmoniously has become more important than advocating on behalf of an individual client is.

Interestingly, however, going rates tend to vary across courtroom workgroup settings, and the variations apply to many forms of offenses. In the case of minors in possession of alcohol mentioned previously, few other California jurisdictions impose the loss of a driver's license on underage drinkers. The loss of a driver's license is a minor consequence compared to other case outcomes. Carol Steiker observed that death verdicts were almost twice as common per 1,000 murders in Harris County (Houston), Texas, compared to Dallas County (Dallas) despite the fact that crime rates were very similar in these two cities.[15] Such issues can make the public question the fairness of criminal justice systems. If justice is not perceived to be fair, some scholars have suggested that citizens will be less likely to follow the law.[16]

On the other hand, it can be argued that local officials have a legitimate interest in problems that have a significant impact on the quality of life in their community. As a result, in a college town, the city council as well as the police and prosecuting attorneys might support harsh sentences for underage drinking. Some people believe it is important for local political and justice system officials to have the ability to target specific offenses and tailor punishments these officials believe will reduce behavior seen as threatening or problematic. To the extent that politics and criminal justice are local, there will be some variation in the severity of sentences across jurisdictions.

Plea Bargaining

In the previous chapter, we explained that few defendants actually proceed to trial and that most cases get resolved through a plea agreement. More than 90 percent of all criminal cases are resolved through plea bargaining. The basis for plea bargaining is that the defendant agrees to plead guilty to a lesser offense (or fewer offenses) in return for a less severe sentence. Accepting a plea has a number of advantages for both the criminal justice system and the defendant. Trials are expensive undertakings, and a prosecuting attorney might not have enough evidence to ensure a conviction. Even if the prosecutor does have a "solid case," juries may be reluctant to convict some defendants. **Jury nullification** occurs when a jury refuses to convict a defendant despite the fact that the person is guilty (e.g., it may be difficult for a prosecuting attorney to convict a person of possession of a small amount of marijuana in a college town because some of the jurors may perceive that the punishment is worse than the crime). Moreover, the district attorney might be reluctant to force a witness or victim to testify. Last, once a plea agreement has been reached, the offender is unlikely to appeal the conviction successfully, thereby ensuring that the conviction stands.

Although plea bargaining allows cases to be disposed more quickly and at less financial cost, there are a number of potential problems. There is the possibility that some innocent defendants will be pressured into accepting a guilty plea as an alternative to going through an expensive trial that they might lose and risking a more severe sentence. To encourage defendants to plead guilty, a prosecutor might overcharge them—either charging them with a more severe crime (e.g., attempted murder instead of assault) or charging them with a number of offenses and dropping all but one in return for a plea of guilty. Some court observers refer to these practices as vertical and horizontal overcharging.[17] Wealthier defendants may also be able to retain an attorney who is more skilled at negotiating these matters, or who has more time and resources to invest in fact finding (to better understand the prosecutor's case) before the plea negotiations are initiated. Even if a wealthy defendant is convicted, that person may benefit from the services of sentencing consultants (who are usually former correctional officials) hired to assist the offender in arranging alternatives to incarceration or minimizing the severity of a sentence.[18]

Criticisms of plea bargaining span the political spectrum. Walker describes how the process is hated by some conservatives who perceive it as soft on crime, whereas many liberals are critical that plea bargaining results in unjust outcomes.[19] On the other hand, there appear to be few sound alternatives to plea bargaining. Even when alternatives are implemented, the courtroom work group often undermines reforms in favor of the going rate.[20]

Although plea agreements are negotiated between the prosecution and the defense, they still must be approved by the court. In some cases, the judge might not agree to the negotiated plea. Scheb and Scheb point out that if "the court is unwilling to approve the plea bargain, the defendant must choose between withdrawing the guilty plea (and thus going to trial) and accepting the plea bargain with such modifications as the judge may approve."[21] The intervention of judges in this process is rare in most jurisdictions and probably occurs when the defense and prosecution stray too far from the going rate, although as we outlined earlier, the going rate does vary between jurisdictions.

Corrections and Policy

In *Bordenkircher v. Hayes,* the U.S. Supreme Court allowed prosecutors to threaten a defendant with more serious charges if the defendant will not plead guilty to the initial charge.[22] The Court rejected the claim of Paul Hayes that his prosecution was motivated by prosecutorial vindictiveness. Hayes was a two-time offender in Fayette County, Kentucky, who was offered a choice of accepting a five-year prison term for a minor felony (forging a check for $88) or going to trial, and if convicted, being sentenced to life imprisonment under Kentucky's habitual offender legislation. After not taking the plea deal, Hayes was convicted, found to be a habitual offender, and sentenced to life in prison. William Stuntz argues that the policy precedent established in this case has given prosecutors an unfair advantage in plea agreements.[23] Prosecutors can overcharge to extract a guilty plea. Some claim that such policies have even led innocent persons to plead guilty to avoid worse outcomes if convicted.[24]

Racial, Gender, and Class Disparities at Sentencing

One of the most important challenges that justice systems confront is disparities in the severity of sanctions or sentences based on race, class, and gender. Blacks and Latinos, for instance, are at higher risk of arrest and incarceration. In fact, a black male born in 2001 has an almost one in three likelihood of going to prison in his lifetime, whereas a white male born in the same year is one-fifth as likely to be imprisoned. In Chapter 2, we provided an overview of statistics about arrests and correctional populations. These statistics reveal that blacks have a higher arrest rate than their representation in the population. These statistics translate into higher levels of imprisonment as well. Recent federal government studies reveal that at midyear 2005, there were 4,682 black males incarcerated in a state or federal prison for every 100,000 black males in the population. By contrast, there were 1,856 Hispanics imprisoned for every 100,000 Hispanics, and 709 whites incarcerated for every 100,000 whites in the population.

A number of questions arise from these sentencing disparities. Are the differences a consequence of differential involvement in crime or of discriminatory practices of law enforcement and the courts? Further, do these differences reflect the fact that many members of minority groups live in neighborhoods that are economically disadvantaged and tend to have higher levels of law enforcement patrol and interventions? It is possible that sentencing disparities are a result of higher rates of involvement in crime by members of minority groups, but other factors could also be responsible. For example, there may be a greater number of police patrols in poor neighborhoods where minority populations tend to live and a greater reliance upon calling the police in minority communities. Either factor might result in a greater number of arrests of blacks or Latinos. To a large extent, arrests influence jail populations, and poor individuals (regardless of race) may have a more difficult time making bail. Although a percentage of these cases is ultimately dropped, these arrests may result in an overrepresentation of blacks or Latinos in local jails.

One controversial extralegal factor is the socioeconomic status of the defendant. Most people would probably agree that wealthier defendants have a greater likelihood of avoiding or minimizing legal consequences than do their poor counterparts. Indigent defendants, for example, rely on public defenders (discussed in Chapter 5) who tend to be overworked and have few resources to mount an aggressive legal defense.[25] A wealthier defendant, by contrast, might be able to afford a more experienced private attorney who can devote more time to a case and has the resources to pay for the independent examination and testing of evidence. Moreover, a wealthy defendant might be able to make restitution prior to sentencing, something that a judge might consider a mitigating factor when imposing a sentence. Despite these advantages, some rich or famous defendants are punished. Sentencing Paris Hilton to 45 days in jail and Martha Stewart to five months in federal prison are exceptions to the norm.

Women are also underrepresented in the correctional population compared to their numbers in the general population. Recent federal statistics reveal that there are 1,366 men incarcerated for every 100,000 men in the population, compared to 129 women for every 100,000 women in the population.[26] At the same time, there is concern that the growth in women's incarceration has outpaced male increases in recent years. Again, there are a number of explanations. In Chapter 2, for example, we suggested that women historically benefited from the paternalism of criminal justice agencies. They were less likely to be prosecuted for some offenses and were seen to be in need of protection (the chivalry hypothesis). We also have to acknowledge that this paternalistic treatment resulted in some girls being harshly punished by juvenile justice systems when they experimented with their sexuality. Also, some women received equally harsh punishments for their involvement in prostitution, while the males who bought their services were seldom arrested.

As women gained more political, social, and political power—and moved toward more equality—it has been suggested that they were treated more equitably by justice systems as well. This might explain why the prevalence of female juveniles and women in arrests has increased in the past two decades. Other scholars suggest that the greater number of women placed in jail or prison is a consequence of the war on drugs.[27] It is also possible that women are more comfortable assuming male roles, including engaging in crime, and this may be one reason why rates of female juveniles convicted of driving under the influence and assault have increased significantly over the past decade.[28]

To make sure that justice systems in a jurisdiction apply punishments consistently, a number of states and the federal government have established **sentencing commissions** to review sentencing outcomes. These agencies make recommendations about sentencing in a jurisdiction. Recommendations are based on studies conducted in the jurisdiction about sentencing outcomes.

One of the problems with conducting research after a person has been sentenced, however, is that sanctions tend to be fairly uniform *within* a given jurisdiction for offenders with similar offenses and criminal histories. When the judge sentences an offender, the process tends to be fairly transparent, meaning that the public can attend sentencing or learn about it through the media. The process can thus be scrutinized. We have comparatively less understanding of the factors that affect sentences. It is difficult to determine whether a defendant received a "break" at some point in the process (e.g., at the arrest or at the prosecutor's decision to charge or dismiss the case), whether there is horizontal or vertical overcharging, and the ability of the defendant to plea bargain. These processes tend to be less transparent; they occur behind closed doors, and it is difficult to know the extent to which various biases influence any of these decisions.

One observation about sentencing, sentencing reform, and disparities is that it is almost impossible to determine whether small amounts of bias throughout the entire process may result in a cumulative disadvantage for a defendant.

Moreover, sentencing based on prior convictions may also contribute to a disadvantage. Mitchell notes, "Over time, as offenders are recycled through the criminal justice system, the small disadvantages suffered in prior sentencing episodes may grow and accumulate into substantial disadvantages."[29]

Individualized Justice versus Consistency

In U.S. justice systems, a tension often exists between (1) the desire for **individualized justice**, where sentencing is tailored around the offender's unique combination of strengths and weaknesses (including potential for rehabilitation), and (2) the desire for consistency in treating everybody who commits a certain offense the same. **Mandatory minimum sentences**, where the same minimum sentence is imposed on all offenders convicted of a specific type of offense, provide a case in point. Treating all offenders the same is one principle underlying such sentences. Although mandatory minimum sentences are often associated with drug crimes, they were implemented before the start of the war on drugs. Mandatory sentencing approaches, however, can lead to problems because policies are rarely implemented in a consistent manner throughout different jurisdictions, and justice systems have often resisted reforms.

In some jurisdictions, there are mandatory sentencing enhancements for using a firearm in an offense, committing a crime while wearing a mask or a bullet-resistant vest, or being a member of a gang. These enhancements can sometimes add a decade or more to an offender's original sentence, and they are often intended to deter crimes that contribute to higher rates of violence (such as using a gun in a robbery). A criticism of sentencing enhancements, or any other type of mandatory minimum sentence, is that decisions to proceed with these sanctions are made by prosecutors, and once the defendant enters a plea of guilty, the judge historically had little discretion in the length of sentence imposed.

Recent U.S. Supreme Court decisions, such as *United States v. Booker,* have affirmed that judges can use more discretion in sentencing and that mandatory minimum guidelines are advisory rather than rigid.[30] Such decisions have moved the pendulum slightly away from consistency and toward individualized justice, giving some leeway to judges in sentencing (see **Figure 6–1**). Scholars who study the courts are closely observing the long-term impact of these decisions, and it is too early to know whether judges will simply follow the original mandatory sentencing guidelines or whether they will exercise more discretion. Because judges and prosecutors in many jurisdictions are elected officials who must stand for reelection, they are usually reluctant to exercise discretion in a way that can be interpreted as "soft on crime." Thus, while some judges might exercise discretion in a lenient direction, others might do so in a direction that is more harsh than the mandatory minimum.

Besides the potential for excessive judicial leniency or harshness, there is another problem with the discretion inherent in the individualized approach to

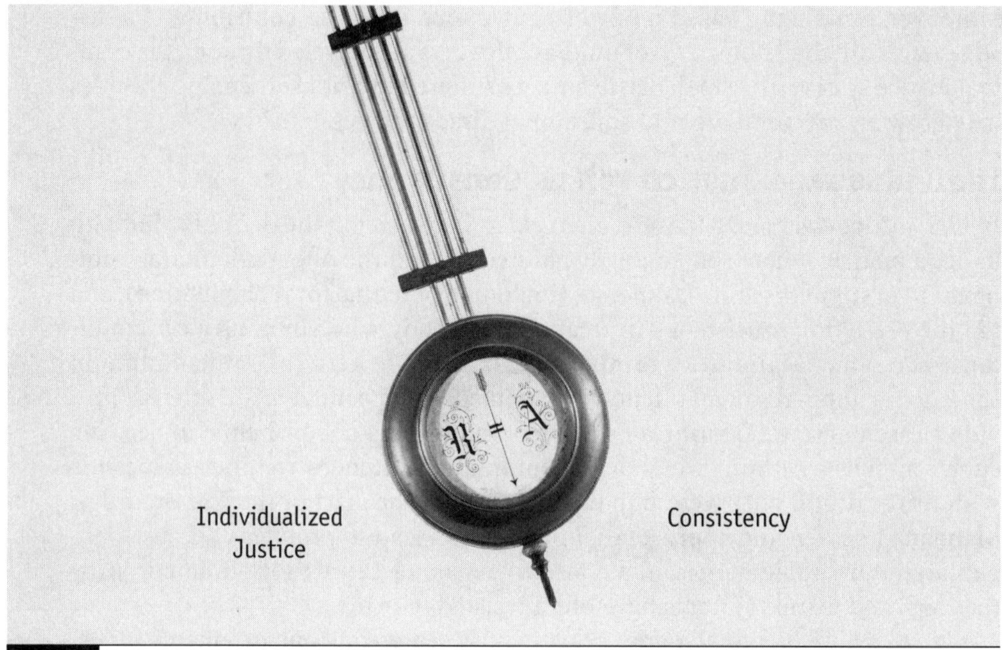

Figure 6–1 The sentencing pendulum.
Source: © *Krzysztof Gorski/ShutterStock, Inc.*

justice. Some offenders—usually those with middle-class characteristics—are considered to be at lower risk of recidivism and better candidates for rehabilitation than others convicted of similar crimes. In other words, the individualized approach has potential to perpetuate extralegal biases.

Presentence Investigations

PSIs—also discussed in Chapter 8—were originally designed as a way to assess amenability to rehabilitation and individualize sentences. PSIs are typically conducted by probation officers following conviction and prior to sentencing. These investigations result in a report that provides the judge with information about the offender and offers recommendations for sentencing. Although the format of PSIs differs across jurisdictions, they are generally composed of the components listed in **Figure 6–2**, which provide the judge with a comprehensive overview of the offender, the offender's strengths and weaknesses, as well as the success and failure of prior criminal justice interventions.

Because they consume time and money, PSIs are not ordered for every case. Judges in some jurisdictions request them only in very serious cases or when deciding whether a community-based sentence such as probation is appropriate. Judges use these reports to better understand the offender, including the offender's degree of remorse and amenability for rehabilitation. The probation officers

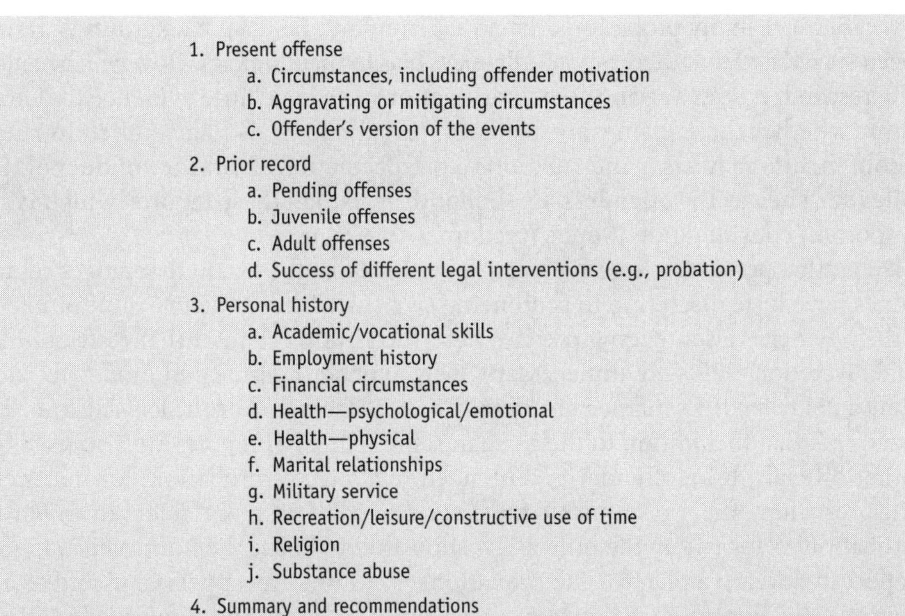
1. Present offense
   a. Circumstances, including offender motivation
   b. Aggravating or mitigating circumstances
   c. Offender's version of the events
2. Prior record
   a. Pending offenses
   b. Juvenile offenses
   c. Adult offenses
   d. Success of different legal interventions (e.g., probation)
3. Personal history
   a. Academic/vocational skills
   b. Employment history
   c. Financial circumstances
   d. Health—psychological/emotional
   e. Health—physical
   f. Marital relationships
   g. Military service
   h. Recreation/leisure/constructive use of time
   i. Religion
   j. Substance abuse
4. Summary and recommendations
   a. Officer's impressions of the subject
   b. Sentencing alternatives
   c. Officer's recommendations
5. Sources of information

Figure 6–2 Components of a presentence investigation report.

who conduct the investigations and write the report collect information about the offense and the offender. In many cases, the investigators interview the police officers who arrested the offender, trying to get more insight into the offense and the offender's role (e.g., if the offender was a leader or a follower). It is also common for the offender, family members of the offender, the victim(s), and employers to be interviewed.

These investigations are not without controversy. PSI reports have been criticized because the probation officer may make recommendations based on knowledge of a judge's sentencing expectations and practices (see Chapter 8). Judges have also been criticized because they often follow the probation officer's recommendations; one study found that judges accepted the recommendations for sentences in a high percentage of cases.[31] Another controversial factor is that some offenders have the financial resources to have an independent (also called defense-based) investigation completed, and these reports may provide a greater emphasis on rehabilitative community-based treatment that might help the offender avoid jail or prison.

Perhaps a greater criticism of these reports is that probation officers may unconsciously favor offenders with backgrounds and histories similar to their own. Because many probation officers are from middle-class backgrounds, it has been speculated that there is a middle-class bias to these reports. However, because the research is so scant on these investigations, we have little evidence to determine whether these claims are accurate. Jurisdictions also have different rules about including hearsay information (e.g., speculation about the conduct of the offender, such as the offender's alcohol or drug use) in these reports, which is an important consideration if one's freedom is in jeopardy.[32]

A rather persuasive argument can be made that to the extent that judges sometimes have little discretion in sentencing (e.g., as with mandatory minimums or other changes to sentencing policies described in this chapter that restrict judicial discretion), PSIs are unnecessary. Why expend the time and money to conduct a PSI when the sentence has already been determined by the legislature? The answer is that in addition to their use in sentencing, PSI reports are also used by other officials. If an offender is sentenced to a term of probation, for instance, the probation officer responsible for that *caseload* (a term that refers to all of the probationers for whom the officer is responsible) will use the information in the report to develop a plan for the probationer's community supervision and treatment (see Chapter 8). Likewise, these reports follow the offender to jail or prison and may be used to place the inmate in the most appropriate housing unit. In addition, these reports can also be used to develop an institutional case plan, which is a set of goals for the offender's treatment while incarcerated. PSI reports may also be used for release planning, parole decisions, and even to conduct research. Altogether, these investigations represent an important document that is likely to be used by many different officials for many years.

Changes in Sentencing Policy

One of the most interesting aspects of corrections involves recent changes in our sentencing practices and policies. As shown in **Figure 6–3**, incarcerated populations were more or less stable from 1920 to 1970. During times of economic crises, wars, increasing and decreasing rates of crime, as well as periods of domestic instability and social change, rates of imprisonment hovered around 100 persons per 100,000 residents in the U.S. population. In fact, correctional populations were so stable that they were thought by some to be self-regulating.[33] Yet, rates of incarceration increased dramatically starting in the mid-1970s. The rates have kept increasing for the past three decades, despite the fact that crime has decreased since the early 1990s. The United States locks up five to seven times as many prisoners per capita as most other industrialized nations.[34] At midyear 2006, there were 2,245,189 persons incarcerated in state and federal prisons and local jails across the United States.[35]

Perhaps the most fundamental change in sentencing policy over the last two to three decades has been the shift from **indeterminate sentences** to **determinate**

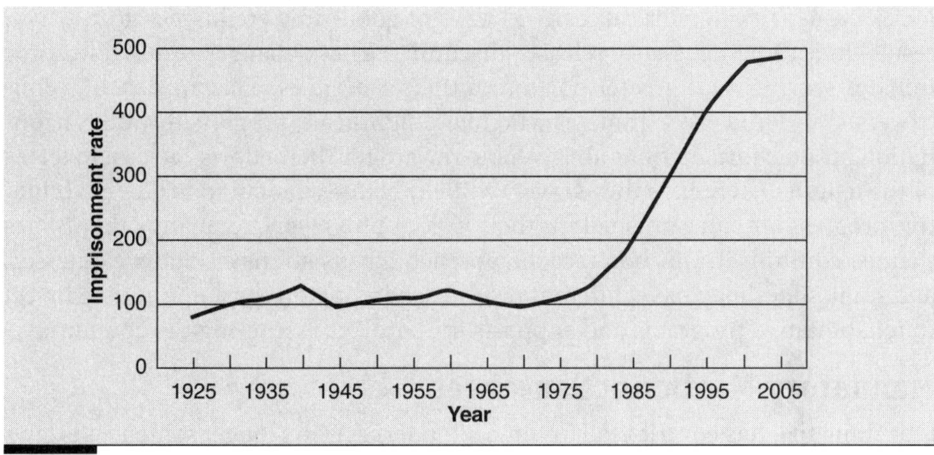

Figure 6–3 Federal and state imprisonment, 1920–2005.
Source: *Reproduced from Bureau of Justice Statistics Imprisonment Data.*

sentences. With indeterminate sentences, the judge specifies a minimum and maximum time frame within which the offender serves the prison term. With determinate sentencing, the exact number of years the offender has to serve is specified by the judge at sentencing. For example, a prisoner convicted for residential burglary might receive an indeterminate sentence of three to five years. Although prisoners could apply for parole after serving some portion of the sentence (usually some percentage of the lower limit), the parole board in that state would ultimately determine the prisoner's release date (see Chapter 13).

Indeterminate sentences were based on the assumptions that offenders could rehabilitate themselves. Early release was a seen as a reward for positive behavior in the institution and for making some form of rehabilitative change, such as finishing a general equivalency diploma, attending Alcoholics or Narcotics Anonymous programming, or learning a trade or job skill while in prison. By contrast, prisoners who were disruptive or uncooperative, or who engaged in criminal activities in the institution, spent a longer time in prison before they were released. Yet, indeterminate sentencing was not popular with legislators, who thought that parole boards were too forgiving and that releasing offenders early contributed to higher crime rates.

The first state to abandon parole release was Maine in 1976, and eight other states adopted similar policies by 1980. The federal government also ended parole with the Sentencing Reform Act of 1984, although it did not come into effect until 1987. In many respects, the federal government is a leader in corrections, and many jurisdictions followed its lead. As a result, determinate sentences have become the norm in most states.

In states that eliminated parole release, the primary way that a prisoner could shorten the length of his or her sentence was though earning good-time credits, which amounted to time off for good behavior. In the Federal Bureau of Prisons, for example, an offender can earn 54 days of good-time credits per year, which results in a 15 percent early release "discount." These changes resulted in some inmates serving a far greater portion of their sentences. This approach is controversial in some ways. Inmates who make significant progress toward rehabilitation in some jurisdictions are given no reward for their efforts (at least in terms of the length of sentence they serve), while inmates who fail to make rehabilitative progress are not punished for their lack of progress. Like many other issues that we confront in this book, each approach has its strengths and weaknesses, and some states may have different results depending on the amount they invest in rehabilitative programs and supports for parolees reentering the community.

Mandatory Minimum Sentences

One thing that has contributed to some offenders serving longer sentences is laws (also discussed earlier in the chapter) that specify a certain minimum penalty for a crime. These types of sentences were first enacted more than 30 years ago with the Bartley–Fox law in Massachusetts that specified that persons illegally carrying a firearm would be subject to a one-year term of incarceration regardless of mitigating factors. The cornerstone of mandatory minimum sentences is that the minimum sentence is fixed. Judges have little discretion to lower the severity of the sentence; departures from the minimum sentence have to be supported with a written rationale, and most jurisdictions forbid them. Mandatory sentencing was popular with legislators and the public who believed that offenders were not being held accountable for their crimes. Such sentences were also intended to have a strong deterrent effect on potential offenders.

Most people associate mandatory minimum sentences with the war on drugs and the lengthy sentences imposed for some offenders convicted of possession or trafficking of drugs. The federal government, for instance, enacted a number of strict mandatory sentences for drug offenders in 1986. The end result of this war on drugs is that more than one-half of all prisoners in the Federal Bureau of Prisons were sentenced on drug offenses in May 2007,[36] and approximately one-fifth of state prisoners were being held on drug offenses on December 31, 2003.[37]

Although mandatory minimum sentences have popular appeal, there are a number of limitations to this sentencing scheme. First, to have a strong deterrent effect, an offender has to know that these policies exist, and this is not always the case. A second concern is that discretion is not eliminated in the courtroom; discretion simply shifts from judges to prosecutors, because prosecutors can charge the accused with an offense that falls within or outside a mandatory sentencing scheme. If the defendant is convicted of a crime falling inside the scheme, the judge is required to sentence him or her to the mandatory sentence. Although judges can depart from the mandatory sentence, these departures are rare, even though judges have more discretion today given the *Booker* decision discussed earlier. A third limitation of this approach is that mandatory minimum sentences might not have a significant crime control impact on some offenses, such as street-level drug sales, where there is a large supply of potential offenders.

Perhaps the biggest concern about mandatory sentencing is that in jurisdictions that do not have parole, some offenders are sentenced to long periods of incarceration without any hope of early release. Although such an approach may be appropriate for violent criminals, mandatory sentences are often imposed on drug offenders. Imprisoning these offenders for long terms is thought to yield a marginal reduction in crime, is expensive, does not promote rehabilitation, and does not provide an incentive to avoid misconduct in prison. In recent years, there has been a grassroots movement initiated by the families of these prisoners. The movement seeks to modify these laws, although there seems to be little political willingness to enact less severe sentences.

Sentencing Guidelines

We described earlier how researchers discovered differences in sentence severity based on geography, as well as on such individual factors as race, class, or gender. One approach that is used to reduce sentencing disparities is to impose sentences that consider only the severity of the current crime and the offender's prior record so that judges are less subject to bias at sentencing. **Sentencing guidelines** (also called structured sentencing in some states) were first introduced in Minnesota and had two main goals: (1) ensuring uniformity in sentencing to reduce differences based on race, gender, or class, and (2) achieving proportionality such that offenders are punished according to the severity of the offense (**Figure 6–4**). Since the introduction of guidelines, some additional goals have been added, but the underlying purpose is to preserve confidence in the justice system by sentencing in a comparable manner offenders who commit similar crimes. Although some states provide sentencing guidelines for misdemeanor offenses, most jurisdictions use these guidelines for felony offenses.

Although sentencing guidelines place restrictions on the judge at sentencing, the court does have some discretion to sentence within a narrow range. Thus, the judge can consider aggravating or mitigating factors. **Aggravating factors** are circumstances that warrant a more punitive sentence. Examples include unusual cruelty, a helpless victim (such as an infant or elderly person), or membership in a criminal organization or gang. **Mitigating factors**, by contrast, include evidence about the character of the offender or facts about the offense that would warrant a less harsh sentence. Some mitigating factors include being a follower in the commission of a crime (rather than a leader), providing help to the police in solving the offense (as with the use of informants), immature age, or making restitution to a victim for the victim's losses prior to sentencing. Since the *Booker* decision, judges can more easily justify departing from the sentencing guidelines as long as they provide a justification for the departure. One problem, however, is that departures may actually increase the disparities that sentencing guidelines were created to reduce.

SENTENCING TABLE
(in months of imprisonment)

	Offense Level	Criminal History Category (Criminal History Points) I (0 or 1)	II (2 or 3)	III (4, 5, 6)	IV (7, 8, 9)	V (10, 11, 12)	VI (13 or more)
	1	0–6	0–6	0–6	0–6	0–6	0–6
	2	0–6	0–6	0–6	0–6	0–6	1–7
	3	0–6	0–6	0–6	0–6	2–8	3–9
	4	0–6	0–6	0–6	2–8	4–10	6–12
Zone A	5	0–6	0–6	1–7	4–10	6–12	9–15
	6	0–6	1–7	2–8	6–12	9–15	12–18
	7	0–6	2–8	4–10	8–14	12–18	15–21
	8	0–6	4–10	6–12	10–16	15–21	18–24
	9	4–10	6–12	8–14	12–18	18–24	21–27
Zone B	10	6–12	8–14	10–16	15–21	21–27	24–30
Zone C	11	8–14	10–16	12–18	18–24	24–30	27–33
	12	10–16	12–18	15–21	21–27	27–33	30–37
	13	12–18	15–21	18–24	24–30	30–37	33–41
	14	15–21	18–24	21–27	27–33	33–41	37–46
	15	18–24	21–27	24–30	30–37	37–46	41–51
	16	21–27	24–30	27–33	33–41	41–51	46–57
	17	24–30	27–33	30–37	37–46	46–57	51–63
	18	27–33	30–37	33–41	41–51	51–63	57–71
	19	30–37	33–41	37–46	46–57	57–71	63–78
	20	33–41	37–46	41–51	51–63	63–78	70–87
	21	37–46	41–51	46–57	57–71	70–87	77–96
	22	41–51	46–57	51–63	63–78	77–96	84–105
	23	46–57	51–63	57–71	70–87	84–105	92–115
	24	51–63	57–71	63–78	77–96	92–115	100–125
	25	57–71	63–78	70–87	84–105	100–125	110–137
	26	63–78	70–87	78–97	92–115	110–137	120–150
	27	70–87	78–97	87–108	100–125	120–150	130–162
Zone D	28	78–97	87–108	97–121	110–137	130–162	140–175
	29	87–108	97–121	108–135	121–151	140–175	151–188
	30	97–121	108–135	121–151	135–168	151–188	168–210
	31	108–135	121–151	135–168	151–188	168–210	188–235
	32	121–151	135–168	151–188	168–210	188–235	210–262
	33	135–168	151–188	168–210	188–235	210–262	235–293
	34	151–188	168–210	188–235	210–262	235–293	262–327
	35	168–210	188–235	210–262	235–293	262–327	292–365
	36	188–235	210–262	235–293	262–327	292–365	324–405
	37	210–262	235–293	262–327	292–365	324–405	360–life
	38	235–293	262–327	292–365	324–405	360–life	360–life
	39	262–327	292–365	324–405	360–life	360–life	360–life
	40	292–365	324–405	360–life	360–life	360–life	360–life
	41	324–405	360–life	360–life	360–life	360–life	360–life
	42	360–life	360–life	360–life	360–life	360–life	360–life
	43	life	life	life	life	life	life

Figure 6–4 Sentencing guidelines.
Source: *Reproduced from the Federal Public Defender for the District of Columbia, www.dcfpd.org/sentencing/sentencing_ grid.pdf.*

Truth-in-Sentencing

In jurisdictions that choose to retain parole release, there is dissatisfaction because prisoners are released after serving only a portion of their term. In California, for instance, an average prisoner in 2006 was sentenced to 46.7 months, but served only 24.1 months.[38] To provide an incentive for more severe sentences for violent offenders, the federal government provided funding to states that enacted **truth-in-sentencing** laws, where offenders who had been convicted of certain violent offenses would serve 85 percent of their sentence before being eligible for release. By 1998, 27 states had implemented truth-in-sentencing guidelines.[39]

The federal government has a long history of providing financial incentives to states that adopt criminal justice policies that the federal government supports. There are a number of hazards when states take this funding. First, the federal government provided funds only for prison construction, but the highest costs occur when the prison is actually staffed and incarcerating offenders. The second challenge is that states are now holding offenders for longer terms of imprisonment, and this creates institutional overcrowding.

Three Strikes

Washington State first enacted three strikes sentencing guidelines in 1993 to impose lengthy prison sentences for repeat felony offenders. Like many other seemingly novel criminal justice policies, other jurisdictions were quick to copy Washington's approach, but these sentencing schemes were not imposed in the same manner. In California, for instance, almost 8,000 offenders had been sentenced to minimum terms of 25 years for their third felony by December 31, 2006 (there are another 33,296 inmates who are serving prison sentences for their second strike, which doubles the prisoner's term of incarceration).[40] In other states, by contrast, the sanction is rarely used. A report by the Justice Policy Institute in 2004 found that judges in 14 states (of the 23 states that adopted these policies) had sentenced fewer than 100 offenders to terms of 25 years to life for a third strike.[41]

Similar to other controversies in criminal justice, there appears to be a great deal of discretion in the use of these sentences. A much-publicized California case, for example, resulted in an offender receiving a 25-year term of imprisonment based on the strongarm robbery of a piece of pizza from a group of youngsters. Yet, the media reports of these cases do not generally acknowledge that some of these offenders may have been arrested dozens of times for misdemeanors and felonies prior to the offense triggering the third strike. Repeat offenders or career criminals have always been a problem for justice systems, and there is a long history of using incarceration to incapacitate them.

Although the name of the legislation has popular appeal ("three strikes and you're out"), habitual offender legislation (also mentioned in Chapter 3) is not a new approach to sentencing repeat offenders. Some states had habitual offender statutes as early as the 1870s. Moreover, offenders did not have to be sentenced to lengthy periods of incarceration based on violent offenses. In 1980, the U.S. Supreme Court upheld the life imprisonment of a Texas offender for three property crimes that resulted in a loss of a total of less than $300 to the victims of his crimes.[42] Consistent with that earlier decision, the Supreme Court recently upheld the third strike sentence of a California man convicted of a nonviolent "wobbler" (a wobbler is an offense that can be filed by the prosecutor as either a felony or misdemeanor).[43]

Repeat offenders are an ongoing problem for justice systems. The Lincoln, Nebraska, police arrested a 41-year-old man for the 226th time in August 2006, and the police reported that at least one person in that city had more than 500 arrests before he died.[44] Clearly, this report suggests that a small number of offenders are long-term or career criminals, and their conduct places the community at risk. As a result, habitual offender laws such as three strikes enable a prosecutor to sanction serious or high-risk offenders. Policy analysts, scholars, and critics, however, sometimes disagree about who should receive these lengthy sentences.[45] For instance, should a person's prior involvement in a juvenile offense (when the offender was less mature and may have received fewer due process protections) count as a strike? Another problem surrounding these sanctions is justice by geography. Even in California, there are significant differences in the application of three strike laws; some counties have a much higher use of three strike sentences than others do.[46] Such observations challenge our ideas of fairness because how much time an offender serves in prison is sometimes a consequence of where the offender was arrested, rather than the offender's role in the offense or prior criminal history.

Controversies

Incarceration rates in North and South Dakota—what factors contribute to the difference? A review of the Bureau of Justice Statistics (BJS) incarceration statistics for midyear 2005 shows that for every 100,000 residents in the state of North Dakota, 199 persons were in prison, compared to 430 inmates for every 100,000 residents in South Dakota.[47] Crime rates, economic conditions, and population characteristics for these two states are generally similar. Given these facts, can you think of some reasons why South Dakota uses prison at twice the rate of its northern neighbors?

Death Penalty

Perhaps the most controversial sanction, and one that can be imposed at the federal level and in most states, is the death penalty. Once widely used in the United States and in other countries, capital punishment is gradually falling out of favor in many nations. Even in the United States, there are now fewer offenders who can be executed; the Supreme Court has recently made it unlawful to execute persons who were juveniles or mentally retarded at the time of the offense. In the past, persons who were convicted of serious felonies—such as sexual assaults—were eligible for the death penalty, but since the 1960s nobody in the United States has been put to death for any offense other than murder. Despite the fact that the death penalty is rarely imposed, it remains popular with many members of the public.

In 2005, the Federal Bureau of Investigation reported that there were 16,692 murders and nonnegligent manslaughter offenses.[48] Yet, the BJS reported that 128 persons were sentenced to death in 2005 for murders that they had committed, and 60 offenders were legally executed.[49] The same BJS study reported that the average person on death row spent approximately 144 months before the sentence was carried out, and in some jurisdictions, such as California, the average time on death row prior to execution is almost twice that length. Thus, the likelihood of being sentenced to death is very low for any given person convicted of murder, and once sentenced the prisoner will likely reside on death row for a very long time before execution.

One common question is why persons sentenced to death will serve such a long time in prison before the sentence is actually carried out. With the sole exception of South Carolina, states automatically review the conviction and sentence of anybody sentenced to death. These reviews are important to ensure that the individual liberties of the offender are protected and that innocent persons are not executed. Many prisoners have been released from prison in the past decade as a result of wrongful convictions. By May 2007, for example, the Innocence Project had documented 201 cases where prisoners were proven innocent (generally by DNA evidence).[50] The Death Penalty Information Center reports that 124 death row inmates had been exonerated by May 2007 because they had been wrongfully convicted.[51]

Although it is generally agreed that offenders should be held accountable for serious crimes, many people are not convinced that the death penalty has been applied fairly. The sentence is sometimes said to be imposed in an arbitrary manner. Serial killers such as Jeffrey Dahlmer, Kristin Gilbert, or Herbert Mullin were all sentenced to serve terms of life imprisonment without the possibility of parole, but some offenders who have killed one victim, such as Scott Peterson (although his wife was pregnant at the time of the murder), were sentenced to die.

Some factors that might contribute to these outcomes are laws in different jurisdictions, changing legislation (some states have reintroduced the death penalty in recent years), or the quality of representation if some defendants have more skilled attorneys.

One of the most problematic aspects of death penalty sentencing is that African Americans are greatly overrepresented on death row. According to the BJS, there were 1,805 whites under a sentence of death at the end of 2005. The 1,372 blacks condemned to die made up about 42 percent of the total death row population.[52] Given that blacks represent approximately 13 percent of the U.S. population, they are greatly overrepresented in the condemned inmate population. Latinos, by contrast, accounted for 362 condemned prisoners, or about 13 percent of the death row population, which is approximately the same as their prevalence in the general population.

A number of scholars have tried to explain why blacks are overrepresented in death row populations, and again there are no easy answers. A recent study of California death sentences between 1973 and 2003, for instance, found that the death penalty was more likely to be imposed when the victim was white, after controlling for other offense-related factors such as multiple victims.[53] In addition to the race of the victim, there may be other extralegal factors that influence the use of the death penalty. It is possible that media attention, such as that devoted to the death of Lacey Peterson, may contribute to the imposition of the death penalty in certain celebrated cases.

Death penalty trials tend to be expensive undertakings, and they are often related to some high-profile, outrageous, or egregious circumstances (a particularly violent attack, a vulnerable victim, a murder that occurs as part of another crime, or a crime with many victims). Such cases usually draw a significant amount of media attention, and as described in Chapter 1, celebrated cases often shape our ideas about justice systems. Death penalty trials tend to be much longer than other homicide trials are and are made up of two components, including establishment of guilt followed by a sentencing phase where the fate of the convicted offender is decided. Because these trials are broken into two parts or stages, they are called **bifurcated proceedings**. (Other civil, family, or criminal trials can also be split into two different components.)

Capital sentencing may be undertaken by a judge or jury, depending on the jurisdiction. The judge or jury examines the aggravating and mitigating factors and decides whether a sentence of death is appropriate given the offender's past record, the current offense, and the circumstances surrounding that offense. In some states, the jury must unanimously decide that the death penalty is a more appropriate sanction than life in prison without the possibility of parole.

Because the stakes are so high in capital punishment cases, the Supreme Court has placed boundaries on the execution of offenders. In the landmark 1972 case of *Furman v. Georgia*,[54] the Court found that executions were being applied in an arbitrary and capricious manner and directed states to develop procedures to reduce the discretion of judges and juries in recommending the death penalty. Until states had corrected these procedures, there was a halt to executions. In the *Gregg v. Georgia*,[55] decision in 1976, the Court found that states had developed mechanisms to ensure that capital punishment was not implemented in a capricious manner. These changes included (1) considering the mitigating and aggravating circumstances and (2) instituting a bifurcated trial process with a determination of guilt and a subsequent penalty phase. Also, in most states, there is an automatic appellate review of capital cases.

Although executions resumed after the *Gregg* decision, the Supreme Court has restricted who can be executed. The *Atkins v. Virginia*[56] decision in 2002 made it unlawful to execute offenders who were mentally retarded. Three years later, the *Roper v. Simmons*[57] decision precluded the use of capital punishment for persons under the age of 18 years at the time of the offense. These decisions have reduced the number of offenders who are legally able to be executed. Advocates for the abolition of the death penalty argue that offenders who are immature, have developmental problems, or are not sophisticated offenders should not be put to death.

Craig Haney examined the trial and sentencing process of U.S. death penalty cases and is critical of the jury selection and sentencing instructions to the jury.[58] Haney argues that jurors likely to oppose the death penalty are excluded from the jury pool. He also contends that the judge's instructions to juries about sentencing options do not always adequately address the issue of mitigation (and a sentence of life without the possibility of parole), making the death penalty the likely outcome. Despite Professor Haney's claims that the justice system is set up to execute offenders, legal executions of condemned offenders are rare. In 2006, for instance, there were 53 executions, down from a high of 99 in 1999.[59]

The mechanisms of legal execution have changed over time, and executions today are considered more humane than executions were in the past. This is probably why the United States is one of the few wealthy democratic nations that still uses the death penalty. Until the 1930s, in some U.S. jurisdictions, executions were public events, and townspeople witnessed the condemned offender being put to death (e.g., being hanged or electrocuted). These events are more private today, and only a few dozen people (including the family of the victims in some states) might witness the execution. In addition to making executions private, the methods of execution are thought to be more humane. Lethal injection is the usual method for execution of most U.S. prisoners and is usually less violent than are electrocution, the firing squad, gas chamber, and hanging. These latter methods were commonly used until the past few decades, and some are still authorized in certain states. Still, there has been recent criticism over the use of lethal injection because of the difficulty in ensuring that the inmate has been properly anesthetized.

As alluded to earlier in the chapter, there is intense debate over the general deterrent value of capital punishment. A number of scholars have examined whether the execution of a condemned offender deters other citizens from committing murder. More than 20 years of studies, however, have failed to conclusively decide this question. Some researchers have found support for capital punishment reducing homicide.[60] Some of the research claims that every execution saves a particular number of lives, such as five.[61] But other investigators have found that far from having a deterrent impact, capital punishment has a **brutalization effect**, actually causing murders to increase after an offender is put to death.[62]

Conclusion

To achieve the unprecedented increase in the U.S. incarceration rate, judges sentenced more offenders to jail or prison and sentenced them to longer terms. Abandoning rehabilitation as the leading goal of punishment was a driving philosophical force behind the increased use of incarceration. The public lost confidence in the ability of prisoners to reform themselves, and politicians enacted legislation that shifted the goals of punishment to incapacitation, retribution, and deterrence. The end result is that the United States now has the highest rate of incarceration in the world[63] and is one of the few wealthy nations that still sentences offenders to death.

Correctional officials had little say in these changes. In the previous chapters, we outlined how local, state, and federal corrections officials often find themselves reacting to the activities of law enforcement and the courts. In the remaining chapters, we describe the challenges of carrying out the sentences imposed by courts—from supervising a probationer's sentence to executing condemned inmates. These tasks are made more difficult when correctional agencies are understaffed, overcrowded, and underfunded. Perennial problems confront community corrections as well as jail and prison operations.

READY FOR REVIEW

- The degree to which an offender is punished depends on the philosophies of punishment (also called goals of sentencing) in a jurisdiction and include deterrence, incapacitation, retribution, rehabilitation, restoration, and restitution. None of these philosophies exists in a pure form, and these goals are often blended.
- A shift in the philosophy of punishment from rehabilitation to incapacitation and deterrence occurred in the 1970s and 1980s, and this is one factor that has contributed to stricter sentences and correctional overcrowding.
- Deterrence, incapacitation, and retribution focus on making the criminal accountable for the offense, whereas rehabilitation, restoration, and restitution focus on helping the offender make a positive change in attitude and behavior.
- Judges have a variety of sentencing options, from fines to the death penalty, although these sanctions are often limited by the actions of the legislature (such as imposing sentencing guidelines) and parole boards that can release an offender.
- Extralegal factors such as the race, class, or economic status of the offender are thought to influence sentencing and result in disparities. Some researchers have speculated that small bits of bias from arrest to sentencing might result in a cumulative disadvantage.
- Presentence investigations completed by probation officers or other authorities result in a presentence report that outlines the strengths and weaknesses of the offender and outline a number of recommendations about sentences.
- Forms of sentencing, such as three strikes, truth-in-sentencing, and mandatory minimum sentences, have made it easier to impose lengthy prison sentences on offenders.
- The shift from indeterminate (e.g., where the prisoner has to serve from 3 to 5 years) to determinate sentences (e.g., where the prison term is fixed at 5 years) was seen as a significant change in sentencing policy because indeterminate sentences encouraged offender rehabilitation.
- The federal government and some states have eliminated parole but give prisoners good-time credits, which translates into time off for good behavior.
- Capital punishment is a controversial sanction because research has produced conflicting evidence of whether it deters or actually encourages more murders.
- The number of offenders sentenced to death and the number of executions actually carried out have decreased over the past decade.
- The Supreme Court has limited the number of offenders in jeopardy of this punishment (e.g., forbidding the execution of persons who were mentally retarded or juveniles at the time of their offense).

KEY TERMS

aggravating factors Relevant circumstances—such as unusual cruelty, a helpless victim (such as an infant or elderly person), or membership in a criminal organization or gang—that warrant a more punitive sentence

bifurcated proceedings Trials that are broken into two parts or stages (e.g., one part of a capital murder trial establishes guilt, and the other stage determines the punishment)

brutalization effect Refers to an increase in murders as a result of offenders being executed

circle sentencing A restorative justice intervention that brings together justice officials, victims, and community members to develop a sanction or punishment for an offender

determinate sentences Where the exact sentence the offender has to serve is specified by the judge at sentencing (e.g., four years for a residential burglary)

deterrence The philosophy of preventing future criminal behaviors by punishing and threatening to punish offenders (see *general* and *specific deterrence*)

extralegal factors Factors that are taken into account at sentencing that are not related to the law (such as gender, race, or social class)

family group counseling A restorative justice intervention that brings together justice officials, the offender, and the victim to develop a response to a crime that repairs the damage or harm done to the victim and community

general deterrence The belief that after witnessing an offender being punished, the onlookers who observe the punishment or other members of society who hear about this punishment will learn from these experiences and be less likely to violate the law

goals of sentencing Also known as philosophies of punishment (such as deterrence, incapacitation, or rehabilitation)

going rate The beliefs shared by the courtroom work group about the appropriate sentence levels for defendants charged with given crimes in a jurisdiction

incapacitation A philosophy of punishment that separates an offender from society so that the offender cannot commit any further crimes

indeterminate sentence A sentence where the judge specifies a minimum and maximum time that the offender would have to serve in prison (e.g., "You will serve not less than three years, nor more than five years for the residential burglary")

individualized justice When decisions about sentencing are made on the basis of knowledge of the offender's strengths and weaknesses, including factors that might mitigate the severity of punishment

jury nullification When a jury refuses to convict a defendant despite the fact that the defendant is guilty (often believing that the punishment that the person would receive is worse than the crime committed)

just deserts Punishing an offender to achieve a state of fairness or equity
justice by geography When different sentencing outcomes occur based on where the court processes the case (e.g., urban and rural differences in sentencing)
lex talionis The principle or law of retaliation that specifies "an eye for an eye, and a tooth for a tooth"
mandatory minimum sentences When the same minimum sentence is imposed on all offenders convicted of a specific offense and mitigating factors are not considered
mitigating factors Evidence about the character of the offender or facts about the offense that warrant a less harsh sentence (e.g., juveniles have often received a mitigated sentences because of their immaturity)
philosophies of punishment The sentencing goals of a justice system, such as rehabilitation, incapacitation, or deterrence
plea agreement When a defendant pleads guilty to an offense to receive a less serious sentence after an agreement has been made between the offender's attorney and the district attorney
presentence investigation (PSI) report A report typically prepared by a probation officer that highlights the offender's history and amenability for rehabilitation
rehabilitation The philosophy of punishment that proposes that engaging in educational, counseling, or other treatment opportunities will improve the ability of the offender to avoid future involvement in crimes
restitution Occurs when an offender makes repayment for the damages that occurred in a crime
restoration Extends the philosophy of rehabilitation to include supervised contact between the offender and victim to develop a sanction that repairs the damage to the victim and community
retribution A philosophy of punishment that is based on revenge (also called retributive justice)
selective incapacitation A philosophy of punishment that is based on the belief that identification and incarceration of high-risk or high-rate offenders will increase public safety
sentencing commissions Government bodies established to ensure that sentences are fair and do not discriminate on the basis of race, gender, class, or geographical biases
sentencing disparity Variation in sentences for similar crimes
sentencing guidelines Sentencing policies that are intended to preserve confidence in the justice system by sentencing in a similar manner offenders who commit similar crimes
sentencing hearing After a trial is completed, a hearing is conducted and the judge hears evidence about the offense, the offender, reviews information from a presentence investigation report, and imposes a sentence
specific deterrence The proposition that the pain of being punished will deter the person punished from committing future offenses

three strikes laws Laws that were enacted to levy lengthy prison sentences (e.g., 25 years to life) for repeat felony offenders

truth-in-sentencing Sentencing schemes that make offenders who commit violent offenses serve 85 percent of their sentence before being eligible for release

YOU ARE THE CORRECTIONS PROFESSIONAL SUMMARY

1. How would you reassure the public that the probation officer has been held accountable? Or should the focus be on the actions of the probationer who was involved in the six murders? A case like this actually took place in Florida and has been widely reported as the "Deltona x-box murders." In all likelihood, the probation officer would have been fired, and this is what took place in the actual case. Although you could point out that, ultimately, the probation officer does not choose behaviors for probationers, you also need to reassure the public that such negligence among officers in your department will not be tolerated.
2. What is your view on whether first-time offenders convicted of violent crimes should be placed on probation? Correctional officials such as you are not responsible for assigning offenders to probation; judges do this. Statistics show that about half of all probationers are on probation for having committed a felony. You might mention this and say that it is not at all unusual for felons to be on probation, even felons who have committed violent offenses. The responsibility of correctional officials is to supervise these offenders as well as possible. This did not happen in the present case, and the problem has been addressed.
3. What are you going to say if you are asked about your views on capital punishment in this case? Your own personal views on capital punishment are irrelevant. If pressed by the media, you might mention that although it is important to hold offenders accountable for serious crimes like this and ensure that they do not offend further, there is much scientific controversy about whether death sentences deter murder.

NOTES

1. M. R. Durose and P. A. Langan, *Felony Sentences in State Courts, 2002* (Washington, DC: Bureau of Justice Statistics, 2004), 3.
2. A. Ashworth, "Responsibilities, Rights, and Restorative Justice," *British Journal of Criminology* 42(2002): 578–595.

3. See D. Fogel, *We Are the Living Proof: The Justice Model for Corrections*, 2nd ed. (Cincinnati, OH: Anderson, 1979).
4. F. E. Zimring and G. J. Hawkins, *Deterrence: The Legal Threat in Crime Control* (Chicago: University of Chicago Press, 1973).
5. D. S. Nagin and G. Pogarsky, "Integrating Celerity, Impulsivity, and Extralegal Sanction Threats into a Model of General Deterrence: Theory and Evidence," *Criminology* 39(2001): 865–889.
6. T. H. Cohen and B. A. Rainville, *Felony Defendants in Large Urban Counties, 2002* (Washington, DC: Bureau of Justice Statistics, 2006), 29.
7. L. E. Glaze and T. P. Bonczar, *Probation and Parole in the United States, 2005* (Washington, DC: Bureau of Justice Statistics, 2006), 1.
8. R. Ruddell and G. L. Mays, "Rural Jails: Problematic Inmates, Overcrowded Cells, and Cash-Strapped Counties," *Journal of Criminal Justice* 35(2007): 251–260.
9. P. W. Greenwood and A. Abrahamse, *Selective Incapacitation* (Santa Monica, CA: RAND Corporation, 1982).
10. E. Powell, "Settlement of Disputes by Arbitration in Fifteenth-Century England," *Law and History Review* 2(1984): 21–43.
11. D. Roche, *Accountability in Restorative Justice* (New York: Oxford University Press, 2003).
12. Federal Bureau of Investigation, *Crime in the United States, 2005* (Washington, DC: FBI, 2006).
13. J. Eisenstein and H. Jacob, *Felony Justice: An Organizational Analysis of Criminal Courts* (Boston: Little, Brown, 1977).
14. J. Casper and D. Brereton, "Evaluating Criminal Justice Reforms," *Law and Society Review* 18(1984): 122–144.
15. C. S. Steiker, "Capital Punishment and American Exceptionalism" (paper presented at the International Law, Human Rights and the Death Penalty Conference, Geneva, Switzerland, July 19, 2002).
16. T. R. Tyler, *Why People Obey the Law* (Princeton, NJ: Princeton University Press, 2006).
17. K. Guzik, "The Forces of Conviction: The Power and Practice of Mandatory Prosecution upon Misdemeanor Domestic Battery Suspects," *Law and Social Inquiry*, 32(2007): 41–74.
18. R. Scherer and S. B. Miller, "Martha Stewart Preps for Prison, Consultant in Tow," *Christian Science Monitor*, July 15, 2004, www.csmonitor.com/2004/0715/p01s01-usju.html.
19. S. Walker, *Sense and Nonsense about Crime and Drugs* (Belmont, CA: Thompson, 2006).

20. Walker, *Sense and Nonsense.*
21. J. M. Scheb and J. M. Scheb, *Criminal Procedure* (Belmont, CA: Thompson, 2003), 153.
22. *Bordenkircher v. Hayes*, 434 U.S. 357 (1978).
23. W. J. Stuntz, "*Bordenkitcher v. Hayes:* The Rise of Plea Bargaining and the Decline of the Rule of Law," in *Criminal Procedure Stories: An In-Depth Look at Leading Criminal Procedure Cases*, ed. C. Steiker, 351–180 (New York: Thompson, 2006).
24. O. Bar-Gill and O. Gaza, "Plea Bargains Only for the Guilty," *Journal of Law and Economics* 49(2006): 353–364.
25. American Bar Association, *Gideon's Broken Promise: America's Continuing Quest for Equal Justice* (Chicago: American Bar Association, 2004), www.abanet.org/legalservices/sclaid/defender/brokenpromise/fullreport.pdf.
26. P. M. Harrison and A. J. Beck, *Prisoners at Midyear*, 2005 (Washington, DC: Bureau of Justice Statistics, 2006), 1.
27. M. Mauer, C. Potler, and R. Wolf, *Gender and Justice: Women, Drugs, and Sentencing Policy* (Washington, DC: The Sentencing Project, 1999), www.sentencingproject.org/pdfs/9042.pdf.
28. H. Snyder and M. Sickmund, *Juvenile Offenders and Victims, 2006 National Report* (Washington, DC: Office of Juvenile Justice and Delinquency Prevention, 2006).
29. O. Mitchell, "A Meta-Analysis of Race and Sentencing Research: Explaining the Inconsistencies," *Journal of Quantitative Criminology* 21(2005): 439–466.
30. *United States v. Booker*, 543 U.S. 220 (2005).
31. M. D. Norman and R. C. Wadman, "Probation Department Sentencing Recommendations in Two Utah Counties," *Federal Probation* 64(2000): 47–51.
32. G. W. Carman and T. Harutunian, "Fairness at the Time of Sentencing: The Accuracy of the Presentence Report," *St. John's Law Review* 78(2004): 1–14.
33. A. Blumstein and J. Cohen, "A Theory of the Stability of Punishment," *Journal of Criminal Law and Criminology* 64(1973): 198–207.
34. R. Ruddell and N. E. Fearn, "The Stability of Punishment Hypothesis Revisited: A Comparative Analysis," *International Journal of Comparative Criminology* 5(2005): 1–28.
35. W. J. Sabol, T. D. Minton, and P. M. Harrison, *Prison and Jail Inmates at Midyear 2006* (Washington, DC: U.S. Department of Justice, Bureau of Justice Statistics, June 2007).

36. Federal Bureau of Prisons, "Weekly Population Report," www.bop.gov/locations/weekly_report.jsp (accessed May 26, 2007).
37. Harrison and Beck, *Prisoners at Midyear*, 9.
38. California Department of Corrections and Rehabilitation, "*Second Quarter 2006 Facts and Figures*," www.cdcr.ca.gov/DivisionsBoards/AOAP/FactsFigures.html (accessed May 13, 2007).
39. P. M. Ditton and D. J. Wilson, *Truth-in-Sentencing in State Prisons* (Washington, DC: Bureau of Justice Statistics, 1999).
40. California Department of Corrections and Rehabilitation, *Second and Third Strikers in the Adult Institution Population, December 31, 2006* (Sacramento, CA: California Department of Corrections and Rehabilitation, 2007).
41. V. Schiraldi, J. Colburn, and E. Lotke, *Three Strikes and You Are Out: An Examination of the Impact of 3-Strike Laws 10 Years after Their Enactment* (Washington, DC: Justice Policy Institute, 2004).
42. *Rummel v. Estelle*, 445 U.S. 263 (1980).
43. *Lockyer v. Andrade*, 538 U.S. 63 (2003).
44. "226 Arrests Aren't a Record in Nebraska," *Detroit News*, August 16, 2006, www.detnews.com.
45. F. Zimring, G. Hawkins, and S. Kamin, *Punishment and Democracy: Three Strikes and You're Out in California* (New York: Oxford University Press, 2001).
46. Schiraldi et al., *Three Strikes and You Are Out*.
47. Harrison and Beck, *Prisoners at Midyear*, 9.
48. Federal Bureau of Investigation, *Crime in the United States, 2005*.
49. T. L. Snell, *Capital Punishment, 2005* (Washington, DC: Bureau of Justice Statistics, 2006), 1.
50. Innocence Project, homepage, www.innocenceproject.org.
51. Death Penalty Information Center, homepage, www.deathpenaltyinfo.org/article.php?did=412andscid=6.
52. Snell, *Capital Punishment*, 6.
53. G. L. Pierce and M. L. Radelet, "The Impact of Legally Inappropriate Factors on Death Sentencing for California Homicides, 1990–1999," *Santa Clara Law Review* 46(2005): 1–47.
54. *Furman v. Georgia*, 408 U.S. 238 (1972).
55. *Gregg v. Georgia*, 428 U.S. 153 (1976).
56. *Atkins v. Virginia*, 536 U.S. 304 (2002).
57. *Roper v. Simmons*, 543 U.S. 551 (2005).

58. C. Haney, *Death by Design: Capital Punishment as a Social Psychological System* (New York: Oxford University Press, 2005).
59. Snell, *Capital Punishment*, 11.
60. P. R. Zimmerman, "State Executions, Deterrence, and the Incidence of Murder," *Journal of Applied Economics* 7 (2004): 163–193.
61. H. N. Mocan and R. K. Gittings, "Getting Off Death Row: Commuted Sentences and the Deterrent Effect of Capital Punishment," *Journal of Law and Economics* 46(2003): 453–478.
62. J. M. Shepherd, "Deterrence versus Brutalization: Capital Punishment's Differing Impacts among States," *Michigan Law Review* 104(2005): 203–256.
63. R. Walmsley, *World Prison Population List*, 6th ed. (London: King's College, 2007).

7

Jails and Detention Facilities

Chapter Objectives

- Describe the origins of the American jail.
- Describe the different types of jails in the United States today.
- Understand the challenges that "special-needs" inmates create for jail administrators.
- Explain the advantages of the new-generation jail design.
- Describe jails operated by different levels (or forms) of governments.
- Explain the impact of courts on the functioning of local jails.
- Describe the advantages and disadvantages of regional jail operations.
- Explain why community-based jail programs were developed and some of their characteristics.

CASE STUDY

You are working in a small rural jail, and it is close to the midnight shift change. One of the inmates in your unit is complaining of minor chest pains, but this person often complains of discomfort or problems to get attention from the nursing staff (who only work during the day shift) or to avoid unpleasant work details, a condition medical staff call **malingering**. This inmate is regularly admitted to the jail and has a long history of alcohol abuse. The inmate has some minor mental health problems but is otherwise in good physical condition. There will be no medical staff on duty until the next day. If this inmate is to receive medical care, you will have to transport him to the hospital, and you might wait several hours before the medical assessment and treatment are completed and you return to the jail.

1. What is your best response to this problem, especially given that this inmate is prone to malingering?
2. Whatever your decision, what are the risks of the action you choose? (Think about all of the potential risks—inmate health as well as the risks of transporting an inmate—including the possibility of escape).

Introduction

On December 31, 2005, there were 747,529 inmates residing in U.S. jails,[1] and another 71,905 participated in community-based jail programs.[2] Originally intended for the short-term detention of arrestees or defendants waiting for their court dates, jails have gradually transformed into institutions that hold a diverse range of inmates temporarily for other elements of justice systems. This includes growing numbers of sentenced offenders bound for state prisons and other persons who do not seem to fit into other places or systems. Jails have been called an institution that was created to control the "rabble" or "underclass," and a tour through a local jail will show that many inmates are mentally ill, drug and alcohol addicted, homeless, and poor.[3] Yet, the number of serious offenders held in jail has also increased, and in some jurisdictions, most inmates are felons.

Very few academics study what occurs in jails, and that is unfortunate because jail operations are among the most interesting in criminal justice. According to the Federal Bureau of Investigation, there were approximately 14.1 million arrests in 2005, and most of these persons passed through some 3,300 U.S. jails or juvenile halls.[4] As a result, jails have been described as the **gatekeeper** or transfer point for the entire criminal justice system—the place through which everybody must pass. Minor offenders are held in jails, but alleged serial killers also spend time in jail after their arrest, awaiting court dates, and pending transfer to prison if convicted.[5]

People sometimes confuse jails and prisons. The main factor that differentiates these two operations is that jails are intended to hold inmates temporarily, while prisons hold felons sentenced to at least one year of incarceration or death. Like many other examples outlined in this text, however, there are some exceptions to this observation, and we discuss them later in the chapter. The second factor that distinguishes jails from prisons is the diversity of inmates. Among the millions of persons admitted to jails are men and women, the rich and famous, the poor and homeless, as well as elderly inmates and juveniles in some places. There is also an ethnically diverse population in most facilities, and racial minorities tend to be overrepresented in jail populations.

There is a wide variety of jails, from the local detention facility in rural America that holds three or four inmates to the jail systems in Chicago, Los Angeles, or New York that hold upward of 20,000 inmates and employ thousands of correctional officers and staff. Regardless of the size of the facility, a constant theme is that jails have almost always been in some form of crisis as a result of overcrowding, resource limitations, litigation, suicide and violence, staff turnover, or the influence of local politics. Part of the problem is that most American jails are operated by elected sheriffs and funded by local counties, and many county administrators or boards of supervisors would rather spend their scarce budget dollars on anything but the jail. One interesting question is: How did we adopt these arrangements, and why do they persist? To better understand jails today we need to know the historical arrangements that lead to the development of city- and county-operated jails.

History of Jails

Like other elements of U.S. criminal justice systems, local jails in the United States closely resembled those established in England. As a result, the operations of today's jails in the United States have been shaped, in large part, by decisions and arrangements that were made centuries ago across the Atlantic Ocean. Unlike most nations that have national, state (provincial), or regional jails that are operated according to national-level standards and receive funding from federal governments, many U.S. jails are dependent upon local funding and overseen by local politicians. In many cases, problems identified in English jails, also called **gaols**, hundreds of years ago are still apparent in U.S. jails today.

English Gaols

Conditions in early English jails were characterized by indifference toward the inmates and brutal conditions of confinement, especially for the poor. In many places, jails operated as businesses, and prisoners were expected to pay for their incarceration. If detainees were wealthy, they could afford to purchase better meals, and alcohol use in these jails was common. Jails were violent places, and aggressive inmates preyed upon the weaker prisoners, including women and children. Inmates of all ages and both genders were held together, and in some cases, they would sleep among the animals stabled in the facility. Moreover, criminals were also mixed with debtors, and defendants were mixed with sentenced offenders.

Before the establishment of prisons, defendants who were found guilty were subject to corporal punishment (physical punishments such as whippings), sometimes transported to a colony such as Australia, or executed soon after their sentence was imposed. (The practice of penal transportation is discussed in further detail in Chapter 13.) Although transportation was a humane alternative to capital punishment, some offenders were not permitted to return to their homeland, and conditions on these ships were harsh.[6] Moreover, after the convicts arrived at their destination, they were often placed in work camps or leased out to individuals or other enterprises.

The alternative to corporal punishments or transportation, however, was capital punishment. Executions often took place at the jail, or in the surrounding courtyards, and these events were shocking spectacles that were attended by large crowds. After these executions, an offender's corpse would sometimes be displayed to deter potential lawbreakers. The body of the pirate Captain Kidd, for instance, was covered in tar after his execution, placed in an iron cage, and was hanged for several years alongside the Thames River in London.[7]

The origin of county- and city-level jail systems in early America can be traced directly to English gaols that were first established in the twelfth century. In 1166, King Henry II ordered that the sheriff of each county establish a jail. These places of detention were often "make-do" facilities. Smaller communities set up their gaols in courthouse cellars, stables, castle dungeons, under bridges, or other places not intended to house offenders. Larger cities, on the other hand, constructed facilities designed to hold inmates.

Often these early jails were used temporarily, and then abandoned when more appropriate places were built. Pugh notes how lockups or "some kind of prison seems likely to have been tucked away in the cellar or attic of every fifteenth-century guildhall."[8] This pattern of holding inmates in temporary, and often inappropriate, structures persisted in England. Authorities even held prisoners in decommissioned naval or merchant ships, called **hulks**, pending their penal transportation. Conditions in these ships were filthy and foul, and inmates were often chained to the hulls when they were not engaged in hard labor ashore.[9] Although hulks were intended as a temporary alternative to incarceration, these ships were moored to the docks along the Thames River in London for decades.

Reliance upon "make do" gaols resulted in a number of problems for an inmate that increased with the length of detention. Few standards for confinement existed in those days, but some officials were concerned about the quality of detention facilities. Pugh reports that the operator of one early English gaol was ordered five times up to the year 1233 to repair that facility.[10] Local politicians, even in that day, were reluctant to spend much on incarcerating the accused, and the jailers managing London's infamous Newgate Prison (it was actually a jail for most of its history) were under more or less constant pressure to repair and rebuild this facility for almost 200 years, from the mid-1200s to 1432. These investments in repairs paid off over the long term, because this facility was not closed until 1904; it had been transformed during that time from a gaol to a prison. Some English jails proved to be almost indestructible, and some of these facilities operated for hundreds of years, a pattern that would be repeated in the United States.

The earliest jails were places of disrepute, and **keepers** (a historical term for jail or prison staff) had little compassion for their inmates. Griffiths observes:

> *Their chief aim was to make their office profitable. Salaries that were often infinitesimal had to be eked out by the iniquitous system of "gaol fees." Innocent men, acquitted in court, were hauled back to prison till they paid the gaoler's dues. Extortion in every shape was openly practiced. The keepers sold spirituous liquors.*[11]

Because conditions in the jail were so bad, sickness and illness were widespread, and many inmates contracted jail fever (which was actually typhus), and others starved to death awaiting court appearances.

Early American Jails

In Chapter 3, we described how America was settled primarily by English colonists who imported the common law. In the first few decades of the colony, populations were small and clustered in villages, and there was little crime or delinquency. Formal law enforcement was rare. Parents disciplined their children, and few serious or violent offenses occurred. Cases of deviance were often dealt with informally and harshly, primarily through corporal punishments, although some delinquents and adult criminals were put to death.

There were few formal places of incarceration in the colony. As a result, corporal punishments were common. The stocks, pillories, dunking stool, and whipping were used instead of confinement. Although capital punishment was justified for most felonies, many scholars today believe that harsh punishments in the colony were used less frequently than they were in Europe.[12] Though jail was not a punishment in the strictest sense, conditions of confinement were harsh.

As colonial populations grew and **informal social control** (the ability of the community to regulate behavior through its approval of positive behavior or shunning those who did not conform) decreased, there was more crime and disorder. As we outlined in Chapter 4, law enforcement also became more formal, and over time it became necessary to detain persons accused of offenses until their appearances in court. As a result, the first formal American jails emerged in the 1600s, and like their British counterparts, a wide variety of "make-do" structures were used. Jails were often built close to the courthouse or occupied the same structure. These facilities were often very small, and in many cases the sheriff or jailors and their family would live in a house attached to a jail (sometimes called a jail annex); this arrangement continues to the present day in some small or rural jails.

Despite the fact that American correctional systems started with a clean slate, they were quick to adopt the worst of British gaols. Classification of inmates (defined later in the chapter) did not exist. Male inmates were mixed with women, children with adults, felons placed with minor offenders, and the mentally and physically ill put with healthy inmates. Jailors often confined everybody in a common area or room. This increased the transmission of disease and attacks on vulnerable inmates. Historical documents of the era suggest that the quality of jailers in America was not much better than their counterparts in England. Greenberg notes that keepers were often corrupt:

> *Sheriffs and constables took bribes to release prisoners in their custody or to fix juries, extorted money from prisoners in exchange for preferential treatment, assaulted innocent citizens without cause, charged excessive fees, committed a variety of other crimes and used their office as a protective device to advance their private interests.*[13]

A lack of professionalism, high levels of corruption, and inadequate or makeshift jails resulted in many escapes from these rudimentary facilities.

There were several important reasons for establishing places of detention in every county several hundred years ago. Transporting an arrestee through the countryside increased the chances that the prisoner would escape or that somebody might interfere with the escort. This was a serious problem in the American South and West because arrestees or defendants were sometimes killed by angry mobs of citizens before they ever reached the jail.

One of the most important descriptions of the conditions in American jails in the early 1900s was written by Joseph Fishman, a federal jail inspector, in his book titled *Crucibles of Crime*. Fishman's job was to ensure that federal inmates held in local jails were confined in humane conditions (the federal government paid counties to hold these offenders, as they do today). He labeled jails "human dumping grounds" and places of debauchery, dirt, disease, and dependency.[14] The poor conditions in these jails were often a result of untrained or incompetent jailors operating these facilities. In many jurisdictions, jailors were paid according to a fee system, where the county paid the sheriff a set rate per day to hold an inmate, and the sheriff kept any funds that remained after food and housing were provided. Thus, if the sheriff was paid a dollar a day, and only spent five cents feeding and housing the inmate, he kept the remainder, and some sheriffs became rich depriving inmates of food.[15]

Each region of the nation developed criminal justice systems in a slightly different manner, and jails were often a product of the local culture, history, and economic conditions. In the South, for instance, inmates sentenced to a term of jail or prison incarceration were often leased out to farmers or businesses, after being detained in a county stockade. Later, misdemeanor offenders were sentenced to work on county-operated road camps and chain gangs. On the East Coast, houses of detention closely resembled English gaols. Sentenced misdemeanants were often held in county prisons, workhouses, reformatories, and houses of correction, and until the early 1900s northern states also placed convicts on road crews. As a result, the jails that exist today developed in a different manner throughout the nation, and some of this is a result of their history and the influences of the nations that first colonized that region.

Types of Jails

A recent federal government survey found that there are currently about 3,300 jails in the United States, and there is at least one in most U.S. counties.[16] There is also a wide range of jail facilities operated by different forms of governments or corporations, and these include the following:

- Locally operated jails
- Privately run jails and detention centers
- Regional jails
- Jails operated by tribal governments
- Consolidated jail–prison systems managed by state governments
- Federally operated jails

These jails go by a wide range of names, including parish jails in Louisiana, houses of correction, detention centers, county prisons or penitentiaries, correctional centers, and regional jails. However, most of us know these facilities as county jails. Though the names differ, the main goals of these operations are to provide temporary custody for detainees awaiting their court appearances, to house inmates serving periods of incarceration of less than one year, and to hold detainees or prisoners awaiting transfers.

Locally Operated Jails

Local jails range in size from 3 to 20,000 beds. The 50 largest jails, mostly in urban areas, held almost one-third of all U.S. jail inmates at midyear 2004.[17] Some of these facilities house the population of a small town, and the administrators are responsible for thousands of employees who work in a wide variety of professional, health care, correctional, and support roles. As a result, these large jails are often professionally operated organizations that are led by correctional administrators, many of whom are certified jail managers (a designation that is awarded by the American Jail Association to jail administrators who have passed a series of courses) or hold advanced degrees in business administration, criminal justice, public administration, or a related area.

Large jails have a unique set of challenges, and a review of the National Institute of Correction's (NIC) *Large Jail Network* publications reveals that one of the most significant problems these agencies face, in addition to working with limited budgets, is health care, including management of costs for mental and physical health services. These problems are common to all jails, but the sheer size of these facilities (or systems) leads to a tremendous health care demand. Approximately 1.8 percent of state prison inmates have been diagnosed with human immunodeficiency virus (HIV)/acquired immune deficiency syndrome (AIDS).[18] Applying this estimate to a jail population of 10,000 inmates suggests that 190 inmates would have this disease in a single jail. In addition to HIV/AIDS, jails also house a high number of persons who have tuberculosis, sexually transmitted infections, hepatitis, and a host of chronic conditions, such as diabetes. Jail populations are constantly changing, and inmates with severe illnesses create a medical care burden on these agencies. Neglecting to treat these diseases or illnesses, however, places the community at higher risk when these inmates are released.

Most of us are familiar with smaller local jails, and they come in a wide variety of sizes and designs. These jails' operations are shaped by the social, cultural, political, and economic characteristics of the community or county where a jail is located. A review of the American Jail Association's *Who's Who in American Jails*, for instance, shows that some jail facilities operating in 2003 were more than 200 years old.[19] In some places, the jails are professionally operated, and a tour will show that they are clean, well-run operations with few problems. In other places, however, the jail operations are noisy, dangerous, and poorly run. What separates these two types of facilities seems to be the professionalism and expertise of the jail staff. One long-standing problem with jails is that the political nature of the sheriff's position sometimes results in people being hired on the basis of patronage (such as a reward for political support) rather than being hired because they are the best people for the job.

Jail managers closely track expenditures. The largest budget item is staff salaries, which represent about three quarters of the total operational cost of any correctional facility. Recruiting and retaining good employees is a challenge for any jail, especially when salaries are low in poor cities or counties. In addition, there are challenges to retaining jail officers over the long term, especially because they work unsocial shifts (such as evenings, nights, weekends, and holidays) and supervise inmates who may act unpredictably, abusively, or violently. In addition, many of the arrestees being admitted to a facility are under the influence of alcohol or drugs, have been living on the street, are experiencing the stress of incarceration (some for the first time), or have been arrested on serious charges and their futures are uncertain. These inmates are likely to "act out," which adds to the stress of working in a jail.

In contrast to their larger counterparts, there are about 1,775 jails of less than 100 beds in the United States (**Figure 7–1**). These facilities are often located in

Figure 7–1 The disappearing small jail.
Source: *© Dee Golden/ShutterStock, Inc.*

rural areas, and we do not know a lot about these smaller jail operations, except that they are gradually disappearing. **Table 7–1** shows that there were 1,000 fewer jails of less than 100 beds in 2003 than there were in 1982. Some of the smallest facilities are closing, while others are expanding their operations to hold more than 100 inmates. The disappearance of these facilities may be a positive development in rural justice systems because rates of suicide in smaller facilities are five times greater than in larger jails[20] and rates of violence have historically

Table 7–1 Disappearing Small Jails: 1982 to 2003

	10 Beds or Less	50 Beds or Less	Less Than 100 Beds
1982	355	2,469	2,821
1991	320	1,988	—
1994	—	1,739	—
2003	171	1,202	1,775

Source: R. Ruddell and G. L. Mays, "Expand or Expire: Jails in Rural America," *Corrections Compendium* 31(2006): 1–5, 20–21, 27.

been higher as well.[21] Last, these small jails tend to offer fewer rehabilitative opportunities for inmates.

Ken Kerle, a well-known jail scholar and editor of *American Jails* magazine, reports that in the past many small jails were operated by a sheriff and his spouse (until recently most sheriffs were men), and Kerle named these facilities "mom and pop" jails.[22] Often, the sheriff's wife cooked inmate meals, washed the jail laundry, and sometimes acted as county dispatcher, all without pay. One reason for relying on this help was that these small jails were often located in poor counties, and the jail had little priority in the county budget. Although things have progressed, the last national jail census found that smaller facilities operate less efficiently than do larger jails; they have a higher percentage of unoccupied beds, and the ratio of inmates per employee in small jails tends to be higher, which also increases costs.[23] Further, small jails cannot take advantage of bulk purchasing, or employee specialization, and they draw fewer resources from their communities.

Jails Operated by Tribal Governments

Tribal governments operate approximately 70 different jails in the United States, and most of them are located in western and southwestern states. These facilities tend to be fairly small, with the largest such jail having only 152 beds, and altogether these institutions hold less than 2,000 inmates.[24] As a result, they share the same problems as the smaller jails mentioned earlier. One additional difficulty is recruiting qualified jail officers from a very small population base. Few people are suited for a career as a jail officer, and the remote areas where many of these tribal jails are located make it hard to find qualified staff.

Some tribal jails have come under scrutiny because these operations have fallen short of health and safety standards. A recent report by the U.S. Office of the Inspector General (OIG) found that jails in Indian Country were neither safe nor secure. Many of these facilities were unsanitary, had high rates of fatalities and escapes, and were generally a hazard to both staff and inmates.[25] The OIG found that many of these facilities had poor leadership, and jail officers were poorly trained. Such problems are not confined to jails in Indian Country, however; they are present throughout the nation.

Race and Gender in Corrections

When discussing issues of race in corrections, we often neglect the fact that American Indians have high rates of arrests and jail admissions. Can you think of some social, historical, economic, or cultural reasons for these differences?

Privately Operated Jails

One type of specialized jail is operated by private firms who are in business to make a profit. There were only approximately 50 of these jails in 1999.[26] In the 1990s, there was a movement to reinvent government and make the delivery of government services more cost-efficient, responsive to the public, and accountable.[27] Part of this movement related to corrections, and **privatization** of correctional facilities became a popular issue in the criminal justice literature. It was proposed that corporations could provide correctional services more efficiently than government and reduce taxes. Some operators even argued that they could reduce recidivism. Private operators became involved in the delivery of corrections, from supplying goods and services for a growing inmate population to delivering correctional services (e.g., food or medical services), and in some cases private operators took responsibility for running the entire jail or prison.

Despite the initial excitement over privatization, Sabol, Minton, and Harrison note that 7.2 percent of adult prisoners were held in these operations at midyear 2006, and this total has increased over 20 percent since 2000.[28] Although the percentage of state prisoners increased by over 70 percent, the number of federal prisoners held in privately operated facilities has almost doubled in 6 years. It is likely that most of these federal inmates are short-term immigration detention cases, and these inmates would receive few rehabilitative services.

One of the reasons why the populations of state prisoners held in private prisons have increased slowly is that private prisons have not reduced problems for correctional administrators (e.g., the state governments are often listed as a co-defendant if an inmate sues the private operator), and these operators never delivered on the reduced recidivism that some promised.[29] While acknowledging that governments are often inefficient, one of the few ways that private operators can save money is by paying their staff lower salaries or benefits (staffing represents the greatest cost of corrections). The downside to this approach, however, is that it is difficult to retain professional staff paid low wages, and this increases employee turnover and all the problems of running a correctional facility with less-senior staff (e.g., more disorder and violence).

Private operators are, however, more efficient at undertaking some tasks, such as facility construction (because they don't have to engage in costly and time-consuming government contracts), financing agreements (for jail or prison construction), and maintaining efficiency by ensuring full facilities (without crowding) by transferring inmates. This issue is somewhat controversial because large national corporations sometimes transfer inmates to private prisons in other states. Inmates generally oppose such transfers because they are removed from their families and homes.

The issue of privatizing corrections has generated considerable debate, with proponents arguing that if private operators can deliver equal services while saving the taxpayer money, it is a good deal for everybody involved. Opponents, by contrast, argue that companies should not try to profit from the punishment of offenders. Although this issue is complex (and the subject of entire books), privatization has some positive aspects such as financing, construction, and flexibility. It also has downfalls such as the potential to reduce officer professionalism or provide fewer goods and services to inmates to save money.

Regional Jails

Regionalization refers to building one jail to serve a number of counties. Regionalized jails can operate more cost efficiently, and these larger operations can reduce some of the problems that we identified with smaller jails. The Ohio and West Virginia State governments, for example, have offered financial incentives to encourage counties to consolidate a number of small jail operations into a single facility. Some county administrators feel that it is more cost effective to pay another jail to house their inmates. Regionalization may be a promising approach when jails in adjacent counties are overcrowded, require extensive renovations, or need to be replaced. For example, it might be practical to build a single 200-bed facility to replace several aging or crowded 20- or 30-bed facilities from surrounding counties.

These changes are not always easy to implement. Jail operations are the hub of the local justice system, and changing the jail's location has an impact upon local stakeholders. Attorneys and family members visiting jail inmates may have to travel farther, and bail bondsmen and attorneys may be forced to relocate their offices.[30] In addition, transferring inmates from the local police station or sheriff's office to a regional jail also increases transportation costs and reduces the amount of time that a deputy or officer is "on the street," which in turn, may reduce public safety.

Corrections and Policy

Federal and state governments often provide financial incentives if a jurisdiction will adopt policies that they promote. In some states, counties are given state funding for jail construction if they build regional jails. The federal government, on the other hand, has provided states with funding for prison construction if they enact truth-in-sentencing policies. States or counties that choose not to participate in these initiatives have to live with the economic consequences of these decisions.

Like other examples given in previous chapters, local government officials do not always pursue the most rational solution to criminal justice problems. Although regional jails are a cost-efficient strategy, it is difficult to convince local sheriffs to reduce their political power and influence by surrendering their jail operations to another county. The NIC lists a number of barriers to regionalization, including differences in philosophies, management styles, inconvenience, and increased transportation costs.[31] Local government officials have to consider these factors and how a regional jail would affect the crime-fighting abilities of the police and courts. Moreover, when the local jail closes, it is a signal of decline in a small town. Although the sheriff may lose flexibility, job positions, and prestige when a jail closes, a local jail closure may also represent the deterioration of the community. Thus, the decision to keep operating an inefficient small jail may have little to do with the effectiveness of the local justice system.

State-Operated Jail Systems

The United States is one of the few wealthy nations where local governments assume the cost of housing **pretrial detainees** or those sentenced to short terms of incarceration. In most countries, state, provincial, or federal governments are responsible for these tasks. There are numerous advantages to such arrangements, and six states (Alaska, Connecticut, Delaware, Hawaii, Rhode Island, and Vermont) have integrated jail–prison systems (also called consolidated systems). This means that a facility might hold pretrial detainees as well as a person sentenced to a term in the state prison. State-operated jails have been found to increase correctional officer professionalism, reduce political influence, and standardize services (conditions are similar throughout the state), and jail conditions are often better.[32] The main advantage, however, is financial because the costs of jail operations can be borne by a larger tax base. Despite these advantages, state governments have been hesitant to assume these additional costs. Local communities are often unwilling to give up the autonomy of having their own facility, and sheriffs are reluctant to give up their influence.

State-operated jail systems are found in small or sparsely populated states. The fact that the number of states that have these systems has not increased in several decades suggests that the likelihood of further change is small. Perhaps it is positive that local corrections are not state operated. State involvement might actually result in higher jail incarceration rates. The actions of the courtroom work group are sometimes governed or constrained by the bed capacity of the local jail, and judges might mete out harsher sentences if there were a greater number of jail beds available. Moreover, state-operated jails would add to the burden of the Department of Corrections and result in a larger bureaucracy and the delays and inefficiency that go along with larger organizations. Such examples illustrate that each time a criminal justice policy changes, there is the possibility of unanticipated or unforeseen outcomes.

Federally Operated Jails

The Federal Bureau of Prisons (BOP) operates detention facilities in a number of urban areas, including Chicago, Los Angeles, and New York. These jails are known by a number of different names including federal detention centers, metropolitan correctional centers, and metropolitan detention centers. These facilities are often located beside federal courthouses and resemble large office towers. These structures are the preferred jail design in urban areas where building space is scarce and expensive. From a distance, these facilities are almost identical to other office buildings, and most passersby would not recognize these buildings as correctional facilities.

The federal government had historically paid local sheriffs to hold federal prisoners in their county or city jails. Fishman outlines the problems with this approach because inmates received very poor care in some jails.[33] Starting in the 1970s, the BOP constructed a number of these facilities because it was becoming difficult to lease enough detention beds in large urban counties. In addition, by building these correctional centers, the BOP could deliver community-based pretrial services (such as supervised release on bail) to provide alternatives to jail detention.

The federal government still leases beds from counties for federal inmates. In addition, the BOP and the U.S. Citizenship and Immigration Services also provide payment to privately operated detention centers that hold offenders and illegal immigrants. In fact, the increased number of inmates living in privatized facilities over the past 10 years is largely a result of the expansion of federal incarceration. Conditions in these facilities have been called substandard, and critics have observed that the detainees have been held in underfunded and understaffed facilities that are subject to very little oversight.[34] Altogether, whether jails are operated by local authorities, state agencies, tribal governments, private operators, or the federal government, the lack of resources and difficulty in retaining and recruiting staff seem to be significant challenges.

Characteristics of Jail Inmates

We noted earlier that jails hold varied inmate populations; these inmates range from material witnesses, who are held to ensure their attendance at court, to felony offenders, who will serve their prison term in the local jail. Many pretrial detainees remain in jail because they cannot obtain bail. In most jurisdictions, more than one-half of all jail inmates are awaiting their court appearances, from their initial arraignment to sentencing. Although the public tends to hold fairly punitive attitudes when it comes to jail inmates, it is important to remember that a vast majority have not been convicted of any offense and are innocent until convicted.

A federal study found that the following types of defendants or inmates are held in jails:

- Individuals pending arraignment and awaiting trial, conviction, or sentencing
- Probation, parole, and bail-bond violators and absconders
- Juveniles pending transfer to juvenile authorities
- Mentally ill persons pending their movement to appropriate mental health facilities
- Individuals for the military, for protective custody, for contempt, and for the courts as witnesses
- Inmates awaiting transfer to federal, state, or other authorities
- Inmates for federal, state, or other authorities because of crowding of their facilities
- Inmates sentenced to short terms (generally less than 1 year)[35]

Some of these groups present significant challenges for jail administrators and the officers who work with these inmates on a day-to-day basis. The following pages outline the demographic and offense-related characteristics of jail inmates as well as the different groups of special-needs jail inmates.

Jails were intended to hold pretrial detainees and misdemeanants sentenced to terms of incarceration of less than one year. Harrison and Beck, however, report that nearly 10 percent of all U.S. jail inmates are actually sentenced state or federal prisoners.[36] These inmates make the job of correctional officers and administrators more difficult because jails were never intended for holding long-term prisoners. It has been observed that jails have experienced mission creep where the tasks expected of these facilities have gradually changed to the point where jail roles and functions today bear little resemblance to the jail operations several generations ago.[37]

Demographic Characteristics of Jail Inmates

Historically, most jail inmates were misdemeanants—persons arrested for relatively minor public order offenses such as trespassing, disorderly conduct, or being drunk in public. These jail inmates were overwhelmingly poor, and many were substance abusers. Although these populations still represent a high percentage of jail inmates, there have been some long-term changes in the characteristics of jail populations. In the past 20 years, there has been a significant increase in the number of serious or violent offenders detained in jail, or those who actively try to undermine the jail operations, such as gang members. Such changes have had a long-term impact upon jail disorder, misconduct, and violence because the demographic and offense characteristics of the inmate population often shape what occurs in the jail. A young male charged with a violent offense, for instance, is likely to be a greater concern (and threat) to the jail than a middle-aged woman detained for driving while intoxicated.

The main difference between jails and prisons is that most jails house both genders, whereas most prisons hold only either males or females (although there are some exceptions to both statements). Males are arrested at much higher rates than are their female counterparts, and they tend to be arrested for more serious offenses. According to the latest federal government study of jail populations, at midyear 2005, women represented about 12.7 percent of all inmates.[38] The percentage of women in jail has been increasing in the past decade, and this change was a result of changing law enforcement practices, a greater involvement in some types of crime (such as drug offenses), and changing social attitudes toward the role of women in society.

Women pose special challenges for jail operations. Recent studies show that they have a greater need for psychological services.[39] The stress of an arrest and the uncertainty of upcoming court dates also add to the mental health problems that these women face. Another challenge is providing medical care for these jail inmates, especially given their reproductive health care needs. A large percentage of women jail inmates are pregnant, or have given birth within the previous year.[40] Many jails are unable to provide these women with the proper prenatal, medical, or psychological care.

One of the most challenging problems facing criminal justice systems is the disproportionate percentage of minority populations that come into contact with law enforcement, are arrested, or are incarcerated. According to the U.S. Census Bureau, African Americans and Latinos represented approximately 12.1 and 14.5 percent of the U.S. population in 2005, yet at midyear 2005 they made up 38.9 and 15 percent of all jail populations, while other races made up approximately 1.7 percent of the population.[41] Whites, by contrast, represented approximately 44.3 percent of the jail population, making them underrepresented in jail and blacks highly overrepresented.

Like the observations made about the female jail population, there are no easy explanations for disproportionate minority contact and incarceration. These findings could be a consequence of patterns of differential policing and prosecution, racist criminal justice system interventions, differential involvement in crime, or a combination thereof. All of these explanations are troubling and contribute to perceptions among minority groups that the justice system has little legitimacy. Blacks consistently report, for instance, that they have less confidence in the criminal justice system than do whites.[42]

It is also plausible that minority populations are more likely to be poor and have a greater difficulty in arranging bail than their white counterparts do. It has long been argued that one of the primary factors leading to jail incarceration is poverty and the inability of many poor defendants to obtain bail.[43] In the past few decades, there has been a movement to implement release on recognizance policies that allow poor defendants with some ties to the community (such as housing or a job) to simply promise to appear, and they are not required to post bail (see Chapter 5). As a result, it is possible that ROR policies have reduced the number of poor defendants in jails. The latest Bureau of Justice Statistics (BJS) data reveal that the percentage of African Americans in jail has decreased from 43.5 to 38.9 percent in the past 10 years, which is an encouraging trend.

Despite the fact that less than 1 percent of the entire U.S. jail population was under 18 years of age at midyear 2005, the vast majority of them had been charged or sentenced as adults (a total of 5,750 persons under the age of 18 years), although there were at least 1,000 held as actual juveniles.[44] Although the number of juveniles held in jails is relatively low, they pose significant problems for jail administrators. Youngsters are at higher risk of suicide and victimization by other inmates. In addition to being at greater physical risk, placing an impressionable juvenile with hardened offenders makes little sense if the goal is to discourage juveniles from engaging in further crimes.

Incarcerated adolescents tend to be more disruptive and unsettled than adults are. Moreover, juveniles are impulsive and apt to engage in risk-taking, criminal, or risky behaviors in the jail because they are bored, need stimulation, or want attention. Some juveniles vandalize jail property, taunt or fight with other inmates, cause disturbances, or attempt escape. In the chaotic environment of the jail, such behaviors are apt to result in serious consequences, including further criminal charges. Moreover, juveniles usually have more energy than their adult counterparts do, and they contribute to higher levels of "wear and tear" on a facility. As a result, most jail officers prefer that youths be placed in juvenile facilities (see Chapter 14). But, in some jurisdictions, there are no alternatives, and the care of a juvenile becomes the problem of the jail.

Some arrestees younger than 18 years of age are charged as adults and are placed in county jails. In some states, juveniles are legally defined as adults at age 16 or 17 years, and upon arrest they are admitted to the jail instead of a juvenile hall. Other juveniles are temporarily held in an adult jail as delinquents because there might not be a juvenile detention facility in the county. Regardless of the cause, there is considerable concern about what happens to these youths because their placement in jails increases the risks of victimization or self-harm.[45]

Offense Characteristics of Jail Populations

No national data outline the offense characteristics of all jail inmates. We know that of the approximately 14.1 million arrests (this total includes juvenile arrests) made in 2005, approximately one-half million were for the most serious violent offenses (murder, rape, and aggravated assault), and another 1.3 million were misdemeanor assaults. Otherwise, the vast majority of arrests are for nonviolent crimes, and most inmates do not serve a lengthy time in jail for property or public order offenses (the exception are those who cannot obtain bail).

The state of California publishes quarterly statistics that describe jail populations; at the end of August 2006, 79 percent of all inmates in California jails were being held on felony charges. More than two-thirds of these inmates had not been sentenced, which is slightly more than the national average of 62 percent.[46] From these statistics we can infer that most jail offenders are felons and have not been sentenced. This finding has some implications for jail operations because a vast majority of these inmates are innocent according to the law. Many believe that these defendants should not be mixed with sentenced offenders.

Special-Needs Jail Inmates

There are a number of special-needs inmate populations that pose special challenges for jail officers and administrators, and they include persons with mental illness, **frequent fliers** (inmates with multiple jail admissions), long-term jail inmates, gang members, and those with severe health problems. These groups increase disruptive behaviors, medical care costs, criminal behaviors, or high-risk behaviors (such as self-harm or suicide). The following paragraphs present the characteristics of these inmates.

Inmates with mental illness are overrepresented in local jails. A recent national study of jail inmates found that 64 percent of jail detainees reported having a mental health disorder, and 21 percent had received mental health treatment in the previous year.[47] Inmates with mental illnesses have a greater likelihood of being involved in disruptive behaviors, including violence, self-harm, and suicide. Larger jails are usually able to manage these populations more effectively using special housing units. Creating a separate mental health unit is not an option for smaller jails, and inmates with mental illnesses in these facilities are more likely to be placed in the **general population** (inmates who are not living in specialized housing units) and are therefore at higher risk of victimization.[48]

Jail officers also have to manage a group of inmates that is unique to local corrections, the frequent flier. Also called **habitual misdemeanor offenders** or frequent users, these inmates have dozens of (and in some cases more than 100) jail admissions. A study of a small sample of frequent fliers in a Florida county jail revealed that their impact on jail operations was significant: One frequent flier had 151 separate jail admissions, and the average person in the sample had 56.9 admissions.[49] A national-level study of jails found that more than 20 percent of inmates were thought to be frequent fliers, with more than 20 jail admissions in a five-year period.[50] Even one of these offenders can have a significant impact on community resources; a single offender cost the Reno, Nevada, community approximately a million dollars in law enforcement, social service, and medical care costs.[51]

Seriously ill jail inmates, such as persons with HIV/AIDS, tuberculosis, or hepatitis, pose additional problems for jails. Correctional facilities do not always provide the type of medical care these inmates require. In fact, 47.6 per-

Controversies

The city of Seattle has recently created an apartment building for 75 "chronic public inebriants." Most of these hard-core alcoholics were homeless, and many have long histories of involvement with the law, much like frequent fliers. By containing these alcoholics in a common building, it is hoped that the city will save money in reduced medical care, including fewer ambulance trips to emergency rooms. These residents can continue drinking in their new apartments, but the program is controversial. This group receives free housing, while some Seattle citizens who work at full-time jobs cannot afford their own apartments.[52]

cent of all deaths in American jails in 2002 were attributed to illness.[53] Jail inmates are often at higher risk of poor health because of the long-term effects of addictions, a lack of regular health care, poor lifestyle choices, and prior involvement in risky behaviors. Providing medical care for these severely ill inmates creates a significant economic burden. The medication needed to treat one inmate with HIV/AIDS, for instance, costs taxpayers $13,900 to $36,500 per year.[54] One issue (addressed more thoroughly in later chapters) is that when the government incarcerates someone, it also assumes responsibility for all his or her basic needs, including health care.

Jails are also holding more inmates who have been incarcerated for a year or longer. A recent national-level study found that a small subgroup of inmates serves very long periods of detention or incarceration.[55] These long-term inmates are held awaiting trial, are sentenced to a period of jail incarceration, are actually serving their state prison sentence in a local jail, are awaiting transfer to a state or federal prison, or some combination of these factors. These individuals can create problems because jails are neither designed nor intended to incarcerate long-term inmates. The average stay in a county jail is measured in days. As a result, few jails have the educational, vocational, recreational, or rehabilitative programs to keep a long-term inmate constructively occupied. In addition, when inmates are held for a long period of time in jail, they are unlikely to develop vocational, life, or social skills that will help them when they return to the community.

There are a number of possible causes of long-term jail incarceration. First, some jail inmates serve long periods of detention while awaiting various court dates. As the seriousness of a defendant's case increases, that person's period of detention usually becomes longer. A study of felony sentences in urban counties, for instance, found that the median case-processing time for a homicide offender was 361 days from arrest to sentencing.[56] Second, jails hold sentenced offenders for terms of incarceration of up to one year, and 58 percent of those sentenced to jail in 2002 in urban counties were misdemeanants, while 37 percent were felons.[57] Third, a large number of prisoners are held in jail awaiting transfer to state penitentiaries. A recent BJS study reported that approximately 10 percent of all U.S. jail inmates are awaiting such transfers and that numbers of these prisoners increased greatly from 2000 to 2005.[58] The inability to transfer these inmates has resulted in overcrowding in some jails. Last, some state prison inmates may serve their entire sentence in a local jail. In Tennessee, for instance, judges can sentence felony offenders who would otherwise go to prison to local jails under some circumstances.

One additional group of inmates presents special challenges for jail officers and administrators. Members of gangs, also known as **security threat groups**, cause problems for jail operations because they are more violent than other inmates and they sometimes undermine jail operations (e.g., arranging for an assault, importing contraband into the facility, or some other illegal act; see **Figure 7–2**). Gangs are also thought to increase racial tensions, challenge or undermine rehabilitative programming, and contribute to the failure of community reentry if these gang-involved inmates continue their criminal conduct in the community.[59]

Figure 7–2 Persons involved with gangs can cause difficulties for jails.
Source: *© Tim Harman/ShutterStock, Inc.*

A recent national survey of jails found that approximately 13 percent of all jail inmates were thought to be gang-involved.[60] The researchers discovered that gang members were more likely to be found in larger communities and that almost half of jail administrators thought the gang problem had gotten worse in the past five years. When asked about the problems that these inmates caused, jail administrators reported that gang members were less disruptive than inmates with severe mental illnesses, but were more likely to assault other inmates. These inmates can cause serious difficulties, and many jail administrators and gang investigators in jails work closely with other law enforcement agencies to share information and intelligence about security threat group members.

Impact of Courts on Jails

The nature of jail incarceration has changed over time, in part because of changes in court systems, both in the day-to-day relationships between criminal courts and jails, as well as the impact of litigation on jail operations. Jails were initially intended to hold defendants temporarily until their court appearances and hold those who could not arrange bail. Most arrestees booked into a jail are able to obtain bail or are released on recognizance. Of the remaining inmates, most are held because they are thought to be a flight risk, represent a threat to witnesses, pose a risk to community safety, or could not make bail; this last group is usually the largest.

Court Case Processing and the Jail

In California, approximately two-thirds of all jail inmates in 2006 had not been sentenced.[61] This statistic is close to the national average, and recent national studies also note that the percentage of sentenced jail inmates is decreasing over time.[62] Jails often have a strained relationship with local courts because they have little influence over what will happen with a given inmate. All court jurisdictions have a unique culture, and in some places it takes a comparatively long time for a case to work its way through the justice system. In some jurisdictions, a problem called "jail bloat" occurs, where delays in the time that it takes for a case to be processed result in jail overcrowding.[63] We have previously described how the courtroom work group attempts to process cases quickly. When relationships between these officials are strained or the work group is inefficient, cases move through the system slowly, and this is one factor that contributes to jail overcrowding.

Inmate Litigation

Jails have been increasingly vulnerable to lawsuits over the past two decades. When jail or prison staff members admit an inmate, the county, state, or federal government assumes responsibility for that person's care, safety, and well-being. In cases where the jail has not provided appropriate care and inmates have been harmed, these inmates can initiate a lawsuit in federal courts (we discuss correctional litigation more comprehensively in Chapter 11). Schlanger conducted a survey of large jails and prisons and found that rates of litigation were much lower in jails compared to prisons.[64] This is understandable because most jail inmates spend a relatively short period of time in jail. According to the California Corrections Standards Authority, the average jail inmate spent 21.3 days in jail in 2006.[65] By contrast, the average prison inmate in the same state serves 24.8 months.[66] As a result, a prison inmate has a much greater likelihood of being harmed by some type of negligence. In addition, prison inmates have more time on their hands to initiate a lawsuit.

Schlanger's study is important because it sheds light on the main reasons why jail inmates initiate lawsuits. Jail administrators reported that the following were the major causes of inmate litigation in large jails:

- Medical care
- Use of force
- Personal injury
- Loss or damage to property
- Inmate-on-inmate violence[67]

A lack of medical care is the foremost cause of jail litigation, and officials have sometimes been negligent in providing basic care for inmates. For example, a Tampa jail inmate, Kimberly Grey, gave birth over a toilet after officers refused to transport her to a hospital, despite the fact that she had been in labor for several hours.[68] This admittedly extreme case resulted in the death of the infant, but the incident also underscores the types of health care treatment that some inmates receive (or do not receive).

One of the challenges of admitting arrestees directly from the "street" is that many are under the influence of drugs and alcohol, and some are at risk of overdose. Moreover, many arrestees are suffering from unintentional injuries, have been assaulted (or were injured while being arrested), or suffer from chronic health conditions. Last, some arrestees are at high risk of harming themselves, and families have initiated lawsuits after jail staff failed to properly assess and supervise at-risk inmates who later committed suicide.

There are two types of lawsuits that can be initiated by inmates: individual suits that address a particular incident, and class-action lawsuits where a number of prisoners who have suffered a similar harm file a case together. Some lawsuits are resolved in a financial award to the inmate, while others result in a **consent decree**, where the jail or prison administrators agree to change policies or procedures under the supervision of the court. Welsh observes:

> *Correctional officials have been ordered to make sweeping changes to comply with court directives: to improve medical care and recreational services, to reduce chronic overcrowding, to increase staffing levels and improve training, to make use of various pretrial and post-conviction release mechanisms, even to build new facilities.*[69]

In some lawsuits, courts have appointed special masters to ensure that the conditions of a consent decree are met. Agency administrators who do not make a good faith effort to correct the problems identified in a consent decree might find themselves in court and may be in jeopardy of being placed in jail for being in contempt of court. Inmate litigation in the 1980s and 1990s had a significant impact on the way that correctional agencies operated: agencies were forced to deliver a broader range of services (such as better medical or dental care), or change policies relating to strip searches, use of force, or responding to inmate grievances.[70]

Although courts forced local jails to spend millions of dollars to improve services, some long-term consequences were positive. Jail administrators in Schlanger's study thought that court orders had positive impacts on inmate behavior, the jail's **physical plant** (the buildings or grounds), and inmate health; but less positive effects were reported as regards staff morale or overall jail functioning. This finding has a commonsense appeal: when jail inmates have better access to health care or improvements in living conditions, they are likely to behave better.

New-Generation Jails

Between 1983 and 2005, jail populations increased nearly threefold, from 261,556 to 747,529 inmates, and jail populations continue to increase. To accommodate this growth, counties had to renovate facilities, add beds to existing jails, and sometimes construct entirely new jails. In some places, the local jail may be several hundred years old. One problem with these aging facilities is that most were built using a linear design, with cells in long rows or hallways (the type of structure portrayed in old prison movies). These antiquated designs are inefficient for supervising inmates. They only allow for **intermittent supervision** because inmates are supervised only when officers make rounds, and this contributes to higher rates of disruptive and illegal behavior.

Most new jail and prison construction is based on **new-generation jail** designs. This design features cells that are located on the exterior walls, leaving a large interior space or common area that inmates use for dining, recreation, or education. In some facilities, a single correctional officer can supervise several hundred inmates on a number of different **living units** (units that house the inmates and utilize a common area for recreation, education, and meals). This is done from a control room overlooking the inmates. Other jails place an officer at a work station located in the unit. Either approach provides higher levels of inmate and officer safety through **direct supervision**. Although many jurisdictions have utilized new-generation architecture, they have been criticized for failing to implement the new-generation philosophy that aims to improve the offender's skills through interacting with jail officers.[71]

Ken Kerle is a critic of jurisdictions that fail to implement the entire new-generation philosophy. This philosophy has nine goals, including the following:

- Effective control
- Effective supervision
- Competent staff
- Safety of staff and inmates
- Manageable and cost-effective operations
- Effective communications
- Classification and orientation
- Justice and fairness
- Staff ownership of operations[72]

Although new-generation jails are able to facilitate more effective supervision, the original goal underlying these programs was to place an officer in each living unit to help inmates engage in problem solving and to increase safety by resolving minor problems before they escalated. As a result, this approach requires jail officers to have excellent communication and interpersonal skills. One of the challenges of implementing any type of major innovation, however, is that staff members might not support this change, and it is possible that jail officers in some places undermined the implementation of the entire new-generation philosophy.[73]

A cornerstone of jail safety is inmate **classification**. In its simplest definition, classification refers to the placement of an inmate in a facility or living unit that best matches his or her need for security. For instance, a gang member charged with a violent offense and who has a previous history of escapes has a different security requirement than does a middle-aged first-time offender charged with driving under the influence. The first inmate would be assigned to a living unit that has a higher degree of security. He may, for instance, be placed in a single cell in a living unit and may rarely leave the unit, and never without an officer's supervision. The second inmate, by contrast, may be placed in a dormitory set-

Corrections in the Real World

In May 2007, the *New York Times* reported the California practice of jails allowing inmates to purchase upgraded jail conditions for $75 to $127 per day. Inmates who "upgraded" are housed in cleaner and quieter cells and are allowed a greater number of possessions, including iPods and cell phones in some jails. In addition to being more comfortable, violent offenders cannot participate in these programs, so these wealthier inmates are safer as well.

ting where unsupervised inmates regularly leave the unit for recreation or vocational programs (such as working in the jail's kitchen or laundry).

We address classification, including the instruments that officers use to assess the degree of risk that an inmate poses to others, in Chapter 12. Generally speaking, inmates are classified based on prior criminal record (e.g., including convictions for violent crimes or escaping custody), current offense (e.g., whether or not it is violent), prior behaviors in jail, and involvement in gangs.

The federal government recently conducted a national-level study of homicides in jails. This research found that the jail homicide rate decreased by 40 percent from 1983 to 2002 to three deaths per every 100,000 jail inmates.[74] Two possible reasons for less violence were: (1) the increased ability to supervise inmates, and (2) the use of classification to place inmates in the most appropriate living units. Larger jails have a better capability to separate inmates, and this may also lead to a safer environment. A jail with a few dozen beds might place all of its inmates, regardless of classification, in a common living unit, and this is one reason that smaller facilities have historically been more dangerous places for inmates.

Community-Based Jail Programs

The threefold growth in jail populations forced many jurisdictions to make costly expansions to their facilities. To reduce the impact of overcrowding, jail administrators are increasingly dependent on community-based programs that provide an alternative to jail incarceration. A recent BJS study reported that some 71,905 persons were serving their jail sentence in the community.[75] Because jail populations grew more rapidly than officials could add beds, sheriffs and jail administrators developed community-based alternatives to jail incarceration. The population of jail offenders in the community more than doubled between 1995 and 2005, and the most common programs are listed in **Table 7–2**.[76] (Of the programs in Table 7–2, electronic monitoring, home detention/confinement, day reporting, and community service are covered in Chapter 9.)

Of jail inmates living in the community, the largest group is serving pretrial supervision. These programs reduce jail overcrowding by releasing detainees who have positive social supports and are likely to show up for their court dates. Jail employees working in these units gather information about defendants (including criminal history information from other jurisdictions) and determine whether there are family or community resources not considered by the judge (e.g., a stable residence). Persons on pretrial supervision must abide by a number of conditions, including meetings with a pretrial supervision worker on a regular basis, submitting to drug or alcohol testing, and abiding by a curfew, or attending substance abuse counseling.

Table 7–2 Jail Inmates Supervised Outside the Facility

Weekender programs	16.4%
Electronic monitoring	16.6%
Home detention	1.7%
Day reporting	9.4%
Community service	18.7%
Pretrial supervision	20.4%
Other work programs	10.2%
Treatment programs	3.1%
Other	3.6%

Source: P. M. Harrison and A. J. Beck, *Prison and Jail Inmates at Midyear, 2005* (Washington, DC: Bureau of Justice Statistics, 2006).

One popular alternative sentencing option is the weekender or **intermittent jail sentence**. Designed for offenders who work during the week, these programs incarcerate offenders during the weekends, often admitting inmates Friday afternoons and releasing them Monday mornings. This continues until their jail sentence has been served. These programs are popular with inmates but create extra work for jail officers because of frequent admissions and discharges and the difficulty of placing these inmates in appropriate housing units each weekend. In addition, weekends tend to be busy times in jails because of the high number of arrestees. Despite these challenges, weekender incarceration is cited as an example of a win-win approach to justice, where offenders keep their jobs and are held accountable for their behavior.

Some sheriffs departments offer community service programs where unemployed inmates spend the day doing community service (e.g., road maintenance or county chores such as collecting litter from roadways) and go home every evening. These programs have different names, such as Sheriff's Work Alternative Program, but the common element is that such programs provide low-risk inmates with work to keep them constructively occupied, and they do not fill a jail bed. Other inmates may be given a furlough from the jail (also known as a temporary release—see Chapter 13) to participate in a residential drug or alcohol treatment program.

These community-based programs are often operated out of necessity because the alternative is a costly jail expansion. These programs are also examples of jail administrators having to respond to forces beyond their control because jails are often at the mercy of the entire justice system. The activities of the police determine jail populations, and the number of inmates also depends on the case processing time of the courts. Moreover, jail overcrowding is a result of state prison officials refusing to accept prisoners because of overcrowding.

Conclusion

Jails are an integral part of the criminal justice system and have been described as both the gatekeeper (the place through which most arrestees will pass) and the transfer point for law enforcement, juvenile corrections, mental health agencies, and state or federal prisons. There are some 3,300 jails in the United States, and with the exception of some larger facilities in cities such as Chicago, New York, and Los Angeles, researchers have little data on what occurs in many of these places. Local sheriffs operate more than 70 percent of these facilities. These sheriffs are elected officials who have, historically, used their power to hire political supporters to manage or work in the jail. These practices resulted in serious problems because political appointees often had no experience or interest in corrections. The threat of litigation, however, has forced jail administrators to hire more professional staff, establish written policies, and amend procedures that were discriminatory or abusive to inmates. In addition, about two-thirds of states have statewide standards that facilities are required to follow, while other jails are accredited through agencies such as the American Correctional Association, American Jail Association, and the National Commission on Correctional Health Care (for medical services).

In addition to the political nature of some jail operations, one of the greatest obstacles to jail operations is local funding. Increasing jail populations raised the total cost to local counties from $3 billion in 1982 to $18.6 billion in 2003, an increase of 519 percent.[77] In a recent study of small jails, for instance, budget problems were cited as the foremost problem confronting jail administrators.[78] There are a number of remedies to this challenge, including creating regional jail operations that take advantage of economies of scale to reduce costs. In addition, state-operated jail systems might be a positive change, but this alternative, along with increasing the number of regional jails, would require local governments and sheriffs to surrender political and economic power, something that is unlikely to occur in the near future.

Because jail inmates tend to reside in these facilities for only a short period of time, conditions are apt to be more volatile than they are in prisons. Rates of interpersonal violence are liable to be high, although these facilities report fewer murders than they did two decades ago.[79] The challenges confronting jail administrators today are similar to the problems facing all elements of local criminal justice systems. The influence of local government politics; being dependent on local municipalities, cities, or counties for funding; the need for more professional staff; and the growth in litigation were experienced in all elements of criminal justice systems. These changes may, however, have had a greater impact on local jails because they were so dysfunctional several decades ago. Today, many of these facilities are more professionally operated and staffed, and this has improved conditions.

READY FOR REVIEW

- The practice of placing a jail in each county was established in England in 1166, and many of these early facilities were makeshift structures that were operated by corrupt keepers who charged inmates for their incarceration.
- Jail development in the United States closely followed the English model, and until the First World War most were operated on the fee system (where sheriffs would be paid a daily rate to keep an inmate and were allowed to keep any funds remaining after the inmate was fed and housed).
- There is a jail in almost every U.S. county, and the vast majority are operated by local sheriffs. These facilities range in size from 3 to almost 20,000 beds, and the smaller facilities are gradually disappearing.
- There were about 50 privately operated jails in 1999, and these operations contract with a city or county to provide all of the jail services.
- There is a long history of private operators delivering goods and services in corrections. Private operators can deliver some services more efficiently than government can, but there is concern that punishment should not be delivered by corporations whose sole interest is making a profit.
- Regionalization occurs when a number of smaller jails consolidate their jail operations in a single facility. Larger jails can operate more cost effectively, take advantage of new-generation jail designs, and are often safer for inmates and staff.
- There are about 70 jails operated by tribal governments in Indian Country. These jails are usually small and suffer from the same disadvantages of other small jails (e.g., funding and overcrowding) and find it difficult to retain and recruit good officers.
- The states of Alaska, Connecticut, Delaware, Hawaii, Rhode Island, and Vermont operate integrated jail–prison systems where jail services are delivered by the state rather than counties.
- The federal government operates detention facilities in urban areas to hold federal detainees. In the rest of the nation, the federal government pays local jails to hold its detainees temporarily.
- There are increasing numbers of persons with mental illness, severely ill inmates, frequent fliers, long-term inmates, and security threat groups being held in jail. Each of these populations poses challenges for jail administrators and officers—either because of the costs of holding these special-needs inmates or their disruptive or aggressive behaviors in the jail.
- The primary causes of jail litigation are medical care, use of force, personal injury, loss or damage to property, and inmate-on-inmate violence.
- Most jails today are built using the new-generation design, which enables a greater degree of supervision. Although most jurisdictions have used this jail design, they have often failed to implement the new-generation

philosophy (which strives to improve the problem-solving skills of the inmates).

- As a result of jail overcrowding, many jurisdictions now provide community-based jail programs to inmates thought to be at a low risk of offending.

KEY TERMS

classification An analysis of the risks that the inmate poses (and in some jurisdictions, the inmate's needs). Classification enables jail staff to place an inmate in a facility or housing unit that best meets the inmate's risk to the staff and other inmates and level of functioning

consent decree An agreement where a correctional agency avoids formal litigation by agreeing to change its policies or procedures under the supervision of the court

direct supervision New-generation jails enable an officer to supervise offenders continuously either from a control room overlooking the living unit or a desk located in the unit

frequent fliers Inmates who are admitted to local jails more than 20 times in the previous five years—these arrestees are usually charged with minor crimes and many have mental illness or alcohol or drug problems (also called habitual misdemeanor offenders)

gaols The historical English name for local jails

gatekeeper Jails are said to be gatekeepers (the entry point through which everybody passes) and transfer point for the criminal justice system

general population Refers to the jail or prison population that is not in a specialized housing or living unit

habitual misdemeanor offenders Another term for frequent fliers

hulks Jail inmates awaiting transportation from England were held in the hulls of warship and merchant vessels on the Thames River

informal social control The ability of the community to regulate behavior through encouragement or informal sanctions (such as scorn)

intermittent jail sentence Also called a weekender sentence where jail inmates are employed or attend school during the week, and then return to jail each weekend to serve their sentences

intermittent supervision When officers observe inmates on an infrequent or intermittent basis because of the architectural design of a facility

keepers The historical term for jail officers

living unit A self-contained area in a new-generation jail where inmates reside, eat their meals, engage in recreational activities, and sometimes attend classes

malingering A term used to describe when somebody pretends to be ill to receive attention or to escape an unpleasant task

new-generation jail An architectural design that enables officers to supervise inmates more efficiently

physical plant A term used to describe the physical structure and grounds of a correctional facility

pretrial detainees Someone held in jail until the person's next court appearance

privatization When a jail or prison system sells part or all of its operation to private enterprise (e.g., some private operators provide all of a jail's medical services, while other jails are entirely operated by corporations)

regionalization When a number of smaller jails combine their operations into a single facility to consolidate their operations

security threat groups A term used to describe a group of inmates (usually three or greater members) who have a common name, symbols, or signs. Similar to a jail or prison gang

YOU ARE THE CORRECTIONS PROFESSIONAL SUMMARY

1. What is your best response to this problem, especially given that this inmate is prone to malingering? By ignoring the inmate's concerns, even though you might believe that he may be malingering, the inmate may have a heart attack and die (which is a worst-case scenario). Not only would you be morally accountable for the inmate's death, both you and the agency could potentially be held civilly responsible for damages. Without an accurate assessment of the inmate's condition by a medical professional, you are betting your career on your judgment.
2. Whatever your decision, what are the risks of the action you choose? (Think about all of the potential risks—inmate health as well as the risks of transporting an inmate—including the possibility of escape). Every time an officer takes an inmate into the community, there is a risk of escape—even though the prisoner would be transported in handcuffs and possibly shackles (leg irons)—or involvement in traffic accidents (especially at night). Many officers are reluctant to take inmates to the emergency room because these visits often take several hours. By not taking action, however, you risk the inmate dying.

NOTES

1. P. M. Harrison and A. J. Beck, *Prisoners in 2005* (Washington, DC: Bureau of Justice Statistics, 2005), 1.
2. P. M. Harrison and A. J. Beck, *Prison and Jail Inmates at Midyear, 2005* (Washington, DC: Bureau of Justice Statistics, 2006), 1.

3. J. Irwin, *The Jail: Managing the Underclass in American Society* (Berkeley, CA: University of California Press, 1985).
4. Federal Bureau of Investigation, *Crime in the United States, 2005* (Washington, DC: FBI, 2006).
5. G. L. Mays, "Jails: Purposes, Populations, and Problems," in *Encyclopedia of Criminology*, ed. R. A. Wright and J. M. Miller (Hampshire, UK: Taylor and Francis, 2004).
6. E. C. Casella, "Prisoner of His Majesty: Postcoloniality and the Archaeology of British Penal Transportation," *World Archaeology* 37(2005): 453–467.
7. P. B. Nutting, "The Madagascar Connection: Parliament and Piracy, 1690–1701," *American Journal of Legal History* 22(1978): 202–215.
8. R. B. Pugh, *Imprisonment in Medieval England* (Cambridge, England: Cambridge University Press, 1968), 101.
9. G. Durston, "Magwitch's Forbears: Returning from Transportation in Eighteenth-Century London," *Australian Journal of Legal History* 9(2005): 137–158.
10. Pugh, *Imprisonment in Medieval England,* 79.
11. A. Griffiths, *Memorials of Millbank* (London: Chapman and Hall, 1884).
12. N. King, "The Origins of Felony Jury Sentencing in the United States," *Chicago Kent Law Review* 78(2003): 937–993.
13. D. S. Greenberg, *Persons of Evil Name and Fame: Crime and Law Enforcement in the Colony of New York, 1691–1776* (doctoral dissertation, Cornell University, 1974), 270.
14. J. F. Fishman, *Crucibles of Crime: The Shocking Story of the American Jail* (Glen Ridge, NJ: Patterson Smith, 1923), 13.
15. L. N. Robinson, *Penology in the United States* (Philadelphia: John C. Winston, 1921).
16. J. J. Stephan, *Census of Jails, 1999* (Washington, DC: Bureau of Justice Statistics, 2001).
17. P. M. Harrison and A. J. Beck, *Prison and Jail Inmates, 2004* (Washington, DC: Bureau of Justice Statistics, 2005).
18. L. M. Maruschak, *HIV in Prisons, 2005* (Washington, DC: Bureau of Justice Statistics, 2007), 1.
19. American Jail Association, *Who's Who in Jail Management* (Hagerstown, MD: American Jail Association, 2003).
20. C. J. Mumola, *Suicide and Homicide in State Prisons and Local Jails* (Washington, DC: Bureau of Justice Statistics, 2005), 5.
21. G. L. Mays and J. A. Thompson, "Mayberry Revisited: The Characteristics and Operations of America's Small Jails," *Justice Quarterly* 53(1988): 421–440.

22. K. E. Kerle, *American Jails: Looking to the Future* (Boston: Butterworth-Heinemann, 1998).
23. J. J. Stephan, *Census of Jails.*
24. T. D. Minton, *Jails in Indian Country* (Washington, DC: Bureau of Justice Statistics, 2005).
25. Office of the Inspector General, *Neither Safe nor Secure: An Assessment of Indian Detention Facilities* (Washington, DC: Office of the Inspector General, 2004).
26. Stephen, *Census of Jails.*
27. D. Osborne and T. Gaebler, *Reinventing Government: How the Entrepreneurial Spirit Is Transforming the Public Sector* (New York: Plume, 1993).
28. W. J. Sabol, T. D. Minton, and P. M. Harrison, *Prison and Jail Inmates at Midyear 2006*, 2007, 4.
29. C. W. Thomas, "Recidivism of Public and Private State Prison Inmates in Florida: Issues and Unanswered Questions," *Criminology and Public Policy* 4(2005): 89–100.
30. R. K. Davis, B. K. Applegate, C. W. Otto, R. Surette, and B. J. McCarthy, "Roles and Responsibilities: Analyzing Local Leaders' Views on Jail Crowding from a Systems Perspective," *Crime and Delinquency* 50(2004): 458–482.
31. National Institute of Corrections Information Center, *Regional Jails* (Boulder, CO: National Institute of Corrections Information Center, 1992), 1, www.nicic.org/pubs/1992/010049.pdf.
32. Mays and Thompson, "Mayberry Revisited," 435.
33. Fishman, *Crucibles of Crime.*
34. M. Welch, *Detained: Immigration Laws and the Expanding I.N.S. Jail Complex* (Philadelphia: Temple University Press, 2002).
35. Harrison and Beck, *Prison and Jail Inmates at Midyear,* 7.
36. Harrison and Beck, *Prisoners in 2005.*
37. D. Leach, "Mission Creep and the Role of the Jail in Public Health Policy," *National Institute of Corrections Large Jail Network Exchange*, 2004, 37–44.
38. Harrison and Beck, *Prison and Jail Inmates at Midyear.*
39. D. J. James and L. E. Glaze, *Mental Health Problems of Prison and Jail Inmates* (Washington, DC: Bureau of Justice Statistics, 2006).
40. K. Parker, "Pregnant Women Inmates: Evaluating Their Rights and Identifying Opportunities for Improvements in Their Treatment," *Journal of Law and Health* 19(2006): 259–295.

41. Harrison and Beck, *Prison and Jail Inmates at Midyear.*
42. A. L. Pastore and K. Maguire, *Sourcebook of Criminal Justice Statistics, 2003*, www.albany.edu/sourcebook/pdf/t211.pdf.
43. J. Reiman, *The Rich Get Richer and the Poor Get Prison* (Boston: Allyn & Bacon, 2007).
44. Harrison and Beck, *Prison and Jail Inmates at Midyear,* 8.
45. P. Elrod and M. T. Brooks, "Kids in Jail: 'I Mean You Ain't Really Learning Nothing [Productive],' " in *Convict Criminology,* ed. S. C. Richards and J. I. Ross, 325–346 (Belmont, CA: Thompson/Wadsworth, 2003).
46. Harrison and Beck, *Prison and Jail Inmates at Midyear,* 8.
47. James and Glaze, *Mental Health Problems.*
48. R. Ruddell, "Jail Interventions for Inmates with Mental Illnesses," *Journal of Correctional Health Care* 12(2006): 118–131.
49. M. Chandler Ford, "Frequent Fliers: The High Demand User in Local Corrections," *Californian Journal of Health Promotion* 3(2005): 61–71.
50. R. Ruddell, "Mayberry Revisited: The Challenges Confronting America's Small Jails" (paper presented at the annual meeting of the American Society of Criminology, Toronto, Canada, November 17, 2005).
51. M. Gladwell, "Million-Dollar Murray," *The New Yorker*, February 13, 2006, 96, www.newyorker.com/fact/content/articles/060213fa_fact.
52. S. Eskenazi, S. "75 Hard-Core Alcoholics to Be Offered Apartments," *Seattle Times,* December 15, 2005, www.seattletimes.nwsource.com.
53. Mumola, *Suicide and Homicide,* 1.
54. Infectious Diseases in Corrections Report. "News and Literature Reviews: HIV Costs Have Decreased," *Infectious Diseases in Corrections Report* 9(2006): 8.
55. R. Ruddell, "Long-Term Jail Populations: A National Assessment," *American Jails,* March/April, 22–27, 2005.
56. T. H. Cohen and B. A. Rainville, *Felony Defendants in Large Urban Counties, 2002* (Washington, DC: Bureau of Justice Statistics, 2006).
57. Cohen and Rainville, *Felony Defendants,* 36.
58. Harrison and Beck, *Prisoners in 2005.*
59. R. Ruddell, S. H. Decker, and A. Egley, "Gang Interventions in Jails: A National Analysis," *Criminal Justice Review* 31(2006): 33–46.
60. Ruddell et al., "Gang Interventions in Jails."
61. Corrections Standards Authority. *Jail Profile Survey Report: 2006 Third Quarter Results* (Sacramento, CA: Corrections Standards Authority, 2006).

62. Harrison and Beck, *Prison and Jail Inmates at Midyear.*
63. A. R. Beck, *Jail Bloating: A Common but Unnecessary Cause of Jail Overcrowding* (Kansas City, MO: Justice Concepts Incorporated, 2001), http://justiceconcepts.com/jail%20overcrowding.pdf.
64. M. Schlanger, "Inmate Litigation: Results of a National Survey," *National Institute of Corrections Large Jail Network Exchange*, 2003, 1–12.
65. Corrections Standards Authority, *Jail Profile Survey Report.*
66. California Department of Corrections and Rehabilitation, "Second Quarter 2006 Facts and Figures," www.cdcr.ca.gov/divisionsboards/csa/2002constructionauditguide6thed/aoap/FactsFigures.html.
67. Schlanger, "Inmate Litigation," 3.
68. M. Crawley and J. King-Greenwood, "Mother Asks Why After Her Baby, Born in Jail, Dies," *Tampa Tribune*, March 13, 2004.
69. W. N. Welsh, *Counties in Court: Jail Overcrowding and Court-Ordered Reform* (Philadelphia: Temple University Press, 1995), 3.
70. Schlanger, "Inmate Litigation."
71. C. Tartaro, "Watered Down: Partial Implementation of the New Generation Jail Philosophy," *Prison Journal* 86(2006): 284–300.
72. K. Kerle, *Exploring Jail Operations* (Hagerstown, MD: American Jail Association, 2003), 240.
73. L. L. Zupan, *Jails: Reform and the New Generation Philosophy* (Cincinnati, OH: Anderson Publishing, 1991).
74. Mumola, *Suicide and Homicide,* 1.
75. Harrison and Beck, *Prisoners in 2005,* 1.
76. Harrison and Beck, *Prison and Jail Inmates at Midyear,* 7.
77. K. A. Hughes, *Justice Expenditure and Employment in the United States, 2003* (Washington, DC: Bureau of Justice Statistics, 2006), 3.
78. R. Ruddell and G. L. Mays, "Rural Jails: Problematic Inmates, Overcrowded Cells, and Cash-Strapped Counties," *Journal of Criminal Justice* 35(2007): 81–101.
79. Mumola, *Suicide and Homicide.*

8

Diversion and Probation

Chapter Objectives

- Define the terms *diversion* and *probation*, and summarize the characteristics of offenders found in these programs.
- Summarize the historical origins of both diversion and probation.
- Describe the roles of staff members in diversion and probation, with particular attention to the potential for conflict between the law enforcement and rehabilitative roles; state how the balance between these roles has changed over time.
- Demonstrate understanding of the following stages in the diversion process: (1) case and service selection, (2) initiation and implementation of programming, and (3) program termination.
- Illustrate the advantages and disadvantages of diversion, including potential for net widening.
- Summarize the effectiveness of diversion at saving money and controlling recidivism.
- Describe the process by which people are placed on probation, including understanding of the presentence investigation and the probation order.
- Understand the process by which probationers are assessed, supervised, and receive services.
- Describe the process by which probation orders are terminated, including understanding of probation revocation.
- Summarize important changes that have been taking place in probation.

- Provide illustrations of and critically analyze controversial issues related to probation.
- Summarize the relationship between probation and recidivism.

CASE STUDY

You are a probation officer working in an agency where the chief probation officer (your supervisor) has recently indicated at a staff meeting that the state's governor wants a statewide "crackdown" on probation violations, especially violations involving substance abuse. She explains that this is a result of recent negative publicity surrounding two probationers who, while under the influence of drugs, committed murders; these probationers were under supervision in separate parts of the state. At the same meeting, your supervisor indicates that she is getting pressure from the state Department of Corrections to hold off on as many revocations as possible because of prison overcrowding and limited space to house probation violators.

A 25-year-old probationer on your caseload has now completed 3.5 years of a 4-year probation sentence that was initially handed out for felony theft. During this period, the probationer has, to your knowledge, complied with all conditions of probation except one. The exception took place at the one-year mark in the sentence when the probationer tested positive for drug use. At that time, based on your recommendation to the judge, the probationer was continued on probation, with increased reporting requirements and mandatory substance abuse counseling. The probationer successfully completed the counseling and reporting as required. However, the probationer has just tested positive again at his most recent drug test. No other problems (e.g., family, job) are apparent; except for drug use, the probationer is displaying signs of success. Should you recommend revocation of probation followed by incarceration?

1. Can revocation and incarceration in jail or prison be expected to improve this problem? Could it possibly make matters worse?
2. What alternatives to revocation are available, given all the political rhetoric and pressure?

Introduction

This chapter deals with two of the most frequently used responses to adult and juvenile offenders—probation sentences and diversion from system processing. It is unnecessary to place most offenders (including many felons) in jail or prison, and even if it was desirable to do so, there are not enough resources available.

Were it not for diversion, probation, and the other community-based placements discussed in Chapters 9 and 13, judges would be confronted with far more cases than they could ever hope to handle solely by sentencing offenders to institutions. In fact, over the last 25 or so years, increases in both jail and prison crowding have led to placement of more and more felons on probation, something known as **felony probation**. Concomitantly, more and more misdemeanor offenders, who might formerly have been put on probation, have been diverted from the system.

This point was illustrated earlier in Figure 2–2 (see Chapter 2), which displayed the large increases between 1980 and 2005 in both adult institutional (prison and jail) and community correctional (probation and parole) populations. At the end of 2005, approximately 70 percent of the more than 7 million adults under one of these four forms of correctional supervision were either on probation or parole.[1] These figures do not even take into account the large numbers of persons diverted from the system. Clearly, then, community programs are essential to the operation of the entire criminal justice system.

Definitions and Numbers

Persons who are convicted of misdemeanors and less-serious felonies are frequently sentenced to probation, especially if they have no or minimal prior legal record. **Probation** is a sentence in which the offender is permitted to remain in the community for all or most of the term as long as he or she complies with certain rules and conditions imposed by the court. A probation agency supervises the extent of compliance and makes recommendations to the court in the event of noncompliance. The agency also delivers or coordinates services meant to promote offender betterment (e.g., counseling, remedial education, employment assistance). As noted in Chapter 2, more than 4,162,500 adults were on probation by the end of 2005, representing an average increase of 2.5 percent per year since 1995.[2] Half of these probationers had been convicted of committing felony offenses (refer to Chapter 2 for a breakdown by specific type of offense).

Probation is sometimes thought of as a type of diversion because probation involves channeling offenders away from incarceration and into alternative programming in the community. Unlike probation, diversion is not a formal sentence. In fact, diversion often involves channeling offenders away from formal sentences altogether, including probation sentences, and offenders are commonly placed on probation if they fail to complete diversion agreements.

Diversion refers to systematic efforts to move less-serious cases away from any further penetration into the criminal justice process than they have already experienced, often into alternative programs that are meant to address identifiable needs better (e.g., drug treatment, mental illness, family counseling). A case can be diverted anytime after initial contact with the police up to formal sentencing by the judge. As with probation, a jurisdiction's laws or policies often specify which offenders and offenses are eligible for diversion. Also, persons who are diverted must comply with certain rules and conditions; otherwise, criminal justice processing can resume at the point where it was suspended when the case was diverted.

Although probation agencies are administered at the federal level, as well as at the state and local levels, it is common for diversion programs to be locally operated. This, combined with the fact that diversion is not a formal sentence, means it is much more difficult to obtain data on the number of diverted cases and their characteristics. Instead of the type of probation statistics reviewed earlier that are furnished by the federal government, for diversion we have to rely on studies of individual programs.

Such studies show that diversion is quite common, especially in juvenile justice. For instance, a study of a Missouri juvenile court found that nearly 92 percent of all new cases referred to court over a six-month period received some kind of diversion in lieu of formal prosecution. Younger juveniles, and those referred to court for nonfelony offenses, were more likely to receive diversion.[3] A study of drug courts (which are also discussed in Chapter 5) reported that 1,727 individuals were diverted into drug court programs for felons between 1995 and 2002 in Virginia.[4] There were more than 1,500 drug court programs operating across the country at the end of 2005.[5] Similarly, specialized courts are becoming increasingly popular for persons with mental health problems.[6]

Historical Overview of Probation and Diversion

As an informal practice, diversion is as old as criminal processing itself. Milder alternatives to formally imposed sentences have been in use for centuries.[7] However, when viewed as concepts that are formally recognized and systematically implemented in the system, the history of probation is longer than that of diversion.

Probation evolved out of various English common law practices. One especially relevant practice, **judicial reprieve**, permitted a convicted person to petition the court to suspend the main sentence (e.g., corporal punishment or execution) so that the person could seek a pardon for the offense. Eventually reprieves came to be granted contingent upon whether the person's future behavior complied with the wishes of the court, which for practical purposes is very similar to modern probation. A less direct forerunner of probation was the common law practice of **benefit of clergy** (also discussed in Chapter 10). Originally, during medieval times, this practice allowed members of the clergy to ask judges to be lenient in sentencing them. These requests came in the form of biblical scripture readings to demonstrate that the person was in fact a cleric, for most nonclerics were illiterate. Over time, however, the benefit of clergy was extended to nonclerics.[8] Other antecedents of probation include bail and recognizance (see Chapter 5), both of which entail a person remaining free from confinement depending on behavior (e.g., appearing for scheduled court hearings and avoiding illegal activity).

The direct origins of modern probation are found in the United States, and specifically, with the efforts of Bostonian shoe manufacturer and philanthropist, **John Augustus** (1785–1859), who is considered the founder of probation. Augustus made it a practice to regularly visit the lower courts in Boston and observe proceedings. In the 1840s, he began putting up bail for persons convicted of relatively minor crimes, especially drunkenness. Augustus convinced judges to defer sentences and allow time for him to supervise and work with offenders in the community. He would monitor offenders' activities, assist them with housing and employment, and take whatever other measures he deemed necessary to help them avoid return to crime. After a period of time, he would report back to the court on the person's progress so that the judge could decide whether to proceed with formal sentencing. Augustus discovered that few of these early "probationers" failed to follow the conditions he and the court set.[9]

Citizen volunteers eventually began assisting Augustus and replicating his activities. These efforts ultimately resulted in Massachusetts passing the first probation legislation in 1878. This particular legislation was geared toward juveniles, but the stage was set for the development of probation in adult corrections as well. New York enacted the first adult probation legislation in 1901, and probation at the federal level was organized in the 1920s. By the mid-1950s, all states had legislation allowing for both juvenile and adult probation.[10] Since the mid-1800s, then, probation in the United States has evolved from the pioneering volunteer work of Augustus into the most commonly used sentence for adult and juvenile offenders, with agencies existing at all levels of government and employing thousands of staff.

By contrast, the history of formal, systematic diversion dates only to the 1960s. During this era, some groups expressed concern that traditional court and correctional processing might make some offenders more, rather than less, inclined toward criminal behavior by labeling them as criminals and leading them to see themselves as such.[11] Additionally, many units of government had an incentive to curtail the financial costs of criminal justice because of revenue expenditures associated with the Johnson administration's War on Poverty and Great Society initiatives.[12] Hence, the idea took hold of diverting relatively minor offenders away from more stigmatizing and often more expensive procedures into programs potentially more capable of helping them avoid further criminal behavior.

Many early diversion programs focused on providing employment assistance and substance abuse treatment. One illustration of the latter is Treatment Alternatives to Street Crime (TASC). The federal government launched this program in the early 1970s to divert arrested substance abusers away from traditional criminal processing into appropriate treatment where their progress toward recovery could be monitored. Designed for offenders whose crimes were thought to arise from drug use, many TASC programs were funded by the federal Law Enforcement Assistance Administration (LEAA). In fact, one source estimated that LEAA spent more than $12 million during the 1970s funding diversion programs for adults and juveniles, including but not limited to TASC. However, with the advent of the "get tough" movement in sentencing and corrections during the 1980s, federal spending on such programs was curtailed.[13]

Staff Roles in Probation and Diversion Programs

Although outnumbered by staff working in correctional institutions and jails, thousands of professional staff members work in the nation's numerous probation and diversion programs for adult and juvenile offenders. Probation officers typically must hold a bachelor's degree or, in some places, have appropriate work experience to substitute for such a degree. The qualifications for working in diversion programs are much more variable across jurisdictions and depend on the nature of the program (e.g., a program providing substance abuse counseling might require staff with appropriate certifications). Additionally, in many jurisdictions, both probation and diversion staff must undergo a period of in-service training prior to starting duties and then train periodically thereafter. Although entry pay levels vary tremendously, depending on the cost of living in a region, these levels generally fall in the low to moderate range. For instance, across the nation it is pretty uncommon to find entry pay levels for probation officers that exceed $40,000 per year; by the same token, it is quite common for experienced officers to earn considerably more, especially if they hold administrative positions.

Probation and diversion staff work with diverse types of people and perform diverse tasks. These tasks can range from routine information collection, report writing, and counseling to periodic crisis intervention and investigation of serious criminal activity. A fundamental task is to ensure that offenders are following rules and participating in services in the manner they have been ordered and agreed to do. Another fundamental task is to communicate information about the offender's progress or lack thereof to judges and other officials so that appropriate decisions can be made about whether the offender should be deemed successful or unsuccessful. Effective performance of such duties requires knowledge and skills in several key areas, including (1) the ability to communicate professionally both orally and in writing, (2) counseling-related interpersonal skills, (3) the ability to problem solve and make objective and ethical decisions that are substantiated by evidence rather than conjecture, as well as (4) organizational and time management skills. The ability to respond effectively to diversity is also necessary because, across the United States, the clients of diversion and probation programs are increasingly diverse along gender, racial, socioeconomic, and cultural lines.

It is common for probation and diversion program staff members to face tension between two potentially competing roles. First, these staff are responsible for what is, in effect, a law enforcement role, which can promote suspicion and cynicism among offenders. Staff have to make sure offenders comply with the conditions of the probation or diversion agreement. This involves supervising and monitoring offenders' activities, investigating possible instances of noncompliance, and enforcing consequences for noncompliance. Second, however, staff have a helping or rehabilitative role to assist offenders in refraining from further legal problems. This role involves establishing the rapport needed to provide counseling and other services directly to offenders. It also entails functioning as a resource broker to (1) accurately identify offenders' problems and needs and (2) see that offenders are matched with and referred to community agencies that can best address problems and needs. West and Seiter describe this potential tension as existing on a "casework to surveillance continuum," implying that staff have to be able to move between and balance the two functions. West and Seiter found that although staff in their study spent more time performing casework duties, staff perceived themselves as more concerned with surveillance.[14] This self-perception is indicative of a change in goal orientation that many community correctional programs have experienced over the last two to three decades.

The roles just described reflect two important goals of diversion, probation, and other similar community programs. The first goal is to protect the community from crime, which implies the law enforcement role. The second is to assist offenders in overcoming problems, such as substance abuse and unemployment, believed to undermine law-abiding behavior patterns. This goal implies the rehabilitative role. Traditionally, community correctional programs emphasized rehabilitation. However, with the broader movement toward a "get tough" approach in criminal justice starting in the 1970s, and faced with ever-increasing numbers of cases, these programs began emphasizing other goals such as community protection, deterrence, and punishment. A study published at the end of the 1980s found that the attitudes of community corrections staff had undergone a shift away from rehabilitation and toward law enforcement.[15]

One can argue that diversion and probation staff should have a single, well-defined role (or a least compatible roles if there are multiple ones) so that the potential for role conflict and stress is reduced, and so that it is easier to accomplish something worthwhile with offenders. Realistically, however, it seems unlikely that these jobs will ever involve roles that are entirely compatible. As such, it seems important to strive toward balancing the roles in a consistent and complementary manner when working with offenders so that supervision and enforcement functions support and facilitate helping activities, which reduce the need for enforcement.

A Closer Look at Diversion

As mentioned earlier, a wide variety of diversion programs exist in the United States. Nevertheless, diversion can be divided into two general types. These types are discussed in detail below.

Types of Diversion

Diversion is used most frequently in juvenile justice, but it is also used regularly in adult criminal justice. Irrespective of whether it occurs in the juvenile or adult system, there are two basic types of diversion. **Pretrial diversion** takes place sometime prior to a formal plea or finding of guilt. Instead of making an arrest, the police might divert a drug addict to a detoxification center, or they might divert a juvenile to his or her home and notify the parents that more formal action will be taken if the juvenile causes further problems. Or a prosecutor might divert a domestic disorder case to family counseling in lieu of filing charges; alternatively, charges could be filed and then later dropped if the conditions of diversion are met. In the adult system, a pretrial diversion agreement is frequently incorporated as part of a bail or release on recognizance arrangement, such as when a person suffering from mental problems is diverted from jail into treatment. Diversion agreements are common at the intake stage of the juvenile justice process (see Chapter 14).

Posttrial diversion takes place following admission or finding of guilt and is used in lieu of formal sentencing. For example, a shoplifter could be ordered to complete a counseling program, make restitution, or perform community service rather than being sentenced to a brief jail term, or perhaps with the jail term imposed but suspended contingent on behavior. A juvenile who has admitted guilt could be diverted to teen court to receive a "sentence" from his or her peer group.[16]

As the preceding illustrations reveal, diversion can be administered by a wide variety of agencies. Examples include police departments, prosecutorial offices, defense agencies, courts, jails, and probation agencies, as well as both publicly and privately operated social service agencies that are not part of the justice system *per se*. Indeed, one of the advantages to this variation is the potential for matching the service to the specific needs of the case.

Diversion programs are used mainly with offenders who do not have serious charges or extensive histories of offending (especially first offenders). These programs are particularly common with cases involving domestic disturbance, substance abuse, and mental illness. However, there are many varieties of diversion programs besides those emphasizing family counseling, substance abuse intervention, and psychological or psychiatric treatment. Additional examples include programs emphasizing remedial education, vocational training, job placement and retention, restitution payments to the victim, service work performed for the community, and reconciliation of disputes with the victim or community.

The Diversion Process

No matter which type of diversion program is used, the process of diversion generally involves three steps: selection of cases and services, program intiation and implementation, and program termination. These stages are described in detail in the following section.

Selection of Cases and Services

Two of the most important considerations in the diversion process involve correctly identifying cases for diversion and matching these with appropriate programming. Typically, cases must meet certain criteria defined by statute or policy to be eligible for diversion (e.g., nonviolent offense, minimal criminal background, voluntary agreement to participate). This is done to try to minimize risks to community security. By the same token, the offense and risks posed should not be so trivial that charges would simply have been dismissed had it not been for the availability of a diversion program (see the later discussion of net widening). In other words, cases should present at least minimal risk, and diversion should represent an alternative to further penetration into the system.

Additionally, it is important that the offender actually possess the needs being targeted by the program services selected. For instance, all too often in the past, offenders without documented substance abuse problems (especially juveniles) have been diverted into drug counseling programs on the erroneous assumption that "a little counseling cannot hurt," thereby ignoring potentially negative labeling effects.[17] An offender who lacks a substance abuse problem but needs to learn responsibility and accountability for actions might benefit more from a program emphasizing restitution or community service. So, as in the case of probation (see the next section), a crucial first step in the diversion process is careful screening and assessment of cases to promote correct classification of risk and need in relationship to program criteria.[18]

Program Initiation and Implementation

Some diversion programs, typically private ones and ones functioning outside the system, have discretion over which referrals are accepted. But others (such as one that is an arm of a court) have much less discretion. In addition, it is quite common for offenders to volunteer for participation as an alternative to undergoing the next stage of system processing. When a referral enters a program, the person usually undergoes an orientation concerning program expectations and conditions. The person must agree to follow these within a specified time period; written contracts are common. After programming begins, staff are responsible for monitoring the offender's compliance with conditions, reviewing progress toward goals, and seeing that services are delivered as intended. Persons who fail to meet expectations may receive warnings or have modifications made to their diversion agreements in an effort to encourage successful completion.

Program Termination

All officials involved in a particular diversion case (e.g., judge, prosecutor, program service provider) must eventually collaborate to determine whether the person has satisfactorily completed the program. Persons who are successful usually have their charges dropped without moving further into the system. Those who are deemed unsuccessful usually reenter the system to resume processing from whatever point they were when diversion took place. Those who commit new crimes while participating in diversion can have charges filed and be prosecuted anew.

Some Issues in Diversion

Diversion has a number of attractive features. It can help avoid the stigma and other possible negative effects of moving the offender deeper into the system. Likewise, there is potential for offenders to receive interventions geared toward their problems and needs more rapidly, interventions that might not have been provided had the case gone further into the system. Diversion thus gives officials such as police and prosecutors options and needed flexibility. If used properly, it can conserve time and money for more serious cases.

But diversion can be criticized on grounds that it often amounts to a "slap on the wrist," thereby negating any deterrent lesson the system's response to crime is supposed to teach. It can also be argued that diversion expands official discretion and threatens constitutional rights, such as the rights to counsel or trial. Diversion programs require admission of guilt (either explicitly or implicitly), and typically, program operations are subject to relatively few due process protections.

A related critique is that officials might sometimes use diversion as an excuse to intervene in cases where the circumstances really warrant no intervention. Officials may see some control over a case as preferable to none. When diversion programming is used with a case in which no further official action would have been taken had diversion been unavailable, **net widening** occurs. When the net of the system is cast wider in this fashion, more cases are pulled

Corrections in the Real World

Any police, court, or corrections professional who is responsible for helping make determinations about which cases to place in a diversion program, and who wishes to avoid net widening, should always ask himself or herself the following question: "What action would have been taken on this case had diversion not been an option?" If the answer is no or minimal action, the case is not appropriate for diversion. If the answer is that the case would have been processed further into the system, diversion will not promote net widening.

under its reach. Some of these will penetrate further into the system as a result of noncompliance with diversionary agreements. The danger is that diversion programs can become filled with cases that should simply have been dismissed, while cases that are truly appropriate for diversion are processed traditionally from the outset.

There are also questions about whether diversion really promotes cost savings. Obviously, cost savings are replaced by expenditures if net widening transpires. Potential savings also vanish when diversion is not successfully completed and cases reenter the system as a consequence; costs then expand to include both diversion and traditional processing. Even when diversion programs avoid net widening and are successful at curtailing violations of conditions, the costs of services, such as counseling, must be taken into account.

To What Extent Is Diversion Effective?

Evidence is mixed on the effectiveness of diversion at containing both costs and reoffending or **recidivism**. This is not surprising considering the enormous variation across programs and studies. Despite the caveats mentioned earlier about the cost savings of diversion, a careful cost-effectiveness study of an Oregon drug court concluded that the program resulted in a savings over alternative processing of $4,800 per participant.[19] But estimates of cost-effectiveness were more varied in a study of diversion from jail into substance abuse and mental health intervention.[20]

The effectiveness of diversion has also been studied by looking at recidivism, generally as measured by violations of program conditions, new arrests, or new convictions following diversion. Here too results have been mixed. A study of a juvenile diversion program in Lexington, Kentucky, reported that while most participants successfully completed the program, about two-thirds of them were rearrested within one year; program completers and noncompleters were equally likely to recidivate.[21] By contrast, building upon earlier research on the Treatment Alternatives to Street Crime initiative, a study of a pretrial diversion program targeting substance abuse in Birmingham, Alabama, demonstrated that program participants had significantly fewer rearrests than did a comparison group of nonparticipating offenders.[22] It could be argued that because many diversion programs engage in "**creaming**"—that is, they select low-risk cases that show relatively good prognoses for success—these programs should, as a general rule, have high rates of successful completion and low rates of recidivism. However, the Birmingham program was limited to felons and did not cream cases. Furthermore, another study reported that persons jailed for violent crimes and then diverted showed no greater likelihood of recidivism than did a comparable group of diverted nonviolent offenders, suggesting that eligibility for diversion need not necessarily exclude people charged with violent crimes.[23] Again, a key to successful diversion seems to be assessing correctly what cases have risk and need patterns that warrant services and then accurately matching these cases with the appropriate services.

Probation

It was stated earlier that probation agencies have the dual goals of promoting community protection and offender betterment. Because they tend to handle more-serious cases than diversion programs do, including many felons, probation agencies usually put greater emphasis on community protection and supervision than diversion programs do. In addition, as will become apparent, probation agencies have as a third very important goal of conducting investigations, providing information, and making recommendations to assist the court in reaching decisions about sentences.

Types of Probation

Probation is the single most frequently handed out sentence in corrections. But probation sentences are not all the same. *Probation* is an umbrella term that can be divided into five categories.

1. **Straight probation**—This is also referred to as regular or standard probation. Under straight probation offenders are sentenced directly to supervision in the community without placement in a jail, prison, halfway house, or other residential facility being part of the arrangement.
2. **Suspended sentence probation**—The offender is sentenced to incarceration, but execution of that sentence is suspended or withheld contingent on the offender's performance during a probation sentence. This differs from a **deferred sentence** in which the judge withholds formal sentencing of a guilty person contingent on the person staying out of legal trouble. It also differs from a straight suspended sentence where a sentence is imposed without placement on probation and not carried out unless additional illegal activity occurs. But probation officers occasionally supervise persons with deferred or straight suspended sentence arrangements, a practice known in some places as **informal probation**.
3. **Shock probation**—In jurisdictions where it is used, shock probation involves the offender first being sent to prison and shortly thereafter (e.g., three or four months) being returned to court and sentenced to probation. The idea is that a brief sample of prison life might decrease the likelihood of probation violations.
4. **Split sentence probation**—The judge divides or splits a total sentence of, say, 4.5 years between a jail term (e.g., the first 6 months) and a probation term (e.g., 4 years after release). Split and suspended sentence probation can be combined. For example, an offender might serve the first three months of a six-month jail term and have the remaining three months suspended contingent on performance on probation.

Split sentence probation differs from shock probation in that the former involves a jail term, while the latter involves prison time.

5. **Residential probation**—The probationer is ordered to serve part of the probation term in a halfway house setting that provides relatively open but still structured living. Such residential placements are sometimes used for the early part of a probation sentence (e.g., the first 6 months or year) or for persons who have violated their probation.

Any of these types of probation can be combined with other sentence elements, such as fines owed the government, restitution orders owed the victim, or community service. Offenders can be denied discharge from probation until such other elements are completed. This is why probation is sometimes described as an umbrella sentence under which other sanctions are carried out.

A distinction can also be drawn between federal probation and state/local probation. The Administrative Office of the U.S. courts operates federal probation, and federal probation officers oversee persons who have been sentenced by the federal courts for violating federal law. By contrast, persons put on probation for violations of state or local statutes are handled by agencies that, depending on the jurisdiction, are administered at the state level (such as a Department of Corrections) or local level (such as a county or city probation department).

There are other ways that professionals in the field distinguish types of probation. A distinction is sometimes drawn between probation that is administered out of the judicial branch of government (as with federal probation) versus probation administered out of the executive branch (as with a state that administers probation out of its Department of Corrections). Likewise, probation is sometimes categorized according to the caseloads that officers carry or the functions they perform. In some jurisdictions, caseloads may be specialized by (1) kind of offense—felony versus misdemeanor probation, or (2) age of the probationer—adult versus juvenile probation. Additionally, in some places caseloads consist entirely of probationers, whereas in other places caseloads include both probationers and parolees (see Chapter 13 for a discussion of parole). Finally, some jurisdictions have begun to separate probation staff according to function. For example, certain staff may work entirely at preparing presentence investigation (PSI) reports and related reports, while other staff supervise probationers and deliver services. Or some staff may be assigned to supervise probationers and enforce conditions while other staff perform the rehabilitative role; this is one way to reduce the potential for role conflict.

The Probation Process

Probation is a sentence regularly handed down by judges in jurisdictions throughout the United States. Nevertheless, the process of probation does not end with the sentence; in fact, the probation process only begins at the sentence. In the following section, we discuss that process and the various decisions involved in the probation process.

Eligibility and Placement

The statutes and policy guidelines of a given jurisdiction define which offenders are eligible for probation. Eligible cases are considered to determine those that are actually suitable for placement in the community. This is a discretionary decision that has been shown to vary across locations and individual judges.[24] In reaching decisions, judges consider a variety of factors such as (1) recommendations from the prosecution and defense, which often reflect plea negotiations; (2) the wishes of the victim and/or community, which may be expressed in a formal victim impact statement; (3) whether the defendant was jailed in the time frame leading up to sentencing; (4) availability of jail bed space; (5) recommendations from the probation agency itself; (6) the seriousness and circumstances of the offense; and (7) the offender's prior legal record. The last two of these are especially important considerations for most judges.

As covered in Chapter 6, probation agency recommendations usually take the form of a PSI report. Though not ordered in all cases, judges tend to order the probation office to conduct PSIs in felony cases and even in more serious misdemeanor cases within the interval between a plea or finding of guilt and sentencing. Although some are far more extensive than are others, a PSI is a narrative report on the circumstances of the offense as well as the offender's legal, psychological, and social (e.g., family, educational, and employment) background and current situation. Officials (usually probation officers) preparing a PSI get most of the information from conducting interviews, reviewing documents, and searching legal background databases. The PSI usually concludes with a recommendation to the judge regarding what sentence is appropriate. In some jurisdictions, PSIs are accompanied by a recommended sentence based on structured sentencing guidelines designed to promote greater consistency in sentences across cases (see Chapter 6).

Race and Gender in Corrections

Considerable research has been conducted on whether a convicted defendant's race is related to the severity of the sentence received, such as being sentenced to probation instead of prison. As might be expected, findings are mixed across studies and jurisdictions. But on balance, studies show that African Americans and Hispanics often receive more severe sentences for some violent crimes and for drug offenses. According to Stephanie Bontrager and her colleagues, this may be because minorities are stereotyped by some persons and groups as representing a greater social threat, and as such, are sometimes targeted for tighter control.[25]

Although sentences chosen by judges are typically (though certainly not always) consistent with PSI recommendations, there is debate about the reasons for this. Are judges influenced by the report to try to tailor the sentence to the unique circumstances of the case? Or are probation officers just skilled at selectively structuring information in the report and recommending what they know judges would deem a proper sentence, given the nature of the offense and legal history of the offender?[26] In fact, it has been suggested that the PSI has minimal, if any, utility for predicting the future behavior of offenders and should be discontinued.[27]

Offenders are officially placed on probation by issuance of a formal **probation order**. The probation order used by the Administrative Office of the U.S. Courts at the federal level appears in **Figure 8–1**. The order specifies the defendant and the length of the sentence; length is ordinarily shorter in misdemeanor cases (e.g., a year or two) than it is in felony cases (e.g., three or four years). Length is established by the judge working within the guidelines of the law, and in some instances, the law makes it possible for the judge to reduce the probation term later for positive behavior, or increase it for failure to comply with all conditions in a timely fashion.

A very important component of the probation order is the conditions the offender must follow to receive a successful discharge and remain free from incarceration; progress is evaluated by reference to these conditions. Two basic categories are **standard conditions** (sometimes called general conditions) and **special conditions** (sometimes called specific or individualized conditions). Standard conditions are illustrated in Figure 8–1. As the name suggests, standard conditions are applied to all probated cases in the jurisdiction and are often established by statute. Special conditions are tailored to the individual case at the discretion of the judge and/or probation officials. Another difference is that standard conditions typically deal more with control and supervision of the case. Special ones tend to be more concerned with addressing offender needs, such as substance abuse counseling or rectification of educational deficits, but these conditions can also address victim needs through restitution or compensation requirements, or community needs through service orders. Finally, special conditions can later be modified to make them either more or less restrictive depending on the probationer's progress.

Assessment/Classification, Supervision, and Delivery of Services

Not everyone assigned to probation will need the same kind of supervision and services. Thus, after an offender is placed on probation, frequently one of the first steps is to assess the levels of risk and need posed (known as a **risk/needs assessment**). This is done so that both the type and amount of supervision and services can be determined. In part, information for doing this assessment is conducted as part of the PSI report, but in recent decades, more probation agencies have been supplementing PSI data with data derived from risk and needs assessment scales to promote greater objectivity and consistency in decision making. These scales assess risk by examining factors regarding the present offense and legal history. Need is assessed by considering variables such as education and employment background, substance abuse, and psychological functioning. (See **Figures 8–2** and **8–3** for

AO 245B (Rev. 06/05) Judgment in a Criminal Case
Sheet 4—Probation

DEFENDANT: Judgment—Page ______ of 3
CASE NUMBER:

PROBATION

The defendant is hereby sentenced to probation for a term of:

The defendant shall not commit another federal, state or local crime.

The defendant shall not unlawfully possess a controlled substance. The defendant shall refrain from any unlawful use of a controlled substance. The defendant shall submit to one drug test within 15 days of placement on probation and at least two periodic drug tests thereafter, as determined by the court.

☐ The above drug testing condition is suspended, based on the court's determination that the defendant poses a low risk of future substance abuse. (Check, if applicable.)

☐ The defendant shall not possess a firearm, ammunition, destructive device, or any other dangerous weapon. (Check, if applicable.)

☐ The defendant shall cooperate in the collection of DNA as directed by the probation officer. (Check, if applicable.)

☐ The defendant shall register with the state sex offender registration agency in the state where the defendant resides, works, or is a student, as directed by the probation officer. (Check, if applicable.)

☐ The defendant shall participate in an approved program for domestic violence. (Check, if applicable.)

If this judgment imposes a fine or restitution, it is a condition of probation that the defendant pay in accordance with the Schedule of Payments sheet of this judgment.

The defendant must comply with the standard conditions that have been adopted by this court as well as with any additional conditions on the attached page.

STANDARD CONDITIONS OF SUPERVISION

1) the defendant shall not leave the judicial district without the permission of the court or probation officer;
2) the defendant shall report to the probation officer and shall submit a truthful and complete written report within the first five days of each month;
3) the defendant shall answer truthfully all inquiries by the probation officer and follow the instructions of the probation officer;
4) the defendant shall support his or her dependents and meet other family responsibilities;
5) the defendant shall work regularly at a lawful occupation, unless excused by the probation officer for schooling, training, or other acceptable reasons;
6) the defendant shall notify the probation officer at least ten days prior to any change in residence or employment;
7) the defendant shall refrain from excessive use of alcohol and shall not purchase, possess, use, distribute, or administer any controlled substance or any paraphernalia related to any controlled substances, except as prescribd by a physician;
8) the defendant shall not frequent places where controlled substances are illegally sold, used, distributed, or administered;
9) the defendant shall not associate with any persons engaged in criminal activity and shall not associate with any person convicted of a felony, unless granted permission to do so by the probation officer;
10) the defendant shall permit a probation officer to visit him or her at any time at home or elsewhere and shall permit confiscation of any contraband observed in plain view of the probation officer;
11) the defendant shall notify the probation officer within seventy-two hours of being arrested or questioned by a law enforcement officer;
12) the defendant shall not enter into any agreement to act as an informer or a special agent of a law enforcement agency without the permission of the court; and
13) as directed by the probation officer, the defendant shall notify third parties of risks that may be occasioned by the defendant's criminal record or personal history or characteristics and shall permit the probation officer to make such notifications and to confirm the defendant's compliance with such notification requirement.

Figure 8–1 A probation order.

INDIANA ADULT RISK ASSESSMENT INSTRUMENT

Probationer's Name ______________________ Cause No.__________

Probation Officer's Name ______________________ Date Completed__________

1. AGE AT FIRST CONVICTION OR ADJUDICATION

24 or Greater	0
20 – 23	3
19 or Less	6

2. NUMBER OF PRIOR CONVICTIONS

0	0
One	3
Two or More	6

3. NUMBER OF PRIOR COMMUNITY SUPERVISIONS

0	0
One or More	4

4. NUMBER OF PRIOR VIOLATIONS OF COMMUNITY SUPERVISIONS

0	0
One or More	3

5. NUMBER OF PRIOR COMMITMENTS

None	0
One or More	4

6. HISTORY OF SUBSTANCE USE

No Known Interference	0
Some Disruption	1
Serious Disruption	2

7. TIME EMPLOYED/FULL-TIME STUDENT IN LAST 12 MONTHS

9 months or More or N/A	0
5 to 8 months	1
Less than 5 months	2

8. RESIDENCE CHANGES IN LAST 12 MONTHS

None	0
One	2
Two or More	4

9. EDUCATIONAL ATTAINMENT

College and Post-College	0
High School/GED	1
Not Graduated from High School	2

10. EXPECTATION OF COMPLIANCE

Reasonably Certain	0
No Opinion	2
Serious Concern	4

TOTAL ______

AUTOMATIC POLICY OVERRIDE/UNDERRIDE

Specify policy: ______________________

OVERRIDE/UNDERRIDE

Must explain: ______________________

Supervision level needed: ______________________

Supervisor's signature______________________

CUT OFF SCORES: 0-10 LOW 11-19 MEDIUM 20 OR *MORE* HIGH

Figure 8–2 Sample adult risk assessment instrument used in Indiana.
Source: *Reproduced from the Indiana Judicial Center (1993).* Probation Case Classification & Workload Measures System for Indiana. *Indianapolis, IN: Indiana Judicial Center. Available: www.nicic.org/pubs/1993/011273.pdf.*

INDIANA JUVENILE NEEDS ASSESSMENT INSTRUMENT

Probationer's Name____________________ Cause No.________________

Probation Officer____________________ Date Completed________________

	Factor Score		Relationship To Criminal Behavior		Total
1. School/Employment 0 Not Applicable; None 1 Moderate 2 Serious	_____	+	_____	=	_____
2. Substance Use 0 None 1 Experimental/Some disruption 2 Serious	_____	+	_____	=	_____
3. Family Relationships 0 No problems 1 Moderate 2 Serious	_____	+	_____	=	_____
4. Peer Relationships 0 No problems 1 Some delinquents 2 Mostly delinquents	_____	+	_____	=	_____
5. Emotional Stability 0 No problems 1 Moderate problems 2 Serious problems	_____	+	_____	=	_____
6. Health and Hygiene 0 No problems 1 Illness or physical condition interferes with functioning 2 Serious physical condition or chronic illness interferes with functioning	_____	+	_____	=	_____
7. Learning Ability 0 No problems 1 Some need for assistance 2 Serious interference with functioning	_____	+	_____	=	_____
8. History of Abuse/Neglect 0 No known history of victimization 1 Evidence of victimization but no outward manifestation 2 Evidence of victimization and juvenile is exhibiting rated behavior	_____	+	_____	=	_____
9. Other, Must Explain: (0, 1, or 2) ________________	_____	+	_____	=	_____

Comments____________________________________

Supervisor's signature: ____________________________
(Optional)

Behavior

If one or more individual categories total score is 3 or above, then the 3 most serious needs should be prioritized and a supervision plan developed.

Scores

Relationship to Criminal
0 – No relationship
1 – Moderately related
2 – Directly related

Figure 8–3 Sample juvenile risk assessment instrument used in Indiana.
Source: *Reproduced from the Indiana Judicial Center (1993).* Probation Case Classification & Workload Measures System for Indiana. *Indianapolis, IN: Indiana Judicial Center. Available: www.nicic.org/pubs/1993/011273.pdf.*

examples from Indiana.) The scales are meant to help ensure that cases with higher risk and need scores receive the most supervision and/or services, while those with lower scores receive less. Risk and need scores are sometimes classified as maximum, medium, and minimum so that supervision and services can be calibrated according to level. Furthermore, levels frequently change as the probation sentence unfolds; for instance, an offender could be reclassified from maximum to medium risk level following a year with no violations.

A main goal of probation supervision is to monitor compliance with both general and special conditions. This involves maintaining periodic contact with the offender by having the offender visit the probation agency. Phone calls, visits to the offender's home or place of work, and electronic monitoring (see Chapter 9) can also be used. Supervision can involve periodic conversations with people able to judge the offender's progress (e.g., employers or providers of treatment services). The frequency of such contacts (e.g., monthly versus weekly visits to the agency) varies with the level of risk and also with the progress demonstrated. Random drug testing and regular checks of police records for criminal activity are other common features of a supervision plan.

Similarly, service delivery usually reflects the special conditions of the sentence. Services are delivered either directly by the probation agency, as in the case of an officer counseling a client, or through referral of the offender to another community agency that provides the service, like when an offender seeks a general equivalency diploma or job placement or participates in drug rehabilitation. When such referrals are made, probation officers are responsible for monitoring and documenting that offenders are participating as intended.

In some jurisdictions across the United States, probation caseloads are excessively high. There are too many probationers relative to the number of probation officers that the agency has resources to hire. For instance, a single officer might be responsible for 150 felony cases that require moderate to high supervision/services. Although this number would not be considered high (and could be considered low) for lower risk misdemeanor cases requiring minimal supervision/service, it is probably excessive for moderate- to high-risk felons. As such, the usual result is that many (or most) cases receive minimal supervision and services, while those few cases that cause the most problems consume a disproportionate amount of the officer's attention.

Probation officers file follow-up reports as scheduled by the court so that judges and other court officials can be apprised of a probationer's performance in the community. Considered in conjunction with the PSI, which represents important baseline information, follow-up reports are the foundation for an ultimate recommendation to the court about when and how probation should be terminated.

Termination of Probation Orders

When probation officers possess evidence that probationers have fulfilled the conditions of their sentences within allotted time frames, officers recommend discharge.

Some agencies use gradations of discharge to connote the amount of offender success (e.g., discharged with minimal progress).

By contrast, when officers have evidence of failure to comply with conditions, they have two basic choices. One choice is to recommend to the judge that the probationer be continued on probation with a tightening of conditions (e.g., jail time, a curfew, mandatory employment). This choice is often made when the violation is not very serious and the probationer has not accumulated many prior violations. The other choice is to recommend **revocation** or repeal of the probation sentence. This choice is common when the violation is relatively serious, such as a felony, or when the probationer has a pattern of continued petty violations and has been unresponsive to previous warnings and condition modifications.

Officers may recommend revocation of probation in reaction to two kinds of violations: (1) new misdemeanor or felony crimes or (2) **technical violations**. Though technical violations are not law violations per se, they nevertheless constitute violations of technical probation conditions, such as not reporting as scheduled to the probation officer or not completing counseling as ordered.

When revocation is recommended by the probation agency and the court decides to follow the recommendation, a preliminary inquiry is conducted. At the inquiry, the probation agency must establish probable cause that a violation(s) took place, which is the same standard needed by the police to make an arrest. Given probable cause, the probationer can either admit to the violation or proceed to a formal revocation hearing. At such a hearing, the judge will decide whether the evidence (typically a preponderance thereof, which is the civil law standard of proof) substantiates the alleged violation and, if so, whether the violation warrants revoking the probation sentence. The preliminary inquiry and revocation hearing may be conducted at separate points or combined as two stages in one setting. At both, probationers have the right to advance notice both of the charges and the hearing. They are also entitled to attend the hearing and present evidence on their behalf, including cross-examination of witnesses. Rather than being required to provide all indigent probationers with counsel, courts are permitted to make decisions regarding the provision of counsel on a case-by-case basis.[28] Counsel is often furnished, particularly when the circumstances and legal issues are such that a probationer may experience difficulty representing his or her interests.

The typical outcome of a revocation decision is incarceration in jail or prison, but this is not inevitable in all cases. In select cases, the offender may be warned to avoid future violations and be placed back on probation with tighter conditions.

The Changing Face of Probation in the Twenty-First Century

The field of probation has undergone important changes in recent decades. One change was alluded to earlier. Compared to the traditional emphasis on individual casework and rehabilitation, in many jurisdictions probation has become more oriented toward surveillance and risk management; these efforts are often directed toward entire groups rather than individual probationers.[29] Three other developments reflect this broader shift in orientation and are illustrative of the changing face of probation in the present century.

One development involves **police–probation partnerships**, or collaborations between probation agencies and police agencies. Such partnerships are most common at the local level but may also include state and federal officials. A first area of collaboration is the sharing of data about individuals under community supervision (e.g., information about legal histories, probation violations). Such information could, for instance, be highly useful in locating probation **absconders**, or offenders who have fled the jurisdiction in which they were under supervision. Second, some probation and police agencies work together to perform surveillance and investigate illegal activities that involve probationers. Police often consider this arrangement advantageous. In many places, probationers account for a disproportionately high amount of crime, so partnering with probation officials is seen as a way to improve crime control. In addition, due process protections governing areas such as self-incrimination and privacy are not normally as stringent for persons on probation. For instance, a warrantless search can be conducted of a probationer's residence based only on reasonable grounds, a lesser standard than probable cause.[30] Additionally, information that probationers divulge about illegal actions during counseling sessions can be conveyed to the police.[31] For their part, probation officers (especially those employed in jurisdictions where probation agents do not carry firearms) may feel safer when entering high-crime areas accompanied by the police.[32]

A second development in probation stems directly from the community policing movement, specifically the "broken windows" approach espoused by Kelling and Wilson in the early 1980s.[33] This approach argues for expanded citizen participation in helping the police to address various social problems surrounding the crime problem. In particular, the contention is that citizen support is needed to help prevent communities and neighborhoods from being inundated with symbols of high crime such as abandoned and vandalized buildings, dilapidated housing, disorderly behavior, and public fear. These areas invite more crime. Around the turn of the last century, this logic was applied to formulate the concept of **broken windows probation**.[34] Among other things, the concept suggests that probation agencies should prioritize public safety above other objectives and be held accountable for doing so. This means probation conditions must be rigorously enforced, and agency performance must be assessed. In addition, probation officers should spend less time in their offices and more time in the field supervising offenders, connecting with the public, and building support for and familiarity with probation operations among various community organizations. Some see this approach as overly simplistic and as downplaying empirical evidence that supports rehabilitative approaches to probation. But in an era of prison/jail crowding and high probation caseloads, combined with higher than desired levels of probation recidivism and public fear, the broken windows approach will continue to find appeal among agencies.

A final development to be mentioned here is the growing importance of computer-based technology, a trend that might be called **automated probation**. Just as people are growing increasingly comfortable with (and dependent on) automated banking, bill paying, reservation booking, product ordering, and so forth, probation agencies have been automating more and more of their functions. For instance, many agencies have software designed specifically to expedite the preparation of PSIs and risk/needs assessments. Given that proper computer security is maintained, much of the information needed for progress monitoring and preparation of reports can be communicated electronically through e-mail and various file attachments and linkages. Also, agencies are increasingly using technology to facilitate supervision and enforcement of conditions. One development allows select lower risk probationers to report to the agency via kiosk machines that resemble ATMs; these machines let officials identify offenders through fingerprints, identification cards, or voice verification. During image-recorded interactions, offenders can make payments or perform other simple activities required of them (e.g., answering routine questions or making appointments).[35] Related areas being affected by automation include drug testing and DNA sample collection with offenders, as well as in-service training of probation staff.[36]

Advocates of automating probation point to the efficiency involved, especially the potential for time and cost savings. Others caution, however, that excessive reliance on technology can lessen the quality of probation services by depersonalizing the experience and can even threaten to displace staff positions. There is concern about decisions and policies being driven less by a desire for quality probation services and more by technological needs and capabilities; and because private vendors provide so much of the technology, there is concern about the role of commercial interests in shaping the future of probation.

Issues in Probation

By this point, it should be apparent that probation is a field with many controversial questions and issues. Based on the foregoing material in the chapter, the following list presents examples of such questions. The questions have no easy answers and require critical, innovative thinking.

- Given the various factors that can affect judicial decision making, how do we help make sure that decisions are made in a consistent and fair way about who is placed on probation, as opposed to those given more lenient or more severe sanctions?
- Is the PSI process worth the time and effort it consumes? How can it be made more useful?
- How can probation agencies better address: (1) the potential for conflict between the law enforcement and rehabilitative roles of officers and (2) the reality of high caseloads and limited resources?

- To what extent should people on probation have the same due process rights as people not under correctional supervision?

All of these issues are presently under debate in the field. Additional issues are discussed in the following paragraphs.

A question that has been debated for a long time is whether probation is sufficiently punitive or, like some claim about diversion, it amounts to a mere "slap on the wrist." There is little question that probation costs less money than jail or prison does, assuming of course that probation is successful in helping avert incarceration. Community supervision avoids the costs of residential placement, and many probationers hold jobs, pay taxes, and otherwise participate in the local economy. Probation also avoids disintegrating offenders from the community (e.g., breaking ties with family) and placing them in what many consider to be a **criminogenic** jail or prison environment. Plus, available community services (e.g., vocational training or drug treatment) can be drawn upon. But does probation provide sufficient punishment for crime?

There are really two parts to this question—whether probation is punitive enough to deter crime and whether it is sufficient retribution for criminal behavior to exact justice. The first part is fairly easy to address because research has consistently shown two things: (1) formal criminal justice sanctions of all types have limited capacity to deter crime; that is, offenders and would-be offenders are much more influenced by informal groups (e.g., friends, family, church);[37] and (2) net of other relevant factors, probation is associated with less recidivism than imprisonment is.[38] The second part is trickier because research has shown important differences between people in their perceptions of the severity of various sanctions.[39] What one individual finds very punitive, another might find considerably less punitive or not punitive at all. Thus, whether probation constitutes adequate retribution depends on how severe a given person perceives probation to be and also on exactly what a particular probation sentence entails.

Controversies

It is hard to imagine anything more controversial in the field of corrections than the placement of convicted sex offenders on probation. There is enormous public fear that if retained in the community, even under close supervision, most of these offenders are sure to "prey" on innocent victims. Although there have certainly been isolated (and media-sensationalized) instances of such predatory behavior, public fears are likely exaggerated. One study found that only 4.5 percent of sex offenders placed on probation committed a new sex crime, and only 16 percent had any kind of reoffending. Other studies have reached similar conclusions.[40]

Though probation is less costly than incarceration, increases in the probation population have been associated with escalating costs in many jurisdictions. In response, some jurisdictions require probationers to pay monthly **probation fees** (e.g., $50) to help defray supervisory and even rehabilitative costs. These fees, which become a probation condition, are distinct from other financial penalties such as fines, restitution payments, and court costs. Those who favor fees argue that fees are a practical way to contain costs, that fees help teach accountability for behavior, and that fees increase the likelihood probationers will actually use services made available to them. Critics counter that fees are an undue financial burden for economically disadvantaged persons and question why people should have to pay for supervision and services they are being required to receive by the court. There is also concern that various bureaucratic costs associated with collecting fees can diminish cost savings.[41]

Another issue is the degree to which probation services should be privatized or contracted to private vendors. Over time, jurisdictions have contracted with the private sector for preparation of PSI reports, supervision, and various rehabilitative services. The justifications have been that (1) privatization is a viable and efficient means of handling growing workloads in organizations plagued by understaffing and limited budgets, and/or (2) the private sector can provide certain specialized services and expertise that the government is ill equipped to provide. But some contend that much caution must be exercised when contracting probation services because government officials relinquish a certain amount of control, while ultimately remaining accountable for outcomes. Other critics see inherent problems with having probation policy and practice shaped by the stereotypical motif of the private sector—driving up profit through a combination of high sales volume and cost-cutting measures.

To What Extent Is Probation Effective?

As with all sanctions and programs in corrections (including diversion), there is debate over how the success or effectiveness of probation should be defined. We discuss the nature of that debate in the following paragraphs.

Defining Success and Demonstrating Effectiveness Because control of recidivism is obviously a goal of any probation agency, most studies have focused on recidivism. However, recidivism can be measured in various ways, as illustrated in **Table 8–1.** Because the influence of discretion grows cumulative across the arrest, court referral, conviction, revocation, and incarceration decisions, many researchers opt for new arrests as the least biased measure; however, data are often collected on the other measures of recidivistic events as well. Likewise, though researchers and agency officials are often most concerned with felony recidivism, they tend to collect data on misdemeanors and technical violations as well. Finally, when feasible, researchers try to measure recidivism that takes place during the probation sentence, and for offenders

Table 8–1 Common Approaches to Defining Probation Recidivism

Recidivistic Event
- New Arrest
- New Court Charges
- New Conviction
- Revocation
- Incarceration

Category of Recidivistic Behavior
- Felony
- Misdemeanor
- Technical Violation

Timing of Recidivism
- Time to Recidivism During Probation Sentence
- Recidivism After Successful Probation Discharge

who complete the sentence, recidivism during a specified time frame (e.g., a year) after discharge. Sometimes the number of days to the recidivistic event is measured as well.

Even though recidivism is a commonly used measure of probation effectiveness, sole reliance on this measure is limiting. Probationers can have accomplishments and better their lives in a variety of ways without refraining entirely from illegal activities. Examples include obtaining and maintaining employment, making educational gains, improving social skills, and eventually completing the probation sentence successfully. Others have suggested that the success of probation agencies should be assessed through the use of specific agency performance measures, such as supervisory contacts and violations detected.[42]

Even if agreement were reached on a measure(s) of probation effectiveness (e.g., reduced recidivism or enhanced employment), another difficulty remains. This difficulty is demonstrating that it was the probation experience per se, rather than some outside factor (e.g., improvement in the local economy), that produced any positive outcome observed. For practical reasons, most studies have not used the kinds of rigorous experimental research designs needed to make unambiguous conclusions about the effects of probation on offenders. Related to this, probation is susceptible to the creaming critique mentioned earlier in regard to diversion; favorable statistics about program outcomes may be caused by conservative selection criteria and the fact that these programs often start with "safer bet" cases from the outset. These caveats must be kept in mind when considering probation recidivism research.

Probation Recidivism Since the 1950s, numerous studies have examined recidivism among probationers. Much of the earlier research, conducted when probation had more of a casework, rehabilitative emphasis, looked at whether the size of officer caseloads was associated with recidivism. The idea was that with smaller caseloads (25 to 40), officers could devote more individualized attention and service to each probationer, thus achieving lowered recidivism. This issue remains paramount today, given the record high caseloads in many locations.

Earlier studies yielded no evidence of a link between lower caseloads and reduced recidivism. In fact, some evidence shows that lowered caseloads allow officers to monitor probationers more closely, thereby increasing detection of recidivism, especially technical violations.[43] More recent research has confirmed these results.[44] From a financial standpoint, this is important. If an offender has his or her probation revoked and is incarcerated for technical violations detected because of closer surveillance, taxpayers end up supporting the cost of both closer community supervision and the subsequent period of incarceration.

Aside from studying the relationship between caseloads and recidivism, researchers have sought to discover the volume of probation recidivism and the factors related to or predictive of it. Two widely known studies were published in the 1980s and 1990s.[45] In one, Joan Petersilia and her colleagues in California tracked approximately 1,700 probated felons for 40 months and reported that two-thirds were rearrested, more than half were reconvicted, and about a third were incarcerated. In another study, the U.S. Justice Department tracked 79,000 felons who had been placed on probation in 1986 in 17 states. Within 36 months, 43 percent had been rearrested for new felonies, including 8.5 percent for violent crimes. Understandably, these studies caused some alarm because they implied that felony probation poses a serious threat to public safety. But the findings of other studies have provided less reason for alarm. Vito, for instance, reports a rearrest rate of about 22 percent among felony probationers in Kentucky; another study of federal probationers (81% of whom were sentenced for felony convictions) reports that nearly 70 percent had no official sentence violations whatsoever within two years of placement on probation.[46] In fact, Spohn and Holleran demonstrate that, net of other factors, imprisoning felons (especially drug offenders) is associated with considerably higher recidivism, and thus a greater risk to public safety, than placing them on probation is.[47]

Given that roughly half of the persons serving probation sentences at any given time have been convicted of nonfelonies (mostly misdemeanors), a reasonable question is whether offense category predicts the likelihood of recidivism. Researchers have examined this and other variables as potential predictors. For instance, Morgan conducted an extensive review of the research and concludes that failure on probation was best predicted by the factors shown in **Table 8–2**; notice that the felony versus nonfelony distinction is not among these variables.[48] The study of

Table 8–2 Predictors of Probation Failure

- Property conviction offense
- More prior offending
- Unemployed
- Younger
- Male
- Not residing with spouse or children
- Residential instability (frequent moving)
- Use of heroin

Source: Adapted by the author from Morgan (1993).

federal probationers mentioned earlier found that official sentence violations were significantly more likely among nonwhites, younger probationers, those with prior convictions, and those ordered by the court to undergo mental health treatment. Here again, offense type (felony versus misdemeanor) was not predictive of recidivism.[49]

It would be convenient if the question of probation effectiveness could be answered in simple yes or no terms, but like most questions in the fields of corrections and juvenile justice, this is just not realistic. Research does suggest that, in general, probation contains recidivism just as well as, and in some instances better than, incarceration, and it does so at considerable cost savings.[50] Still, probation is not always and everywhere as effective as we might desire. The rather wide variation in findings across studies is likely to reflect several key factors, captured in the following guidelines to which probation programs should conform.

- There should be evidence that the offenders selected for placement on probation are really the ones who can be expected to benefit from such a sentence.
- Probationers who show evidence of greater risks of reoffending should get higher levels of supervision.

Race and Gender in Corrections

David Olson and his colleagues report that among the Illinois probationers in their study, women were significantly less likely than men were to be rearrested. However, gender was not related to the likelihood of technical violations. The researchers also found that the predictors of recidivism among women were always the same factors that predicted recidivism among men.[51]

Corrections and Policy

In an effort to improve probation outcomes, some jurisdictions have moved toward a policy of tailoring rehabilitative services to the special needs of specific subgroups, such as sex offenders or substance abusing offenders. There is some evidence to support the efficacy of this trend. For example, Pamela Lattimore and her colleagues demonstrate that, among approximately 134,000 drug-involved probationers in Florida, those receiving substance abuse treatment were less likely to experience new arrests than were those not getting treatment.[52] Another study suggests that treatment is more important than supervision is in accounting for such positive results.[53]

- Probationers who show evidence of greater needs for rehabilitative services should receive more services. Furthermore, services should be tailored to addressing criminogenic needs.
- The supervision and services being delivered should be monitored and adjusted as necessary to help ensure high quality. Efforts should be made to structure supervision and service modes so that offenders are likely to benefit and respond in a positive manner.[54]

Application to Criminal Justice and Corrections

The material covered in this chapter has clear application to criminal justice in general, and other aspects of the correctional system in particular. Given the sheer volume of cases diverted and probated each year, these two programs are, in many respects, centerpieces of the criminal and juvenile justice systems. Certain key linkages and relationships are illustrated in the following paragraphs.

Law is the foundation for diversion and probation programs because law authorizes jurisdictions to establish and operate these programs and lays out the parameters in which the programs must function. For instance, laws normally stipulate which persons are eligible for diversion or probation, and jurisdictions sometimes alter their laws and policies based on their experiences with diversion and probation programs. There is also a body of case law that lays out the due process protections of those on probation.

Police agencies are involved with diversion and probation. Diversion programs can provide the police alternatives to arrest and traditional processing. In some jurisdictions, the police actually operate diversion programs; such programs are frequently designed for juveniles. Even when diversion is operated by another agency, such as a court, the police may participate in the provision of services or work with diversion officials to deal with people who have not complied with program conditions. Likewise, the police often work with probation officials to provide information pertinent to case investigations, assessments, and status reports as well as to help monitor the activities of persons on probation and locate absconders. As pointed out earlier, some jurisdictions have initiated formal police–probation partnerships, wherein the agencies collaborate and try to coordinate certain activities. It is also important to bear in mind that to the degree diversion and probation are effective in controlling recidivism, there will be that much less criminal behavior for the police to address on the streets; of course, the converse is also true.

Court personnel (i.e., prosecutors, defense attorneys, judges, and various other court staff) are integral to probation and to most diversion programs. These personnel determine which cases, out of those eligible, actually receive diversion or probation. This determination commonly entails negotiations between the prosecution and defense, as overseen by the judge. As with the police, diversion and probation provide court personnel with alternatives to more restrictive sanctions. One or more of these personnel may be directly involved in the implementation of diversion. Court staff frequently operate diversion programs, and some prosecutorial and defense offices do so as well. Court personnel also work with diversion and probation staff to monitor the behavior of those in programs and also to determine whether an offender has ultimately been successful at complying with conditions. As with the police, the success of diversion and probation at controlling recidivism has implications for the workload of court personnel.

As alluded to near the outset of this chapter, those officials who operate jails, prisons, and other correctional facilities are heavily reliant on diversion and probation to keep facility populations to manageable levels and help contain crowding. For those offenders who are incarcerated, such officials are also reliant on whatever background investigation and other reports a diversion or probation agency might have completed; this information is helpful for determining how to handle new prisoners. By the same token, those staff running diversion and probation programs rely on the threat of incarceration to encourage compliance with conditions.

Some jails participate in diversion efforts to channel offenders for whom jail confinement is deemed inappropriate (e.g., the mentally ill, alcoholics) into more appropriate programs. Also, jails house many persons who either (1) eventually will be placed on probation once their presentence court hearings are completed, or (2) are serving jail time as part of an extant probation arrangement. The latter category consists of people serving split sentences and those jailed for probation violations, and both can contribute significantly to jail crowding. In some jurisdictions, offenders may be transitioned into the community on probation following a prison stay, as with shock probation programs or shock incarceration (boot camp) programs.

Jails and prisons hold many persons who have recidivated following diversion and/or placement on probation. Building more effective diversion and probation programs can help alleviate the crowding strains placed on correctional facilities and thereby help stem the steadily increasing flow of ex-prisoners reentering the community (see Chapter 13).

READY FOR REVIEW

- Diversion and probation are two of the most frequently used responses to illegal behavior; these programs are essential to the operation of the correctional system, because the majority of people who commit offenses cannot (and should not) be incarcerated.
- Probation is a sentence in which the offender remains in the community for all or most of the term, with the probation agency engaged in the following: (1) supervising compliance with rules and conditions imposed by the court, (2) making recommendations to the court in the event of noncompliance, and (3) delivering or coordinating services meant to promote offender betterment. More than 4 million adults are on probation, and about half of them were placed on probation for felony offenses.
- Diversion refers to systematic efforts to move less-serious cases away from any further penetration into the criminal justice process than they have already experienced, often moving cases into alternative programs that are meant to better address needs. Laws or policies often specify which offenders and offenses are eligible for diversion, and persons who are diverted must comply with certain rules and conditions. Diversion is used frequently, especially in juvenile justice.
- Probation evolved from English common law practices and has its formal beginnings in the United States during the 1840s with the work of John Augustus in Boston.
- The history of formal diversion dates to the 1960s when the idea of moving less-serious offenders away from system processing into potentially less stigmatizing, less expensive, and more helpful programs became popular.
- Staff who work in diversion and probation programs need certain professional qualifications and competencies. These staff face potential conflict between their law enforcement and rehabilitative roles. Studies show that many diversion, probation, and related community correctional programs have moved away from a rehabilitative orientation toward a more punitive, law enforcement orientation since the 1970s.
- Diversion can take place before or after a plea or finding of guilt, can be administered by a wide variety of agencies, and can assume a variety of programmatic forms (e.g., drug treatment, vocational training, family counseling). Offenders selected for diversion must meet the legal criteria for diversion and should actually possess the needs targeted by the particular program.
- Persons who enter diversion agreements are expected to follow certain rules and complete diversion requirements within a specified time frame. People who fail to do so usually reenter the system to resume processing.

- Diversion is controversial. Some critics contend it is too lenient, while others argue that it is not governed by sufficient due process. Other critiques are that diversion promotes net widening and may not result in cost savings.
- Though the results of studies are mixed about the effectiveness of diversion at controlling recidivism, research does suggest that some diversion programs are effective at controlling both costs and recidivism.
- Probation agencies typically handle more-serious cases than diversion programs, and there are several different types of probation.
- Judges consider diverse factors in determining which offenders are placed on probation. Many such factors are summarized in the presentence investigation into the offender's legal, psychological, and social background conducted by the probation agency.
- Offenders placed on probation are expected to follow both the standard and special conditions of the sentence. In many jurisdictions, both the amount and types of supervision and services probationers receive are guided by risk/needs assessment data.
- When probation officers become aware that offenders have violated probation conditions, officers may, depending on the number and/or seriousness of violations, recommend to the judge that probation be revoked. Given probable cause to think a violation occurred, the probationer may admit to the violation or proceed to a formal revocation hearing where the probationer has certain due process rights.
- Probation has experienced important changes in recent years. Examples include the development of partnerships between police and probation agencies, a broken windows probation movement to prioritize public safety and public relations, and a movement toward greater automation.
- There are numerous controversial issues in probation that require critical thinking. For example, debates ensue over whether probation is a sufficiently punitive sanction for criminal behavior, whether probationers should pay fees to defray the cost of their supervision and services, and whether governments should contract certain probation functions out to private vendors.
- There are various ways to define the effectiveness of probation. Control of recidivism is one of the most common approaches, but there are different ways to measure recidivism.
  - Research is fairly conclusive that simply reducing probation officers' caseloads does little to reduce recidivism; if anything, detection of violations may increase.
  - Studies have found rearrest rates among probationers to range from roughly a quarter to two-thirds.
  - Research makes it clear that probation can control recidivism as well as incarceration, and in some cases better; and probation is considerably less costly than prison is.

KEY TERMS

absconders Offenders who have fled the jurisdiction in which they were under probation supervision

automated probation The trend toward probation functions, such as investigation and supervision, becoming dependent upon and driven by, advances in computer technology

benefit of clergy Common law forerunner of probation that allowed members of the clergy (and eventually nonclergy) to ask judges to be lenient in sentencing them. These requests came in the form of biblical scripture readings

broken windows probation Approach to probation that suggests probation agencies should prioritize public safety, rigorously enforce conditions, and actively build support for and familiarity with probation operations among various community organizations

creaming Selecting low risk cases for a correctional program that show relatively good prognosis for success

criminogenic A factor or set of factors conducive to criminal behavior

deferred sentence The judge withholds formal sentencing of a guilty person contingent on the person staying out of legal trouble

diversion Systematic efforts to move less-serious cases away from any further penetration into the criminal justice process than they have already experienced, oftentimes into alternative programs that are meant to address offender needs better. Cases can be diverted any time after initial contact up to formal sentencing. Laws or policies often specify which offenders and offenses are eligible for diversion, and persons who are diverted must comply with certain rules and conditions

felony probation The practice of giving felons probation sentences and supervising them in the community

informal probation An arrangement wherein probation officers supervise persons with deferred sentence arrangements (see earlier definition)

John Augustus Considered the founder of probation

judicial reprieve An antecedent of probation whereby a convicted person could petition the court to suspend the main sentence (oftentimes corporal punishment or execution) so that the person could seek a pardon for the offense. Reprieves came to be granted contingent upon whether the person's future behavior complied with the wishes of the court

net widening When diversion programming is used with a case in which no further official action would have been taken had diversion been unavailable; the effect is that more cases are pulled into the system even though the goal of diversion is supposed to be to pull in fewer cases

police–probation partnerships Collaborations between police and probation agencies that entail sharing of information and joint efforts to accomplish surveillance and investigation of illegal activities

posttrial diversion Diversion taking place following admission or finding of guilt in lieu of formal sentencing

pretrial diversion Diversion taking place sometime prior to a formal plea or finding of guilt

probation A sentence in which the offender remains in the community for all or most of the term, with the probation agency: (1) supervising compliance with rules and conditions imposed by the court, (2) making recommendations to the court in the event of noncompliance, and (3) delivering or coordinating services meant to promote offender betterment

probation fees Fees charged to offenders to help defray the costs of their supervision and services

probation order The legal document by which an offender is officially placed on probation. The document specifies the length of the sentence as well as the conditions

recidivism Return to criminal behavior following criminal justice system intervention

residential probation The probationer is ordered to serve part of the probation term in a halfway house setting that provides relatively open but still structured living; residential placement is sometimes used for the early part of a probation sentence and also for persons who have violated their probation

revocation Repeal of a probation sentence

risk/needs assessment Gathering of data to determine both the type and amount of supervision and services that a probation case requires

shock probation When the offender is first sent to prison and shortly thereafter returned to court and sentenced to probation; not used in all jurisdictions

special conditions Probation conditions tailored to the individual case by the judge and/or probation agency; tend to be more concerned with addressing offender treatment needs

split sentence probation The judge divides or splits a total sentence between a shorter jail term and a longer probation term

standard conditions Probation conditions applicable to all probated cases in the jurisdiction, often as established by statute; typically deal more with control and supervision of the case

straight probation A direct sentence to supervision in the community without any kind of residential placement (jail, prison, or halfway house) as a component of the probation sentence

suspended sentence probation The offender is sentenced to incarceration, but carrying out of that sentence is suspended contingent on the offender's performance under community supervision

technical violations Though not law violations per se, these represent violations of technical probation conditions, such as not reporting as scheduled to the probation officer

YOU ARE THE CORRECTIONS PROFESSIONAL SUMMARY

1. Can revocation and incarceration in jail or prison be expected to improve this problem? Could it possibly make matters worse? Incarceration is unlikely to assist the person in overcoming a substance abuse problem unless the facility to which he is sent operates a high-quality treatment program. This is improbable in an era of limited resources and declining emphasis on rehabilitation. However, incarceration certainly has potential to worsen the situation. The probationer would almost certainly lose his job and undergo separation from his family and community. There is a good chance that he would be confined in a criminogenic environment that would do more to socialize him toward greater crime and deviance rather than greater conformity.
2. What alternatives to revocation are available, given all the political rhetoric and pressure? Political rhetoric tends to be highly fickle, such that next week or next month, the governor will likely be preoccupied with another issue. You could recommend that the person be continued on probation and that substance abuse treatment and increased reporting resume. If further drug use is evident during the six-month duration of the probation term, it might be possible for the judge to extend the probation sentence for an additional six months or year while treatment and monitoring continue. The probationer might benefit from referral to a drug court if your jurisdiction has such a program.

NOTES

1. L. E. Glaze and T. P. Bonczar, *Probation and Parole in the United States, 2005* (Washington, DC: U.S. Department of Justice, Bureau of Justice Statistics Bulletin, 2006).
2. Data presented on this topic derive from Glaze and Bonczar, *Probation and Parole.*
3. K. I. Minor, D. J. Hartmann, and S. Terry, "Predictors of Juvenile Court Actions and Recidivism," *Crime and Delinquency* 43(1997): 328–344.
4. Office of the Executive Secretary, Supreme Court of Virginia, and Virginia Department of Criminal Justice Services, *Summary Report on Virginia's Drug Court Programs,* 2003, www.spa.american.edu/justice/publications/evalreportva.pdf.

5. National Institute of Justice, *Drug Courts: The Second Decade* (Washington, DC: U.S. Department of Justice, National Institute of Justice, 2006).
6. H. J. Steadman, A. D. Redlich, P. Griffin, J. Petrila, and J. Monahan, "From Referral to Disposition: Case Processing in Seven Mental Health Courts," *Behavior Sciences and the Law* 23 (2005): 215–226.
7. B. R. McCarthy, B. J. McCarthy Jr., and M. C. Leone, *Community-Based Corrections*, 4th ed. (Belmont, CA: Wadsworth, 2001). See also: J. Q. Whitman, *Harsh Justice: Criminal Punishment and the Widening Divide between America and Europe* (New York: Oxford University Press, 2003).
8. Whitman, *Harsh Justice.*
9. J. Augustus, *A Report of the Labors of John Augustus: First Probation Officer, 1784–1859* (Lexington, KY: American Probation and Parole Association, 1984). See also: L. Gesualdi, "The Work of John Augustus: Peacemaking Criminology," *ACJS Today* 18, no. 3 (1999): 1, 3–4.
10. J. Petersilia, *Reforming Probation and Parole in the 21st Century* (Lanham, MD: American Correctional Association, 2001). See also: B. Foster, *Corrections: The Fundamentals* (Upper Saddle River, NJ: Pearson/Prentice Hall, 2002); H. E. Allen, E. W. Carlson, and E. C. Parks, *Critical Issues in Adult Probation: A Summary* (Washington, DC: National Institute of Law Enforcement and Criminal Justice, 1979); D. J. Rothman, *Conscience and Convenience: The Asylum and Its Alternatives in Progressive America* (Boston: Little, Brown, 1980); McCarthy et al., *Community-Based Corrections.*
11. E. M. Lemert, *Human Deviance, Social Problems, and Social Control* (Englewood Cliffs, NJ: Prentice Hall, 1967).
12. A. T. Scull, *Decarceration: Community Treatment and the Deviant—A Radical View* (Englewood Cliffs, NJ: Prentice Hall, 1977).
13. McCarthy et al., *Community-Based Corrections.*
14. A. West and R. P. Seiter, "Social Worker or Cop: Measuring the Supervision Styles of Probation and Parole Officers in Kentucky," *Journal of Crime and Justice* 27(2004): 27–57.
15. P. M. Harris, T. R. Clear, and S. C. Baird, "Have Community Supervision Officers Changed Their Attitudes toward Their Work?" *Justice Quarterly* 6(1989): 233–246.
16. K. I. Minor, J. B. Wells, I. R. Soderstrom, R. Bingham, and D. Williamson, "Sentence Completion and Recidivism among Juveniles Referred to Teen Court," *Crime and Delinquency* 45(1999): 467–480.
17. E. M. Lemert, "Diversion in Juvenile Justice: What Hath Been Wrought?" *Journal of Research in Crime and Delinquency* 18(1981): 34–46.

18. H. J. Steadman, S. M. Morris, and D. L. Dennis, "The Diversion of Mentally Ill Persons from Jails to Community-Based Services: A Profile of Programs," *American Journal of Public Health* 85(1995): 1630–1636.
19. S. M. Carey and M. W. Finigan, "A Detailed Cost Analysis in a Mature Drug Court Setting: A Cost-Benefit Evaluation in the Multonomah County Drug Court," *Journal of Contemporary Criminal Justice* 20(2004): 315–338.
20. A. J. Cowell, N. Broner, and R. Dupont, "The Cost-Effectiveness of Criminal Justice Diversion Programs for People with Serious Mental Illness Co-occurring with Substance Abuse: Four Case Studies," *Journal of Contemporary Criminal Justice* 20(2004): 292–314.
21. J. J. Kammer, K. I. Minor, and J. B. Wells, "An Outcome Study of the Diversion Plus Program for Juvenile Offenders," *Federal Probation* 61(1997): 51–56.
22. A. Harrell, O. Mitchell, A. Hirst, D. Marlowe, and J. Merrill, "Breaking the Cycle of Drugs and Crime: Findings from the Birmingham BTC Demonstration," *Criminology and Public Policy* 1(2002): 189–216.
23. M. Naples and H. J. Steadman, "Can Persons with Co-occurring Disorders and Violent Charges Be Successfully Diverted?" *International Journal of Forensic Mental Health* 2(2003): 137–143.
24. B. D. Johnson, "The Multilevel Context of Criminal Sentencing: Integrating Judge- and County-Level Influences," *Criminology* 44(2006): 259–298.
25. S. Bontrager, W. Bales, and T. Chiricos, "Race, Ethnicity, Threat and the Labeling of Convicted Felons," *Criminology* 43(2005): 859–622.
26. J. Rosecrance, "Maintaining the Myth of Individualized Justice: Probation Presentence Reports," *Justice Quarterly* 5(1988): 235–256.
27. M. Weinrath, "Are New Directions Warranted for the Presentence Report? An Empirical Assessment of Its Predictive Utility in the Adult Court System," *Journal of Crime and Justice* 22(1999): 113–129.
28. *Morrissey v. Brewer*, 408 U.S. 271 (1972); *Gagnon v. Scarpelli*, 411 U.S. 778 (1973).
29. M. M. Feeley and J. Simon, "The New Penology: Notes on the Emerging Strategy of Corrections and Its Implications," *Criminology* 30(1992): 449–474. See also: D. G. Evans, "Actualizing Probation in an Actuarial Age," *Corrections Management Quarterly* 4(2000): 17–22.
30. *Griffin v. Wisconsin*, 483 U.S. 868 (1997).
31. *Minnesota v. Murphy*, 465 U.S. 420 (1984).
32. J. Petersilia, ed. *Community Corrections: Probation, Parole, and Intermediate Sanctions* (New York: Oxford University Press, 1998), chap. 7. See also:

D. Parent and B. Snyder, *Police–Corrections Partnerships* (Washington, DC: National Institute of Justice, 1999).

33. G. Kelling and J. Q. Wilson, "Broken Windows: The Police and Neighborhood Safety," *Atlantic Monthly* 249(1982): 29–38.

34. T. Arola and R. Lawrence, "Broken Windows Probation: The Next Step in Fighting Crime," *Perspectives* 24(2001): 26–33. For a critique, see: F. Taxman and J. Byrne, "Fixing Broken Windows Probation," *Perspectives* 25(2001): 22–29.

35. M. Renzema, "Reporting Kiosks: A Logical Idea Meets Resistance," *Journal of Offender Monitoring* 11, no. 3(1998): 13–14.

36. M. Jones, *Community Corrections* (Prospect Heights, IL: Waveland Press, 2004).

37. For a good summary of such research, see R. L. Akers and C. S. Sellers, *Criminological Theories: Introduction, Evaluation, and Application,* 4th ed. (Los Angeles: Roxbury, 2004).

38. See C. Spohn and D. Holleran, "The Effect of Imprisonment on Recidivism Rates of Felony Offenders: A Focus on Drug Offenders," *Criminology* 40(2002): 329–358.

39. C. M. Flory, D. C. May, K. I. Minor, and P. B. Wood, "A Comparison of Punishment Exchange Rates between Offenders under Supervision and Their Supervising Officers," *Journal of Criminal Justice* 34(2006): 39–50. Also, D. C. May, P. B. Wood, J. L. Mooney, and K. I. Minor, "Predicting Offender-Generated Exchange Rates: Implications for a Theory of Sentence Severity," *Crime and Delinquency* 51(2005): 373–399.

40. M. Meloy, "The Sex Offender Next Door: An Analysis of Recidivism, Risk Factors, and Deterrence of Sex Offenders on Probation," *Criminal Justice Policy Review* 16(2005): 211–236. Also, C. Kruttschnitt, C. Uggen, and K. Shelton, "Predictors of Desistance among Sex Offenders: The Interaction of Formal and Informal Social Controls," *Justice Quarterly* 17(2000): 61–87.

41. See T. R. Clear and H. R. Dammer, *The Offender in the Community,* 2nd ed. (Belmont, CA: Thompson/Wadsworth, 2003). McCarthy et al., *Community-Based Corrections.*

42. Refer to Arola and Lawrence, "Broken Windows Probation."

43. See J. S. Albanese, B. A. Fiore, J. H. Powell, and J. R. Storti, *Is Probation Working: A Guide for Managers and Methodologists* (New York: University Press of America, 1981).

44. F. S. Taxman, "Supervision—Exploring the Dimensions of Effectiveness," *Federal Probation* 66(2002): 14–27. Also, J. Petersilia and S. Turner,

"Intensive Probation and Parole," *Crime and Justice: A Review of Research* 17(1993): 281–335.

45. J. Petersilia, S. Turner, J. Kahan, and J. Peterson, *Granting Felons Probation: Public Risks and Alternatives* (Santa Monica, CA: Rand. Bureau of Justice Statistics, 1985). Bureau of Justice Statistics, *Recidivism of Felons on Probation,* 1986–89. (Washington, DC: U.S. Department of Justice, 1992).
46. G. F. Vito, "Felony Probation and Recidivism: Replication and Response," *Federal Probation* 50(1986): 17–25; K. I. Minor, J. B. Wells, and C. Sims, "Recidivism among Federal Probationers: Predicting Sentence Violations," *Federal Probation* 67(2003): 31–36. See also: T. R. Clear, P. M. Harris, and S. C. Baird, "Probationer Violations and Officer Response," *Journal of Criminal Justice* 20(1992): 1–12.
47. Spohn and Holleran, "The Effect of Imprisonment."
48. K. Morgan, "Factors Influencing Probation Outcome: A Review of the Literature," *Federal Probation* 57(1993): 23–29.
49. Minor et al., "Recidivism among Federal Probationers."
50. Spohn and Holleran, "The Effect of Imprisonment."
51. D. E. Olson, M. Alderden, and A. J. Lurigio, "Men Are from Mars, Women Are from Venus, but What Role Does Gender Play in Probation Recidivism?" *Justice Research and Policy* 5(2003): 33–54.
52. P. K. Lattimore, C. P. Krebs, W. Koetse, C. Lindquist, and A. J. Cowell, "Predicting the Effect of Substance Abuse Treatment on Probationer Recidivism," *Journal of Experimental Criminology* 1(2005): 159–189. For similar results, see M. L. Hiller, K. Knight, and D. D. Simpson, "Recidivism Following Mandated Residential Substance Abuse Treatment for Felony Probationers," *Prison Journal* 86(2006): 230–241.
53. D. Banks and D. C. Gottfredson, "The Effects of Drug Treatment and Supervision on Time to Rearrest among Drug Treatment Court Participants," *Journal of Drug Issues* 33(2003): 385–412.
54. See D. A. Andrews, I. Zinger, R. D. Hoge, J. Bonta, P. Gendreau, and F. T. Cullen, "Does Correctional Treatment Work? A Psychologically Informed Meta-analysis," *Criminology* 28(1999): 369–404.

9

Community Corrections and Intermediate Sanctions

Chapter Objectives

- Define the concepts of community corrections and intermediate sanctions.
- Discuss the circumstances leading to the development of intermediate sanctions.
- List the common goals for intermediate sanctions.
- Describe the major types of intermediate sanctions available today.
- Discuss the findings from available research on intensive supervision probation/parole (ISP), electronic monitoring, and boot camps.
- Summarize the major conclusions that have been drawn from the available research on intermediate sanctions.
- Discuss four strategies for improving intermediate sanctions.

CASE STUDY

You have recently been appointed chief probation officer of Charity County Probation Department in a midwestern state. The probation department is midsize and operates a restricted range of sanctions, including regular probation, a community service program, and a small electronic monitoring program. The county and the state are experiencing severe fiscal problems, and major budget cuts are forthcoming. The county commissioner, your supervisor, has asked you to develop a proposal for a continuum of sanctions that will help to divert offenders from the county jails and state prisons, and thus save money.

1. What type of intermediate sanctions would you recommend? Why?
2. What steps would you take to ensure that these sanctions achieve the goals of prison diversion and cost savings?

Introduction

"Get tough" policies contributed to an unprecedented growth in the U.S. prison population between 1970 and 1980. Most prisons were operating beyond their stated capacity; this contributed to increases in violence and to litigation by prisoners claiming that the crowded conditions violated their constitutionally protected right against cruel and unusual punishment. With forecasts calling for continued increases in the prison population, federal and state governments began looking for alternatives to prison as a way to cut costs. The answer was found in tough new community corrections programs called "alternatives to incarceration" or "intermediate sanctions." These programs were well received by the public, politicians, and practitioners because they "shared the rhetoric of punishment but offered to accomplish crime control at a reduced cost."[1]

This chapter describes the concepts of community corrections and intermediate sanctions. The most common types of intermediate sanctions are described, and findings from studies of their effectiveness are reviewed. The chapter closes with discussions about lessons learned from more than 20 years of experience with intermediate sanctions and reviews important steps that must be taken to achieve their stated objectives.

History and Current Status of Community Corrections and Intermediate Sanctions

Community corrections refers to agencies and programs at the local, state, or federal level that are responsible for the punishment and management of adult offenders within their local communities.[2] Community corrections include a range of programs from regular probation to ISP. Some consider local jails to be part of community corrections; others, including the authors of this text, limit their conceptualization of community corrections to programs and services provided in a nonsecure environment.[3]

Traditional forms of community corrections, such as probation and parole, have existed since the mid-1900s. The term *community corrections*, however, did not come into vogue until the early 1980s when alternatives to incarceration (e.g., intensive supervision probation, electronic monitoring) emerged in response to prison crowding. Along with these new forms of community-based punishments came **Community Corrections Acts**, legislation that provided a structure for generating, coordinating, funding, and evaluating community corrections programs.[4] In addition to promoting more effective sentencing, these acts were designed to promote more effective use of public resources and more extensive involvement of local communities in the development and implementation of associated programming.[5]

In recent years, intermediate sanctions have emerged as an important subset of community corrections. **Intermediate sanctions** are sanctioning options that fall on a continuum between traditional probation and traditional incarceration.[6] Although initially promoted as a less costly alternative to prison, intermediate sanctions are now recognized as important sentencing options that enable judges to better match the sentence to the offender's situation and the seriousness of the crime.[7] They are used as a standalone sanction, a condition of probation supervision, and a sanction for probation/parole violations. Offenders who require more supervision than that provided by traditional probation but who do not require the restrictions imposed by prison are placed in these options that are, for the most part, administered in the community by federal, state, or local probation and parole agencies.

The goals for intermediate sanctions are many and varied. The most common goals include the following:

- Diverting offenders from prison
- Saving money
- Providing judges with sanctioning options that are proportionate to the crime
- Enhancing public safety through increased surveillance and control
- Promoting rehabilitation and reintegration

There is a variety of intermediate sanction programs including economic sanctions, community service, ISP, home confinement/electronic monitoring (EM), day reporting centers (DRCs), boot camps, and halfway houses. Although this list is not exhaustive, it reflects the most common types of intermediate sanctions that are available across the United States.[8]

Many jurisdictions have developed a **continuum of sanctions** from least restrictive (e.g., regular probation) to most restrictive (e.g., prison) (**Figure 9–1**). Some jurisdictions have even incorporated this continuum into sentencing guidelines or a sentencing matrix as a means to facilitate more equitable decision making.[9] Theoretically, offenders are placed in the least restrictive option necessary for their safe management. As their risk level and severity of offense increase, so too does the restrictiveness of the sanctioning options. Then, based on progress in the program and their reduction in risk, offenders are transitioned to a less restrictive option before being completely released from correctional supervision.

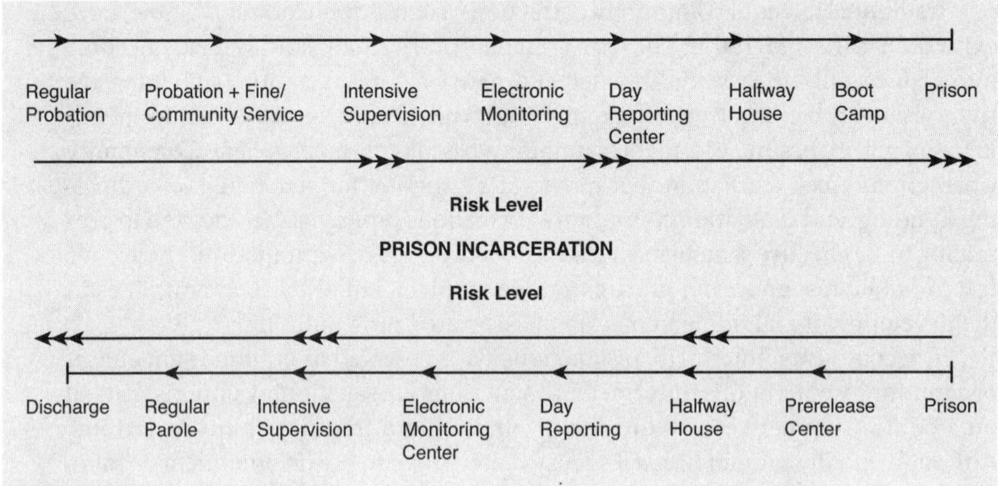

Figure 9–1 Continuum of sanctions.
Source: *Reproduced from D. Wagner, (2001).* Risk and Needs Assessment for Juvenile Justice *(Madison, WI: National Council on Crime and Delinquency).*

Corrections in the Real World

Mary Anne is a 26-year-old white female. She was placed on probation for child endangerment after leaving her 3-year-old daughter alone in their apartment for six hours. Mary Anne's past criminal history includes a charge of disorderly conduct and two prior convictions for driving under the influence. This is her first felony conviction and her first time on probation. Since being placed on probation approximately one year ago, she has regained custody of her daughter. Reports from Children Services indicate that both have made great strides. Mary Anne has been employed as a cashier at Wal-Mart and seems to be doing well. She also has been participating in outpatient treatment, attending regular Alcoholics Anonymous meetings, and remaining sober.

On Monday, you receive a call from her substance abuse counselor who indicated that Mary Anne showed up drunk for treatment last week. She also missed your regularly scheduled weekly appointment. During a home visit, you confront Mary Anne about this behavior. She admits to drinking but says that she now has things back on track. Other than this violation, Mary Anne has complied with all other conditions of probation, but you are concerned that she is heading for a serious relapse. Describe how you would use the continuum of sanctions depicted in Figure 9–1 to encourage Mary Anne's compliance with the conditions of her probation and ensure the safety of her daughter and the community.

The scaling of sanctions based on how restrictive or punitive they are is not as simple as it sounds. Criminal justice personnel often disagree about which sanction is more punitive.[10] Furthermore, some research suggests that offenders do not necessarily agree that prison is the most severe sanction.[11] When asked to rate the severity of sanctions, several alternative sanctions were rated as more severe than prison is, challenging the conventional wisdom of criminal justice policymakers who view prison as the most punitive and intrusive sanction.

Types of Intermediate Sanctions

As mentioned earlier, there is a wide variety of intermediate sanctions available. Some of the more common intermediate sanctions are discussed below.

Economic Sanctions

There are three general types of economic sanctions: restitution, supervision fees, and fines. Economic sanctions are not especially innovative. Restitution, in particular, has been used to punish criminal transgressions since biblical times.[12] The use of economic sanctions, however, has increased dramatically since the early 1980s. This increase has been attributed to financial constraints experienced by community corrections agencies.[13] Fines and fees can offset the cost of offender programs, removing some of the burden from taxpayers and making the agency appear fiscally responsible. Additionally, requiring offenders to shoulder the financial burden of their penalty and to pay restitution to victims supports the quest for offender accountability that has accompanied the "get tough" approach to crime.

Economic sanctions are not without controversy, however.[14] One area of controversy centers around the potential for economic sanctions to violate offenders' constitutional right to equal protection. Are offenders who cannot afford to pay economic sanctions subjected to harsher forms of punishment? And how do community corrections agencies respond to violations when nonpayment is clearly related to the offenders' economic status? Other areas of controversy are more pragmatic. For example, how can probation and parole officers provide quality supervision to offenders and, simultaneously, serve as bill collectors? And will expectations regarding the collection of fines and supervision fees lead legislatures to decrease the level of funding to community corrections agencies? Following are brief discussions that further explore the controversies and benefits associated with each type of economic sanction.

Restitution

Restitution requires offenders to compensate victims for losses incurred as the result of the crime. All 50 states have statutes relating to the collection of restitution.[15] Judges typically order offenders to pay restitution at the time of sentencing based upon information reported in the victim impact statement. Commonly, restitution payments are made to the clerk of court, probation department, or some other intermediary who then disburses the payment to crime victims.[16] The amount of the monthly payments is based on the offenders' ability to pay.

One purpose of restitution is to deter the offender's future involvement in crime. Meta-analyses that have examined the effects of restitution on recidivism conclude that restitution programs have a positive, but modest, effect on recidivism; offenders who were ordered to pay restitution had slightly lower rates of recidivism than did offenders in comparison groups who were not required to pay restitution.[17]

The primary purpose of restitution is to compensate victims of crime for their losses. Research reveals, however, that only a small amount of the restitution ordered by the court is actually collected.[18] For example, a 1999 study of Colorado courts found that of the $26 million offenders were ordered to pay their victims in 1996, $20 million had still not been paid.[19] Many states are taking measures to improve the enforcement of restitution orders, including improved monitoring of restitution payments, attachments of wages earned, attachments of any state payments that the offender receives (tax returns, lottery prizes), and the use of sanctions or revocation when an offender fails to pay.[20]

Supervision Fees

Since 1970, 28 states have enacted legislation that authorizes probation and/or parole agencies to collect **supervision fees** from offenders under their supervision.[21] These fees are used to cover the cost of supervision, EM, drug testing, and treatment. Supervision fees are not intended to punish offenders; they were introduced as a means of generating revenue for correctional programs that were experiencing dire financial constraints.[22] Research reveals that supervision fees have generated a significant amount of revenue for probation and parole agencies. Probation departments in Texas collected more than $57 million in probation fees in 1990.[23] Additionally, supervision fees are often put back into correctional services that directly benefit offenders (e.g., substance abuse treatment, life skills training).

As the practice of fee collection has evolved, two concerns have been raised about the potentially punitive aspects of supervision fees.[24] First, it has been argued that these fees place additional financial strain on offenders who are, in the majority of cases, already experiencing financial difficulties, and that this strain can have a negative impact on offender behavior. Probation and parole officers have speculated that offenders who are behind on their payments fail to report as directed to avoid confrontation, and some fear that offenders may resort to criminal activity to pay the required fees. This concern has been refuted by several writers who claim that failing to pay supervision fees and failing to report to a probation officer are signs of deeper problems that should trigger intervention.[25]

Second, although not intended as a punitive measure, supervision fees become punitive when offenders are sanctioned for nonpayment. It is unlikely that offenders' supervision would be revoked for nonpayment, but other types of sanctions are often imposed such as a short-term stay in jail or community service. Indigent offenders are protected by the precedent established in *Beardon v. Georgia* (1983). In this case, the Supreme Court ruled that revoking an offender's probation for nonpayment when the offender is genuinely unable to pay would be a violation of the equal protection clause in the Fourteenth Amendment. These concerns about the punitive aspects of supervision fees are based on anecdotal reports from probation and parole officers. Until they are confirmed through systematic research, supervision fees are likely to remain an important component of community corrections programming.

Fines

Fines are monetary penalties, the amount of which is generally determined by the severity of the offense.[26] In the United States, fines are used more frequently in limited jurisdiction courts (e.g., courts that impose sanctions for misdemeanors). Additionally, as stand-alone sanctions, fines have traditionally been reserved for traffic and other noncriminal law violations. A recent report, however, reveals that fines were imposed on 25 percent of convicted felons sentenced by state courts in 2002, sometimes as a stand-alone sanction, but more often as a condition of probation.[27] Other countries use fines far more often for a broader range of crimes. For example, a study conducted in the mid-1980s reports that in West Germany, 81 percent of adult crimes and 73 percent of violent crimes were punished by fines as the sole penalty.[28]

Problems associated with fines are similar to those associated with restitution and supervision fees. The two major problems with fines are (1) difficulties in determining the amount of money that is both equitable and punitive and (2) obstacles to enforcement. Recently, several courts have implemented "day fines" as a way to overcome these problems and make fines a viable sentencing option.

Day fines, or structured fines, are "means-based." That is, in addition to basing the amount of the fine on the severity of the crime, it is also based on the offender's daily income.[29] The objectives of day fines are as follows:

- Lower the cost of punishment
- Achieve equity
- Enhance the severity of punishment
- Expand the range of offenders for whom fines are used
- Enhance collection rates[30]

Three steps are involved in the calculation of day fines. First, offenses are ranked according to their severity level. Second, offenses are assigned units of punishment. Third, units of punishment are converted into dollar amounts based on offenders' ability to pay.

Several studies have examined the impact of fines on offender recidivism. A quasi-experimental study in Los Angeles County reported lower recidivism rates for offenders who received a fine in addition to a probation sentence (25%) as compared to offenders who received only a probation sentence (36%).[31] The results of a structured fines demonstration project sponsored by the Bureau of Justice Assistance show that offenders sentenced to day fines had significantly lower rates of technical violations (9%) and rearrests (11%) than did offenders sentenced to conventional fines (22% and 17%, respectively).[32] A study of Milwaukee's Municipal Court Day-Fine Pilot Project comparing offenders who received a day fine with a group of offenders who received traditional fines reports no differences in the percentage of offenders who violated municipal ordinances during a nine-month follow-up period.[33]

Despite the controversies, economic sanctions are likely to continue as a staple in the continuum of sanctions. Innovative practices, such as day fines, appear to promote equity for offenders and increase the level of offender compliance. Economic sanctions appear to add value to a system of sanctions by generating revenue for strapped community courts and corrections agencies and, in the case of restitution, by compensating crime victims. More research is needed, however, to explore the relationship between economic sanctions and offenders' future criminal behavior.

Community Service

Offenders who are sentenced to **community service** are required to provide free labor to a public or nonprofit agency in the community. Community service is used as a stand-alone sanction or as a condition of probation. It has traditionally been used with low-risk and/or misdemeanor offenders. Statistics compiled by the National Judicial Reporting Program reveal that in 2002, in 300 nationally representative counties, only 4 percent of convicted felons received a community service sanction.[34] Although initially introduced as a punishment to deter offenders from future crime, community service is closely aligned with the **restorative justice** philosophy that suggests that an important goal for criminal sanctions is to repair the harm to the community that was caused by the crime.

Many advantages are associated with community service.[35] First, the community gains the tangible benefit of free labor for financially strained public service organizations. Second, it satisfies the public's demand for punishment. Third, it enhances community safety as the result of the structure and supervision provided while the offender is performing the community service. Last, the offender learns responsibility and new job skills and has the opportunity to develop positive community ties. The extent to which these benefits are realized, however, is often dependent on the quality of the supervision provided at the community service site.[36] An inappropriate placement creates frustration for all involved and wastes valuable time and resources.

Community service could potentially reduce an offender's risk of recidivism by teaching offenders attitudes and skills conducive to prosocial behavior and by exposing them to positive role models in the community.[37] At this time, however, the effect of community service on offender recidivism is unknown. Most of the research on community service programs focuses on its restorative aspects such as the number of hours worked. For example, the Center for Alternative Sentencing and Employment Services, a community service program operated by New York courts, enrolled 1,617 participants from July 1995 to July 1996, who worked 87,120 hours.[38] Furthermore, over a five-year period, 754 offenders placed in the Federal Community Service program in the northern district of Georgia performed services valued at more than $2 million.[39]

Intensive Supervision Probation/Parole

Intensive supervision probation/parole (ISP) is the most common form of intermediate sanction in the United States. Instead of meeting a probation officer once a month, as is typical of traditional probation, offenders on ISP are seen up to five times a week through a combination of office and field visits. Although some ISP programs also include the provision of a more intensive level of treatment than that provided to offenders on traditional probation, ISP programs, in general, are designed to be more incapacitative than rehabilitative. That is, through intensive monitoring, ISP programs are expected to restrain offenders' movements in the community and, as a result, limit their opportunity to engage in crime. ISP programs are also designed to deter offenders from future crime through an increased threat of detection and sanctions for noncompliance.

Intensive supervision dates back to the early 1960s when the California Special Intensive Parole Unit and the San Francisco Project were implemented to experiment with smaller caseloads of offenders.[40] The first wave of ISP programs in the 1960s and early 1970s was designed primarily as a probation management tool to examine the effectiveness of various caseload sizes.[41] The experimentation with smaller caseloads was based on the assumption that smaller caseloads would allow for increased contact and lead to greater success.[42] Probation and parole programs during this era operated under the "rehabilitative ideal" that focused on individual offenders and sought to reduce recidivism through interventions aimed at changing offenders' attitudes and behaviors.[43] Rehabilitative interventions were the primary focus of these programs; punishment and community protection were viewed as secondary goals. These early ISP programs met their demise when research revealed that offenders in ISP had similar or marginally lower arrest rates and more technical violations than offenders under regular supervision.[44] More contemporary ISP programs have retained small caseloads in their design, but they differ in most other ways.

The Georgia Department of Corrections reintroduced the concept of ISP in 1982 in response to prison crowding and a shrinking budget. Most states followed suit throughout the 1980s, developing ISP programs as alternative sanctions for offenders who would have otherwise gone to prison. Program designs reflected the leading penal principles of the time—deterrence and incapacitation.[45] The emphasis was on controlling the offender in the community through the use of punishment and surveillance-oriented measures. Treatment components and other service-oriented components generally received a lower priority. Today, most ISP programs target high-risk offenders for participation and include the following elements:

- Frequent contact with offenders
- Smaller caseloads
- Curfews, house arrest, or electronic monitoring
- Drug and alcohol testing
- Performance of community service work
- Graduate sanctions in response to violations
- Treatment and other interventions
- Required employment, employment-seeking activities, or schooling

Two 1987 evaluations of this new model of ISP found that they were diversionary and resulted in cost savings and lower rates of recidivism.[46] Both evaluations, however, suffered from major methodological flaws that threatened the validity of the findings.[47] Later evaluations revealed that the research results for this vastly different approach to ISP were no more favorable than those of early ISP programs.[48] Of particular note is the multisite study conducted by the RAND Corporation.[49]

Corrections and Policy

Intermediate sanctions have changed the way probation and parole officers do business. As discussed in Chapter 8, probation was originally introduced as a social service for offenders geared at helping them become law-abiding citizens. It was believed that offenders needed treatment to overcome circumstances that led them to crime. The focus was on rehabilitation, not punishment. With intermediate sanctions came an emphasis on surveillance and enforcement. The new tongue-in-cheek slogan for what probation and parole officers did with offenders was "tail 'em, nail 'em, and jail 'em." With this new approach to offender supervision, agencies wanted probation and parole officers with a different skill set. Instead of someone who had counseling or social work skills, they looked to hire persons with a law enforcement background who are familiar with investigation and surveillance techniques.

RAND, an independent research agency, conducted an evaluation of ISP programs operating at 14 sites. Between 1986 and 1991, RAND researchers tracked the activities of the programs and the behaviors of offenders who had been randomly assigned to either ISP or regular probation. A review of program activities revealed that the ISP programs did provide a more intensive form of punishment; as compared to regular probation, officers had more contact with offenders, offenders were subject to more stringent supervisory conditions, and violations of program conditions were more likely to lead to incarceration. Other key findings were not so favorable. ISP programs failed to significantly affect the recidivism rates of offenders, with the number of new arrests on ISP being equal to or more than the number of new arrests of offenders on regular probation. Additionally, increased supervisory conditions coupled with closer surveillance led to an increased rate of technical violations for offenders on ISP. As a result of these higher recidivism rates for ISP offenders, the programs failed to alleviate prison crowding and were more costly than originally thought. Last, the researchers reported that the ISP programs that focused more on treatment and services appeared to produce better results than did strict surveillance-based ISP programs in terms of addressing offender needs and reducing recidivism.

The latter finding regarding the potential relationship between ISP programs with a stronger treatment orientation and lower rates of recidivism has been reported in several other studies.[50] In response to this finding, and a growing concern regarding the safe management of certain offender groups in the community, specialized caseloads or courts have emerged that integrate the stringent monitoring associated with ISP and intensive treatment regimens. Specialized programs for sex offenders, drug offenders and domestic violence offenders, can be found across the United States.[51] A recent study supports this direction, finding that it is the addition of the drug treatment component and not the intensive supervision alone that contributes to lower rates of failure for offenders involved in a drug court program.[52]

Despite less than favorable research results, ISP continues to garner a high level of support and commitment from a broad constituency. Judges enjoy the sentencing options it provides; administrators appreciate the many resources ISP programs have generated for probation and parole; and line staff believe that ISP represents the way probation and parole should have been conducted all along. The popularity of ISP has extended to other countries and to use with juvenile offenders. Some variation of ISP has been implemented in Canada, England, Wales, and New Zealand.[53] In concert with the more rehabilitative mission of juvenile justice programs, most juvenile ISP programs appear to include a stronger treatment component.[54] There are too few methodologically sound evaluations, however, to assess the effectiveness of ISP with juvenile offenders.

Home Confinement/Electronic Monitoring

Home confinement and **electronic monitoring** (EM) are other popular intermediate sanctions. Both are designed to incapacitate offenders by restricting them to their homes for specified periods of time. Typically, offenders subject to home confinement or EM are allowed to leave their homes at preapproved times to look for employment, go to work, or participate in treatment programs. The fundamental difference between the two sanctions is that the surveillance for offenders on home confinement is conducted manually by probation or parole officers through home visits and telephone calls, whereas the surveillance for offenders on EM, as the name implies, is accomplished through electronic means. The earliest recorded use of electronic means to monitor offenders' compliance with home confinement was recorded in Boston, Massachusetts, in 1964.[55] During the late 1980s, the use of EM tripled, leading many to predict that EM would become the dominant form of offender supervision.[56] As recently as 2002, however, it was estimated that only 3 percent of the correctional population in the United States was supervised electronically.[57]

There are two traditional types of EM equipment.[58] The passive system involves random telephone calls to offenders' homes that are generated from a preprogrammed machine, at which time offenders must verify their presence with a coded wristlet or anklet or voice verification. The active system involves the emission of a constant signal from a miniaturized transmitter that is strapped to the offender's ankle/wrist to a receiver placed in the offender's home (**Figure 9–2**). If the offender leaves his or her home, the receiver dials a central computer and a violation is recorded. Newer technology that is used to monitor offenders includes field monitoring devices and Global Positioning System (GPS) devices that allow probation and parole officers to monitor offenders' presence in places other than the offenders' homes.[59] The field monitoring devices, or "drive-by units," consist of a portable receiver carried by the supervising officer and a transmitter worn by the offender. When within 2 to 800 feet of the offender's transmitter, the receiver emits a signal verifying the offender's presence. The field monitoring device is helpful in verifying attendance at treatment and at work. GPS, the newest technology, uses satellites and mapping technologies to track the actual movements of the offender in real time.

Technological malfunctions have created a special set of challenges for community corrections agencies. Problems experienced include interference with the signal, tampering with the wristlets/anklets, false positives (a signal alerting officials of a violation when one has not occurred), and false negatives (the failure of the equipment to alert officials of a violation). Most of the problems have been resolved with improved technology. Identity verification techniques are often used with EM equipment to ensure that it is the offender who is responding to random calls and pages.[60] These techniques include voice verification and transmission devices that send images of fingerprints and faces to the monitoring unit.

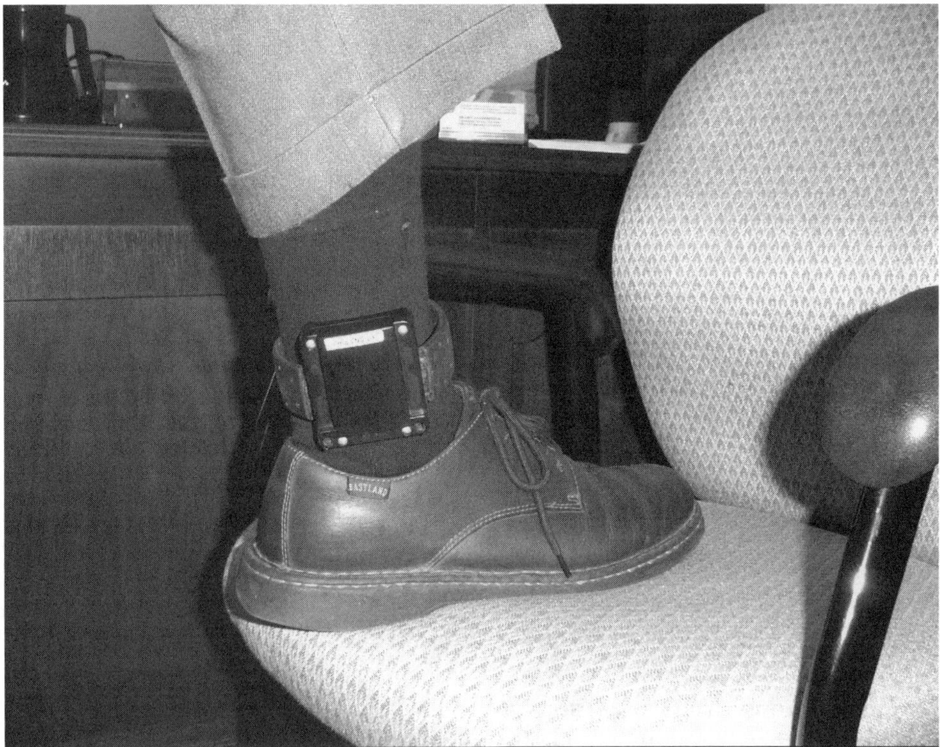

Figure 9–2 Offenders on electronic monitoring must have a transmitter strapped to their ankle or wrist at all times.
Source: *Courtesy of Shawn Satterfield and Paula Bryant, Florida Department of Corrections.*

Electronic surveillance was introduced to enhance offender monitoring and provide a less labor-intensive option to manual surveillance. Research suggests, however, that neither of these benefits has been realized.[61] A comparison of EM and manual supervision found that 42 percent of offenders in both groups had unauthorized absences. Furthermore, EM was found to replace the burden of field work for probation and parole with other labor-intensive duties such as program installation and maintenance, offender scheduling, coordination with the EM service provider, and the verification of and responses to violations.

As the technology continues to evolve, so do discussions regarding ethical concerns associated with its use.[62] Among those concerns is the movement toward overregulation and the potential threat to offenders' right against cruel and unusual punishment. According to a 1986 article by Del Carmen and Vaughn, EM does not violate the standard against cruel and unusual punishment because (1) offenders, at least theoretically, are given a choice between EM and imprisonment, making their participation in EM voluntary; and (2) EM is far more humane and less intrusive and oppressive than imprisonment.[63] Another concern has to do with the escalating infringement on offenders' right to privacy that is often discussed in the context of the Fourth Amendment (see Chapter 3). Previous challenges to the use of warrantless searches as part of traditional probation and parole supervision have been overruled because of the diminished constitutional rights associated with the offenders' legal status; as a condition of their supervision, offenders must agree to submit to warrantless searches.[64] More recent analysis of this issue as it relates to EM concludes that because most EM techniques are less intrusive than are the warrantless searches conducted by probation officers (e.g., rooting through offenders' closets and refrigerators), they do not constitute a violation of offenders' rights against unreasonable searches and seizures. Although EM has not yet been successfully challenged on either of these grounds to date, the advent of GPS is sure to evoke more challenges of this nature.

Another ethical issue associated with the use of EM revolves around the clause in the Fourteenth Amendment guaranteeing all persons equal protection under the law. Offenders may not have equal access to the technology if they cannot afford the supervision fees frequently required of EM participants to cover a portion of the costs of the program or the telephone needed to operate the equipment.[65] Most agencies address the first issue by creating a sliding scale fee to accommodate low-income and indigent offenders.

EM is typically reserved for low-risk, nonviolent offenders, including drug offenders, drunk drivers, and property offenders. Studies of these programs have revealed low arrest rates ranging from 3.3 to 16 percent.[66] Recidivism rates reported for these programs range from 6 to 30 percent with most failures a result of technical violations and absconsions (i.e., escapes).[67] The few studies available on EM programs that target high-risk offenders, however, report no significant differences in rates of recidivism between the EM group and matched groups of offenders who were not subjected to EM.[68] One study that compared the outcomes of violent male parolees who were subject to EM to a control group of similar offenders who received traditional parole supervision found that although there were no overall differences in the rates of recommitment, EM reduced the likelihood of and postponed the return to prison for sex offenders.[69] A study on the use of EM with 99 juvenile offenders revealed that only 58 percent of the youth successfully completed their term of supervision.[70] Of the failures, two-thirds of them were a result of the youth tampering with or removing their anklet. Researchers speculated that these youth experienced difficulties coping with their home environment and the extensive limits on their freedom.

As with ISP, better results have been found for those offenders who participated in treatment in conjunction with EM.[71] It is speculated that EM increases the length of offenders' participation in treatment, which in turn increases the likelihood of positive therapeutic outcomes.[72] Additionally, studies suggest that EM strengthens offenders' social bonds by promoting better job attendance and performance and more involvement with family.[73]

Because EM frequently targets low-risk offenders for participation, and because it is associated with higher rates of technical violations that lead to subsequent incarceration, it has limited potential as a strategy for diverting offenders from prison. Its best use may be as a stabilizing force for offenders on probation or parole while they receive necessary treatment. The continued evolution of EM suggests that it will persist as an important correctional option.

Day Reporting Centers

The first known DRC was implemented in Massachusetts in 1986.[74] Within eight years, 114 DRCs were operating in 22 states.[75] **Day reporting centers (DRCs)** are often referred to as "one-stop shops" for offenders who live at home and are required to report to the center on a daily basis. Although program foci vary, the most common model of DRCs serves both treatment and surveillance functions. During the day, offenders participate in a variety of services including individual and group counseling, substance abuse treatment, life skills training, literacy training, job skills training, and general education diploma preparation classes. They may also meet with a probation officer and submit to drug testing. At night, offenders are monitored through various surveillance strategies including random telephone calls, drug testing, curfews, field visits, and EM.[76]

Most studies of DRCs have been limited to descriptive analyses.[77] A national survey of DRCs conducted in 1995 revealed varied rates of success with 14 to 86 percent of offenders being unsuccessfully terminated from the programs.[78] Several studies have reported that longer participation in the DRC is associated with a lower likelihood of being rearrested.[79] An evaluation of DRCs operating in Ohio found that a slightly higher percentage of DRC participants were rearrested (52.8%) than were nonequivalent comparison groups of offenders on regular probation (45.7%), intensive supervision (49.1%), and prison (50.5%).[80] Aside from these few studies, there is very little basis for assessing the effectiveness of DRCs in reducing offender recidivism.

Boot Camps/Shock Incarceration

Boot camps, or shock incarceration programs, were first implemented by the Georgia Department of Corrections in 1983. With their strong focus on discipline and their association with the powerful cultural tradition of military boot camps, they garnered strong political support that led to their proliferation across the nation.[81] By 1998, adult correctional boot camp programs were operating in 28 states and the Federal Bureau of Prisons.[82] Boot camps for juvenile offenders have also been developed.

Nonviolent, first-time offenders are the primary target population for most boot camps. In many cases, however, offenders with a prior criminal record have been selected for participation. Boot camps range from 90 to 180 days in length. The distinguishing elements of boot camps include military drills and ceremony, physical training, strict discipline, and hard labor. As boot camps have evolved, they have placed more emphasis on rehabilitative components.

Research on boot camps has focused on their effect on inmate attitudes, recidivism, prison crowding, and costs. A study comparing 27 juvenile boot camps to 22 traditional secure facilities found that boot camp inmates reported more favorable perceptions of the institutional environment; they perceived less danger and viewed program elements as more conducive to change.[83] At the same time, a study involving an experimental design and pre- and postprogram measures of antisocial attitudes found no differences between the two groups in terms of changes in antisocial attitudes.[84]

There is no evidence that boot camps reduce offender recidivism. A meta-analysis on 44 quasi-experimental and experimental studies available in 2000 revealed no effect on recidivism when comparing boot camp participants with offenders sentenced to more traditional correctional sanctions.[85] Additional results from this meta-analysis suggested that the programs that did reduce recidivism among boot camp participants had a stronger therapeutic component and a period of aftercare. A recent study tested a more rehabilitative model of boot camp and found that, although there were no significant differences in recidivism rates between boot camp participants and offenders released from prison overall, the boot camp was more effective at reducing the recidivism of offenders with a prior criminal record.[86] This latter finding highlights an area for further research, given that boot camps typically target lower risk offenders for participation.

The back-end approach to boot camps may offer the most hope for saving prison bed space and money. That is, rather than a sentencing option for judges, boot camps frequently operate as a back-end option that transfers offenders directly from prison into the program. This and the shorter length of stay in the boot camp enhance the chance of cost savings. The high failure rates, however, minimize the extent to which these cost savings can be achieved over the long term.

Halfway Houses

Halfway houses are among the oldest type of community corrections programs. They are nonsecure, community-based residential facilities for probationers or parolees. **Halfway houses** provide a structured setting that gives offenders the op-

Controversies

Correctional boot camps are controversial. Proponents argue that this sanction instills discipline and respect for authority, while also saving money as a result of shorter sentences. Opponents argue that there is much potential for abuse of inmates and that military regimentation does little to reduce recidivism. What do you think are the merits of these arguments?

portunity to gain some stability in their lives. Most halfway houses require residents to maintain employment during their stay and to assume responsibility for cooking and caring for the residence. Halfway houses are both a front- and back-end sanctioning option: Probationers are often placed in halfway houses as a condition of their probation, and parolees are often placed in halfway houses upon their release from prison to facilitate their reentry to the community. Some halfway houses are administered by government agencies, such as a Department of Corrections, but the majority are run by private, nonprofit agencies that have a contract with the state or county that funds offenders' stays.

The types of services provided by a halfway house are varied. Many provide in-house services, but most refer offenders to services in the community. Some provide generalized services including job readiness, life skills training, and cognitive skills training while others focus on a particular area of need such as drug and alcohol abuse. The most common area of focus for halfway houses is employment. Halfway houses are staffed by counselors who oversee the daily activities of offenders, work individually with offenders, and conduct support groups.

One of the biggest problems facing halfway houses is the not in my backyard syndrome. Community members resist having halfway houses in their neighborhood for two key reasons. First, they are fearful that having offenders in the neighborhood will expose them to a greater risk of crime. Second, they are concerned that having a halfway house close by will reduce their property value.[87] Early research on halfway houses refutes these concerns.[88]

Because halfway houses vary markedly in their purpose and approach to working with offenders, it is difficult to draw conclusions regarding their effectiveness. Several studies report slightly lower recidivism rates for offenders who complete the halfway house program as compared to offenders subjected to other forms of sanctions (e.g., regular probation or parole supervision),[89] but the majority of studies reveal no significant differences in outcomes.[90]

Race and Gender in Corrections

A study that followed probationers for seven years following release from a Michigan halfway house program found that, in comparison to whites, African-Americans were significantly (1) less likely to be discharged successfully from the program, and (2) more likely to experience a new arrest during the seven-year period. Moreover, the study found that African Americans who were *successfully* discharged from the program had a *higher* likelihood of recidivism than did whites who were *unsuccessfully* discharged.[91] What factors do you think might explain these findings?

Summary of Research on Intermediate Sanctions

A review of the literature on intermediate sanctions leads to several conclusions regarding their effectiveness in achieving their stated goals. It is important, however, to recognize the limitations of this research. A comprehensive review by the University of Maryland found that most studies on intermediate sanctions suffered from methodological problems, including small sample sizes, lack of comparison groups, and short follow-up periods. Nonetheless, there are five core conclusions that have been drawn, and each is discussed in the following subsections.

No Significant Reductions in Recidivism Have Been Achieved

The results of evaluations clearly indicate that intermediate sanctions, in their present form, have no significant impact on offender recidivism as measured by new arrests, technical violations, and a return to prison. A comparison of arrest rates shows that offenders in the more control-oriented intermediate sanction programs have similar or higher rates of arrest than do offenders in traditional sanctioning options. Furthermore, ISP, EM, and DRCs appear to drive up the rate of technical violations among participants. In many cases, placement in one of these intermediate sanctions is seen as the offender's last stop before prison. Thus, the response to techical violations is far more stringent and offenders are much more likely to be revoked for technical violations than are offenders in traditional options.

Increased Rates of Technical Violations

As stated earlier, ISP, EM, and DRCs have all been found to be associated with increased rates of technical violations among their participants. One source of this problem has been referred to as "sanction stacking." It is rare that any of these intermediate sanctions is imposed as a stand-alone sanction. Normally, an offender is placed on probation with community service or EM as a condition of that probation. Along with that comes drug testing, increased reporting requirements, and a host of other conditions of supervision. It is speculated that the high expectations placed on offenders, and the multiple conditions of their probation, create a situation in which the likelihood of their success is low. The higher level of surveillance naturally contributes to increased detection of both technical and law violations. Some argue that this increased ability to detect violations is a positive benefit of intermediate sanctions; it contributes to public safety by getting noncompliant offenders off the street.[92] What is still unknown, however, is whether technical violations are truly proxies for future criminal behavior that would threaten public safety. For example, does not complying with a curfew mean that offenders are also engaging in crime? What we do know is that the increased rate of technical violations combined with more stringent responses to these violations undermines the ability of intermediate sanctions to achieve the goals of diversion from incarceration and cost savings.

No Significant Diversionary Impact

A critical review of the research conducted on intermediate sanctions to date suggests that they are not relieving prison crowding because of net widening and high revocation rates.[93] Net widening (see Chapter 8) refers to the practice of placing in intermediate sanctions offenders who would have gotten a lesser penalty (e.g., regular probation) had the intermediate sanction not been available. Judges are sometimes reluctant to place higher-risk offenders in new programs that are likely to be closely scrutinized. This caution leads them to pull from low-risk populations, essentially increasing the level of punishment and supervision received by offenders rather than decreasing it as originally intended.

More Expensive Than Originally Thought

Simple cost comparisons such as those reported in Georgia's Department of Corrections Annual Report 2000 imply that intermediate sanctions save money (**Figure 9–3**). Certainly the daily cost of even the most expensive community-based sanction is less expensive than the daily cost of incarceration. Net widening, however, limits the extent of cost savings, and if offenders fail and are ultimately

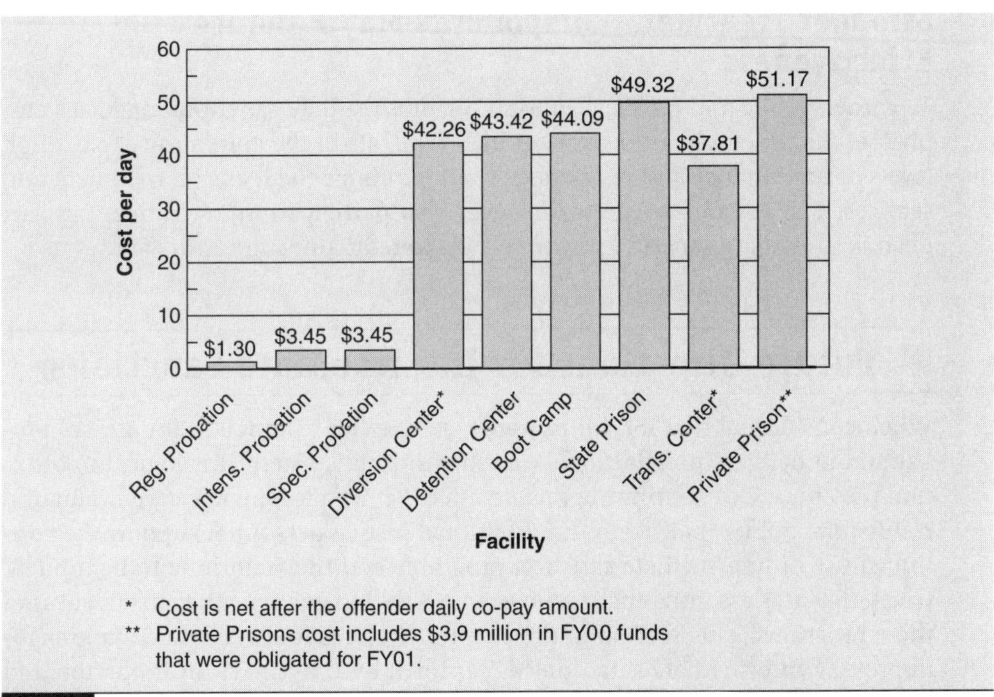

Figure 9–3 The daily cost of intermediate sanctions.
Source: *Reproduced from Georgia Department of Corrections 2002 Annual Report.*

incarcerated, taxpayers have to pay for the cost of the intermediate sanction, expenses associated with the revocation process, and the cost of incarceration, making these sanctions more expensive than originally thought.[94]

Provide an Intermediate Form of Punishment

The most promising result of experimentation with intermediate sanctions is that they appear to provide an intermediate form of punishment that is proportionate to the crime.[95] Having a variety of sanctioning options at their disposal provides prosecutors with more bargaining chips during plea negotiations. More important, it enables judges to better match the sentence to the seriousness of the crime and allows judges to tailor the punishment to the offender's situation (e.g., family, employment).

Having intermediate sanctions available is also beneficial for community corrections agencies that must respond to violations of probation and parole. Prior to the implementation of intermediate sanctions, officers could respond in one of two ways. They could let the offender slide with no more than a verbal reprimand, or they could pursue revocation. Neither of these responses is optimal. The first reduces the credibility of probation and parole and sends offenders the message that they will not be held accountable for their behavior. The second, in many cases, is an overreaction that leads to unnecessary incarceration.

Stronger Treatment Components May Enhance Effectiveness

To garner public and political support for intermediate sanctions, agencies emphasize the development of "prison-like" controls in the community (e.g., high levels of surveillance and restrictions) and minimize the focus on treatment and services. The research suggests, however, that participation in treatment as part of an intermediate sanction program leads to reductions in recidivism.

Future Directions for Intermediate Sanctions

When considering that ISP, for example, costs twice as much as traditional probation but delivers no additional measure of public safety, it is logical to question the efficacy of continuing this practice. Yet despite the negative evaluation results, the public, politicians, and criminal justice personnel support the continued use of intermediate sanction programs and they continue to be funded. Given this, it is essential that agencies find a way to enhance the effectiveness of these programs. This section of the chapter explores four recommendations for improving intermediate sanctions as set forth by the American Probation and Parole Association.[96]

Agreeing on Goals

The goals of intermediate sanction programs are often conflicting. For example, the more stringently they impose punitive conditions (as a means of providing an intermediate punishment and increasing public protection), the more likely they are to exacerbate prison crowding and to approach the costs of imprisonment.[97] Additionally, the claim of reduced costs underestimates the increased level of staffing required for intermediate sanctions and the costs of services needed to achieve rehabilitative aims.[98] It is difficult, but important, to balance system goals (e.g., diversion, cost savings) with sanctioning goals (e.g., deterrence, rehabilitation). Differentiating short- and long-term goals reflects the importance of both in-program crime control and long-term behavioral change and clarifies what the program can be expected to accomplish within specified time frames.

Avoid Net Widening

Defining, identifying, and selecting the target population for participation in intermediate sanctions is one of the most problematic areas of program development. Most intermediate sanction programs exclude certain types of offenders (i.e., violent offenders, repeat offenders) to the extent that the target populations have been reduced to low-risk offenders. This inappropriate selection of offenders leads to net widening and the ineffective allocation of resources. Placing low-risk offenders in high-intensity programs has been found to increase their recidivism.[99] Additionally, some high-risk offenders who are being excluded from the programs by policy are placed under regular supervision rather than sent to prison (something known as "net narrowing") and are not getting the level of services and supervision they need. To avoid net widening, agencies should use a reliable risk/need assessment instrument (see Chapter 8) and place offenders in the least intrusive sanctioning option that is needed to ensure public safety. Offenders with a low risk of recidivism can be placed under minimal probation supervision, while offenders with moderate to high risk of recidivism can be placed on ISP or EM.

Revisit the Near-Exclusive Focus on Surveillance and Control

As stated previously, most intermediate sanctions emphasize intensive surveillance, control, and punishment. Considering the research findings that suggest correlations between participation in rehabilitative programs and recidivism reduction, it may be more important to provide intensive assistance and services that address those needs most strongly related to the offender's criminal behavior. A balanced approach to supervision is recommended with equal emphasis placed on intervention, surveillance, and enforcement. The key, however, is the order in which these components are applied. Current practices center around surveillance and enforcement activities and occasionally interject treatment. A more proactive approach is the identification of and intervention with offenders' needs as a means of promoting behavioral change. If successful with intervention as ascertained through monitoring and surveillance, the application of expensive enforcement strategies becomes unnecessary. This multifaceted approach to offender intervention is congruent with public opinion polls revealing that, in addition to punishment, citizens view rehabilitation as an important correctional goal.[100]

Allocate Sufficient Resources

Joan Petersilia, an expert in the area of intermediate sanctions, calls for a "reinvestment in community corrections."[101] She cites statistics suggesting that although three-fourths of convicted offenders are under some form of community supervision, community corrections agencies get only one-tenth of the correctional budget. Probation and parole populations have grown at a faster rate than prison populations have, yet the budget has not kept pace. This inadequate level of funding is contributing to high caseloads, a lack of program integrity, and a lack of offender services, all of which impede any meaningful programming capable of reducing recidivism. Petersilia proposes that policymakers allocate a sufficient level of resources to allow programs to be implemented as designed and provide a sufficient level of treatment to an increasingly serious probation and parole caseload.

Impact of Intermediate Sanctions on Corrections

Intermediate sanctions clearly did not deliver on their promise of prison diversion. Although growth in the prison population has slowed, it continues to increase year by year. Prisons continue to operate from 1 to 18 percent above capacity and although the majority of prisoners are violent offenders (51%), there are still many nonviolent, low-level offenders behind bars who might be better served in the community.[102]

The advent of intermediate sanctions initially brought attention and status to probation and parole, agencies that were previously viewed as the stepchildren of the criminal justice system. It opened the door for innovation. Federal agencies supported pilot projects to test the feasibility of these new programs and allocated funds for their evaluation. The level of funding, however, was never sufficient, and many intermediate sanction programs are now understaffed or abandoned altogether.

Because of intermediate sanctions, a lot has been learned about what does not work in reducing offender recidivism. Increasingly, through this and other work, more also is being learned about what does work. Fortunately, agencies are beginning to recognize the importance of a balanced approach to supervision that includes a strong treatment component, enhancing the chance that intermediate sanctions will become a valued entity in the corrections continuum.

READY FOR REVIEW

- Community corrections programs are those programs designed to treat and control offenders within the community.
- Community Correction Acts were passed to encourage the development of community-based correctional options that would divert offenders from crowded prisons and reduce correctional costs.
- Most jurisdictions have developed a continuum of intermediate sanctions that fall somewhere between regular probation and incarceration.
- Three common economic sanctions include restitution, supervision fees, and fines. Day fines were introduced as a way to achieve equity in sentencing.
- Community service is a popular sanction that involves offenders providing free labor to nonprofit and government agencies. Although community service provides economic benefits to communities, its effect on recidivism is unclear.
- Intensive supervision programs (ISP programs), the most popular form of intermediate sanctions, are designed to restrict offenders' movement in the community and deter future criminality through the provision of intensive surveillance and threats of incarceration. Research suggests that ISP programs have not been successful in reducing offender recidivism.
- Home confinement and electronic monitoring are popular sanctions that restrict offenders to their homes for specified periods of time. The only difference between the two sanctions is that one is monitored manually by probation officers, and the other is monitored through electronic means. Studies suggest that both can be effective with low-level offenders.
- DRCs are one-stop shops where offenders report for treatment and monitoring. Other than a few descriptive studies, no research is available for assessing their effectiveness in reducing offender recidivism.
- Boot camps are characterized by military drills and ceremony, physical training, strict discipline, and hard labor. Because boot camps operate as a back-end option that transfers offenders directly from prison, they may provide the most hope for saving prison bed space and money. They have not, however, been successful at reducing offender recidivism.
- Halfway houses provide probationers and parolees with an opportunity to gain stability in their lives. Services offered by halfway houses include practical life skills training. As with other community corrections programs, there is no evidence that halfway houses lead to reduced recidivism.
- Because of the variation in the design and integrity of intermediate sanctions programs, the results of research on specific intermediate sanctions programs are mixed. Overall, however, the intermediate sanctions movement has failed to divert offenders from prison, save money, or reduce recidivism.

- To achieve their stated goals, intermediate sanctions programs should avoid net widening by carefully targeting offenders for participation who would have otherwise been incarcerated, provide intensive services that are designed to address offenders' risks and needs, and allocate a sufficient level of resources to ensure the proper implementation of the programs.

KEY TERMS

boot camps Residential programs for nonviolent, first-time offenders that are characterized by military drill and ceremony, physical training, strict discipline, and hard labor

community corrections Agencies and programs at the local, state, or federal level that are responsible for the punishment and management of adult offenders within their local communities

Community Corrections Acts Legislation that provided a structure for generating, coordinating, funding, and evaluating community corrections programs

community service A sanction that requires offenders to provide free labor to a public or nonprofit agency in the community

continuum of sanctions A list of sanctions that are ordered from least restrictive (e.g., regular probation) to most restrictive (e.g., prison)

day fines Fines that are "means-based," that is, in addition to basing the amount of the fine on the severity of the crime, it is also based on the offenders' daily income

day reporting centers (DRCs) Centers where offenders who live at home are required to report daily to receive treatment services, check in with their probation officers, and submit to drug testing

electronic monitoring (EM) A period of home confinement that is monitored through electronic means

fines Monetary penalties, the amount of which is generally determined by the severity of the offense

halfway house Nonsecure, community-based residential facilities that are designed to provide structure and stability for probationers or parolees

home confinement An intermediate sanction that restricts offenders to their homes except during preapproved times when they are allowed to leave their homes to look for employment, go to work, or participate in treatment programs

intensive supervision probation/parole (ISP) A form of probation/parole designed to restrain offenders' movement in the community through a more intensive level of intervention, including frequent office and field visits and more stringent conditions of supervision

intermediate sanctions Sanctioning options that fall on a continuum between traditional probation and traditional incarceration

restitution A sanction that requires offenders to compensate victims for losses incurred as the result of the crime

restorative justice A philosophy of justice that seeks to restore the offender, victim, and community to their state of being prior to the crime being committed by providing services and sanctions that build offenders' competencies, hold offenders accountable, restore losses incurred by the victim, and involve the community in sentencing

supervision fees Fees collected from offenders under some form of community supervision that are used to cover the cost of supervision, electronic monitoring, drug testing, and treatment

YOU ARE THE CORRECTIONS PROFESSIONAL SUMMARY

1. What type of intermediate sanctions would you recommend? Why? Evaluations suggest that fines and community service are promising approaches to keeping offenders in the community, but research on other sanctions suggests that they have not had a strong diversionary impact or reduced offender recidivism. More important than the choice of specific sanction, however, are the choices made about the design and implementation of various sanctions.
2. What steps would you take to ensure that these sanctions achieve the goals of prison diversion and cost savings? The research suggests that to save money and have any diversionary effect, agencies must target offenders who would have otherwise gone to prison and provide intensive treatment services in addition to intensive controls.

NOTES

1. F. T. Cullen, J. Wright, and B. Applegate, "Control in the Community: The Limits of Reform," in *Choosing Correctional Interventions That Work: Defining the Demand and Evaluating the Supply*, ed. A. T. Harland, 6, 9–116 (Newbury Park, CA: Sage, 1996).
2. Center for Community Corrections, "Why Community Corrections?" www.communitycorrectionsworks.org/why%20corrections.htm.
3. M. K. Harris, *The Goals of Community Sanctions* (Washington, DC: U.S. Department of Justice, National Institute of Corrections, 1986).
4. Center for Community Corrections, "Why Community Corrections?"

5. American Bar Association, *Model Adult Community Corrections Act* (Washington, DC: American Bar Association, 1992).
6. National Institute of Corrections, *Survey of Intermediate Sanctions* (Washington, DC: Office of Justice Programs, 1990).
7. M. Tonry and N. Morris, "Between Prison and Probation: Toward a Comprehensive Punishment System," in *Criminal Justice System: Politics and Policies*, 7th ed., eds. G. F. Cole and M. G. Gertz, 407–419 (Belmont, CA: Wadsworth, 1998).
8. Bureau of Justice Statistics, *State Court Organization, 1998* (Washington, DC: U.S. Department of Justice, 2002).
9. M. Tonry, "Intermediate Sanctions in Sentencing Guidelines," in *Crime and Justice: A Review of the Research,* ed. M. Tonry, 199–253 (Chicago: University of Chicago Press, 1996).
10. A. Harland, "Sanctions vs. Punishments," in *The Intermediate Sanctions Handbook: Experiences and Tools for Policymakers*, eds. P. McGarry and M. M. Carter (Washington, DC: State Justice Institute, 1993).
11. P. B. Wood and H. G. Grasmick, "Inmates Rank the Severity of Ten Alternative Sanctions Compared to Prison," *Journal of the Oklahoma Criminal Justice Research Consortium* 2(August 1995): 30–42.
12. F. G. Mullaney, *Economic Sanctions in Community Corrections* (Washington, DC: U.S. Department of Justice, National Institute of Corrections, 1998).
13. Mullaney, *Economic Sanctions*, 2.
14. Mullaney, *Economic Sanctions*, 2; K. D. Morgan, "Officer Attitudes About Supervision Fee Collection in Alabama," *Federal Probation* 59, no. 4 (1995): 62–65; D. Parent, *Recovering Correctional Costs through Offender Fees* (Washington, DC: U.S. Department of Justice, National Institute of Justice, 1990).
15. Office for Victims of Crime, "Restitution: Making It Work," *Legal Series Bulletin*, NCJ 189193 (2002).
16. American Probation and Parole Association, *A Guide to Enhancing Victim Services within Probation and Parole* (Lexington, KY: American Probation and Parole Association, 1994).
17. M. Lipsey and D. Wilson, "Effective Intervention for Serious Juvenile Offenders: A Synthesis of the Research," in *Serious and Violent Juvenile Offenders: Risk Factors and Successful Interventions,* eds. R. Loeber and D. P. Farrington, 313–345 (Thousand Oaks, CA: Sage, 1998); P. Gendreau and T. Little, "A Meta-Analysis of the Effectiveness of Sanctions on Offender Recidivism" (unpublished manuscript, Department of Psychology, University of New Brunswick, St. John, 1993).

18. A. Seymour, "Putting Victims First," *State Government News* 39, no. 9 (1996): 24–25.
19. Colorado Legislative Counsel, *Study of Criminal Restitution in Colorado* (Denver, CO: Colorado Legislative Counsel, 1999).
20. Office for Victims of Crime, "Restitution: Making It Work."
21. John Howard Society of Alberta, "Correctional User Fees," 2001, www.johnhoward.ab.ca/docs/UserFees/cover.htm.
22. D. Parent, *Recovering Correctional Costs through Offender Fees* (Washington, DC: U.S. Department of Justice, National Institute of Justice, 1990).
23. P. Finn and D. Parent, "Making the Offender Foot the Bill: A Texas Program," *National Institute of Justice Program Focus* (Washington, DC: U.S. Department of Justice, National Institute of Justice, 1992).
24. John Howard Society of Alberta, "Correctional User Fees."
25. Finn and Parent, "Making the Offender Foot the Bill"; C. R. Ring, "Probation Supervision Fees: Shifting Costs to the Offender," *Federal Probation* 53, no. 2 (1989): 43–48.
26. F. G. Mullaney, *Economic Sanctions in Community Corrections* (Washington, DC: U.S. Department of Justice, National Institute of Corrections, 1988).
27. M. R. Durose and P. A. Langan, "Felony Sentences in State Courts, 2002," *Bureau of Justice Statistics Bulletin*, NCJ 206916 (2004).
28. S. T. Hillsman, B. Mahoney, G. F. Cole, and B. Auchter, *Fines in Sentencing: A Study of the Use of the Fine as a Criminal Sanction. Final Report* (Washington, DC: U.S. Department of Justice, National Institute of Justice, 1984).
29. Center for Community Corrections, "Why Community Corrections?"
30. Bureau of Justice Assistance, *How to Use Structured Fines (Day Fines) as an Intermediate Sanction* (Washington, DC: U.S. Department of Justice, Bureau of Justice Assistance, 1996).
31. D. Glaser and M. A. Gordon, "Profitable Penalties for Lower Level Courts," *Judicature* 73, no. 5 (1990): 248–252.
32. S. Turner and J. Petersilia, "Day Fines in Four US Jurisdictions," in *Research in Brief* (Washington, DC: U.S. Department of Justice, National Institute of Justice, 1996).
33. C. Worzella, "The Milwaukee Municipal Court Day-Fine Project," in *Day Fines in American Courts: The Staten Island and Milwaukee Experiments*, ed. D. C. McDonald, 61–78 (Darby, PA: Diane Books Publishing Company, 1992).
34. Durose and Langan, "Felony Sentences in State Courts."
35. J. Hudson and B. Galaway, *Criminal Justice, Restitution, and Reconciliation* (Monsey, NY: Criminal Justice Press, 1990); D. Maloney and M. Umbreit,

"Managing Change: Toward a Balanced and Restorative Justice Model," *Perspectives* 19, no. 2 (Monsey, NY: Criminal Justice Press, 1995): 43–46.

36. B. Fulton, *Restoring Hope through Community Partnerships: The Real Deal in Crime Control* (Lexington, KY: American Probation and Parole Association, 1996).
37. S. Levrant, F. Cullen, B. Fulton, and J. Wozniak, "Reconsidering Restorative Justice: The Corruption of Benevolence Revisited?" *Crime and Delinquency* 45, no. 1 (1999): 3–27.
38. Center for Community Corrections, "Why Community Corrections?"
39. R. Maher, "Federal Community Service—Northern District of Georgia," in *Restoring Hope through Community Partnerships: The Real Deal in Crime Control,* ed. B. Fulton, 175–178 (Lexington, KY: American Probation and Parole Association, 1996).
40. M. G. Neithercutt and D. M. Gottfredson, *Caseload Size Variation and Differences in Probation/Parole Performance* (Washington, DC: National Center for Juvenile Justice, 1975); R. Carter and L. T. Wilkins, "Caseloads: Some Conceptual Models," in *Probation, Parole and Community Corrections,* 2nd ed. (New York: Wiley, 1976); J. A. Banks, L. Porter, R. L. Rardin, R. R. Sider, and V. E. Unger, *Evaluation of Intensive Special Probation Projects: Phase I Report* (Washington, DC: U.S. Department of Justice, 1977).
41. J. Petersilia and S. Turner, *Intensive Supervision for High-Risk Probationers: Findings from Three California Experiments* (Santa Monica, CA: RAND Corporation, 1990).
42. Banks et al., *Evaluation of Intensive Special Probation Projects.*
43. L. Sechrest, S. White, and E. D. Brown, *The Rehabilitation of Criminal Offenders: Problems and Prospects* (Washington, DC: National Academy of Sciences, 1979).
44. Neithercutt and Gottfredson, *Caseload Size Variation and Differences*; Carter and Wilkins, "Caseloads"; Banks et al., *Evaluation of Intensive Special Probation Projects.*
45. T. Clear and P. Hardyman, "The New Intensive Supervision Movement," *Crime and Delinquency* 36, no. 1 (1990): 42–60.
46. B. Erwin, *Evaluation of Intensive Probation Supervision in Georgia* (Atlanta: Georgia Department of Corrections, 1987); F. Pearson, *Final Report of Research on New Jersey's Intensive Supervision Program* (New Brunswick, NJ: Rutgers University Institute for Criminological Research, 1987).
47. J. M. Byrne, A. Lurigio, and C. Baird, "The Effectiveness of the New Intensive Supervision Programs," *Research in Corrections* 2(1989): 1–48.

48. J. M. Byrne and L. Kelly, *Restructuring Probation as an Intermediate Sanction: An Evaluation of the Massachusetts Intensive Probation Program* (final report to the National Institute of Justice, Research Program on the Punishment and Control of Offenders, 1989); D. Wagner, "An Evaluation of High Risk Offender Intensive Supervision Project," *Perspectives* (Summer 1989): 22–27; U.S. General Accounting Office, "Intermediate Sanctions: Their Impacts on Prison Crowding, Costs, and Recidivism Are Still Unclear" (Washington, DC: U.S. General Accounting Office, 1990).

49. J. Petersilia and S. Turner, *Evaluating Intensive Supervision Probation/Parole: Results of a Nationwide Experiment* (Washington, DC: National Institute of Justice, 1993).

50. Byrne and Kelly, *Restructuring Probation as an Intermediate Sanction*; E. J. Latessa, "An Evaluation of the Lucas County Adult Probation Department's IDU and High Risk Groups" (unpublished manuscript, University of Cincinnati, Ohio, 1993); Petersilia and Turner, *Evaluating Intensive Supervision Probation/Parole.*

51. See, for example, S. Adams, "Specialized Sex Offender Probation in Cook County Links Supervision, Treatment," *On Good Authority* 4, no. 7 (2001): 1–4; M. J. Junuwine, R. Simmons, and E. Swies, "Community Supervision of Sex Offenders: Integrating Probation and Clinical Treatment," *Federal Probation* 67, no. 3 (2003): 20–27; D. Banks and D. C. Gottfredson, "Effects of Drug Treatment and Supervision on Time to Rearrest among Drug Treatment Court Participants," *Journal of Drug Issues* 33, no. 2 (2003): 385–412; R. R. Johnson, "Intensive Probation for Domestic Violence Offenders," *Federal Probation* 65, no. 3 (2001): 36–39.

52. Banks and Gottfredson, "Effects of Drug Treatment and Supervision."

53. D. Evans, "New Zealand's Approach to a High-Risk Offender Supervision Program," *Corrections Today* 67, no. 2 (2005): 28–31; M. Little, J. Kogan, R. Bullock, and P. Van Der Laan, "Experiment in Multi-Systemic Responses to Persistent Young Offenders Known to Children's Services," *British Journal of Criminology* 44, no. 2 (2004): 225–240.

54. Juvenile Rehabilitation Administration, "Intensive Parole for High-Risk Offenders," 2002, www1.dshs.wa.gov/pdf/ea/govrel/leg1202/ipr.pdf; J. S. Meisel, "Relationships and Juvenile Offenders: The Effects of Intensive Aftercare Supervision," *Prison Journal* 81, no. 2 (2001): 206–245; A. A. Robertson, P. W. Grimes, and K. E. Rogers, "Short-Run Cost-Benefits Analysis of Community-Based Interventions for Juvenile Offenders," *Crime and Delinquency* 47, no. 2 (2001): 265–284.

Chapter Resources

55. A. Crowe, L. Sydney, P. Bancroft, and B. Lawrence, *Offender Supervision with Electronic Technology* (Lexington, KY: American Probation and Parole Association, 2002).
56. R. Corbett, "Electronic Monitoring," *Corrections Today* 74(1989): 78–80.
57. Crowe et al., *Offender Supervision.*
58. Bureau of Justice Assistance, *Electronic Monitoring in Intensive Probation and Parole Programs* (Washington, DC: U.S. Department of Justice, 1989); Crowe et al., *Offender Supervision.*
59. Crowe et al., *Offender Supervision*; J. Hoshen and G. Drake, *Offender Wide Area Continuous Electronic Monitoring Systems: Executive Summary* (Washington, DC: U.S. Department of Justice, National Institute of Justice, 2000).
60. Crowe et al., *Offender Supervision.*
61. T. Baumer, R. Mendelsohn, and E. Rhine, *Executive Summary—The Electronic Monitoring of Non-violent Convicted Felons: An Experiment in Home Detention* (Washington, DC: National Institute of Justice, 1990).
62. See, for example, J. D. Ballard and K. Mullendore, "Legal Issues Related to Electronic Monitoring Programs," *Journal of Offender Monitoring* 15, no. 2 (2002): 5, 16–20; M. Black and R. G. Smith, *Electronic Monitoring in the Criminal Justice System* (Canberra, Australia: Australian Institute of Criminology, 2003); Crowe et al., *Offender Supervision*; B. K. Payne and R. R. Gainey, "Electronic Monitoring: Philosophical, Systemic, and Political Issues," *Journal of Offender Rehabilitation* 31, no. 3/4 (2000): 93–111.
63. R. V. del Carmen and J. Vaughn, "Legal Issues in the Use of Electronic Surveillance in Probation," *Federal Probation* (1986): 60–69.
64. *Griffin v. Wisconsin*, 483 U.S. 868 (1987).
65. Crowe et al., *Offender Supervision*; Ballard and Mullendore, "Legal Issues Related to Electronic Monitoring Programs."
66. M. Brown and S. Roy, "Manual and Electronic House Arrest: An Evaluation of Factors Related to Failure," in *Intermediate Sanctions: Sentencing in the 90s*, eds. J. Smykla and W. Selke, 37–53 (Cincinnati, OH: Anderson, 1995); J. Petersilia, "House Arrest," in *Crime File* (Washington, DC: National Institute of Justice, 1988); L. W. Sherman, D. C. Gottfredson, D. L. MacKenzie, J. Eck, P. Reuter, and S. D. Bushway, "Preventing Crime: What Works, What Doesn't, What's Promising," in *National Institute of Justice Research in Brief* (Washington, DC: U.S. Department of Justice, National Institute of Justice, 1998).
67. Sherman et al., "Preventing Crime."

68. D. Wagner and C. Baird, "Evaluation of the Florida Community Control Program," in *Research in Brief* (Washington, DC: National Institute of Justice, 1993); L. Smith and R. Akers, "A Comparison of Recidivism of Florida's Community Control Program: A Five-Year Survival Analysis," *Journal of Research in Crime and Delinquency* 30(1993): 267–292; A. Jolin and B. Stipak, "Drug Treatment and Electronically Monitored Home Confinement. An Evaluation of a Community-Based Sentencing Option," *Crime and Delinquency* 38(1992): 158–170.
69. M. A. Finn and S. Muirhead-Steves, "Effectiveness of Electronic Monitoring with Violent Male Parolees," *Justice Quarterly* 19, no. 2 (2002): 293–312.
70. T. J. Harig, *Juvenile Electronic Monitoring Project: The Use of Electronic Monitoring Technology on Adjudicated Juvenile Delinquents* (Washington, DC: U.S. Department of Justice, National Institute of Justice, 2001).
71. T. Baumer and R. Mendelson, "Electronically Monitored Home Confinement: Does It Work?" in *Smart Sentencing: The Emergence of Intermediate Sanctions*, eds. J. Byrne, A. Lurigio and J. Petersilia, 54–67 (Newbury Park, CA: Sage, 1992); J. Bonta, S. Wallace-Capretta, and J. Rooney, "Quasi-experimental Evaluation of an Intensive Rehabilitation Supervision Program," *Criminal Justice and Behavior* 27, no. 3 (2000): 312–329; Jolin and Stipak, "Drug Treatment and Electronically Monitored Home Confinement."
72. Bonta et al., "Quasi-experimental Evaluation."
73. Baumer and Mendelson, "Electronically Monitored Home Confinement"; B. K. Payne and R. R. Gainey, "Electronic Monitoring of Offenders from Jail or Prison: Safety, Control, and Comparisons to the Incarceration Experience," *Prison Journal* 84, no. 4 (2004): 413–435.
74. J. McDevitt and R. Miliano, "Day Reporting Centers: An Innovative Concept in Intermediate Sanctions," in *Smart Sentencing: The Emergence of Intermediate Sanctions*, eds. J. Byrne, A. Lurigio and J. Petersilia, 152–181 (Newbury Park, CA: Sage, 1992).
75. D. Parent, J. Byrne, V. Tsarfaty, L. Valade, and J. Esselman, *Day Reporting Centers* (Washington, DC: National Institute of Justice, 1995).
76. McDevitt and Miliano, "Day Reporting Centers"; Parent et al., *Day Reporting Centers*.
77. D. L. McKenzie, "Criminal Justice and Crime Prevention," in *Preventing Crime: What Works, What Doesn't, What's Promising*, eds. L. W. Sherman, D. C. Gottfredson, D. L. MacKenzie, J. Eck, P. Reuter, and S. D. Bushway, chap. 9 (Washington, DC: U.S. Department of Justice, National Institute of Justice, 1998).
78. Parent et al., *Day Reporting Centers*.

79. A. Craddock and L. A. Graham, "Recidivism as a Function of Day Reporting Center Participation," *Journal of Offender Rehabilitation* 34, no. 1 (2001): 81–97; E. Latessa, L. Travis, A. Holsinger, and J. Hartman, "Evaluation of Ohio's Pilot Day Reporting Program: Final Report" (unpublished report, Ohio Office of Criminal Justice Services, 1998); C. Martin, A. J. Lurigio, and D. E. Olson, "Examination of Rearrests and Reincarcerations among Discharged Day Reporting Center Clients," *Federal Probation* 67, no. 1 (2003): 24–30.

80. Latessa et al., "Evaluation of Ohio's Pilot Day Reporting Program."

81. F. T. Cullen, K. R. Blevins, J. S. Trager, and P. Gendreau, "Rise and Fall of Boot Camps: A Case Study in Common-Sense Corrections," *Journal of Offender Rehabilitation* 40, no. 3/4 (2005): 53–70.

82. Bureau of Justice Statistics, *State Court Organization.*

83. D. L. MacKenzie, A. R. Gover, G. S. Armstrong, and O. Mitchell, *National Study Comparing the Environments of Boot Camps with Traditional Facilities for Juvenile Offenders* (Washington, DC: U.S. Department of Justice, National Institute of Justice, 2001).

84. O. Mitchell, D. L. Mackenzie, and D. M. Perez, "Randomized Evaluation of Maryland Correctional Boot Camp for Adults: Effects on Offender Antisocial Attitudes and Cognitions," *Journal of Offender Rehabilitation* 40, no. 3/4 (2005): 71–86.

85. E. L. MacKenzie, D. B. Wilson, and S. B. Kider, "Effects of Correctional Boot Camps on Offending," in *Annals of the American Academy of Political and Social Science*, eds. D. P. Farrington and B. C. Welsh, vol. 578, 126–143 (Thousand Oaks, CA: Sage, 2001).

86. C. A. Kempinen and M. C. Kurlychek, "Outcome Evaluation of Pennsylvania's Boot Camp: Does Rehabilitative Programming within a Disciplinary Setting Reduce Recidivism?" *Crime and Delinquency* 49, no. 4 (2003): 581–602.

87. Kevin Krajick, " 'Not on My Block': Local Opposition Impedes the Search for Alternatives," *Corrections Magazine New York* 6, no. 5 (1980): 15–21, 24–27.

88. R. P. Seiter, E. W. Carlson, H. H. Bowman, J. J. Grandfield, and N. J. Beran, *Halfway Houses* (Washington, DC: Office of Justice Programs, National Institute of Justice, 1977).

89. M. Eisenberg, "Special Release and Supervision Programs: Two Year Outcome Study, Halfway House/PPT Placements" (unpublished report, Texas Department of Criminal Justice, Austin, 1990).

90. E. Latessa and H. Allen, "Halfway Houses and Parole: A National Assessment," *Journal of Criminal Justice* 10, no. 2 (1982): 153–163;

E. Latessa and L. Travis, "Halfway House or Probation: A Comparison of Alternative Dispositions," *Journal of Crime and Justice* 14, no. 1 (1991): 53–75.

91. D. J. Hartmann, P. C. Friday, and K. I. Minor, "Residential Probation: A Seven-Year Follow-up Study of Halfway House Discharges," *Journal of Criminal Justice* 22(1994): 503–515.
92. B. Nidorf, "Nothing Works Revisited," *Perspectives* (Summer 1994): 12–13.
93. J. Petersilia, "Decade of Experimenting with Intermediate Sanctions: What Have We Learned?" *Justice Research and Policy* 1, no. 1 (1999): 9–23; M. Tonry, "Stated and Latent Functions of ISP," *Crime and Delinquency* 36(1990): 174–191.
94. Petersilia, "Decade of Experimenting with Intermediate Sanctions."
95. J. Petersilia, "Evaluating Alternative Sanctions: The Case of Intensive Supervision," *Federal Sentencing Reporter* (July/August 1991): 30–33; M. Tonry, "Stated and Latent Functions of ISP."
96. B. Fulton, M. Paparozzi, and P. Gendreau, "APPA's Prototypical Intensive Supervision Program: ISP as It Was Meant to Be," *Perspectives*, (Spring 1995): 25–41.
97. S. Turner and J. Petersilia, "Focusing on High-Risk Parolees: An Experiment to Reduce Commitments to the Texas Department of Corrections," *Journal of Research in Crime and Delinquency* 29(1992): 34–61.
98. D. Cochran, "A Practitioner's Perspective," *Research in Corrections* 2(1989): 57–63.
99. D. A. Andrews, J. Bonta, and R. D. Hoge, "Classification for Effective Rehabilitation: Rediscovering Psychology," *Criminal Justice and Behavior* 17, no. 1 (1990): 19–52; E. J. Latessa and C. T. Lowenkamp, "Increasing the Effectiveness of Correctional Programming Through the Risk Principle: Identifying Offenders for Residential Placement," *Criminology and Public Policy* 4, no. 2 (2005): 263–291.
100. Bureau of Justice Statistics, *Sourcebook of Criminal Justice Statistics Online*, 2005, www.albany.edu/sourcebook; J. L. Sundt, F. T. Cullen, B. K. Applegate, and M. G. Turner, "The Tenacity of the Rehabilitative Idea Revisited: Have Attitudes toward Offender Treatment Changed?" *Criminal Justice and Behavior* 25, no. 4 (1998): 426–442.
101. J. Petersilia, "A Crime Control Rationale for Reinvesting in Community Corrections," *Spectrum* (Summer 1995): 16–27.
102. Bureau of Justice Statistics, "Prisoners in 2004," in *Bureau of Justice Statistics Bulletin* (Washington, DC: U.S. Department of Justice, Bureau of Justice Statistics, 2005).

Prison Systems and Staff

10

Chapter Objectives

- Describe the origins of the penitentiary.
- Explain the development of the penitentiary in different regions of the United States.
- Understand the primary mission of prison systems.
- Explain how the Department of Corrections headquarters staff supports the operations of individual prisons.
- Describe the different types of prisons, including levels of security and specialized prisons such as the supermax.
- Explain how the training of correctional officers decreases prison disorder and violence.
- Describe some of the major challenges of being a correctional officer.
- Describe elements of the correctional officer subculture.

CASE STUDY

You have been out of the correctional academy for almost one year and are feeling more comfortable working in a prison. During the past year, a long-term inmate helped you maintain control on the living unit and did other small favors that made your shifts go easier (such as telling you when "trouble" was about to occur). You returned the favor by giving him some extra privileges. Last week, you brought in a letter from his wife, even though all mail should have gone through the prison censor to check for contraband. Today, the inmate told you that the letter contained drugs that he shared with his fellow cellmates and that unless you brought in more uncensored letters, they would report your actions to your supervisor. You have heard that officers in similar circumstances have been fired.

1. What alternatives do you have, and what is the best response?
2. What are the pros and cons of the decisions?

Introduction

Prisons operated by state and federal governments and private corrections facilities hold inmates who have been sentenced to a period of imprisonment longer than a year and those condemned to death. Several decades ago, many prisons were characterized by violence, riots, and turmoil, but today they are more settled places. The reduced disorder in correctional facilities today is a result of two centuries of learning about the best ways to manage prisoners.

Useem and Piehl speculate that these conditions are a result of more professional prison management.[1] Correctional professionals and leaders need a comprehensive understanding of both corrections and management. These skills are necessary to coordinate the activities of thousands of staff and manage multi-billion-dollar budgets. The **Federal Bureau of Prisons** (BOP), for example, housed 198,000 inmates in May 2007, and correctional systems in California, Florida, and Texas are almost as large.

Although the management of correctional facilities has improved, perhaps one of the biggest contributions to increased safety and order is the skill and professionalism of today's correctional officers. Even though correctional officers are often portrayed in films in a stereotypical fashion (e.g., as cruel, uncaring, or unethical), officers today are screened carefully before employment and receive more training prior to working with prisoners. Yet correctional work is a tough career, and officers have a unique occupational subculture. This chapter explores the development of the prison, modern prison systems, and prison staff.

History of the Prison

Although jails have existed since the eleventh century (see Chapter 7 for a history of jails), the world's oldest penitentiaries have held prisoners only for about two centuries. You read in earlier chapters that prior to the widespread use of imprisonment, punishments for crimes were often corporal, such as whipping, branding, and placement in stocks or pillories. In addition, offenders could be executed for committing felonies and many earlier methods of execution (e.g., beheading, crucifixion, burning, and even hanging) seem barbaric by contemporary cultural standards. Some countries, such as England and France, also transported offenders to penal colonies. England sent most offenders to Pacific Islands such as Australia and New Zealand (and Virginia and Maryland prior to the Revolution). France, by contrast, established penal colonies in Africa where prisoners served their sentences. Whereas England had abandoned penal transportation in the late 1800s, France was still sending prisoners to Devil's Island in Guyana until it closed in the 1940s.

England and other European countries also developed workhouses or houses of correction; for example, the infamous London Bridewell opened in the mid-1500s. Offenders were sent to workhouses to develop a disciplined lifestyle and productive work habits. Crowding and poor conditions in workhouses helped encourage transportation of convicts to colonies. European workhouses were among the most direct predecessors to the American prison, as was London's Newgate Prison, which was rebuilt in the early 1780s in the aftermath of rioting.

Philadelphia's **Walnut Street Jail** is conventionally regarded as the first American prison. In 1790, the jail was converted from a holding facility into a facility for housing sentenced prisoners. New York opened its first prison (also called Newgate) in 1797. Other states followed the lead of Pennsylvania and New York, thereby launching the penitentiary movement.

At base, penitentiaries were developed to punish, incapacitate, and reform offenders and to deter crime. (See the discussion of philosophies of punishment in Chapter 6.) However, a variety of social and political factors contributed to development of these facilities and the nation's massive penitentiary system that followed.[2]

First, Quakers such as William Penn and reform-minded citizens such as Benjamin Rush (a signatory to the Declaration of Independence) had long advocated for more humane treatment of offenders. Their advocacy was consistent with the ideas of European Enlightenment thinkers, such as Cesare Beccaria, Jeremy Bentham, and John Howard, who regarded the criminal justice system of their day as ineffective, unfair, and inhumane. The penitentiary was presented as an alternative to corporal, capital, and other punishments. The Quakers believed that criminals could be reformed if given time to contemplate their actions, and they lobbied state governments to construct these facilities. The term *penitentiary* was an extension of the word *penance* (a word indicating apology, reparation, and self-punishment).

Another reason for the development of the prison was that offenders were sometimes not punished or were punished lightly. In her study of early American justice systems, King observes that sanctions for many offenses were economic, and penalties were often slight. "Assaults frequently resulted in penny fines; thefts were seldom prosecuted and almost never resulted in a conviction."[3] Even when an offender was convicted, the sentence was not always carried out. State governors often used their ability to pardon criminals using **executive clemency** (the ability of a governor or president to pardon an offender).

As in England, many first-time offenders sentenced to death would have their sentences commuted to warnings, and some convicted offenders were granted benefit of clergy (discussed in Chapter 8 as a forerunner of probation). In medieval times, members of the clergy could argue that their conduct fell outside the jurisdiction of criminal courts, and that they should be judged by church officials. Over time, however, benefit of clergy was extended to most literate first-time offenders. To prevent them from asking for forgiveness more than once, offenders receiving benefit of clergy were branded, usually on the thumb. As a result of these reprieves, relatively few offenders actually received severe punishments even though harsh laws were "on the books."

Given this state of affairs, there was a need for an alternative method of punishing convicted offenders. It was thought that juries and judges would be more likely to convict an offender if the punishment was confinement rather than death or a disfiguring corporal punishment.

Finally, some scholars have suggested that unstable political and economic conditions in the wake of the American Revolution created the need for stronger and more centralized governments at both the federal and state levels. The colonial system of criminal justice was ill-equipped to accommodate the massive population growth, industrialization, and urbanization that followed the Revolution. The penitentiary system was one element in the shoring up of government authority and power.[4]

Controversies

Although presidents have always had the ability to grant an offender a pardon, the use of pardons has decreased significantly in the past five decades. Together, Presidents Roosevelt, Truman, and Wilson granted more than 8,000 pardons. President Clinton, by contrast, granted 456 pardons in his two terms as president, and these created a significant amount of controversy. What factors do you think have contributed to these changes?

Big House Prisons—Silent and Congregate Systems

Pennsylvania developed one the world's first true penitentiary systems, with Philadelphia's Walnut Street Jail (mentioned earlier), the Western State Penitentiary that opened near Pittsburgh in 1826, and the Eastern State Penitentiary (ESP) that opened in Philadelphia in 1829. Although the Walnut Street Jail pioneered some of the approaches that would be introduced at later institutions (such as the segregation of inmates according to offense seriousness and gender), it soon fell into relative obscurity. On the other hand, the other facilities, especially ESP, drew international attention and served as models for prison construction and operation.

At the time of its construction, the ESP was one of the largest and most expensive buildings in the United States. It had indoor plumbing years before this was standard in colleges, hospitals, or private homes. Eastern State Penitentiary was the prototypical **big house** prison and featured walls some 30 feet high and 12 feet thick. The front of the prison has the appearance of a medieval castle, complete with towers, while the interior has been likened to a church. Within the walls, the prison was built like spokes on a wheel, with a central hub, and eight cellblocks radiating outward (something called the "wagon wheel design"). The design and structure of ESP was copied by prisons throughout the world.

Inmates at ESP were housed in solitary confinement and were not allowed to talk with other prisoners, an approach first introduced at the Walnut Street Jail. This approach was labeled the **silent system** and was also called the **separate system**. Although the intent of the silent system was to reform prisoners, the primary mechanism of reform was quiet reflection. Inmates were placed in a 12-foot-by-8-foot cell with a Bible and one other book; no other reading materials were permitted. Given the fact that many prisoners today are illiterate, one can imagine the hardships that these inmates experienced if they could not read (a real possibility at that time). During the day, inmates would engage in solitary work (e.g., weaving, woodwork, shoe repair) in their cells. Inmates were released from their cells into a courtyard for an hour a day to exercise. Photographs of the era show inmates with their heads masked by hoods, so they wouldn't be recognized by other prisoners when they were escorted within the prison.

The innovative approach to correctional treatment drew worldwide attention, and ESP became a tourist destination. Correctional officials, social critics, and philosophers from the United States and abroad visited the penitentiary to learn about this new form of punishment. But there was also some controversy about the silent approach. Even though prison sentences in that era were usually short, prisoners found it difficult to cope with the isolation, and some serving longer sentences developed psychological problems. Despite criticisms of the silent system, it survived until 1913.

Auburn Prison in New York, which opened in 1816, also implemented the big house design, but this institution developed an alternative approach to inmate custody called the **congregate system**. Initial experiences with the silent system at Newgate Prison in New York proved unsuccessful because inmates became mentally ill after long periods of solitary confinement. The congregate system still held inmates in solitary confinement during the evenings and at night, but prisoners dined and worked together during the day. Similar to the silent system, however, inmates were not allowed to converse during these activities.

Although Auburn Prison had a more humane approach to confinement, conditions were very strict. Prisoners worked up to 60 hours a week, wore striped uniforms, and were expected to walk or march in **lockstep** (inmates walked in unison, holding the elbow or shoulder of the prisoner in front of them). Inmates who disobeyed rules were whipped. Despite these conditions, inmates in big house prisons lived in better conditions than their counterparts in southern prisons (see next section).

The congregate system quickly became the model for prisons in the northeastern United States, in part because it was considered more humane but also because economic considerations were important. There were strong desires to use prison labor to turn profit, and the congregate system was far more consistent with the wider trend toward industrialization outside the prison walls.

Plantation-Style Prisons

Much like the development of jails, there was a distinct regional flavor to punishment in the South. Instead of investing in prison construction, many southern states leased convicts to private contractors who paid the state for each prisoner leased. These businessmen were responsible for feeding, clothing, and housing convicts. In return, the prisoners were put to work in labor-intensive jobs, such as farming, forestry, mining, or construction. Most of these prisoners were black, and they labored in conditions closely akin to slavery. Conditions for leased convicts were harsh, and rates of mortality were high.

Lichtenstein observes that the convict lease system supported economic development in the South, but there was opposition to these arrangements. The lease system eventually gave way to county-operated work camps and **chain gangs** in the early 1900s.[5] These prisoners wore striped uniforms and shackles on both ankles attached to a short length of chain and were often chained together so that they could not escape. These inmates were forced to work on labor-intensive county work projects, such as road construction. Conditions in the chain gang were often cruel. Historical accounts of the era report that whipping was common and some inmates were beaten to death by the guards.[6] Life expectancy for inmates sentenced to a chain gang often was short. Despite these conditions, the chain gang persisted for many years and was then reintroduced by Alabama in the 1990s as part of the "get tough on crime" movement.

Plantation-style prisons, such as Louisiana State Penitentiary at Angola, were defined by their farm-like atmosphere. The thick stone fences of the big house, for example, were replaced by wire fences. In addition, some prisons were located near swamps or rivers to act as natural barriers to escape. Instead of large multistory cellblocks, these prisons often were constructed of single-story living units, and the prisoners lived in dormitory settings (often sleeping in rows of bunk beds).

Although one goal of these prisons was punishment, these penitentiaries also created a profit for the state because the costs of inmate care were less than the revenues from the agricultural production of these facilities. To generate profit, prisoners were required to work in the fields surrounding the prisons. Inmates would be supervised by guards on horseback, and they planted, tended, and harvested crops such as cotton. One of the primary concerns of prison administrators was to maximize profits; the state of Arkansas, for example, required that prison wardens hold degrees in agriculture.[7]

One way that southern prison systems could be profitable was to reduce staffing costs, and they developed a unique solution: Inmates were employed as guards. In Arkansas, for example, all of the prison staff members, except the warden and the physician, were **trustees** when Tom Murton (depicted by Robert Redford in the film *Brubaker)* was appointed warden of the Cummings Prison Farm in 1967. In Texas, prison employees relied upon the assistance of **building tenders**, inmates who maintained control in the living units. These arrangements persisted for decades. Angola Prison used trustees from 1916, when almost all of the correctional officers were fired to reduce costs, until the mid-1970s.

The use of inmate trustees and building tenders is viewed as a low-water mark in U.S. corrections. Marquart and Crouch report that building tenders were granted privileges (e.g., separate bathing and recreational periods, better laundered uniforms, open cells, clubs or knives, and "friends" for cell partners) in return for controlling the ordinary inmates in the living areas.[8] Selection of trustees depended on a number of factors. In Texas, for instance, aggressive or dominant inmates were used as building tenders. In other places, murderers, especially those involved in family homicides, were thought to be the most reliable trustees because most of these offenders had not been in trouble prior to their first offense.

The use of inmate trustees and building tenders ended in the 1980s. Judge William Wayne Justice (in the *Ruiz v. Estelle*[9] lawsuit) forced Texas prison officials to terminate the building tender system in 1982. In other states, inmate trustees were phased out as a result of the influence of prison reformers who drew attention to the abuses carried out by these prisoners. Prison staff still rely upon inmate labor to do most of the support work that occurs in a facility (e.g., cleaning, preparing and serving meals, repairing the facility, and maintaining the grounds), but the days of surrendering control of the institution to inmate trustees or building tenders have passed in most prisons.

Reformatories

In the aftermath of the Civil War, discontent with the penitentiaries established in the first half of the nineteenth century was growing. People questioned whether the penitentiary could deter crime, reform offenders, or produce profits from inmate labor. There was also growing concern in some sectors about the threat to free-market jobs, wages, and profit posed by the use of prison labor. One reaction was the development of a different type of prison—the **reformatory**, which was originally designed for younger, less-seasoned offenders who often had committed less serious offenses. The first reformatory opened in Elmira, New York, in 1876. These institutions were designed to instill discipline through military-style regimentation and emphasized both academic and vocational education. They also pioneered the classification of prisoners by progress toward reform, indeterminate sentences, and parole release (see Chapter 13).

In many states today, one still finds both a penitentiary and a reformatory. In Kentucky, for instance, the Kentucky State Penitentiary is located in Eddyville, while the Kentucky State Reformatory is in LaGrange.

Women's Prisons

As the number of incarcerated women increased, and as attention was drawn to their dismal and abusive living conditions, officials began to develop prisons exclusively for women. The first prison that housed exclusively women in the United States, New York's Mount Pleasant Prison, was opened in 1839.[10] In Indiana, a reformatory for women was constructed in 1873 (some scholars have labeled this the first women's prison), and other state institutions for women followed, although female prisons did not emerge in the South until the 1920s.[11] The first federal prison for women, by contrast, was opened in Alderson, West Virginia, in 1927 and is still operational today.

There were, however, a number of problems with this practice. Conditions were grim. Rafter describes the women's unit at New York's Auburn Prison in the 1830s.[12]

Corrections in the Real World

Some people considered correctional "boot camps" to be major innovations when they appeared in the 1980s (see Chapter 9). In reality, military regimentation was used as a strategy to reform offenders at Elmira Reformatory in the 1870s.

> *Female prisoners . . . were confined together in a single attic room above the institution's kitchen. For a number of years they had no matron but rather were "supervised" by the head of the kitchen below. Food was sent up to them once a day, and once a day the slops were removed. No provision was made for privacy or exercise, and although the women were assigned some sewing work, for the most part they were left to their own devices in the "tainted and sickly atmosphere" of their crowded quarters. The wretchedness of their lot came briefly to public attention when one Rachel Welch, impregnated while in prison and severely flogged when she was about five months pregnant, later died.*

In addition to depressing conditions of confinement, women were also vulnerable to sexual assault from male correctional officers and inmates, and some female prisoners engaged in prostitution. As more attention was drawn to these problems, these mixed prisons were replaced by women-only facilities.

Early prisons designed to house females were often cottage-style facilities designed to aid in the rehabilitation of female misdemeanants (some also admitted minor felons) rather than to punish them. At the time, women were thought to be suffering from the effects of poverty and were believed to have been led astray by bad influences. The rehabilitative programming offered to these prisoners reflected these beliefs. Women's facilities taught prisoners how to clean, cook, and sew. Often run by female superintendents, these facilities had better living conditions than did male institutions (e.g., women wore dresses instead of striped uniforms, they slept in individual rooms rather than cells, and dining rooms featured dishes and linen tablecloths). Like the development of local jails, there were also regional differences in the incarceration of women. Females in the South, for instance, were likely to be leased out and work in the same fields as males.

Correctional Institutions, Violent Prisons, and Warehouses

Until roughly 1950, the adult prison system in the United States consisted primarily of a mix of the institutions described previously—big houses and plantation prisons (many of which by 1950 were either more than or nearly a century old and had been expanded in size to accommodate crowding), reformatories, and facilities for women. John Irwin points out that starting in the 1940s a different type of facility, the **correctional institution**, emerged to supplement the other types.[13] Correctional institutions were more modern in design and smaller than big houses. Premised on a medical model of crime (discussed in Chapter 13), these facilities were oriented toward assessing and diagnosing prisoners' problems and then subjecting them to various forms of treatment, mostly counseling and education.

Irwin observes that as correctional rehabilitation increasingly fell into disfavor during the 1960s and 1970s, many prisons did away with certain opportunities for prisoners, such as programs that had helped control prisoners by occupying time and providing them incentives to gain early release. Also, during this era the courts began putting restrictions on the ability of correctional officials to repress prisoner behavior in the manner officials had done in the big house prison. The result was the emergence of what Irwin calls the "violent prison," which was plagued by prisoner idleness and disruptiveness, gang/clique violence, and divisiveness by race. In his more recent work, Irwin describes how, during approximately the last three decades, prison officials implemented various new security designs and measures that transformed the violent prison into the "warehouse prison," where the preoccupation is with holding and controlling prisoners and completely segregating those seen as most troublesome.[14]

Modern Prison Systems

Historically prisons were apt to be stand-alone and relatively self-sufficient facilities, operating with little oversight from the courts, politicians, or media. Many of the goods consumed in the institution were produced there, and the services inmates received were limited. Today, by contrast, prisons are part of complex government systems that are interrelated and interdependent, and their activities are coordinated by a central office or headquarters. Moreover, in most jurisdictions, prisons have formal relationships with community organizations, including agencies that support the primary goal of the prison, which is the safety and security of staff and prisoners.

Federal Bureau of Prisons

Until the Three Prisons Act of 1891, the U.S. government contracted with county jails and state prisons to house federal prisoners. The first three prisons, Atlanta, Leavenworth, and McNeil Island, were constructed using the big house model. By 1930, another 11 federal prisons had been added, and all 14 operations fell under the jurisdiction of the newly formed BOP. By that time, the need for federal incarceration had increased, in part because of the increasing federal law enforcement efforts to control illegal liquor and narcotics after the Volstead and Harrison Acts were enacted in 1914 and 1919, respectively.

The size of the BOP slowly expanded until 1940, and the federal prison population remained more or less constant until 1980. During this era, the federal government opened several famous institutions, such as Alcatraz Prison in San Francisco Bay and the maximum security facility at Marion, Illinois, that replaced Alcatraz in 1963. These facilities were probably more famous for the prisoners that they held, such as Robert Stroud, the so-called Birdman of Alcatraz (although he never actually kept any birds at Alcatraz) and Al Capone, the prohibition-era gangster. Although Alcatraz ultimately was closed because of high operational costs, several older big house prisons, such as Atlanta and Leavenworth, still hold inmates.

In the past few decades, the number of prisoners in the federal system has grown at a fast rate. According to the BOP, in 1980 their population was approximately 24,000 inmates, but they held almost 198,000 prisoners in May 2007.[15] Part of that growth is from incarcerating drug offenders, and in May 2007, approximately 53 percent of BOP inmates were convicted of drug offenses.

To manage a population of 198,000 prisoners, the BOP has developed a diverse range of correctional facilities. Altogether, the BOP houses inmates in 114 minimum to maximum security facilities (we describe the difference in prison security levels in our review of the state prisons later). In addition to prisons, these institutions include operations that serve as medical or detention centers and low-security facilities (often called camps) that are built adjacent to prisons. One of the most unique features of the BOP system is the Federal Transfer Center, located at the Oklahoma City Will Rogers World Airport. This facility held 1,551 inmates in May 2007 and serves as a hub for the U.S. Marshal's Service Justice Prisoner and Alien Transportation System (JPATS). JPATS operates a fleet of airliners to transport 300,000 prisoners and immigration detainees per year.

Coordinating the imprisonment of 198,000 inmates is a large undertaking, and the BOP has a central office that serves as its headquarters in Washington, D.C., and six regional offices that support the managers and staff in the 114 institutions. Altogether, these offices oversee and support the activities of some 35,000 staff members. To accomplish this task, the BOP has established the following divisions:

- Administration
- Correctional Programs
- Health Services
- Human Resource Management
- Industries, Education, and Vocational Training
- Information, Policy, and Public Affairs
- National Institute of Corrections
- Office of General Counsel
- Program Review

Some of these divisions, such as Correctional Programs or Health Services, provide support, supervision, and assistance directly to prison managers and staff. Other divisions, such as Program Review, ensure that prisons are in compliance with federal regulations and BOP policies. Such oversight is important to make sure that inmates are receiving interventions that assist in their rehabilitation, but also so that prison administrators and staff are not sued for delivering programs that do not comply with the agency's policies.

The BOP also provides funding for the National Institute of Corrections (NIC), which is an important resource for U.S. corrections. According to the BOP, "The NIC provides training, technical assistance, information services, and policy/program development assistance to Federal, state, and local corrections agencies."[16] The NIC serves as a clearinghouse for assistance and information, and this agency has helped establish the BOP's leadership role in U.S. corrections.

State Prison Systems

Each state operates its own prison system. These state prison systems range in size from the North Dakota Department of Corrections and Rehabilitation (with 1,385 inmates) to the Texas Department of Criminal Justice that held 169,003 prisoners on December 31, 2005.[17] The development of prison systems in each state has been influenced by that jurisdiction's unique political and correctional history, as well as the public and political opinion in different eras (at certain times in history we have been more forgiving of offenders). As a result, some state prison systems may have a more rehabilitative orientation whereas other systems may focus on incapacitation. One common theme, however, is that the operations of these facilities have become increasingly complex. Moreover, prison administrators are increasingly accountable for what occurs behind the walls of the institution.

On June 30, 2000, there were 1,208 prisons in the United States and 460 community-based programs for prisoners.[18] Modern prisons often have the same population as a small town and must provide all of the basic goods and services for the inmates they house, from food services to medical and psychological care and vocational and rehabilitative programs. To accomplish this task, a large

Race and Gender in Corrections

One of the challenges of correctional work is that prisons tend to be built in rural areas, far away from the inner-city neighborhoods where a disproportionate number of the inmates grew up. Many of the correctional officers from these rural areas have historically been white. Members of minority groups, on the other hand, are overrepresented in prison populations. A recent federal study of prison population characteristics notes that black males in their 20s and 30s made up more than one-third of the entire prison population.[19] Although there has been an increased number of correctional officers who come from ethnic or minority groups, one tension in correctional work stems from the misunderstandings between inner-city minority prisoners and their rural and predominately white male keepers.

prison (such as Sing Sing outside New York City) may have more than 1,000 employees supported by the labor of inmates.

A central state office or headquarters (located in the capital city) coordinates the programs and personnel within individual prisons. In addition to managing the activities of these different operations, these agencies usually manage recruiting and training. The central office also employs policy analysts who examine correctional policies, spending, and issues such as offender rehabilitation (or recidivism). In addition, every prison system has a research division to collect, analyze, interpret, and publish statistics about their offender populations. Last, all prison systems have a division or unit that investigates staff wrongdoing and a legal affairs unit that responds to inmate or staff litigation. Altogether, these support functions are intended to remove administrative burdens from the wardens and other prison leaders so that they can focus on the management of their prisons.

The size of the prison organization increases with the inmate population; as such, the BOP and the Departments of Corrections in California, Florida, and Texas are big enterprises. Expenditures for these prison systems approach $10 billion annually, and running these organizations even 1 percent more efficiently can save millions of dollars. As a result, the leaders of these organizations (usually called commissioners or directors) are increasingly well educated, and surround themselves with the most talented leaders they can find.

In addition to operating efficiently, the primary goal of a state correctional system is to ensure public safety by carrying out the sentences imposed by the courts. A review of a correctional system's **mission statement** (a short description of the agency's major purpose) can help us understand the priorities of the administrators. Although these statements publicize the role of the agency, they also serve as important reminders to the correctional staff about the purpose of the system. The following is the mission statement for the Texas Department of Criminal Justice:

> *The mission of the Texas Department of Criminal Justice is to provide public safety, promote positive change in offender behavior, reintegrate offenders into society, and assist victims of crime.*[20]

In the past decade, there has been an increased emphasis on the role of rehabilitation in corrections, and this theme is present in most agency mission statements and in the formal names of the organization. On July 1, 2005, for instance, the state prison system in California added the term "Rehabilitation" to the agency name (becoming the California Department of Corrections and Rehabilitation).

Types of Prisons

Prison systems usually have multiple institutions located in different areas of the jurisdiction. In any given jurisdiction, most of these will be male institutions, but there is usually at least one facility for women; some states also have cocorrectional facilities that house both men and women separately. In 2000, there were 98 prisons in the United States that were exclusively for women and 93 that mixed male and female populations.[21]

In large prison systems, inmates may be placed in dozens of facilities, from low security **correctional camps** to supermax prisons. A given prisoner can expect to serve time at various facilities during the sentence, depending upon a blend of considerations such as security risk, bed space availability at different institutions, and proximity to family.

A convicted offender's first exposure to prison is typically a relatively short stay at a **reception center** (some states call these reception and diagnostic centers or classification centers), where security needs, psychological functioning, and physical health are evaluated. In addition, information is collected about the inmate's offense and criminal history, employment experience, and education. In some prisons, the inmate's presentence report is used to help reception center staff conduct these evaluations. The product is typically a classification report that helps prison officials decide the facility in which the prisoner will initially be placed, the prisoner's custody level (something distinct from the institutional security levels described later), and the rehabilitative needs. In turn, this report follows the prisoner throughout the sentence, serving as a baseline for later classification and release decisions. (Prisoner classification is covered further in Chapter 12.)

Prisoners who are evaluated to be low security risks—such as first-time nonviolent offenders with short prison terms—are typically assigned a minimum custody level and transferred to low-security facilities. Violent offenders, inmates sentenced to long terms of imprisonment, and disruptive prisoners are often assigned higher custody levels and sent to prisons that have higher levels of security. The following are the characteristics of four levels of prison security in the North Carolina Department of Correction:

Maximum security units are composed of cells with sliding cell doors that are remotely operated from a secure control station. Maximum security units are designated by the Director of Prisons at selected close security prisons. These units are utilized to confine the most dangerous inmates who are a severe threat to public safety, correctional staff, and other inmates. Inmates confined in a maximum security unit typically are in their cell 23 hours a day. During the other hour, they may be allowed to shower and exercise in the cellblock or an exterior cage. All inmate movement is strictly controlled with the use of physical restraints and correctional officer escort.

Close security prisons typically are composed of single cells and divided into cellblocks, which may be in one building or multiple buildings. Cell doors generally are remotely controlled from a secure control station. Each cell is equipped with its own combination plumbing fixture, which includes a sink and toilet. The perimeter barrier is designed with a double fence with armed watchtowers or armed roving patrols. Inmate movement is restricted and supervised by correctional staff. Inmates are allowed out of their cells to work or attend corrective programs inside the facility.

Medium security prisons typically are composed of secure dormitories that provide housing for up to 50 inmates each. Each dormitory contains a group toilet and shower area as well as sinks. Inmates sleep in a military style double bunk and have an adjacent metal locker for storage of uniforms, undergarments, and shoes. Each dormitory is locked at night, with a correctional officer providing direct supervision of the inmates and sleeping area. The prison usually has a double fence perimeter with armed watchtowers or armed roving patrols. There is less supervision and control over the internal movement of inmates than in a close security prison.

Minimum security prisons are composed of nonsecure dormitories that are routinely patrolled by correctional officers. Like the medium security dorm, dorms in minimum security prisons have their own group toilet and shower area adjacent to the sleeping quarters, which contain double bunks and lockers. The prison generally has a single perimeter fence that is inspected on a regular basis but has no armed watchtowers or roving patrol. There is less supervision and control over inmates in the dormitories and less supervision of inmate movement within the prison than at a medium facility. Inmates assigned to minimum security prisons generally pose the least risk to public safety.[22]

Staff members at prison reception centers have become skilled at placing inmates in prisons that best match their need for security. As a result, the use of maximum security has decreased over the years, at the same time that prison violence has also dropped. In their study of prison violence, Useem and Piehl report on the use of prison security levels in the United States (see **Table 10–1**).[23]

Table 10–1 The Decreasing Use of Maximum Security Housing, United States, 1974–2000

	Percentage of Total Prison Population Per Custody Level					
Type of Custody	**1974**	**1979**	**1984**	**1990**	**1995**	**2000**
Maximum security	41.4	51.7	43.7	38.0	38.9	35.7
Medium security	35.5	37.4	44.3	49.2	49.6	47.6
Minimum security	23.0	10.8	11.8	12.7	11.5	16.6

Source: B. Useem and A. M. Piehl, "Prison Buildup and Disorder," *Punishment and Society* 8 (2006): 87–115.

In addition to reducing violence successfully, placing fewer inmates in maximum security prisons is also a cost-saving measure because it is expensive to provide higher levels of security. Despite the fact that the use of maximum security has decreased, some inmates require a high degree of supervision and control. Prison systems have developed a number of options for difficult-to-manage inmates as well as for prisoners requiring other special services.

Specialized Correctional Facilities and Units

Special security or other services may be provided in a specialized unit of a larger prison or at a separate facility. Many larger prisons have special units where certain prisoners can be segregated from the general prison population for various reasons. Some prisoners represent a danger to other prisoners or staff, some are being disciplined for violating prison rules, others are highly vulnerable to assault by other prisoners and are therefore in need of protection (known as protective custody), and still others have special mental health or medical needs.

Prisoners who require long-term placement in secure units are held in **security housing units** (also known as special handling units, segregated housing units, or administrative segregation in some jurisdictions). These inmates seldom interact with other prisoners and usually spend 23 hours per day in their cells. When released from their cells for recreation or hygiene, they are handcuffed (and sometimes placed in shackles) and escorted by correctional officers. In addition to being supervised more closely, these inmates have fewer items in their cells, and rehabilitative programming or privileges are limited.

Corrections and Policy

A famous instance of lethal violence took place in 1983 at the United States Penitentiary at Marion, Illinois (the supermax prison that replaced Alcatraz in the federal system). Two extremely dangerous prisoners, Thomas Silverstein and Clayton Fountain, both members of the Aryan Brotherhood being held in the maximum custody unit at Marion for multiple killings of other prisoners, conspired with other prisoners to get shanks (prison-made knives) and then killed two correctional officers in separate instances on the same day. Fountain also injured two other officers, bragging that he was not going to let Silverstein get ahead of him in the body count.[24] Needless to say, prison officials do not take such matters lightly, and changes in security policies often result. This incident resulted in Marion adopting a 23-hour-per-day lockdown policy and helped encourage the eventual construction of an entirely new supermax prison, the Administrative Maximum U.S. Prison in Florence, Colorado (see Chapter 15). Fountain is deceased, but Silverstein is now housed at Florence.

Most large prison systems have separate **supermax prisons** (also known as supermaximum prisons or Administrative-Maximum in the BOP) to hold long-term high-security inmates. The number of these high-security prisons has increased significantly since the first one was introduced in the mid-1980s, although less than 2 percent of all U.S. inmates are now held in these institutions.[25] Prisoners often are placed in these facilities because of behavioral problems within the prison system or because of gang involvement.[26]

There has been some controversy about placing inmates in supermax facilities for extended periods of time. Haney observes "Because supermax units typically meld sophisticated modern technology with the age-old practice of solitary confinement, prisoners experience levels of isolation and behavioral control that are more total and complete and literally dehumanized than has been possible in the past."[27] Perhaps a more troubling issue is that some of these inmates are held in these special handling units for years and then are released directly into the community. While these prisoners have been living in an isolated environment, their social, problem-solving, and interpersonal skills may have worsened, along with their mental health.[28] As a result, returning these inmates directly to the community without first placing them in the general population or a transitional housing unit places them at high risk of recidivism.

Because prisons are holding large number of inmates with health problems, many departments of corrections have established **correctional medical centers** and other specialized housing facilities. These facilities offer a broad range of correctional medical care and are intended to provide better diagnostic and treatment services for prisoners. With the increasing number of elderly prisoners, some states (i.e., South Carolina and others) have developed housing units specifically for these geriatric prisoners. In addition, the Colorado Department of Corrections has also constructed stand-alone housing units to care for inmates with severe mental illness.

Inmates who have shown stable behavior (e.g., no involvement in incidents or disciplinary reports) might be considered for placement in a facility or housing unit that has less security. Most states, for example, have constructed low-security correctional camps that have little visible security and may not even be enclosed by a fence. Prisoners spend the entire day in the community (e.g., working with a supervised crew that is fighting forest fires or with a community employer) and return to the camp in the evening. If the prisoner is disruptive or noncompliant, the prisoner is transferred back to a higher security facility.

Last, some prison systems have **prerelease centers** or reentry facilities. These programs are intended to help a prisoner prior to release into the community. These facilities offer programs that help inmates make the transition from the prison to community (e.g., vocational skill development, life or social skills development, or addictions relapse prevention). These pre-release centers also take the form of **halfway houses** that are located in the community and provide low levels of supervision for the inmate. (These two types of facilities are discussed further in Chapters 9 and 13.)

Prison Staff

Correctional facilities are bureaucratic organizations that rely upon their staff members to ensure the safety and security of the institution, provide rehabilitative and treatment opportunities (such as educational programs), and support the activities of the facility (such as maintenance). In the following pages, we review these different roles.

The chief executive of an institution is usually called a warden or superintendent. Depending on the size of the institution, there are typically two or more deputy wardens in charge of such areas as security, rehabilitative or treatment programs, and administration or support services. Furthermore, a distinction is usually drawn between custodial and noncustodial staff. Noncustodial staff are divided into areas responsible for such things as business operations, physical plant, health care, industry/agriculture, religion, education, and treatment programs. Custodial staff are organized in a paramilitary hierarchy with a clear chain of command and rules and regulations. Thus, correctional officers report to sergeants, who in turn answer to lieutenants and captains. Despite the fact that correctional officers are at the bottom of the organizational chart, they are the most numerous and most important positions in a prison. They supervise inmates' daily activities and are responsible for the security and safe running of the institution.

Correctional Officers

Correctional work probably was not the first occupational choice of many officers, although after growing accustomed to the rules and routines of a facility, many settle in for long-term careers. Historically, the salary, benefits, and selection standards for officers were low. This led to a number of problems. First, many correctional facilities experienced rapid employee turnover because many employees stayed for a year or two and then moved on to other law enforcement or government jobs; high turnover is expensive and can make institutions unstable. A second problem is that low pay may have drawn persons into the job who might not have been hired if wages and standards were more competitive. Some of these officers have engaged in misconduct, and this produced stereotypes of corrections officers as cruel, apathetic, or unethical. As salaries improved, administrators were able to recruit and retain more professional officers; in 2004, the median yearly earning of correctional officers was $33,600.[29] As a result, correctional officers today are better educated, prepared, and trained than were officers of previous generations (**Figure 10–1**).

In the past, officers learned about the profession "on the job," and they received only minimal formal training. Today's officers are likely to attend several months of full-time academy training before their first shift as a correctional officer; this is followed by periodic in-service training. Although training goals and curriculums vary between states, an important training goal is to give offi-

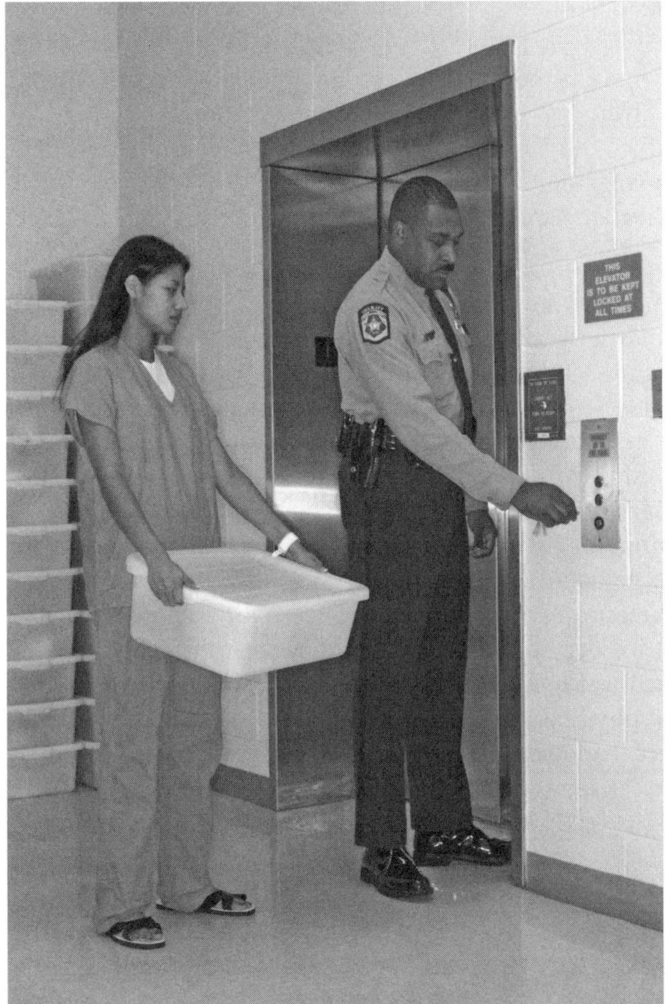

Figure 10–1 Many jurisdictions have moved toward increased professionalism for correctional officers.
Source: *© Ron Chapple/Thinkstock/Alamy Images.*

cers the knowledge, skills, and abilities they need to ensure the safety and security of a facility. The Arizona Department of Corrections, for instance, provides the following training topics in its nine-week academy:

Topic	Main Themes
■ Administrative/Personnel	Rules and regulations, drill and ceremony
■ Ethics and Professionalism	Workplace relationships and professionalism
■ Inmate Management	Supervision, inmate sociology, gangs
■ Legal Issues	Inmate rights, use of force
■ Communication	Radio and written communication
■ Officer Safety	Firearms and chemical weapons training

■ Applied Skills	Searches, contraband, cell extraction, restraints
■ Security, Custody, and Control	Emergency procedures, crime scene preservation
■ Conflict and Crisis Management	Conflict resolution, self-defense training
■ Medical & Mental Health	Communicable disease, suicide risk[30]

Training enhances the correctional cadets' confidence and gives them a better range of skills than previous generations of correctional officers had (**Figure 10–2**). Officers typically serve a probationary period for the first year, and in many states they receive more supervision and support during this time. In some cases they are mentored by correctional field training officers. Providing this training, mentoring, and support are essential parts of preparing officers for their careers. A classic study by Allen and Bosta found that officers in their first year were at high risk of being manipulated by prisoners.[31] Officers quickly learn whether they are suited for correctional careers, and officer turnover during the first few years of employment tends to be quite high.[32]

Officer retention is a significant concern for most prison systems because of investment in recruitment, conducting background checks, and training. Like other law enforcement jobs, correctional officers work in environments with a high risk of physical harm. In addition, working with correctional populations tends to be emotionally draining. Many inmates are not compassionate toward officers and see new ones as targets for manipulation.

Figure 10–2 Correctional officers often receive extensive training in academy-like settings.
Source: *Courtesy of the California Department of Corrections and Rehabilitation.*

As outlined in the next chapter, most prison inmates have relatively long histories of involvement in crime, and only a small percentage see themselves as ordinary citizens. Cornelius suggests that most prisoners scrutinize officers and look for inconsistencies in their behavior, feelings of sympathy toward offenders, signs of fear, or other signs of being "soft" targets.[33] If inmates think that an officer can be manipulated for their advantage, many will test that officer's limits. Correctional officers who fail to abide by institutional rules or who want to "give inmates a break" are at greater risk of being manipulated.

Despite the challenges of being a correctional officer, there are many opportunities for career development and advancement. Throughout an officer's career, there are opportunities for different assignments within the prison and community as a parole officer or agent. Some officers earn a degree and become correctional counselors who assess, classify, and counsel prisoners. Others volunteer for specialized assignments, such as gang investigator, canine handler, corrections emergency response team member, or correctional cadet trainer at the academy. In addition, the growth of correctional systems has created a demand for officers who want to assume supervisory and leadership roles within the prison system.

One of the foremost changes in corrections in the past two decades is the increasing number of women employed as correctional officers. In the last national census of federal and state prisons, Stephan and Karberg found that the number of women working in prisons increased by 41 percent between 1995 and 2000, and women now represent 33 percent of all employees.[34] Although women have worked in corrections for more than a hundred years, most were employed in women's facilities, and few were appointed to managerial positions (notable exceptions were Katherine Bement Davis, who was appointed as Commissioner of Corrections for New York in 1914, and Kathleen Hawk-Sawyer, who served as Director of the U.S. Bureau of Prisons from 1992 to 2003). Correctional work has been described as a "highly masculinized" environment where women faced harassment and discrimination by male officers and supervisors.[35] Today, however, the working conditions for women have improved, and one goal of prison administrators is to retain the female staff members who have been recruited and trained.

Controversies

The *Albany Times Union* newspaper reported that on March 23, 2006, a former New York corrections officer pled guilty to two charges of rape of female inmates in his custody. Controversial cases such as this underscore the importance of doing comprehensive background checks on potential corrections officers to protect the inmates, as well as the legitimacy of the correctional system.

Correctional Officer Subculture

Correctional work is sometimes misunderstood. Historically, officers were stereotyped as "hacks," "screws," and "turnkeys," with little appreciation for differences in the way officers respond to the challenges they face. Sometimes it was mistakenly assumed that officers have unrestrained power over prisoners and that this power led to brutality and corruption. Certainly, some officers do abuse the authority they hold over prisoners by virtue of officers' roles, particularly when prisoners are perceived as uncooperative or belligerent. However, as recognized by Sykes, another source of corruption can be the limits placed on the power of officers.[36] Limits can come from superiors as well as the courts. Furthermore, officers cannot rely entirely on coercion and force to gain compliance from prisoners in mass; they must obtain at least some degree of voluntary compliance to operate efficiently, avoid disturbances, and minimize scrutiny from their superiors, the courts, and the media. Because officers do not directly control all the major incentives that can shape prisoner behavior (e.g., release dates), some officers resort to informal negotiations and exchanges with prisoners. This can result in corruption (e.g., smuggling contraband in exchange for prisoners curbing violence in the cellblock).

Even today, correctional officers are often referred to as "guards," a term that does not accurately reflect the interpersonal skills required for the job or the human service responsibilities of the correctional officer in modern prisons.[37] Despite the fact that officers are better educated, trained, and paid, working in a prison is not a prestigious job, and this may contribute to feelings of alienation or isolation and contribute to the development of a subculture of like-situated peers.

A number of scholars have described the subculture of correctional officers. Robert Johnson writes about the alienation, cynicism, and bitterness that some correctional officers feel.[38] These feelings can both arise from and fuel the negative stereotypes that many people hold toward correctional officers. The job is also stressful, and officers often report feeling misunderstood and underappreciated by the prison administrators. Moreover, correctional officers tend to become suspicious and distrustful, an occupational hazard considering that inmates are constantly attempting to manipulate them. Finally, success as a correctional officer is often hard to measure. Tracy interviewed a correctional officer who reported, "Unlike a carpenter or even a computer worker, at the end of the day you have nothing to show for your work. Here the goal is to do as much as possible to *prevent* incidents."[39]

There is no single "type" of correctional officer. The attitude and behavior of correctional officers may be related to the culture of their place of employment, the quality of supervision, their training, as well as career assignments. Mary Ann Farkas developed a typology of officers based on her interviews of prison staff. She describes the following types of officers:

- **Rule enforcer**—Officers who are rule bound, inflexible in discipline, and have a common bond with officers who share this philosophy.
- **Hard liner**—An extreme type of the rule enforcer who is inflexible with rules, has few interpersonal skills, and is aggressive.
- **People worker**—These officers are perceived as professional, have good interpersonal skills, are responsible, and try their best. These officers tend to have better relationships with inmates and gain compliance through their communication skills rather than by using rule enforcement.
- **Synthetic officer**—These officers are a blend (or synthesis) of the rule enforcer and people worker. They are more flexible than the rule enforcers are, and they are unlikely to deviate far from the rules.
- **Loner**—These officers are perceived as isolated and often work in prison jobs that have less contact with other officers (e.g., in a gun tower). These officers are similar to the rule enforcer, but adopt this role because they are afraid of making mistakes.
- **Residual officers**—These officers have little commitment to the organization, and they are called "officer friendly, lax officers, and wishy-washy officers" by the other correctional officers. These staff members are haphazard in their rule enforcement and are described as putting in time.[40]

It is important to realize that these officer "types" do not exist in a pure form, and some officers likely have attitudes and behaviors that span more than one classification. Nevertheless, Farkas provides us with another way of understanding the private world of correctional officers.

Support and Administrative Correctional Staff

Correctional facilities employ staff in dozens of job classifications that are not directly responsible for enforcing the facility's rules. The buildings, for example, require ongoing maintenance, and many of these tasks are increasingly technical. Security in prisons today relies on video surveillance and technology; most internal and external doors are electronic and can be locked from a central console. This creates the need for highly skilled technicians and tradespersons to repair these systems. There is also a range of less technical jobs within a prison. Much of the day-to-day work in a prison is done by the inmates, but staff members are required to instruct the inmates to do the work properly and supervise their progress. One example is the kitchen staff members who supervise the preparation, delivery, and clean-up for the thousands of meals consumed per day in a large prison.

Prisons also require a number of staff persons who are responsible for budgeting, clerical, and accounting functions. These are important jobs because these employees are accountable for millions of dollars of inmate funds, receipts from canteen or commissary sales, and ensuring that the staff members are paid each month. Ultimately, the support and administrative staff enable the correctional staff to focus on their primary goal, which is ensuring the safety and security of the facility.

Rehabilitative and Treatment Program Staff

One of the most challenging jobs in corrections is the delivery of educational, vocational, and treatment interventions. The staff who deliver these programs include teachers, vocational instructors (e.g., carpentry or welding), staff who offer life skills instruction, career placement staff (also called reentry staff in some systems), and therapists. Therapists include a broad range of professionals, such as social workers and psychologists, while other staff members hold specialized roles, such as the therapists who work one-on-one and in groups with sex offenders. These jobs are challenging because prisoners are not always motivated to participate actively in these programs.

Because the main goal of a correctional facility is safety and security, staff members who provide educational and treatment services sometimes feel undervalued and unappreciated. In some cases, these professionals feel that the focus on security in the prison undermines their hard work. In addition, because these professionals typically cannot earn overtime, an officer with a grade-school education can sometimes earn a larger salary than a psychologist with a doctorate, which might also lead to tension between occupational groups.

Providing medical care to inmates involves many challenges, and one of the most serious problems that correctional administrators face is recruiting and retaining qualified medical staff. Medical professionals often spend much of their time on administrative duties, such as approving amenities for inmates (e.g., extra blankets or sunscreen) rather than providing medical care. They also "endure professional isolation while struggling in 'clinics' that lack sinks, exam tables, common implements for examinations, computers, internet access, PDAs, electronic medical records, medical education, and timely access to diagnostic studies and specialists."[41] It is often difficult to recruit medical professionals who want to work in this environment. As a result, some of the doctors and nurses practicing correctional medicine have actually been barred from practicing in other places, and their licenses restrict them to working only in correctional facilities.[42]

Job Stressors in Corrections

The constant limit and boundary testing by inmates, the unpredictable nature of the job, the austere environment of the prison, and the fact that correctional staff members are surrounded by felons and stimulus overload contribute to a difficult work environment and job-related stress for all correctional staff. Other factors make correctional officer work especially difficult, including working rotating shifts, working unsocial hours (e.g., nights, holidays, and weekends), and being forced to work mandatory overtime. In many places, correctional officers feel underappreciated by their managers. Correctional facilities also tend to have higher than average levels of interpersonal conflict between staff members, especially when there is competition for time off, desirable shift rotations, or job postings.

Over the long term, the combination of conditions can lead to **burnout**, a condition of psychological and physical exhaustion as a result of long-term exposure to stress.

Some staff members become cynical and disgruntled. In recognition of this problem, correctional systems are more proactive in promoting healthy lifestyles. To increase the coping ability of the officers, many administrators are introducing wellness programs and encourage staff members to adopt exercise and healthy living programs. Some Departments of Corrections have also offered counseling programs for officers (and sometimes their family members as well) to enhance their coping skills. Last, after officers are involved in serious incidents (e.g., after witnessing a homicide or suicide) they are now more likely to participate in critical incident stress debriefing, where the events are reviewed (including the officer's role in the incident) in a supportive environment. Such interventions are thought to reduce the likelihood of post-traumatic stress disorder.

Conclusion

Correctional systems are a product of a state's history, political priorities, and the public's attitudes toward punishment and rehabilitation. Fifty years ago, prisons were often stand-alone facilities that operated with little oversight and were more self-sufficient than today's institutions are. Modern prison systems, by contrast, are large, complex organizations that sometimes have multi-billion-dollar budgets and employ well-educated and politically savvy leaders and deploy well-trained and professional correctional staff.

Prison managers and their staff have done an outstanding job of maintaining order in the face of inmates with lengthy prison sentences, high levels of special needs populations, career offenders who have little interest in rehabilitative programs, and increasingly outspoken associations of correctional officers. One significant change for corrections is the increased interest in rehabilitation and helping prisoners to make a meaningful rehabilitative change so that they can be successful in the community. To accomplish these tasks, there will be a long-term need for dedicated and professional staff members working in jails or prisons. These jobs are supported, in most cases, by headquarters staff possessing expertise in business administration, criminal justice, law, policy analysis, and research. State and federal correctional systems want to improve services and increase the professionalism of their staff, and this may also create great opportunities for long-term careers for today's criminal justice students.

READY FOR REVIEW

- Prior to the development of the prison, corporal and capital punishments and penal transportation were used to punish. The penitentiary was developed as a more humane alternative that would reform offenders.
- Prisons such as Eastern State Penitentiary or Auburn were castle-like facilities encircled by walls built of stone. This big house style of prison architecture and design was popular for more than a hundred years, but big house prisons are now seen as inefficient, costly to operate, and less safe than new generation designs are.
- The silent system contributed to high rates of mental illness, so prison authorities in Auburn, New York, adopted the congregate system, which required prisoners to work outside of their cells during the day (but still forced them to be silent).
- Plantation-style prisons evolved in the southern United States, and unlike the big house design, many were housed in one-story dormitory-style living units. Inmates worked in the fields, and agricultural production created a profit for some state prison systems.
- The BOP was established by the federal government in 1930. The BOP is now the largest prison system in the United States, with almost 200,000 inmates. More than half of BOP prisoners in May 2007 were sentenced on drug offenses.
- Prison systems are complex organizations, and divisions based in a central office or headquarters support the activities of the individual prisons.
- State prison systems provide a range of housing units that best match the security needs of the inmates, from minimum to maximum security facilities.
- Most prison systems have reception centers that receive and classify inmates, and some form of supermax prison (or security housing units for difficult-to-manage inmates). Some larger systems have prison hospitals, correctional medical centers, correctional camps (for low-security inmates), and halfway houses to help prisoners reintegrate into the community.
- Higher salaries and recruiting standards have increased the professionalism of correctional officers, and these officers are now better trained to undertake a human service role rather than the stereotypical "guard" role.
- There are a number of opportunities for correctional officers within the prison system, including case counselor, trainer, canine handler, gang investigator, or member of the institution's correctional emergency response team. In addition, the growing size of prison systems has created numerous opportunities for supervisory and managerial positions.
- Women and members of minority groups are playing an increasingly important role as correctional officers.

- In addition to officer positions, correctional systems require a range of other professionals, from accountants to technicians, to support the activities of the correctional officers.
- Psychologists, medical professionals, teachers, addictions counselors, and instructors aid in the rehabilitation and treatment of inmates.
- Working with offenders, rotating shifts, unpleasant conditions, and working mandatory overtime shifts contribute to high levels of stress and burnout for corrections workers, which in turn leads to employee dissatisfaction and job turnover.
- There are a number of types of correctional officers including the rule enforcer, hard liner, people worker, synthetic officer (blend of different types), the loner, and "officer friendly."

KEY TERMS

big house Large imposing prisons utilizing castle-like architecture that featured tall, thick walls and housing units constructed of cells that were stacked on each other

building tenders Inmate trustees who helped run housing units in the Texas Department of Criminal Justice

burnout A condition of psychological and physical exhaustion as a result of long-term exposure to stress

chain gang A group of offenders who are chained together at the ankles and who do manual labor under the supervision of correctional officers. Popular in the southern states prior to World War II, chain gangs were reintroduced to some prison systems in the 1980s

congregate system Also called the Auburn or New York approach, where inmates worked together during the day, but were expected to remain silent, and then returned to their solitary cells in the evening

correctional camp A low-security community-based correctional facility intended for low-risk inmates or prisoners who will soon be released

correctional institution More modern, smaller prisons that started appearing during the 1940s and 1950s and were oriented toward assessing and diagnosing prisoners' problems and then subjecting prisoners to various forms of treatment, mostly counseling and education

correctional medical center A prison that offers advanced hospital and medical care. These are more common in large prison systems (e.g., California, Federal Bureau of Prisons, Florida, or Texas)

executive clemency The ability of the president or state governor to pardon an offender

Federal Bureau of Prisons An agency established by the federal government in 1930 to oversee the imprisonment of federal offenders

halfway house A community-based residence for offenders who are making the transition from prison to the community
lockstep In the early days of the penitentiary, prisoners were often forced to walk in formation, grasping either the shoulder or elbow of the prisoner in front of them
mission statement A short description of the agency's major purpose
prerelease center A residence that helps to prepare inmates for the transition from prison to the community
reception center A facility that receives newly sentenced prisoners and conducts assessment and classifies the inmate (also called a reception and diagnostic center in some states)
reformatory Prisons designed to aid in the rehabilitation of younger, less-hardened male and female offenders
security housing unit A unit that provides long-term housing for difficult-to-manage inmates, usually in individual hardened cells (also known as a special handling unit in some jurisdictions)
separate system Also known as the silent system
silent system Eastern State Penitentiary promoted a model of prisoner treatment where prisoners were kept in isolation (to reflect upon their offenses and rehabilitation) and were not allowed to speak to other inmates or officers (also called the separate system by some scholars)
supermax prison (also supermaximum prison or administrative-maximum) A high-security prison that holds most of the inmates in single cells and provides them with little time outside their cells (e.g., confined to cells 23 hours a day)
trustee An inmate who has been given some authority by the prison staff to supervise or control other inmates
Walnut Street Jail Located in Philadelphia and conventionally regarded as the first American prison, in 1790 the jail was converted from a holding facility into a facility for receiving sentenced prisoners

YOU ARE THE CORRECTIONS PROFESSIONAL SUMMARY

1. What alternatives do you have, and what is the best response? There are only two alternatives: You can comply with the inmate's demands and place yourself at risk of victimization and loss of career, or advise your supervisor of the problem and how it occurred.
2. What are the pros and cons of the decisions? Allen and Bosta found that officers who reported the problems early (before they had knowingly engaged in any criminal misconduct) were generally disciplined and kept their jobs. The sanctions for officers who knowingly brought

contraband into facilities were much greater. Interestingly, the authors found that offenders rarely received much formal punishment for their involvement in manipulating staff members.

NOTES

1. B. Useem and A. M. Piehl, "Prison Buildup and Disorder," *Punishment and Society* 8(2006): 87–115.
2. D. J. Rothman, *The Discovery of the Asylum: Social Order and Disorder in the New Republic* (Boston: Little, Brown, 1971).
3. N. King, "The Origins of Felony Jury Sentencing in the United States," *Chicago Kent Law Review* 78(2003): 976.
4. P. Takagi, "The Walnut Street Jail: A Penal Reform to Centralize the Powers of the State," *Federal Probation* (1975): 18–26.
5. A. Lichtenstein, *Twice the Work of Free Labor: The Political Economy of Convict Labor in the New South* (New York: Verso Press, 1996).
6. J. F. Steiner and R. M. Brown, *The North Carolina Chain Gang: A Study of County Convict Road Work* (Montclair, NJ: Patterson Smith, 1927).
7. T. O. Murton and J. Hyams, *Accomplices to the Crime* (New York: Grove Press, 1969).
8. J. W. Marquart and B. M. Crouch, "Judicial Reform and Prisoner Control: The Impact of *Ruiz v. Estelle* on a Texas Penitentiary," *Law and Society Review* 19(1985): 562.
9. *Ruiz v. Estelle*, 503 F. Supp. 1265 (S.D. Tex. 1980).
10. N. Rafter, "Prisons for Women, 1790–1980," *Crime and Justice: A Review of Research* 5(1983): 129–181.
11. Rafter, "Prisons for Women."
12. Rafter, "Prisons for Women," 135.
13. J. Irwin, *Prisons in Turmoil* (Boston: Little, Brown, 1980).
14. J. Irwin, *The Warehouse Prison: Disposal of the New Dangerous Class* (Los Angeles: Roxbury, 2005).
15. Federal Bureau of Prisons, "Weekly Population Report," 2007, www.bop.gov/locations/weekly_report.jsp.
16. Federal Bureau of Prisons, "About Central Office—The National Institute of Corrections (NIC)," 2007, www.bop.gov/about/co/nic.jsp.
17. P. M. Harrison and A. J. Beck, *Prisoners and Jail Inmates at Midyear 2005* (Washington, DC: Bureau of Justice Statistics, 2006), 1.

18. J. J. Stephan and J. C. Karberg, *Census of State and Federal Correctional Facilities, 2000* (Washington, DC: Bureau of Justice Statistics, 2003), 6.
19. Harrison and Beck, *Prisoners and Jail Inmates at Midyear 2005*,10.
20. Texas Department of Criminal Justice, "Mission Statement," 2007, www.tdcj.state.tx.us/.
21. Stephan and Karberg, *Census of State and Federal Correctional Facilities, 2000*, 14.
22. North Carolina Department of Correction, "Assigning Inmates to Prison," 2007, www.doc.state.nc.us/dop/custody.htm.
23. Useem and Piehl, "Prison Buildup and Disorder," 100.
24. P. Early, *The Hot House: Life Inside Leavenworth Prison* (New York: Bantam Books, 1992).
25. Useem and Piehl, "Prison Buildup and Disorder," 101.
26. J. Petersilia, *Understanding California Corrections* (Berkley, CA: California Policy Research Center, 2006).
27. C. Haney, "Mental Health Issues in Long-Term Solitary and 'Supermax' Confinement," *Crime and Delinquency* 49(2003): 127.
28. Haney, "Mental Health Issues."
29. U.S. Department of Labor, Bureau of Labor Statistics, *Occupational Handbook* (Washington, DC: U.S. Department of Labor, 2006–2007), www.bls.gov/oco/.
30. Arizona Department of Corrections, "Correctional Officer Training Academy," 2006, www.azcorrections.gov/COTA/cotacurric.htm.
31. B. Allen and D. Bosta, *Games Criminals Play: How You Can Profit by Knowing Them* (Sacramento, CA: Rae John Publishers, 1981).
32. G. Armstrong and M. L. Griffin, "Does the Job Matter? Comparing Correlates of Stress among Treatment and Correctional Staff in Prison," *Journal of Criminal Justice* 32(2004): 577–592.
33. G. Cornelius, *The Art of the Con: Avoiding Offender Manipulation* (Lanham, MD: American Correctional Association, 2001).
34. Stephan and Karberg, *Census of State and Federal Correctional Facilities, 2000*, vi.
35. M. L. Griffin, "Gender and Stress: A Comparative Assessment of Sources of Stress among Correctional Officers," *Journal of Contemporary Criminal Justice* 22(2006): 5–25.
36. G. M. Sykes, *The Society of Captives: A Study of a Maximum Security Prison* (Princeton, NJ: Princeton University Press, 1958).

37. L. X. Lombardo, *Guards Imprisoned: Correctional Officers at Work* (Cincinnati, OH: Anderson, 1989).
38. R. Johnson, *Hard Time: Understanding and Reforming the Prison* (Belmont, CA: Thompson, 2002).
39. S. J. Tracy, "The Construction of Correctional Officers: Layers of Emotionality behind Bars," *Qualitative Inquiry* 10(2004): 509–533.
40. M. Farkas, "A Typology of Correctional Officers," *International Journal of Offender Therapy and Comparative Criminology* 44(2000): 437–445.
41. J. Bick, letter from the editor, *Infectious Diseases in Corrections Report* 9, no. 3 (2006): 3.
42. J. J. Gibbons and N. de B. Katzenbach, "Confronting Confinement: A Report of the Commission on Safety and Abuse in American Prisons," 2006, www.prisoncommission.org/report.asp.

11 Prisoners, Prison Life, and Inmate Rights

Chapter Objectives

- Describe the demographic and offense-related characteristics of the U.S. prison population.
- Identify different special needs prison populations.
- Describe the special needs of female prisoners.
- Understand the impact of security threat groups (gangs) on prison operations.
- Explain how the prison environment shapes the experiences of inmates.
- Describe the deprivations of imprisonment.
- Describe the types of litigation that jail or prison inmates might undertake.
- Explain how inmate litigation has changed prison conditions.

CASE STUDY

You are a member of California's Board of Parole Hearings and have to evaluate the parole application of John Rodriguez, a 95-year-old prisoner who was convicted of murdering his wife at the age of 68 years, and who was sentenced to 16 years to life (he has now served more than 25 years). Rodriguez is California's oldest prisoner and has been living in a prison hospital for the past two years. So far, the

Board of Parole Hearings has approved his application for parole six times, but the past two governors (who must approve the parole of any "lifer") have rejected the board's recommendations and denied parole. Rodriguez has committed no serious infractions while in prison, and he regularly attends Alcoholics Anonymous meetings (he was intoxicated at the time of his offense). This prisoner intends to live with his family upon release. The District Attorney who prosecuted this case and spokespersons for the police department that investigated the crime do not support parole based on the risk that Rodriguez poses to the public if he were to drink again.

1. What factors should be considered prior to paroling a prisoner convicted of a serious crime?
2. Rodriguez had a long criminal history before this offense. Should that be considered by the parole board? Should the board also consider that he is a World War II veteran who was awarded a Bronze Star?

Introduction

Over the past two decades, there have been two significant changes in the prison population. First, overcrowding has been a serious problem in most jurisdictions with the number of inmates often exceeding the rated capacity of prisons. Second, the characteristics of these prisoners have also shifted. The percentage of violent prisoners, for instance, has dropped. At the same time, there has been an increased growth in the percentage of minority and women prisoners. Along with these changes in demographic and offense characteristics, there has been an increase in the percentage of "special needs" inmates, including persons with serious mental or medical health needs, long-term prisoners who may be sentenced to terms of imprisonment lasting decades, and condemned offenders who will wait over a decade before their death sentence is carried out. Moreover, growing numbers of security threat groups (STGs; e.g., prison gangs) have also contributed to prison violence and disorder.

The types of behaviors one might expect in prison are also related to the demographic- and offense-related characteristics of the prison population. The correctional population has always been composed of individuals with lower levels of income and education and histories of alcohol and/or drug abuse. As a result, prisons must provide a number of rehabilitative programs (e.g., substance abuse or educational programs) that respond to these needs if they expect to reduce recidivism. Furthermore, incarcerating higher numbers of younger prisoners (who tend to be more aggressive compared to their older counterparts) may increase rates of prison violence. A higher percentage of drug offenders in the prison population, by contrast, may result in decreased violence.

Prison administrators have developed a number of responses to better manage these changing populations. A growing percentage of aging inmates has forced some prison systems to establish housing units for elderly prisoners. Correctional officers also require additional training to work more effectively with these special needs inmates. An increased number of gang members, by contrast, requires that gang intelligence officers be hired and trained and that interventions that will reduce the influence of these groups are developed. In addition, all officers will require some gang awareness training to understand the gang dynamics in a prison.

Characteristics of Prison Populations

As pointed out in Chapter 2, at the end of 2005 there were more than 1.5 million adults imprisoned at the federal and state levels. The figure represents a substantial increase over the past 25 years (refer to Figure 2–2).

Offense Characteristics

Perhaps one of the most important changes in prison populations over the past two decades has been a decrease in the percentage of violent offenders held in state prisons. **Table 11–1** reveals the changes in the offense-related characteristics of state prisoners since 1980. Violent offenders represented almost 59 percent of the state prison population in 1980, but this total decreased to 46 percent in 1990. Since 1990, the percentage of violent offenders in state prisons has increased to slightly more than half of all prisoners in 2005. These changes have an impact on correctional violence because inmates convicted of violent offenses tend to have a greater involvement in prison murders.[1]

Another noteworthy change is the percentage of prisoners convicted of drug offenses, which more than tripled from 1980 to 2005. These changes also have an important impact on prison violence. A recent federal study shows that prisoners convicted of drug offenses tend to have less involvement in suicide or homicide.[2] Further, an increased percentage of drug offenders requires that additional substance abuse treatment programs be developed.

Table 11–1 State Prisoners, 1980 to 2003

Type of Offender	Percentage of Total Prison Population 1980	1985	1990	1995	2000	2003
Violent	58.95	54.93	46.02	46.64	48.96	52.04
Property	30.37	31.26	25.49	22.99	19.82	20.97
Drug	6.46	8.68	21.81	21.59	20.87	20.08
Public Order	4.22	5.13	6.68	8.78	10.35	6.91

Source: Bureau of Justice Statistics, "Number of Persons in Custody of State Correctional Authorities by Most Serious Offense, 1980–2003," Retrieved from: www.ojp.usdoj.gov/bjs/glance/tables/corrtyptab.htm on October 8, 2007.

In addition to shaping the amount of violence or misconduct in a prison, the offense-related characteristics of the inmate population also influence the types of rehabilitative programs that should be offered to the inmates. If one-half of the prisoners are drug offenders, a priority of the prison system should be the development of substance abuse programs. Higher rates of prisoners convicted of sex offenses, by contrast, require an expansion in the number of sex-offender treatment programs offered.

Demographic Characteristics

There has been a number of shifts in the demographic characteristics of state prison populations in the past two decades. In Chapter 7, we reported that the populations of minority jail inmates were increasing, and a similar pattern is present in prisons. In terms of minority populations, for instance, the Bureau of Justice Statistics reports that in 2005, 34.6 percent of all state and federal prisoners were white, 20.2 percent were Latino, and 39.5 percent were black. From 1995 to 2005, there was an increase in the percentage of Latinos (from 17.6%) and a decrease in the black population, from 45.7 percent to 39.5 percent. The percentage of white prisoners was almost unchanged, increasing by only 1.1 percent between 1995 and 2005.[3]

Parallel with the growing population of female jail inmates discussed in Chapter 7, there has also been an increase in the percentage of women housed in state and federal prisons. At year-end 2005, for example, 7 percent of all prisoners were women, which was up 1 percent from 1995.[4] It is noteworthy that although the imprisonment rate for both genders has increased, the rate of female imprisonment has grown faster. Harrison and Beck report that "since 1995 the total number of male prisoners has grown 34%; the number of female prisoners, 57%."[5] These demographic changes will affect prison systems because women require separate housing units and need access to psychological and reproductive health

Corrections in the Real World

More than half of all federal prisoners are convicted drug offenders. This trend has been relatively constant for several years. Drug offenders made up nearly 56 percent of the federal prison population in 1991 and approximately 54 percent in 2007. Other relatively common offenses for which federal prisoners are serving time include robbery, weapons/arms/explosives offenses, and violations of immigration laws; however, with the exception of robbery during the mid-1990s, each of these types of offenses has rarely exceeded 10 percent of all prisoners.

services. Moreover, there has been increasing awareness of the need for different types of vocational and rehabilitative programs for women.

Special Needs Populations

In Chapter 7, we reported that jails were incarcerating a growing percentage of special needs inmates, including persons with mental illness, gang members, and inmates who were suffering from severe health problems. Prison systems have had a similar growth in these populations. In addition, some groups are more prevalent in prisons today, such as elderly prisoners, death row inmates, and inmates sentenced to long-term sentences. All of these groups create special challenges for correctional systems. As such, we outline the characteristics of these groups (and some of the responses that prison systems have developed to manage these populations) in the following pages.

Persons with Mental Illness

A federal government study of state prisoners found that more than 56 percent of these inmates had a prior mental health problem and that 24 percent had a recent history of symptoms.[6] (A *recent history* was defined as "whether in the past 12 months they had been told by a mental health professional that they had a mental disorder or because of a mental health problem had stayed overnight in a hospital, used prescribed medication, or received professional mental health therapy.")[7] **Table 11-2** shows that many of these inmates had a

Table 11-2 State Prisoners with Mental Health Problems

Characteristics	With Mental Problem	Without Mental Problem
Criminal record		
Current or past violent offense	61%	56%
Three or more incarcerations	25	19
Substance dependence or abuse	74	56
Drug use in month before arrest	63	49
Family background		
Homelessness in year before arrest	13%	6%
Past physical or sexual abuse	27	10
Parents abused alcohol or drugs	39	25
Charged with violating facility rules	58%	43%
Physical or verbal assault	24	14
Injured in a fight since admission	20	10

Source: D. J. James and L. E. Glaze, *Mental Health Problems of Prisoners and Jail Inmates* (Washington, DC: Bureau of Justice Statistics, 2006), 1.

number of serious legal and behavioral problems, including (1) a greater likelihood of being involved in violent offenses, (2) more prior incarcerations, and (3) higher levels of drug use compared to state prisoners who did not have mental health problems.

Table 11–2 shows that the past circumstances of prisoners with mental health problems were often chaotic. Inmates with mental illness were more likely to have been physically or sexually abused. Further, their parents were more apt to have abused alcohol or drugs. Last, many of these inmates had been homeless prior to their last arrest. These facts have implications for community reentry: With fewer stable community resources, parolees are more likely to recidivate (see Chapter 13).

It has long been recognized that persons with mental illness are more likely to be engaged in institutional misconduct than other inmates are. Inmates with mental illness may also be at higher risk of being taunted or abused by other prisoners, becoming involved in fights, or being victimized. James and Glaze found that 58 percent of prisoners with mental illness had violated facility rules, a percentage much higher than prisoners who did not have mental health problems. In addition, these inmates were more likely to have been involved in a physical or verbal assault and twice as likely to report being injured in a fight since being admitted to prison.[8]

These findings have three implications for prison systems. First, high percentages of inmates with mental health problems require that more mental health services be provided. Second, these interventions are costly; the federal study of prisoner mental health shows that approximately 27 percent of state prison inmates had received prescription medication for a mental health problem.[9] Third, prisoners with mental health problems are involved in misconduct at much higher levels than other prisoners are. As a result, one promising approach is to place these prisoners in specialized mental health housing units where staff members receive training on how to best manage these inmates.

Long-Term Prisoners

Most inmates serve only a few years in prison before being released. A national study of state court sentences, conducted by Durose and Langan, found that "the average sentence length to state prisons has decreased since 1994 (4.5 years versus 6 years), but felons sentenced in 2002 were likely to serve more of their sentences before release (51% versus 38%)."[10] As a result, the average state prisoner will serve about 27 months. Those investigators also reported that sentences in federal prisons averaged 58 months, and offenders typically served 91 percent of their sentence, or approximately 53 months.

There are some inmates, however, who are serving lengthy terms of imprisonment, and a small percentage will never be released. Three-strikes sentences, mandatory minimums, and truth-in-sentencing policies (discussed in Chapter 6) have increased the prison terms for some offenders. Others are sentenced to life without the possibility of parole (LWOP), and this population creates a number of problems for prison systems. One important question is how prison administrators develop programs that constructively engage a prisoner who may not be released for decades.

One of the gaps in our knowledge about corrections is that we do not know how many offenders are serving lengthy sentences. To predict the size of the long-term prisoner population, we used California as a benchmark for the entire nation. Approximately 2 percent of the California prison population has been sentenced to LWOP. If we applied that 2 percent statistic to the 1.525 million prisoners in the entire nation, there would be an estimated 30,500 prisoners with that sentence. Although we do not know the true population of these inmates, this estimate provides a good starting point for discussions, and the number is probably much higher (e.g., there are more than 150,000 state prisoners who have been convicted of homicide).

Sentencing an inmate to prison for decades raises two important issues for correctional administrators and policy analysts. First, these inmates may feel that they have nothing to lose and therefore engage in higher levels of misconduct. Second, offenders tend to be sentenced to these lengthy terms of imprisonment when they are older. Criminologists, however, have long known that as offenders age their risk of involvement in further crimes typically decreases. As a result of this **aging out** process, these long-term prisoners may be incapacitated during a time in their lives when some may be at their lowest risk.[11]

Jeremy Travis, former head of the National Institute of Justice, describes the iron law of corrections as "they all come back." With the exception of a few prisoners who die behind bars, 98 percent return to their communities.[12] These long-term populations create challenges for correctional administrators. How do correctional systems prepare prisoners who have been incarcerated for most of their lives to return to the community?

Because most long-term prisoners will return to the community, correctional administrators have developed strategies to engage these inmates in productive

Controversies

A Texas offender was sentenced to a term of life imprisonment for three nonviolent felony offenses that netted him less than $300. This sentence was upheld by the U.S. Supreme Court (see *Rummel v. Estelle,* 445 U.S. 263, 1980). Is the incarceration of these types of offenders the most efficient use of prison? Alternatively, how do we best sanction nonviolent repeat offenders?

activities. Most prisons offer academic, vocational, or rehabilitative programs for these offenders (see Chapter 12 for detailed discussion). Moreover, if these prisoners are judged to be low risk and have not engaged in misconduct, they are sometimes moved to less secure housing units that allow more freedom of movement. Coyle suggests that making these rehabilitative opportunities available is important because the amount of family contact for these prisoners often decreases, and these steps help them better prepare for their return to the community.[13]

Elderly Prisoners

One impact of imposing lengthy sentences is that more prisoners grow old behind bars. Aday points out that "tough on crime" policies have resulted in an increasingly large number of elderly prisoners.[14] The costs of imprisoning these inmates are much higher because of their more extensive health care needs, but elderly inmates also pose other challenges. Because they are slower in conducting the basic activities of daily living (such as eating or showering, or keeping up with the groups being escorted within a facility), older prisoners tend to upset the younger inmates. Jail administrators, for instance, reported that elderly persons were more likely to be victimized than were other inmates,[15] and it is possible that the same is true for elderly prisoners.

There are a number of solutions to the problems created by growing numbers of elderly inmates. Many larger prison systems have developed separate units for aging prisoners with medical staff members (and sometimes volunteer inmates) providing care. High numbers of inmates sentenced to terms of LWOP mean that a growing number of inmates will die behind bars. Mumola reports, for instance, that some 3,000 offenders die in state prisons each year.[16] It is likely that the number of offenders who die in prison each year will increase, because there were some 63,500 inmates older than 55 years of age in state or federal prisons on December 31, 2005.[17]

In recognition of the growing number of dying inmates, many prison systems are developing **hospice** (also known as palliative care) programs. Hospice is intended for inmates with terminal illnesses who have six or fewer months to live. Staff and inmates provide care (but no treatment intended to prolong life) for these prisoners. In addition to basic medical care, these dying inmates may be placed in more comfortable surroundings, with a greater number of visits allowed. In some cases, inmate volunteers will stay with these patients round the clock to ensure that they do not die alone.

Most elderly inmates represent little risk to the community. Although some prisoners remain dangerous, comprehensive assessments enable correctional administrators to accurately gauge the risk that older prisoners pose to the community. Yet, there is often opposition to paroling offenders sentenced to life imprisonment or life without the possibility of parole. Similar to the John Rodriguez example that was used at the start of the chapter, correctional officials and boards of parole have to balance the risk to the community against the benefits to society and the offender.

Some states use **compassionate release** policies to release terminally ill prisoners (generally in the last six months of their lives). Kuhlmann reports that 23 states had compassionate release policies in 1997, but this had increased to 43 states by 2001.[18] Although these policies exist, Kuhlmann's research suggests that such releases are granted only infrequently because the application process often takes months, and correctional officials must carefully weigh the risks to the community. During this lengthy evaluation process, some terminally ill prisoners die. Although these policies reduce the demands on prison health care, in most cases these costs are transferred to the community health care system because most prisoners are indigent.

Severely Ill Prisoners

Prison administrators and correctional health care providers are very concerned about inmates with serious health problems. Rates of illness in prisons are much greater than in the community population, and many prisoners have serious health problems, including life-threatening communicable diseases such as human immunodeficiency virus (HIV)/acquired immune deficiency syndrome (AIDS) and hepatitis. In some cases, these conditions are a result of living on the street, addictions, engaging in risky lifestyles, poor health care, homelessness, poverty, unprotected sex, and undiagnosed or untreated chronic diseases. Responding to these health care problems is an expensive proposition.

A recent national-level study found that approximately 1.8 percent of state and 1 percent of federal prisoners were infected with HIV.[19] Newly developed medicines have enabled HIV-infected inmates to live longer, but there is a significant cost: $13,900 to $36,500 for HIV medication per year in 2006 for a single prisoner.[20] Compared to people living in the free community, prison inmates also have higher rates of tuberculosis, hepatitis, sexually transmitted infections, and infections such as methicillin-resistant staphylococcus aureus (an antibiotic-resistant infection that is becoming common in prison).

Providing health care to prisoners is a challenging task. A survey conducted by the American Correctional Association found that up to one quarter of all expenditures in one state prison system was used for health care.[21] Recruiting and retaining qualified health professionals who want to work in a correctional environment are difficult. Prisons, for instance, have to compete with hospitals and community health care providers that sometimes offer better wages, more sympathetic clients, and a more positive (and less austere) work environment.

Regardless of the costs, prison officials have a constitutional obligation to provide appropriate correctional health care. There is also a significant risk to community health if infections or illnesses are not diagnosed, treated, and controlled while the inmate is incarcerated. The Centers for Disease Control and

Prevention observed that there are numerous ties between correctional and community populations, and prisoners are constantly returning to the community, bringing with them infections contracted in jail or prison.[22]

Some of the medical care needs of prisoners are controversial. Few of us, for instance, would support the prison system paying for an inmate's sex change operation. A more practical question, however, is whether taxpayers should cover the costs of organ transplants. Correctional officials in Arizona recently placed an inmate serving a term of life imprisonment on a waiting list for a kidney transplant. The decision was based on economic benefits rather than compassion because the transplant would generate a cost savings (in reduced dialysis costs) after three years.[23]

Death Row Inmates

Federal research shows that executions are decreasing.[24] After the death penalty was reintroduced in the late 1970s, states carried out many executions, but in recent years this number has decreased. In fact, Snell notes that from 1977 to 2005, only 14 percent of the 7,320 persons condemned to death in the United States were actually executed; 4 percent died of other causes (e.g., suicide or natural causes), and 37 percent were released from death row because of other reasons (most occurred when sentences were commuted from death to life imprisonment).[25]

In 2005, the average time an inmate served on death row awaiting the sentence to be carried out was 147 months, and this waiting period has increased. As we explained in Chapter 6, the amount of time between being sentenced to die and the actual execution is a result of the due process protections granted to condemned offenders. Because of these policies, the numbers of prisoners on death row have increased dramatically (see **Figure 11–1**). Housing these condemned inmates represents a significant challenge for correctional administrators. One

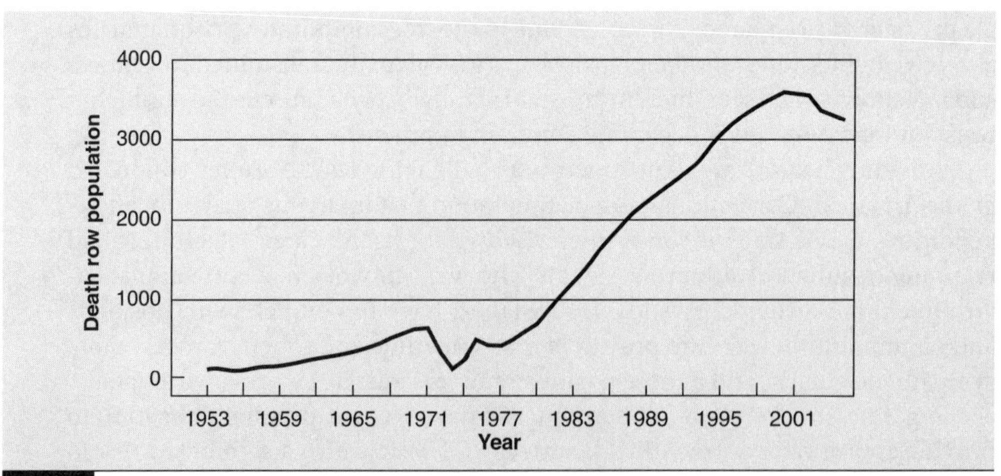

Figure 11–1 U.S. death row population, 1953–2005.
Source: *Reproduced from Bureau of Justice Statistics.*

dilemma is whether these inmates should live on death row or with other maximum security prisoners.

Robert Johnson conducted a study of death row inmates, conditions, and staff. He found that correctional systems use two approaches to manage this population. In the first case, death row inmates are restricted to a housing unit called death row and locked in their cells for most of the day. Johnson observes:

> *Prisoners are confined to their cells 23 hours during weekdays and 24 hours on weekends. The only out-of-cell time is a 15 minute shower three times per week . . . No work, educational or vocational programs are provided.*[26]

Several other states, however, allow inmates on death row to leave their cells. Sometimes there is a living area attached to the cells where the inmates can engage in recreation (e.g., watch television). Further, in at least two states, inmates can actually participate in the prison's vocational programs. When consistent with the safety of the staff and other inmates (e.g., after a careful assessment of the risk that a condemned inmate poses and a lengthy period to observe the inmate's conduct), Johnson advocates placing these prisoners in rehabilitative programs because it is cheaper and the inmates are kept constructively occupied.

Women Prisoners

As mentioned earlier, Harrison and Beck's national study of prison inmates found that women represent 7 percent of the entire U.S. prison population (and 12.7 percent of the jail population)[27] and the growth of female incarceration has outpaced that of men in the past decade. In addition to this growth, the demographic and offense characteristics of women are different than they are for male prisoners. Incarcerated women tend to be better educated than males, although both minority males and females are less educated than their white counterparts. According to Harrison and Beck's research, the largest group of male prisoners is those aged 25 to 29 years, while the largest age group of women prisoners is those aged 35 to 39 years.[28] Another significant gender difference is that only about one-quarter of women prisoners have been sentenced to prison as a result of violent offenses, while twice as many males were convicted of violent crimes. Thus, incarcerated women tend to be less violent, better educated, and older than male prisoners.

Because the population of women prisoners in any state often is small, there are sometimes few alternatives to housing male and female prisoners in a single prison (although they do have separate housing units). Larger prison populations enable administrators to construct gender-specific prison programs. Women-only prisons also have drawbacks, however. First, if there are only one or two women's prisons in a state, there is a likelihood that the prison may be far from the communities where the woman lived prior to her arrest. This makes it difficult and expensive for prisoners to have visits, and this is an especially important consideration because most inmates are mothers with children. Moreover, where there is only one prison in the state, it tends to be a high-security facility. As a result, women who are nonviolent, first-time offenders may be placed in a more secure environment than are males who have committed the same offenses and have the same criminal history.

Another challenge for correctional administrators is to provide women with rehabilitation and treatment opportunities that will lead to economic self-sufficiency.[29] Prisons have historically offered female inmates vocational training programs that were based on gender stereotypes. As a result, women were taught how to be clerks, office support staff, or cosmetologists. The problem is that these jobs are often low-salary careers. Joanne Belknap is critical of these prison-based rehabilitative programs for women, and she raises an important point: How can we expect these women to reintegrate into the community successfully and provide for their families if they can only earn the minimum wage?[30]

Women also have a greater need for correctional health care. Although most prison inmates have received poor community health care, women are also more likely to have engaged in risky behaviors (e.g., alcohol and drug abuse, prostitution, and unhealthy lifestyles or diets). In addition, a number of studies have demonstrated that women tend to have more extensive histories of emotional, physical, or sexual abuse and have a greater need for supportive psychological care than do their male counterparts.[31]

In addition to needing better access to psychological services, women prisoners have a greater need for reproductive health care. Most female prisoners have children or are pregnant at the time of their imprisonment. Either condition is problematic for both the prisoner and the prison authorities. Pregnant inmates must receive specialized prenatal care. This is costly because some 14,000 infants are born to incarcerated women annually.[32] Another dilemma is caring for these infants because prison authorities must decide if the mother and her infant should be able to serve a short term of imprisonment together. Some states, such as California and Washington, have developed prison- or community-based correctional centers where prisoners and their children can be together while the mother serves the remainder of her sentence (provided that it is less than two years, although this time varies depending on jurisdiction). These programs carefully screen potential participants and disqualify those with a history of violent offenses or referrals to child protective services.[33] In most states, however, infants born in prison are placed with family members or in foster care.

There has been considerable interest in how gender role stereotypes have harmed females involved in the justice system. Historically, girls and women were treated in a paternalist fashion (decisions were made on what was thought to be in their best interest regardless of the opinions of these women). Some scholars have called women an "invisible population," and the treatment of women prisoners supports that label.[34] Placed in prisons far away from home, they may receive fewer visits than males do, receive vocational training based on gender role stereotypes, and are often separated from the infants they deliver, despite the fact that prison-based mother–infant programs are cost-effective and successful.[35]

The culture in women's prisons also tends to be different from that in male institutions. Theresa Severance describes how women must cope with imprisonment with few social or family supports. As a result of the isolation that they feel, these women find support from relationships with other prisoners. Female inmates tend to be involved in homosexual relationships (sometimes called **dyads**) at a much higher rate than men are. Severance also describes how some prisoners role-play family relationships (called **pseudofamilies**) where women assume different family roles (such as grandparent, mother, father, or child) to receive "support, affection, belonging, security, and friendship."[36]

Sexual or pseudofamily relationships are seen as problematic by correctional staff. Sexual relationships are perceived as leading to higher levels of misconduct, conflict, or violence as a result of jealousy. Huggins and colleagues surveyed correctional officers in Texas and found that "sexual relationships created anger, suicide attempts, and generally a bad situation for the inmates."[37] Although researchers do not know how many women participate in these types of relationships, some of the Texas officers surveyed in the Huggins study thought that up to 20 percent to 40 percent of inmates living in their units were involved in pseudofamilies, and more than half of officers interviewed believed that 40 percent of women were engaged in sexual relationships.[38]

Security Threat Groups

Over the past two decades, there has been increased concern about gangs, their impact upon crime rates, and developing law enforcement and correctional interventions to control these groups. Gangs are responsible for increased murder rates, and their influence in the drug trade is well documented.[39] One of the problems that correctional administrators have had to confront is that once the police apprehend these individuals, gang members are placed in jails or prisons and their presence changes the culture in many correctional institutions. Gangs are criminal organizations, and most corrections professionals refer to them as security threat groups (STGs), which are defined by the Arizona Department of Corrections as follows:

> *Any organization, club, association or group of individuals, either formal or informal (including traditional prison gangs), that may have a common name or identifying symbol, and whose members engage in activities that include but are not limited to planning, organizing, threatening, financing, soliciting, committing or attempting to commit unlawful acts that would violate the Department's written instructions, which detract from the safe and orderly operations of prisons.*[40]

This definition is comprehensive, but the underlying theme is that members of STGs engage in acts that threaten the safety of the community or correctional facility.

The presence of STGs undermines the authority of correctional officers and the value of the formal rehabilitative programs offered in a prison. Studies of gang-involved jail or prison inmates have generally found that they "contribute to higher rates of prison violence, increase racial tensions within prisons, challenge rehabilitative programming by supporting antisocial values or beliefs, and engage in criminal enterprises within prisons."[41] Being involved in gangs also contributes to higher rates of recidivism after prisoners are returned to the community.[42]

Many prisoners join gangs out of fear of being assaulted. Anecdotal information suggests that inmates serving a short sentence in a minimum security facility might not feel much pressure to join a gang. As the length of sentence and security level of the facility increase, however, it may be difficult to avoid gang recruitment. Ross and Richards observe that "some penitentiaries are literally run by gangs. In those pens, if you don't join one faction or another, you may not be able to defend yourself."[43] Wells and colleagues found that approximately 13 percent of prison inmates were reported to be gang-affiliated.[44] If this figure were applied to the entire U.S. state and federal prison population, there were 188,000 gang members in prisons on December 31, 2005. Despite the fact that large numbers of gang-related offenders are incarcerated, we do not have a good understanding of how attached (or committed) members are to these groups. Some members may be loosely committed to a gang, others want to become members (the "wannabes"), whereas other prisoners are hard-core members who will die for the organization.

A number of strategies have been developed to respond to STGs. Among the most common gang-intervention techniques in prisons are intelligence sharing (between correctional institutions and community law enforcement agencies) and aggressive prosecution of crimes that gang members commit in prison. Other strategies to reduce gang influence include isolating or segregating these inmates and preventing them from communicating with other members. In California, for example, validated or verified gang members are placed in segregated units or facilities and are held in their cells up to 23 hours a day, sometimes for years.[45]

One gang intervention that has been used in some prison systems is called **diesel therapy**.[46] This controversial strategy involves frequent transfers of disruptive prisoners. Although this approach may disrupt an individual from engaging in gang activities, it may actually strengthen the individual's commitment to the gang. Another hazard of diesel therapy is that it may be used for inmates who have legitimate complaints about the correctional system (e.g., those who initiate lawsuits or publicize unethical behavior).

An emerging concern in the gang literature is the possibility that STGs may become recruiting grounds for terrorists. Some prison populations may hold hostile attitudes toward the United States (26.5% of federal prisoners in May 2007, for instance, were not U.S. citizens).[47] The 2005 National Gang Threat Assessment reports that some U.S. gang members have been associated with international terrorists and that "prison gangs seem to be particularly susceptible to terrorism recruitment."[48]

There is a regional element to gang activity in the United States, and jail and prison gangs tend to be more established in western states.[49] But gangs are also a local problem, and some STGs might be limited to a single neighborhood or county.

In addition to established groups, there are some gangs that are becoming well known, including the Nazi Low Riders and the Mara Salvatrucha. Moreover, there is a growing presence of gangs in Indian Country, such as the "Indian Brotherhood, Red Brotherhood, Indian Posse, Warlords, Bear Paw Warrior Society, and similar structures."[50] Last, outlaw motorcycle gangs, such as the Hell's Angels, Bandidos, Outlaws, and Pagans are very active[51] and may have associations with prison gangs.

It is important to acknowledge that each jurisdiction may have a different gang problem. Rural and youth gangs, for instance, tend to have a shorter lifespan than do gangs from urban areas.[52] Other gangs, such as the Black Guerilla Family, have existed for 40 years and may have an intergenerational membership. In addition, some of these groups are involved in the drug trade and individual members are wealthy, while other gangs are transient and not well organized. All of these factors affect the types of interventions that correctional officials can use to respond to these groups.

Race and Gender in Corrections

The following examples of the most commonly encountered prison gangs (from the 2005 National Gang Threat Assessment)[53] reveal that gangs form predominantly along racial lines.

- Aryan Brotherhood
- Black Guerilla Family
- Bloods
- Crips
- Gangster Disciples
- La Nuestra Familia
- Latin Kings
- Mexican Mafia
- Texas Syndicate

Prison Society, Culture, and Experience

A **subculture** is a social group within the larger culture. Prisoners have their own subculture, which has a different set of behaviors, norms, beliefs, and values than those that exist in the free community. To illustrate, rather than status being assigned to people for conventional reasons (e.g., educational or career accomplishments), in the prisoner culture status is often assigned based on criminal accomplishments. Having murdered a police officer typically confers high status, whereas a conviction for raping a child usually brings low status. Displaying signs of fear or weakness, including failure to retaliate when disrespected or not avenging another wrong, can render a prisoner vulnerable to victimization by other prisoners. Signs of too much cooperation with authority figures, especially if this results in a "snitch" or "rat" label, often make prisoners vulnerable as well.

One of the hazards of placing an offender in prison is that the values that offenders adopt while in prison sometimes make it difficult for them to reenter the community successfully. The effects of imprisonment on an individual have long been recognized, and the work of two early scholars, Donald Clemmer and Gresham Sykes, has guided much of our understanding of prison life. We briefly review this work because it provides a foundation to understand the work of current scholars.

Clemmer defined the concept of **prisonization** as the degree to which the experience of imprisonment shapes the individual's norms, values, culture, and language.[54] Prisonization involves the acclimation of a person to prison and assimilation into the prison subculture. Everybody who spends time in what Erving Goffman called a "total institution" (a place where all aspects of the residents' lives are controlled) is influenced by that experience.[55] Prison is often a distinct world. Offenders are removed from society, and conventional norms and rules do not always apply. Prisonization is a process of socialization that starts with the inmate's admission. The longer that a prisoner spends behind bars, the greater influence the institution (and the other inmates) may have on the person. The correctional officers provide inmates with formal rules, but inmates also learn the unwritten or informal rules and expectations. Although these informal rules might differ between facilities, the inmate culture in adult facilities is often oppositional (or antisocial, such as "us versus them").

Walters observes that "prisonization may be an adaptive response that becomes maladaptive once the individual leaves prison."[56] A number of features of prison life, for instance, do not transfer well to the community. Some prisoners might receive "jailhouse tattoos" (crude tattoos made with ink and tattoo guns fashioned from a needle and motors from electronic equipment such as a radio). These tattoos might permanently mark the person as a prisoner and create a lifelong barrier to employment. In addition, inmates may adopt some other "markers" of the penitentiary, such as prison terminology and language, which might also label them after their release. Perhaps the biggest obstacle to community reintegration is the antisocial behaviors that many inmates develop to survive prison. Elliot and Verdeyen report that "deception and manipulation are central features of the inmate culture."[57] Engaging in deceptive, manipulative, or aggressive conduct in the community, however, can be a barrier to success.

Deprivations of Prison Life

Sykes outlines the deprivations of the prison environment and their influence on inmate life in his book titled *The Society of Captives*. The five deprivations (also called losses or pains of imprisonment) include these:

- Liberty
- Goods and services
- Heterosexual relationships
- Autonomy
- Security[58]

These observations provide a framework to help us better understand the inmate subculture. The first deprivation is the loss of liberty. Prisoners are incarcerated apart from society and are further restricted by prison regulations. Sykes believes that inmates see themselves as pariahs (someone who is an outcast from the group), expanding on Clemmer's belief that prisoners adopt an inferior social role.

A second deprivation of prison life is the lack of goods and services. Prison is an austere environment and inmates have little access to goods and services that are often taken for granted. Instead, inmates receive what the prison authorities deem necessary. Although inmates can purchase some types of food at the prison's canteen or commissary (a place that sells items to prisoners), the selection is limited, and inmates must shop on a schedule (e.g., once per week). To understand how difficult it is for a prisoner to adapt to the choices in the community, Chuck Terry describes being in a grocery store after his release from prison:

> *Walking into a supermarket was like entering Disneyland. All those things—lights, products, lines of people—anything you wanted, right there at your fingertips. . . . Compared to the caged-in prison canteen this place was wide open.*[59]

Most prisoners try to make the best of their surroundings, but the correctional environment is drab, and all prisons have rules about the types of items inmates can have in their cells, which further reinforces feelings of deprivation. In many prisons, there is a thriving underground (or *sub rosa*) where prisoners can obtain a variety of items, including **contraband** (items not allowed in a correctional facility) such as drugs and weapons.

The loss of heterosexual contact is difficult for most prisoners to overcome. Inmates are often forced to choose between homosexuality or celibacy, and either may be difficult. A lack of physical contact increases feelings of loneliness and despair. Prisoners often report that they are starved for nonintimate physical contact, such as having a conversation with someone of the opposite sex. **Conjugal visits**, unsupervised overnight family visits that occur in a private setting (such as a trailer on the prison grounds, or a small living unit inside the prison), reduce this pain of imprisonment. These visits are, however, restricted to inmates who are married, have stable institutional behavior, and they may be rare in large prisons. Inmates in only five states had access to these visits in 2002.[60]

Prison authorities structure inmate activities and limit their choices to increase order and security. One limitation of this approach, however, is that prisoners gradually lose the ability to make choices. Depriving inmates of their autonomy makes it more difficult for them to make appropriate decisions when they leave prison.

The last category of deprivation is the loss of security. Sykes observes that inmates live in fear of correctional officers. Prisoners are afraid of being assaulted and of the officer's ability to subject them to internal punishments that place more restrictions on them. Inmates also have to contend with the fear of prison gangs and violence from fellow prisoners. Victor Hassine, a Pennsylvania prisoner serving a term of life imprisonment for murder, has described how the fear of violence shapes most of his day-to-day behaviors.[61]

Women prisoners have different psychological or emotional needs than do men, and they experience a different set of pains of imprisonment. Sykes' description of the deprivations of prison life is based on his observations of male populations. Morton, in contrast, describes a different set of pains for incarcerated women, including the following:

- The loss of family, particularly children
- Being forced to interact with people they do not want to deal with, both other inmates and staff
- Loss of social life, friends, and normal social interaction
- Loss of privacy
- Having to conform to rules made by others that they often see as petty or not warranted[62]

Correctional programs developed for women were historically based on experiences with male prison populations. As a result, programs have sometimes failed to account for these different perceptions of male and female prisoners. This continues to be a shortcoming because understanding these differences can lead to the development of correctional programs that are better able to engage women inmates in rehabilitative pursuits.

Types of Prisoners

The work of Clemmer and Sykes sets the stage for future scholars to develop their interpretations of prison life and culture. John Irwin, a former prisoner, identifies a number of general types of prisoners:

- The thief
- State-raised youth
- The Square John[63]

Writing with Donald Cressey, Irwin suggests that the roles prisoners adopt once incarcerated are heavily influenced by preprison experiences; these experiences are imported into the prison.[64] Further, each of these types of prisoners has an identity that determines how they perceive themselves, serve their time, and relate with other inmates. Irwin's description of the thief, for instance, fits today's definition of a **career criminal**, a person who is committed to a criminal lifestyle (or crime as a form of employment) and who sees a term of imprisonment as a "cost of doing business."

The **state-raised youth** describes persons who grow up in supervised settings, such as foster and group homes, and who are then placed in juvenile detention, juvenile corrections, and adult jails before their admission to prison. These inmates are tough and prone to violence. Irwin describes their worldview as "releases from prison are seen as short vacations. . . . they tend not to see beyond the walls. . . . in prison it is a dog-eat-dog world where force or threat of force prevails."[65] This group of offenders is particularly difficult to work with because their worldview is so tied to crime and prison. Consistent with the career criminal, state-raised youths see themselves as living outside normal society.

The **Square John**, on the other hand, is a prison inmate who holds conventional values, the type of person who lives a normal life and then steals the company payroll. Irwin suggests that these offenders do not share the same values as other inmates, and they have a difficult time adjusting to prison. Richards, an ex-prisoner-turned academic, observes that many of the inmates in federal prisons are "amateur" criminals, a classification much like Irwin's definition of the Square John.[66] Petersilia recently reported that about 24 percent of all U.S. prisoners had not previously been sentenced to probation, jail, or prison.[67]

Our knowledge of corrections has also been enhanced by the contributions of **convict criminologists**, prison inmates who later became college professors and researchers. John Irwin, Daniel Murphy, Greg Newbold, Stephen Richards, Chuck Terry, and Edward Tromanhauser have all written about prison from an insider's perspective. This work often puts a human face on corrections and punishment. As Ross and Richards explain, "Our work is held together by a number of themes, including, but not limited to, understanding the convict experience, forming convict identity, [and] issues of survival in the convict world."[68]

Persons wanting to learn more about life in federal prisons might find it useful to read the work of Michael Santos, who is serving a 45-year term of imprisonment for drug sales made before the age of 23. Santos has written extensively about his two decades as a prisoner. In a series of four books, Santos describes the bureaucracy of the federal prison system, the inmate subculture, types of prisoners, and surviving prison.[69] His work is important because it provides a thoughtful insight into the world of the prison, the pains of imprisonment, and the frustrations of a nonviolent offender serving a four-decade-long sentence.

Inmate Rights and Case Law

When a person is taken into custody, or sentenced to a term of incarceration, the government assumes responsibility for that inmate, including his or her safety and observance of the person's constitutional rights (see Chapter 3 for a description of these rights). This responsibility can place considerable stress on correctional systems. Today appeals courts, especially those at the federal level, require that correctional officials be much more professional than their predecessors were. Historically, the courts showed minimal interest in correctional matters. Minor and colleagues describe three phases that depict the cycle of federal appellate court intervention in corrections over time.[70] These include the hands-off phase (pre-1960s), hands-on phase (1960s through latter 1970s), and due deference phase (latter 1970s until today).

During the **hands-off phase**, courts showed great reluctance to get involved in the day-to-day operations of correctional facilities unless extreme problems existed, such as staff murdering or seriously injuring prisoners. In those cases, courts sometimes intervened on grounds of the Eighth Amendment's prohibition of cruel and unusual punishment.[71] The epitome of the hands-off approach was the 1871 decision of a Virginia court in *Ruffin v. Commonwealth* where prisoners were defined as "slaves of the state" with no constitutional or legal rights or forum to raise grievances except what a jail or prison decided to provide.[72] Judges justified their reluctance to get involved in prison cases on grounds that they did not have expertise in prison management and did not wish to interfere with the work of prison officials.

A departure from the hands-off philosophy came in the 1940s when federal appeals courts ruled that prisoners were entitled to apply to federal courts for *habeas corpus* as a means of challenging the legality and conditions of their confinement.[73] However, until the *Cooper v. Pate* decision some 23 years later,[74] prisoners had little success getting cases into the federal courts, which they generally perceive as more responsive to their cases than state courts are. The *Cooper* ruling allowed prisoners to seek redress or damages on grounds that their civil rights were violated as a result of correctional policies or the actions of individual officers, under Title 42, Section 1983, of the Civil Rights Act of 1871. Prisoners can bring these actions, often called **Section 1983 lawsuits**, directly before the federal courts without first having to exhaust legal remedies at the state level; this can save the prisoner much time and effort. The cases usually allege that the correctional facility or staff members have violated the civil rights of an inmate. But in some cases, a lawsuit is filed on behalf of all (or a group of) inmates in a correctional system, something known as a class action. Some lawsuits seek damages, while others seek some type of remedy, such as a change in policy that mandates better health care for all inmates, for example.

Mainly through using Section 1983 actions, prisoners were awarded a variety of significant rights starting in the 1970s. These cases symbolize the **hands-on phase** mentioned earlier, and some of the main ones are listed in the top panel of **Table 11–3**. The impact of such cases was significant, and in rare cases, federal judges forced correctional administrators to spend hundreds of millions of dollars to reform state correctional systems.

By contrast, the bottom panel of Table 11–3 lists some major cases indicative of the **due deference phase**, which was ushered in by the 1979 *Bell v. Wolfish* decision in much the same way that *Cooper v. Pate* ushered in the hands-on phase 15 years earlier. During the due deference phase, the courts have returned to a more minimalist approach to intervening in prison matters, generally deferring to the discretion of correctional officials—though certainly not always. To illustrate the former point using more recent case law, the Court ruled that the placement of restrictions on prisoner visitation privileges for a limited period of time does not violate the Constitution.[75] To illustrate the latter point, the Supreme Court has ruled that prisoners are entitled to protection from the inappropriate use of force from correctional officers (*Hudson v. McMillian*, 1992) and protection from other prisoners (*Farmer v. Brennan*, 1994).[76] The Court also struck down Alabama's practice of tethering prisoners working on chain gangs to hitching posts when the prisoners presented no immediate threat.[77]

The *Cooper* decision, along with the increased use of imprisonment starting in the mid-1970s, triggered a corresponding growth in the number of inmate lawsuits. A recent national study outlined how the number of Section 1983 lawsuits increased from 3,348 in 1972 (eight years after *Cooper*) to 42,522 in 1996.[78] Because inmates could file these lawsuits in federal courts without paying administrative fees, a large number of cases were dismissed, either because the inmate failed to comply with court rules, their lawsuits were deemed frivolous, or there was no evidence that the prisoner's rights had been violated. Despite the fact that many of these lawsuits ultimately resulted in no action, they tied up the federal courts. In 1980, the federal government enacted the Civil Rights of Institutionalized Persons Act to ensure that inmates would first seek remedy through means other than lawsuits (e.g., institutional grievance procedures) and reserve the courts for the most serious cases. However, most states never

Table 11–3 Some Major U.S. Supreme Court Cases of the Hands-On and Due Deference Phases

Hands-On Phase

1. *Cruz v. Beto* (1972)[a]
   Prisoners must be allowed to practice religions that are unconventional to America, such as Buddhism, when other prisoners are being allowed to practice conventional American religions.
2. *Procunier v. Martinez* (1974)[b]
   Prisoner freedom of speech can be censored only to the extent that restrictions are necessary to promote order, security, and rehabilitation.
3. *Wolff v. McDonnell* (1974)[c]
   Prisoners facing disciplinary actions that threaten their liberties are entitled to (a) an impartial hearing, (b) 24-hours advance notice of the charges, (c) a written statement as to the facts relied upon in reaching decisions, and (d) a limited opportunity to call witnesses and present evidence.
4. *Estelle v. Gamble* (1976)[d]
   Prisoners are entitled to adequate medical care; prison officials may not show deliberate indifference to serious medical conditions.
5. *Bounds v. Smith* (1977)[e]
   Prisoners are entitled either to adequate law library resources or adequate assistance from persons knowledgeable of law.

Due Deference Phase

1. *Bell v. Wolfish* (1979) and *Rhodes v. Chapman* (1981)[f]
   These two cases dealt with related issues; a key ruling was that prison officials may house two persons in a jail or prison cell designed for one as long as such "double bunking" is necessary to address overcrowding.
2. *Hudson v. Palmer* (1984)[g]
   The right to privacy granted under the Fourth Amendment does not apply to a prison cell.
3. *Turner v. Safley* (1987)[h]
   Prison officials can ban mail correspondence between prisoners if justified by "legitimate penological interests."
4. *Wilson v. Seiter* (1991)[i]
   To demonstrate that the prison conditions constitute cruel and unusual punishment, prisoners must show that the conditions exist as a result of deliberate indifference on prison officials' part.
5. *Lewis v. Casey* (1996)[j]
   To prevail when claiming denial of access to the courts, prisoners need to show that problems with the legal resources or assistance provided by the prison caused the prisoner some type of legal injury and hindered his or her efforts to make a case.

[a]*Cruz v. Beto*, 405 U.S. 319 (1972).
[b]*Procunier v. Martinez*, 416 U.S. 396 (1974).
[c]*Wolff v. McDonnell*, 418 U.S. 539 (1974).
[d]*Estelle v. Gamble*, 429 U.S. 97 (1976).
[e]*Bounds v. Smith*, 430 U.S. 817 (1977).
[f]*Bell v. Wolfish*, 441 U.S. 520 (1979); *Rhodes v. Chapman*, 452 U.S. 337 (1981).
[g]*Hudson v. Palmer*, 468 U.S. 517 (1984).
[h]*Turner v. Safley*, 482, U.S. 78 (1987).
[i]*Wilson v. Seiter*, 501 U.S. 294 (1991).
[j]*Casey v. Lewis*, 518 U.S. 343 (1996).

followed through on establishing alternatives, so the number of Section 1983 cases continued to increase.

Although some cases were resolved through monetary settlements to cover damages, in some cases attorneys sued on behalf of groups of inmates to improve the living conditions in a correctional facility (or the entire prison system). Since the 1970s, many correctional administrators (and their attorneys) have entered into court-supervised consent decrees where they agreed to remedy the problem(s) that led to the litigation. In cases involving an entire prison system where it would be almost impossible to determine whether changes were occurring, a special master is appointed by the court to ensure that the terms of these agreements are being upheld. There have been cases where correctional officials have themselves been incarcerated on contempt of court charges if federal judges believed that they were failing to carry out the conditions outlined in the decree.

In many cases, the changes undertaken as a result of these consent decrees have been minor, such as amendments to policies or procedures or increased training for officers. The effects of a few prisoner lawsuits have been more sweeping. Federal Judge William Wayne Justice found in the *Ruiz v. Estelle* (1972) lawsuit (which later became the *Ruiz v. Johnson* matter) that the conditions of confinement in the Texas Department of Criminal Justice (TDCJ) were inconsistent with the Eighth Amendment's prohibition against cruel and unusual punishment. This case lasted for 30 years (it was closed in 2002), and by the time the case was finally resolved, the TDCJ had spent hundreds of millions of dollars (and some say more than a billion dollars) to ensure that conditions of confinement were consistent with constitutional requirements.

In response to the growth in inmate litigation, a number of state attorney generals started distributing lists of the "Top Ten Frivolous Inmate Lawsuits." These lawsuits included suits filed because prisoners' cookies were broken, they had received crunchy instead of smooth peanut butter, or their ice cream was melted when delivered. On the other hand, prisoner advocacy groups also published their lists of nonfrivolous lawsuits; these included cases where correctional officials had sexually abused youngsters in their care, had deliberately placed inmates in jeopardy or harm, or had provided inappropriate (or no) medical care.

In 1995, Congress enacted the **Prison Litigation Reform Act of 1995** (PLRA). This legislation made it more difficult for inmates to file any type of federal lawsuit. PLRA did this by (1) requiring that prior to filing any federal suit, inmates first had to exhaust all administrative remedies available (e.g., prison grievance procedures); (2) making inmates pay to file their lawsuits if they had funds available (such as money in inmate accounts); and (3) restricting inmates from filing suits if they have previously filed two suits judged to be frivolous or malicious. Furthermore, in *Woodford v. Ngo* (2006), the Court held that the PLRA requires *proper* exhaustion of administrative remedies, meaning that prisoners must pursue these remedies in strict adherence to required time lines and procedures.[79]

The impact of the PLRA was almost immediate, and the number of Section 1983 lawsuits decreased significantly. Although the PLRA relieved some pressure on federal courts, prisoner rights advocates argued that it also had the effect of removing a voice from persons who otherwise have few protections from government power.

We noted earlier that inmates have access to the courts through use of writs of *habeas corpus* to challenge their sentence or conviction. Around the same time the PLRA was enacted, Congress also passed legislation titled the **Antiterrorism and Effective Death Penalty Act**. This legislation was intended to limit prisoners' access to the courts through *habeas corpus*, although at least one national study found that the number of *habeas corpus* petitions had not decreased.[80] The authors of that study speculate that these petitions are likely to be filed by prisoners serving lengthy sentences, and these inmates were not deterred by the new legislative requirements intended to create obstacles to these actions.

Schlanger conducted a national-level investigation of prisoner lawsuits in large jails and prison systems and determined the following. First, prisoners tended to file almost four times more lawsuits per 1,000 inmates than jail inmates did. Second, the three main topics of individual and class-action prisoner litigation were medical care, crowding, and use of force.[81] Not only is it expensive to hire attorneys to defend against these lawsuits, but Schlanger found that prison systems and jails paid out millions of dollars in damages to individual prisoners and in class-action lawsuits. In addition to monetary damages, prisons had to make significant changes to their operations, including policies on mental health services, medical services, and use of force, as well as increasing rehabilitative programs for female inmates.

Corrections and Policy

Strip-searching inmates who are admitted to jails to ensure that they have not brought in contraband is an unpleasant activity for the arrestee and the officers involved. Many jails have been sued in recent years for strip-searching misdemeanor offenders without reasonable suspicion that they have contraband, and some have paid millions of dollars of damages. In some cases, these were offenders arrested for minor violations (one Californian alleged that he was strip-searched in front of other naked men after being arrested for driving without a license). What are the risks and benefits of doing these searches?

Perhaps the greatest change resulting from the prisoners' rights movement is that prisoners have a greater voice than they did during the hands-off era. They can raise issues of abuse and prison conditions that threaten their safety, or they may stand up to arbitrary rulings of prison officials. Although the PLRA has reduced the ability of inmates to access the courts, prisoners with legitimate claims can still raise issues. Anecdotal information from correctional officials suggests that when conditions for prisoners improve, prisoners are less likely to be disruptive; this is why some correctional officials have at times supported prisoner suits. Schlanger found that on a scale of 1 (negative impact) to 5 (positive impact), correctional officials reported that court orders on the organization had an overall impact of 3.8 on agency functioning and a 3.5 impact on staff morale.[82] Such findings suggest that correctional officials realize that prisoners who are safer, healthier, treated fairly, and have access to due process protections may be less likely to channel their anger into violence and other forms of acting out.

Conclusion

Understanding the characteristics of prisoners enables us to better appreciate the challenges that they pose for correctional systems. If almost one-quarter of state prisoners have severe mental illnesses, for instance, correctional administrators have to develop strategies to best manage these prisoners while they are incarcerated. Moreover, there is growing concern about how best to prepare these inmates for their return to the community. If prisoners reenter the community without making any rehabilitative changes, they are likely to commit further offenses, and this reduces public safety.

Prisoners have a subculture, and correctional administrators have to understand these values and beliefs to develop prison-based programs that can help prisoners make rehabilitative changes. One challenge, however, is that many inmates are not interested in making these changes. One question that correctional administrators must ask is how they should invest their scarce rehabilitative resources. One of the emerging issues in corrections is the issue of evidence-based corrections, where researchers try to define the most promising correctional interventions and identify what type of inmate receives the most benefit from these interventions. An ongoing consideration in corrections is how to balance prisoners' legal rights against the efforts of prison officials to operate prisons safely and efficiently.

READY FOR REVIEW

- Prison populations have changed over the past two decades. In addition to their growth, prisons are holding fewer nonviolent offenders, including a threefold increase in inmates convicted of drug offenses. The types of offenders held in a prison shape the amount of misconduct in prisons (e.g., drug offenders are less violent) and the types of rehabilitative programs that should be offered.
- Blacks and Latinos represented 59.7 percent of the prison population on December 31, 2005.
- Seven percent of all state and federal prisoners were women, but the rate of growth in women's imprisonment outpaced male imprisonment for the 10 years from 1995 to 2005.
- There is an increasing presence of special needs inmates: prisoners with mental illness, long-term prisoners, elderly inmates, prisoners with severe illnesses, and death row inmates. As the size of these groups increases, there is a greater need for specialized programs, including costly health care interventions.
- The growing number of female prisoners has created a number of challenges because women prisoners have a greater need for psychological support and reproductive health care. One challenge for prison administrators is what to do with infants born in prisons; programs that allow women and infants to be incarcerated together are being introduced in some prison systems.
- Prison gangs, or security threat groups, are an emerging problem in many jurisdictions. These groups undermine the rehabilitative programs of prisons and contribute to higher levels of prison violence. Prisons have established a number of responses to these groups, including segregating gang members or disrupting their activities through frequent transfers.
- Most prisoners undergo some degree of prisonization, which is the influence of the institution on the individual's norms, values, and language.
- Sykes outlines five deprivations of prison life, including liberty, goods and services, heterosexual relationships, autonomy, and security.
- John Irwin identifies a number of prisoner types, including the thief (career criminal), state-raised youth (prisoners raised by the state in foster homes and juvenile facilities), and the Square John (an amateur criminal).
- Convict criminologists are prisoners who became academics; this group of scholars presents an important perspective on inmate life, prison conditions, and debates about the use of imprisonment.
- Courts historically took a "hands-off" approach to inmates, but the *Cooper v. Pate* decision in 1964 enabled prisoners to seek damages if their rights

were violated. This decision led to thousands of inmate lawsuits being launched every year, and although some were legitimate, a large percentage were dismissed.

- The Prison Litigation Reform Act was enacted in 1995 after concern over the number of frivolous inmate lawsuits. This act makes it more difficult for an inmate to launch a lawsuit if the inmate has previously filed two suits judged to be frivolous or malicious.

KEY TERMS

aging out As offenders age, their involvement in crimes tends to decrease

Antiterrorism and Effective Death Penalty Act Signed into law in 1996 by President Clinton, this act makes it more difficult for offenders sentenced to death to file claims of *habeas corpus*

career criminal A person who is committed to a criminal lifestyle (or crime as a form of employment), and who sees a term of imprisonment as a "cost of doing business"

compassionate release The practice of releasing terminally ill prisoners (generally in the last few months of their lives)

conjugal visit An unsupervised overnight family visit that occurs in a private setting, (such as a trailer on the prison grounds or a small living unit inside the prison walls)

contraband An item that is not permitted in a correctional facility (e.g., money, drugs, or weapons)

convict criminologist A former prisoner who becomes a criminologist

diesel therapy The practice of transferring disruptive or gang-involved inmates between prisons to reduce their influence

due deference phase A period that began in the latter 1970s that continues until today during which courts returned to a more minimalist approach to intervening in prison matters, generally deferring to the discretion of correctional officials

dyad A lesbian relationship in prison

hands-off phase Refers to the pre-1960s era when courts showed great reluctance to get involved in the day-to-day operations of correctional facilities unless extreme problems existed

hands-on phase A period from the 1960s through the latter 1970s when courts ruled prisoners were entitled to several important constitutional rights

hospice Care for inmates with terminal illnesses who have six or fewer months to live (also known as palliative care)

Prison Litigation Reform Act of 1995 Legislation that created barriers to prison inmates filing lawsuits

prisonization How imprisonment shapes the individual prisoner's norms, values, and language
pseudofamilies When imprisoned women assume different family roles (such as grandparent, mother, father, or child) to receive support, affection, belonging, security, and friendship
Section 1983 lawsuits Prisoners who feel that their civil rights have been violated can file a Section 1983 lawsuit
Square John A prison inmate who holds conventional values (e.g., a first-time offender admitted to prison)
state-raised youth Prisoners who grew up in supervised settings such as foster and group homes and juvenile facilities prior to their placement in jail or prison
subculture A social group within the larger culture; prisoners have their own culture, which is shaped by their prison experiences

YOU ARE THE CORRECTIONS PROFESSIONAL SUMMARY

1. What factors should be considered prior to paroling a prisoner convicted of a serious crime? Generally, parole boards take into consideration the nature of the offense, the prisoner's prior convictions, risk assessments, the wishes of the victim or the victim's family, institutional conduct (e.g., whether the prisoner has been involved in any disciplinary problems while in prison), prior success or failures while on community supervision, and statements from the prosecuting attorney and sentencing judge. Moreover, parole boards also consider the community supports available to the prisoner and the steps that the inmate has taken to rehabilitate him- or herself.
2. Rodriguez had a long criminal history before this offense. Should that be considered by the parole board? Should the board also consider that he is a World War II veteran who was awarded a Bronze Star? John Rodriguez's application for parole was reviewed by the California Board of Parole Hearings for the sixth time in June 2007 and they recommended his release. Such cases are controversial because it is expensive to care for these inmates, and 95-year-olds do not typically represent much risk to the community. On the other hand, it is important that the public has confidence in the justice system, and returning murderers to the community is not popular, regardless of their age or circumstances. Although the family of Mr. Rodriguez's victim has not formally opposed his release, families of victims do have an opportunity to present their feelings to parole boards. As a member of the California Board of Parole Hearings, your decision to parole a "lifer" ultimately has to be approved by the gover-

nor. During an era when it is politically unpopular to grant mercy, governors are probably less likely to support the decisions of parole boards.

NOTES

1. C. Mumola, *Suicide and Homicide in State Prisons and Local Jails* (Washington, DC: Bureau of Justice Statistics, 2005).
2. Mumola, *Suicide and Homicide.*
3. P. M. Harrison and A. J. Beck, *Prisoners in 2005* (Washington, DC: Bureau of Justice Statistics, 2006), 8.
4. Harrison and Beck, *Prisoners in 2005,* 4.
5. Harrison and Beck, *Prisoners in 2005,* 4.
6. D. J. James and L. E. Glaze, *Mental Health Problems of Prisoners and Jail Inmates* (Washington, DC: Bureau of Justice Statistics, 2006), 1.
7. James and Glaze, *Mental Health Problems*, 2.
8. James and Glaze, *Mental Health Problems.*
9. James and Glaze, *Mental Health Problems*, 9.
10. M. R. Durose and P. A. Langan, *Felony Sentences in State Courts, 2002* (Washington, DC: Bureau of Justice Statistics, 2004), 1.
11. K. Auerhahn, "Selective Incapacitation, Three Strikes, and the Problem of Aging Prison Populations: Using Simulation Modeling to See the Future," *Criminology and Public Policy* 1(2004): 353–388.
12. J. Travis, "Prisoner Reentry: The Iron Law of Imprisonment," in *Key Correctional Issues*, ed. R. Muraskin, 64–71 (Upper Saddle River, NJ: Pearson/Prentice Hall, 2005).
13. A. Coyle, *A Human Rights Approach to Prison Management* (London: International Centre for Prison Studies, 2002).
14. R. H. Aday, *Aging Prisoners: Crisis in American Corrections* (Westport, CT: Praeger, 2003).
15. R. Kuhlmann and R. Ruddell, "Elderly Jail Inmates: Problems, Prevalence and Public Health," *Californian Journal of Health Promotion* 3(2005): 49–60.
16. C. Mumola, *Medical Causes of Death in State Prisons, 2001–2004* (Washington, DC: Bureau of Justice Statistics, 2007), 1.
17. Harrison and Beck, *Prisoners in 2005,* 8.
18. R. Kuhlmann, "Gray Matters: Elderly Jail and Prison Inmates," in *Issues in Correctional Health*, eds. R. Ruddell and M. Tomita, (Richmond, KY: Newgate, 2008), 193–209.

19. L. M. Maruschak, *HIV in Prisons, 2007* (Washington, DC: Bureau of Justice Statistics, 2007), 1.
20. Infectious Diseases in Corrections Report, "News and Literature Reviews: HIV Costs Have Decreased," *Infectious Diseases in Corrections Report* 9(2006): 8.
21. American Correctional Association, "Inmate Health Care," *Corrections Compendium* 29(2004): 10–29.
22. Centers for Disease Control and Prevention, *Drug Use, HIV, and the Criminal Justice System* (Atlanta, GA: Centers for Disease Control and Prevention, 2001).
23. M. Loew, "Should an Inmate Get a Kidney Transplant?" (KOLD News, Phoenix, AZ, December 16, 2005), www.kold.com/Global/story.asp?S=4186951andnav=14RT.
24. T. L. Snell, *Capital Punishment, 2005* (Washington, DC: Bureau of Justice Statistics, 2006).
25. Snell, *Capital Punishment, 2005,* 1.
26. R. Johnson, *Death Work: A Study of the Modern Execution Process* (Belmont, CA: Wadsworth, 1997). 81.
27. Harrison and Beck, *Prisoners in 2005,* 4.
28. Harrison and Beck, *Prisoners in 2005,* 8.
29. J. Belknap, *The Invisible Woman: Gender, Crime, and Justice* (Belmont, CA: Wadsworth, 2001).
30. Belknap, *Invisible Woman.*
31. N. E. Fearn and K. Parker, "Health Care for Women Inmates: Issues, Perceptions and Policy Considerations," *Californian Journal of Health Promotion* 3(2005): 1–22.
32. S. Greene, "Women in Prison," in *Encyclopedia of Crime and Punishment*, ed. D. Levinson (Thousand Oaks, CA: Sage, 2002), 1729–1733.
33. N. E. Fearn and K. Parker, "Washington State's Residential Parenting Program: An Integrated Public Health, Education, and Social Service Resource for Pregnant Inmates and Prison Mothers," *Californian Journal of Health Promotion* 2(2004): 34–48.
34. Belknap, *Invisible Woman.*
35. Fearn and Parker, "Health Care for Women Inmates."
36. T. Severance, " 'You Know Who You Can Go To': Cooperation and Exchange between Incarcerated Women," *Prison Journal* 85(2005): 343–367.
37. D. W. Huggins, L. Capeheart, and E. Newman, "Deviants or Scapegoats: An Examination of Pseudofamily Groups and Dyads in Two Texas prisons," *Prison Journal* 86(2006): 132.

38. Huggins et al., "Deviants or Scapegoats," 132.
39. S. H. Decker and B. Van Winkle, *Life in the Gang: Family, Friends, and Violence* (New York: Cambridge University Press, 1996).
40. Arizona Department of Corrections, "Security Threat Group Unit," 2007, http://adcprisoninfo.az.gov/adc/divisions/support/stg.asp#stg.
41. R. Ruddell, S. H. Decker, and A. Egley, "Gang Interventions in Jails: A National Analysis," *Criminal Justice Review* 31(2006): 33.
42. M. S. Fleisher and S. H. Decker, "Overview of the Challenge of Prison Gangs," *Corrections Management Quarterly* 5(2001): 1–9.
43. J. L. Ross and S. C. Richards, *Behind Bars: Surviving Prison* (Indianapolis, IN: Delta Press, 2002), 127.
44. J. B. Wells, K. I. Minor, E. Angel, L. Carter, and M. Cox, *A Study of Gangs and Security Threat Groups in America's Adult Prisons and Jails* (Indianapolis, IN: National Major Gang Task Force, 2002).
45. J. Petersilia, *Understanding California Corrections* (Berkley, CA: California Policy Research Center, 2006), xi.
46. S. C. Richards, "My Journey through the Federal Bureau of Prisons," in *Convict Criminology,* eds. J. I. Ross and S. C. Richards, 120–149 (Belmont, CA: Wadsworth, 2003).
47. Federal Bureau of Prisons, *Quick Facts,* 2007, www.bop.gov/about/facts.jsp.
48. National Alliance of Gang Investigators Associations, "2005 National Gang Threat Assessment," 2005, www.nagia.org.
49. Ruddell et al., "Gang Interventions in Jails."
50. National Alliance of Gang Investigators Associations, "2005 National Gang Threat Assessment," 12.
51. National Alliance of Gang Investigators Associations, "2005 National Gang Threat Assessment," 12.
52. J. C. Howell and A. Egley, *Gangs in Small Towns and Rural Counties* (Washington, DC: Office of Juvenile Justice and Delinquency Prevention, 2005).
53. National Alliance of Gang Investigators Associations, "2005 National Gang Threat Assessment," 6–7.
54. D. Clemmer, *The Prison Community* (Boston: Christopher Publishing House, 1940).
55. E. Goffman, *Asylums* (New York: Doubleday, 1961).
56. G. Walters, "Changes in Criminal Thinking and Identity in Novice and Experienced Inmates: Prisonization Revisited," *Criminal Justice and Behavior* 30(2003): 413.

57. B. Elliot and V. Verdeyen, *Game Over: Strategies for Redirecting Inmate Deception* (Lanham, MD: American Correctional Association, 2002), 1.
58. G. M. Sykes, *The Society of Captives: A Study of a Maximum-Security Prison* (Princeton, NJ: Princeton University Press, 1958).
59. C. M. Terry, "From C-Block to Academia: You Can't Get There from Here," in *Convict Criminology,* eds. J. I. Ross and S.C. Richards, 95–119 (Belmont, CA: Wadsworth, 2002), 105.
60. C. Hensley, R. Tewksbury, and C. Chiang, "Wardens' Attitudes toward Conjugal Visitation Programs," *Journal of Correctional Health Care* 9(2002): 307–319.
61. V. Hassine, "Prison Violence: From Where I Stand," in *Turnstile Justice,* eds. R. L. Gido and T. Alleman (Upper Saddle River, NJ: Prentice Hall, 2002), 38–56.
62. J. Morton, *Working with Women Offenders in Correctional Institutions* (Lanham, MD: American Correctional Association, 2004), 146. See also: D. Ward and G. Kassenbaum, *Women's Prison: Sex and Social Structure* (Chicago: Aldine Publishing Company, 1965).
63. J. Irwin, *The Felon* (Englewood Cliffs, NJ: Prentice Hall, 1970), 8–32.
64. J. Irwin and D. Cressey, "Thieves, Convicts and the Inmate Culture," *Social Problems* 10(1962): 142–155.
65. Irwin, *The Felon,* 29.
66. Richards, "My Journey through the Federal Bureau of Prisons."
67. Petersilia, *Understanding California Corrections,* 56.
68. J. I. Ross and S. C. Richards, "What Is the New School of Convict Criminology?" in *Convict Criminology,* eds. J. I. Ross and S. C. Richards, (Belmont, CA: Thompson Wadsworth, 2002), 11–12.
69. M. G. Santos, *About Prison* (Belmont, CA: Wadsworth, 2003). See also: M. G. Santos, *Inside: Life behind Bars in America* (New York: St. Martin's Press, 2006).
70. K. I. Minor, J. B. Wells, and I. R. Soderstrom, "Corrections and the Courts," in *Exploring Corrections in America,* eds. J. Whitehead, J. Pollock, and M. Braswell, 60–107 (Cincinnati, OH: Anderson, 2003).
71. D. H. Wallace, "Prisoners' Rights: Historical Views," in *Correctional Contexts,* eds. J. W. Marquart and J. R. Sorensen (Los Angeles: Roxbury, 1997), 248–257.
72. *Ruffin v. Commonwealth*, 62 Va. 790 (Va. 1871).
73. *Ex parte Hull*, 312 U.S. 546 (1941); *Coffin v. Reichard*, 143 F.2d 443, 6th Cir. (1944).

74. *Cooper v. Pate*, 378 U.S. 546 (1964).

75. *Overton v. Bazzetta*, 539 U.S. 126 (2003).

76. *Hudson v. McMillian*, 503 U.S. 1 (1992); *Farmer v. Brennan*, 511 U.S. 825 (1994).

77. *Hope v. Pelzer*, 536 U.S. 730 (2002).

78. F. L. Cheesman, B. J. Ostrum, and R. A. Hanson, *A Tale of Two Laws Revisited: Investigating the Impact of the Prisoner Litigation Reform Act and the Antiterrorism and Effective Death Penalty* (Washington, DC: National Center for State Courts, 2004), 10.

79. *Woodford v. Ngo*, No. 05-146, S. Ct. (2006).

80. Cheesman et al., *Tale of Two Laws Revisited.*

81. M. Schlanger, "Inmate Litigation: Results of a National Survey," *National Institute of Corrections Large Jail Network Exchange*, July 2003, 1–12.

82. Schlanger, "Inmate Litigation," 9.

12

Prison Security, Programs, and Services

Chapter Objectives

- Explain how classification has contributed to lower murder rates in prisons.
- Explain why prison violence has decreased in the past two decades.
- Understand the elements of an effective security program.
- Describe the changes in prison design.
- Describe the different types of prison riots.
- Understand the different elements of correctional rehabilitation.
- Describe the importance of evidence-based corrections.
- Understand the importance of correctional health care.

CASE STUDY

You have been working as a correctional officer for a state prison system for 10 years and have just been promoted to sergeant. Two of the officers on your unit, a male who has been employed in the prison for a number of years and a female officer one year out of the academy, have been getting friendlier over the past few months. Today another officer told you that he saw them embracing and kissing on the night shift while in the staff dining area (on a designated break). Your department doesn't have a nonfraternizing policy, and both of these officers are unmarried.

1. Should you intervene in this situation? If you decided to intervene, how would you approach these officers?
2. What are some possible negative outcomes of this relationship?
3. How will "stepping up to supervisor" affect the relationships with your former coworkers?

Introduction

The previous chapters outline the characteristics of inmates and correctional systems and identify some key features of the inmate and correctional officer subcultures. The primary goal of institutional corrections is the security and control of offenders, and in some cases, the interests of inmates are in opposition to the goals of correctional officers and staff. To achieve the goals of security and control, correctional officers have developed a variety of strategies, a number of which have proven to be effective in reducing violence.

Over the past two decades, murder rates have decreased in state and federal prisons, and prison riots—common during the 1970s and early 1980s—are rare today. The fact that correctional officials achieved these results during an era of institutional crowding and increasingly long sentences while supervising large percentages of difficult-to-manage inmates (such as young gang members serving lengthy sentences) is a very positive outcome.

Over the past four decades, correctional systems have had to respond to a number of problems. The increasing growth of correctional populations since 1975 has led to serious overcrowding problems in some institutions. To respond to overcrowding, some prison systems have had to greatly expand the number of inmates held. This created other problems, because more staff had to be recruited and trained, and these incoming inmates had to be housed. Increasing inmate populations brought about other challenges, such as a greater number of security threat groups. Durham observes that American corrections was driven by crisis and reform.[1] Although violence has decreased, there are other short- and long-term obstacles that correctional administrators have to overcome. Yet, times of crises also create opportunities, and although there is no shortage of criticisms of prison operations—many of which are valid—there has been a meaningful positive change in the way that modern prisons operate.

Inmate Classification

In Chapter 7, we briefly discuss the process of inmate classification in jails and how using this approach to offender management has increased staff and inmate safety. There is no universally accepted method of classifying offender risk, but most prison systems first classify offenders at reception centers where prisoners go after they are sentenced (the exception is condemned inmates). During an inmate's time at a reception center, which can be a period of weeks or months, evaluations are conducted of their physical and psychological functioning as well as their social and legal backgrounds. Some prisons do comprehensive health screening on inmates, including universal testing for some communicable diseases, while other systems have more rudimentary assessments.

Austin observes that in the past, most jails and prison systems used **subjective classification** (gut feelings, intuition, or the best judgment of an officer or placement committee) to place inmates in housing units and establish the appropriate security levels for them.[2] Since that time, however, most prison systems have developed **objective classification**, which bases these judgments on scores on instruments designed to evaluate the risk of escape or misconduct. Factors that are considered in developing a classification score include the current offense and sentence length, prior criminal history, prior institutional history (e.g., disciplinary actions or misconduct in prior incarcerations), age at first arrest, gang involvement, and community factors (including employment history or drug and alcohol use). Some of these factors exert a significant impact on the prisoner's life behind bars; a prior escape, for example, could make the inmate ineligible for community-based programs despite the fact that the escape occurred a decade in the past. Most classification systems utilize a numerical score, and the higher the inmate's score, the greater the likelihood the inmate will be placed in a unit or facility with a higher level of security.

Austin and Hardyman further designate classification into two categories: (1) **external classification** determines a prisoner's custody classification and facility assignment; (2) **internal classification** governs facility-level decisions such as where and with whom the prisoner will be housed, the types of programs and services to which the inmate should be assigned, and the prison industry or work assignment most appropriate for that prisoner.[3]

Objective classification goes beyond the development of an instrument to derive scores for inmates. Essential components of an objective classification system include a "standard set of criteria that have been validated by research, a centralized classification unit that controls transfers between facilities, automated classification systems, annual re-classification of inmates (to determine whether they still require the same level of security), and the ability to override classification."[4]

Classification is a serious business in correctional systems, and an inmate's score can have long-term consequences (see **Figure 12–1**). Classification determines "the level of surveillance, the amount of unrestricted activity, and access to programs . . . [and] the rate of good time that is earned."[5] As a result, it is important that classification be done correctly. Most prisoners would like the lowest score possible, because this represents a faster track to release and less restrictive conditions while in prison.

MALE INMATES
Oklahoma Department of Corrections
Custody Assessment Scale

A. IDENTIFICATION Facility: ____________ **Date:** ___/___/___
Inmate Name (Last, First, Middle Initial): ____________ Inmate DOC# ________
Reception Date: ________ (assessment and reception center) Race/Sex: ___ DOB ______ Date of Last Review: ________

B. CUSTODY EVALUATION Score

1. SEVERITY OF CONVICTION on CURRENT INCARCERATION
   (Use the Offense Severity Scale in Attachment A: rate most serious current charge/conviction, including CC, CS, SS, or rebill cases, detainers, or warrants.)
   - Low = 0 pts.
   - Moderate = 2 pts.
   - High = 5 pts.
   - Highest = 6 pts.

   Offense: ________ Case Number: ________ Case Type: ________

2. SERIOUS OFFENSE HISTORY (From the date of assessment, Use Offense Severity Scale in Attachment A)
   - None or Low (past 5 yrs.) = 0 pts.
   - Moderate (past 5 yrs.) = 1 pt.
   - High (past 10 yrs.) = 4 pts.
   - Highest (past 10 yrs.) = 6 pts.

   Offense: ________ Case Number: ______ Discharge/Conviction: Date: ___/___/___

3. ESCAPE HISTORY
   - No escapes or attempts = 0 pts. ____
   - Escape from community supervision (PPCS, GPS) past 2 years = 6 pts. ____
     Escape or attempted escape from min. custody, com. corrections, juv. AWOL past 5 years = 6 pts. ____
   - Two or more escapes or attempted escapes from minimum, community corrections, juv. AWOL, community supervision w/in 10 years = 6 pts. ____
   - Escapes from any level of security that result in an injury or a felony conviction for a violent crime while on escape status during this incarceration or w/in 10 years of a prior incarceration from the date of this custody assessment = 10 pts. ____
   - Escape or attempted escape from medium (jails, juvenile institutions, detention centers, shu's) or maximum security during this incarceration or w/in 10 years of a prior incarceration from the date of this custody assessment. = 10 pts. ____

   Facility ________ Security Level ______ Escape Date ___/___/___ Apprehension Date ___/___/___
   Facility ________ Security Level ______ Escape Date ___/___/___ Apprehension Date ___/___/___

4. **MAXIMUM CUSTODY SCORE** (Add item 1, 2, 3) **Subtotal**: ______
   SCORE OF 12 OR HIGHER, ASSESS TO MAXIMUM CUSTODY:

5. NUMBER of ACTIVE DISCIPLINARY CONVICTIONS
   (Class A&B – last 6 months, Class X – last 2 years)

		Code	Class	Date
• None	= 0 pts.	________	________	________
• One	= 1 pt.	________	________	________
• Two	= 2 pts.	________	________	________
• Three or more	= 3 pts.	________	________	________

6. MOST SERIOUS DISCIPLINARY CONVICTION (within last 12 months)
   No expiration on current incarceration or w/in 10 years using the date of current assessment for the following Class X offenses: 01-4, 04-1, 04-8 (04-3 prior to September 14, 1989)

		Code	Class	Date
• None	= 0 pts.	________	________	________
• Class B	= 0 pts.	________	________	________
• Class A	= 2 pts.	________	________	________
• Class X	= 4 pts.	________	________	________

7. ASSIGNED PROGRAM PARTICIPATION (since last classification)
   - None, waiting list, enrolled, participating = 0 pts.
   - Completed program within past 2 years (unless has pts. in escape section) = –1 pt.

   Recommended Program: ____________ Completion Date: ___/___/___

8. ADJUSTMENT (indicate earned credit class level assigned)
   - Level 1 = 1 pt.
   - Level 2 (or has points in the escape section) = 0 pts.
   - Level 3 and 4 (unless has pts. in escape section) = –1 pt.

DOC 060103A Section 3 (R 8/07) Page 1

Figure 12–1 Classification form.
Source: *Reproduced from Oklahoma Department of Corrections, www.doc.state.ok.us/Offtech/060103a(m).pdf.*

9. CURRENT AGE
   Age 25 or Younger = 2 pts.
   Age 26 to 31 = 1 pt.
   Age 32 to 39 (or has pts. in escape section) = 0 pts.
   Age 40 to 49 (unless has pts. in escape section) = –1 pt.
   Over 50 (unless has pts. in escape section) = –2 pts.
10. COMPREHENSIVE CUSTODY SCORE (add items 1-9) **Total Score:**

C. SCALE SUMMARY AND RECOMMENDATIONS

1. CUSTODY LEVEL INDICATED by SCALE **Assessed Custody Level**: _______
   - ❑ 6 or fewer points on items 1–9 = Minimum
   - ❑ 7–12 points on items 1–9 = Medium
   - ❑ 12 or more points on items 1–3 = Maximum
   - ❑ 13 or more points 1–9 = Maximum
2. MANDATORY OVERRIDES (No lower than medium security) reason is required
   - ❑ Life without Parole
   - ❑ Time Left to Serve (Highest Crime Category)
   - ❑ Restricted Earned Credits with excess days
   - ❑ ICE Detainer (High or Highest Crime)
3. DISCRETIONARY OVERRIDES FOR HIGHER CUSTODY LEVEL reason is required
   - ❑ Circumstances of the offense
   - ❑ History of Violence
   - ❑ Gang Affiliation
   - ❑ Time left to serve
   - ❑ Management Problem
   - ❑ Escapes
   - ❑ Felony Detainer
   - ❑ Other (specify):_______________
4. DISCRETIONARY OVERRIDES for LOWER SECURITY LEVEL reason is required
   - ❑ Circumstances of the Offense
   - ❑ Time Left to Serve
   - ❑ Outstanding Conduct
   - ❑ Other (specify):_______________
5. RECOMMENDED CUSTODY LEVEL
   ❑ Minimum (< 7300 days) ❑ Medium ❑ Maximum
6. ❑ Community Placement (< 2920 remaining and meet eligibility for community as outlined in OP-060104)
7. **Custody Level Assignment:**_______________
8. Comments: _______________
9. SIGNATURES
   Preparer's Signature _______________ Date: _______
   Committee Member _______________ Date: _______
   Committee Member _______________ Date: _______
   Committee Chair _______________ Date: _______
   Inmate Signature Date:

D. REVIEW AUTHORITY ❑ Concur ❑ Do not Concur **Changed to:** ❑ Min. ❑ Med. ❑ Max.

Reason for Change: _______________

Routine: Case Mgr. Supv: _______________ Date: _______
Non Routine: Facility Head Signature: _______________ Date: _______
(If Changed) Inmate Signature: Date:

E. DIVISION OFFICE ❑ Concur ❑ Do not Concur **Changed to:** ❑ Min ❑ Med. ❑ Max.

Reason for Change: _______________

Deputy Director/designee _______________ Date: _______

F. POPULATION OFFICE: ❑ Concur ❑ Do not Concur **Changed to:** ❑ Min. ❑ Med. ❑ Max.

Reason for Change: _______________

Administrator/Population Coordinator or Population Officer _______________ Date: _______

Distribution:
- White: Population Office (Send back to receiving facility for field file)
- Pink: Population Office
- Canary: Field file before approval
- Goldenrod: Inmate before transfer

DOC 060103A Section 3 (R 8/07) Page 2

Figure 12–1 (continued)

Controversies

Some scholars are critical when "white-collar" criminals are placed in low-security federal prisons that they have labeled "Club Fed." Yet, it can be argued that most of these criminals are often first-time offenders who are sentenced to relatively short periods of imprisonment for nonviolent offenses. Putting these low-risk offenders in a high-security setting would increase costs and would not increase prison safety or security, the primary goals of correctional systems.

It is important for officers to develop methods of properly evaluating the risk of violence or disruptive behavior. Austin found that inmates who are younger, male, have prior histories of violence or mental illness, gang-involved, have had recent misconduct, and are not involved in prison programs are more likely to be disruptive.[6] That same study found that some factors did not have an impact on inmate behavior, including drug and alcohol use, escape histories, length of sentence, severity of the offense, and time left to serve.

Regardless of the approach that is used, two factors are important to consider. First, there must be an ability to override classification scores. A wheelchair-bound elderly inmate admitted to prison for the first time who has been convicted of a sex offense would probably be at lower actual risk than would be a gang-involved young man who has been involved in a series of property offenses, yet the geriatric inmate would probably have a much higher classification score. As a result, classification officers might place the elderly inmate in a lower-security setting that better reflects inmates' true risk. The second issue is that inmates should be classified each year to evaluate changes and to determine whether the inmate is in the appropriate level of housing. This is the process of **reclassification**.

Institutional Security

Considerable changes in institutional corrections have occurred over the past two decades. Prison administrators have made significant changes in increasing the security of prisons, introducing new prison designs (which in turn changes the methods of supervision, staff–inmate interactions, and officer control), and developing violence reduction programs. To a large extent, these changes were a consequence of some well-publicized instances of lethal prison violence and the overcrowding that forced prison systems to expand.

Overcrowding and Correctional Safety

Crowding has a major impact upon federal and state prisons. A national study found that the federal prison system operated at 134 percent above rated capacity on December 31, 2005, while 23 state prisons had inmate populations that were above their capacity.[7] As the overcrowding crisis increased, some facilities began using double bunking (housing two inmates in cells originally designed for one prisoner); others placed inmates in dorms, and dozens or hundreds of beds were placed in common areas such as gyms. In other jails or prisons, inmates slept on cots placed in hallways. The disadvantage to putting inmates in places that were not designed for prisoner housing is that such arrangements are not safe or secure. It is harder to monitor inmate behavior, dangerous inmates may inadvertently be placed in a dorm, the environment is more stressful for inmates (who in turn create stress for officers), and dormitory living arrangements offer less security than traditional rooms do. A recent Office of the Inspector General examination of jails in Indian Country found that;

> *Overcrowding has become a health and sanitary issue. Many inmates sleep on mats on the floor because the jails hold two or three times their rated capacity . . . increasing the potential for altercations and injuries because inmates cannot move without stepping over and around one another.*[8]

Moreover, if the gym is filled with inmates, then the regular inmate population may have fewer opportunities to exercise, and idleness is associated with correctional misconduct.

One of the problems of overcrowding is that the creativity of the correctional system becomes invested in developing alternatives to overcrowding. The orientation of the system shifts from a proactive approach that values planning to one that reacts to crises. To better plan for such contingencies, some prison systems are using computerized modeling systems to forecast population trends. In some jurisdictions, jail or prison authorities have attempted to build enough cells to house all of the incoming inmates, but the more cells constructed, the more the prison population increased, in a seemingly endless cycle. A popular saying about correctional capacity is that "if you build it, they will come."

Rapid expansion in a correctional system tends to dilute the expertise of the staff. New officers are recruited and trained, but in their first months they require considerable support and supervision because there is a steep learning curve on how to interact effectively with prisoners. These supervisory duties often fall upon the shoulders of more experienced staff members, although this increases their levels of stress. But, no matter how difficult it is for staff to cope with overcrowding, it is harder on inmates.

Surviving jail or prison depends on the inmate's ability to fully understand the persons with whom he or she is incarcerated, the risk that the other prisoners pose, and the ability to read situations (e.g., if a fight or disturbance is going to occur). More prisoners in a living unit or facility, however, mean that inmates' knowledge of their counterparts (and their triggers for violence) will be reduced, and this may lead to increased numbers of assaults or murders.[9] In addition, by increasing the number of prisoners in a confined space, there is a greater likelihood that they will interact, disagree, and fight with each other more often. Being placed in a correctional setting that has less physical security (such as a dorm) may also contribute to violence, as will overcrowding in juvenile or young adult populations because this group tends to be more volatile. Another potential problem with facility overcrowding is that when larger numbers of inmates are housed in close proximity, inmates face increased risk for transmission of communicable diseases such as tuberculosis (TB).[10]

Security Programs and Procedures in Modern Prisons

Prisons have become safer places in the past few decades. The number of prison murders in state facilities decreased 93 percent from 1980 to 2002.[11] In addition, riots are rare events. Escapes are also rare, at least from medium or maximum-security facilities. Prisons have become safer places because of the steps that administrators and officers have taken. The reduction in violence and serious events, including inmate suicides, can be attributed to an integrated approach to security where every headquarters, institution, and parole position—whether an employee works as a correctional officer or supports the activity of officers—work together to meet the primary goal of the a prison, which is to provide a safe and secure environment.

Atherton and Phillips outline nine elements of a healthy security system in prisons. These elements are summarized as follows:

1. Fundamental and clear understanding of the agency's and institution's mission by all staff, and sufficient resources to carry out that mission
2. A well-structured and well-staffed headquarters organization (that supports the activities of the prison)
3. A comprehensive institution program, which provides all necessary supporting services to staff and inmates
4. A high-quality personnel-management structure that includes efficient selection, supervision, and staff-development functions
5. Careful matching of the institution (layout, design, age, and level of maintenance) with a given type of inmate and a specific staffing level

6. The availability of appropriate equipment, including locking devices, door and window hardware, perimeter-security devices, and other items used for monitoring and control
7. The availability of programs that enhance security by involving inmates in productive use of their time
8. The availability of information-management systems to assist in security operations (e.g., computerized systems that collect information on inmates, incidents, and costs)
9. The existence of a variety of security systems that act jointly to prevent violence, control contraband, and prevent escapes[12]

Prison security has become an all-encompassing goal and is the foundation for all of the other activities of a prison, from effective rehabilitative programs to prerelease or discharge planning. Certain areas of a prison are considered key to security and are the targets of especially tight control. Examples include the entryway to the facility, the control room (which is central to communication and coordination of activities throughout the institution), the dining or "chow" hall, areas where confidential records are stored, visitation areas, and work sites—especially those with tools and materials that can be converted into weapons. Prisoners are routinely counted by staff, and prisoner whereabouts are monitored throughout the institution (increasingly through the use of computer technology). Standard precautions are also taken any time prisoners must be transported (e.g., to another facility, a courthouse, or a hospital). Prisoners routinely have their persons and housing quarters searched in an effort to control contraband, particularly weapons and drugs. Special precautions are taken to control certain material items found in a prison such as firearms, keys, medication, incoming and outgoing prisoner mail, and even food.

Although security rests in the physical architecture of the prison such as the fences, bars, and individual cells, the activities of the correctional officers and staff are also critical to a safe institution. Correctional officers with excellent interpersonal skills and problem-solving abilities can help inmates develop their social skills, reduce interpersonal conflict, and deescalate problems before they erupt into violence. One significant change in modern prison systems is the recognition of the human service role of correctional officers.

When prisoners violate prison rules, officers have the discretion to issue disciplinary reports (write-ups). In many facilities, officers overlook certain minor rule infractions. But formal disciplinary reports are hardly uncommon and usually lead to a hearing before institutional officials (such as a committee) charged with determining (1) whether the evidence supports the charge(s) in the report and, if so, (2) the sanction to be imposed. Examples of possible sanctions include various restrictions on privileges, placement in the disciplinary segregation unit (known as "the hole"), forfeiture of good time (see Chapter 13), or transfer to a more secure facility. Serious disciplinary violations, such as murders or serious assaults, can result in prosecution in court on new charges.

Prison Design and Security

The understanding most people have of prisons probably comes from watching television or films, and most of these programs depict older big house style facilities that relied upon bars and bricks to keep inmates from escaping or rioting. Some big house prisons are still operational today, but many states are gradually replacing these facilities because of their high operational costs (e.g., staffing, heating, and cooling). These institutions are also less safe because of "blind spots" that make the supervision of prisoners difficult and cell designs that allow only intermittent supervision of inmates.

Newly constructed correctional facilities have a much different look than the traditional big house penitentiary. Gone are certain features that characterized the Pennsylvania, Auburn, and reformatory-style prisons discussed in Chapter 10 (e.g., tall stone walls and buildings that looked like medieval castles). Stone fences have been replaced with rows of wire fences (including electric fences in some jurisdictions). Despite the "less secure look" of modern institutions, prisoners are unlikely to escape. Useem and Piehl report that "in 1981, there were 12.4 escapes per 100,000 inmates in U.S. prisons. By 2001, the ratio had declined to 0.5 per 100,000 inmates."[13]

Inmates living in prisons constructed in the past two decades are likely to be housed in buildings that more closely resemble college dormitories or that were constructed using the new generation design that was described in Chapter 7. This design enables inmates to be housed more efficiently, officers are able to supervise prisoners more closely, and officers can engage in more effective problem solving. Altogether, these approaches are also associated with less disruptive behavior and assaults.[14]

The security of today's prisons is constructed within the actual architecture. Windows are made using thick layers of Lexan or similar glass (what some call bullet-resistant glass) that has reduced the need for iron bars. Furnishings that were once used as weapons—such as tables or chairs—are now bolted to the floor. Fixtures such as sinks and toilets that were once constructed of porcelain are now made of stainless steel. Although inmates still have the ability to flush their toilets or turn on their lights, in many facilities the correctional officers can turn off electricity or water to an individual cell if an inmate is disruptive.

Some of the security in a prison depends on the levels of classification of the inmates. As mentioned earlier, many low-security inmates live in dorms or sleep in bunks in living units that are constantly monitored by correctional officers. Inmates who require a higher level of control are often placed in individual cells and live behind metal doors. Some out-of-control or disruptive inmates are placed in **isolation cells**. Every prison has a few of these single-room "hardened" cells (e.g., bare concrete walls and stainless steel fixtures) for the temporary control of disruptive inmates, and they are usually called **segregation units**.

Violence Reduction Strategies

Even the best-run correctional environments can be crowded, chaotic, and violent places, and as we explored in the previous chapters, a large percentage of the population are persons with mental illness, long-term inmates with little to lose, or gang-involved prisoners. As a result, one of the most significant fears of first-time inmates is that they will be assaulted or killed while trying to serve their sentence. Victor Hassine, an inmate serving a term of life imprisonment without the possibility of parole, describes his relationship with violence;

> *As a convict, I eat prison violence for breakfast, lunch, and dinner. It is with me when I wake in the morning and when I go to sleep at night. It stalks me during the day, and steel-eyed, it glares at me throughout the night. It is the lead actor in my dreams and the villain during my conscious hours.*[15]

These are valid fears, and although a prisoner is still vulnerable to assault or homicide, rates of prison violence have decreased over the past few decades.

A federal study of homicide trends in jails and prisons reported that the 2002 homicide rate in jails was 3 per 100,000 inmates, and in prisons the rate was 4 per 100,000 inmates. The prison murder rate dropped 93 percent in 22 years, and homicide rates in jails decreased by 40 percent during the same time period.[16] These findings are remarkable not only because of the decline, but for the fact that the community murder rate in 2002 for the entire United States was 5.6 persons for every 100,000 residents in the population,[17] meaning that a person was actually safer (at least in terms of homicide) living in a state prison surrounded by felons than he or she would be living in certain communities outside the prison.

Though rates of murder have decreased, there are concerns over the long-term impact of violence on prisoners, both while they are incarcerated and after they return to the community. Exposure to violence and victimization has long-term psychological effects on the individual, including an increased likelihood of depression and post-traumatic stress disorder.[18] These factors may reduce the inmate's ability to benefit from the prison's rehabilitative program. In addition, these conditions might also make it more difficult for the ex-prisoner to return successfully to the community, thus placing people living in the community at greater risk. As a result, we should be concerned about reducing prison violence.

Franklin and Franklin suggest that there are four factors that influence the amount of violence in a jail or prison, and they are the following:

- *Conditions of the prison environment*—As prisons become administratively more fair (as a result of inmate litigation) and conditions improve (less environmental deprivation), inmates have fewer legitimate complaints and are more compliant.
- *Personal characteristics of the inmates*—Prisoners serving sentences for violent crimes tend to be more violent in prison, and this population has decreased, whereas the percentage of drug and property offenders (who tend to be less violent inside prison) has increased.
- *Situational factors*—Changes in the architectural design of jails and prisons have increased the ability of authorities to supervise inmates, and the use of administrative segregation or placement in supermax facilities has helped reduce the influence of the most disruptive prisoners.
- *Prison management techniques*—Inmates are held more accountable for their disruptive conduct, classification techniques have improved, the professionalism and training of correctional officers has improved, and security guidelines are more closely followed.[19]

It is likely that a combination of all these factors is responsible for violence reduction. Although it is likely that increasing percentages of nonviolent offenders contribute to a more settled environment, the overall percentage of violent offenders has decreased only 7 percent since 1980.

Murder is not the only form of prison violence, and there has been increasing concern about the issue of prison rape. In 2003, President George W. Bush signed into law the **Prison Rape Elimination Act of 2003** (mentioned briefly in Chapter 3). This legislation is intended to increase our knowledge about sexual violence behind bars by collecting national statistics about these crimes, developing strategies to prevent these offenses, and providing funding so that correctional agencies can reduce prison rape. One of the problems with sexual violence is that we know it exists, but the estimates of its occurrence vary widely.

In its report titled *No Escape: Male Rape in Prison*, researchers with the Human Rights Watch document the experiences of inmates who have been the victims of sexual violence, and the accounts of prisoners posted on their Web site can help us better understand the powerlessness that these inmates feel:

> *I am a first-time non-violent offender, and committed a white-collar offense. . . . In September, 1994, during the week of Labor Day, I was accosted and raped in the shower. . . . While the entire incident did not last more than a few minutes, it seemed like an eternity. I was certain that I had indeed been sentenced to Hell. I was left badly bruised and crying, with a pretty hopeless outlook on the whole situation.*[20]

Race and Gender in Corrections

In February 2005, the U.S. Supreme Court ruled that the California Department of Corrections and Rehabilitation (CDCR) policy of temporarily segregating inmates by race when inmates were admitted to prison was unconstitutional. The CDCR argued that this policy reduced gang-involved assaults and was a safety measure. Do you think that this practice would contribute to higher levels of safety?

Hochstetler, Murphy, and Simons report that prison violence is particularly stressful because an inmate might have to confront his or her attacker daily, and there is no escaping the memory of the offense.[21]

In addition to the physical and psychological effects of being victimized, prison rape represents a threat to community health as well. Most of the inmates who are victimized tend to be young, nonviolent offenders, and these inmates typically serve short periods of imprisonment.[22] Given the high rates of communicable diseases in prisoner populations, these assaults may transmit human immunodeficiency virus (HIV)/acquired immune deficiency syndrome (AIDS), hepatitis, or other sexually transmitted infections to the victims. As these victims of assaults return to the community, these diseases may then be spread if they are not treated.

Although we have a good understanding of the number of murders in prison, we know less about the extent of assaults, although anecdotal information from correctional officers and inmates suggests that rates are fairly high. In a federal census of state and federal prisons, Stephan and Karberg found that there were 28 inmate-on-inmate assaults for every 1,000 prisoners and 14.6 inmate-on-staff assaults for every 1,000 inmates.[23] Although the rate of correctional officer assaults is fairly high, only five officers were killed in 2000 as a result of assaults.[24] This compares to 51 police officers killed on the job in the same year.[25]

Reducing Prison Riots

One of the worst fears of correctional officials is the possibility of a riot, during which inmates tend to act as a mob and rational thinking seems to disappear. Prison riots were common in the 1970s and 1980s, with two of the worst taking place at Attica in New York in 1971 and Santa Fe, New Mexico, in 1980. Riots have been described as expressive and instrumental events. **Expressive riots** are spontaneous. They occur without much (or any) planning and may be the immediate inmate response to what is perceived as an unfair act by a correctional official. **Instrumental riots**, on the other hand, are ones that have a greater element of planning and generally have some type of goal (such as facilitating an escape or in protest of prison conditions).

In the famous 1971 Attica riot in New York, 11 officers (including 10 who were held hostage) and 29 inmates were killed. This riot had some elements of an instrumental event because protests over conditions in the prison had increased in the days preceding the riot, and there had been rumors that two inmates had been tortured by officers. After the initial disturbance and hostage-taking, the inmates tried for several days to negotiate for better living conditions, inviting a number of prominent officials to speak on their behalf. Governor Rockefeller ordered that the prison be retaken, and an armed force attacked the prisoners. Most of the hostages who were killed, however, died as a result of gunshot wounds (the prisoners did not have access to any firearms).

By contrast, the 1980 New Mexico State Prison riot at Santa Fe had elements of an expressive event. Inmates who were drinking homemade alcohol overcame several correctional officers and then rampaged through the prison, in part because officers had failed to secure the riot gates in one unit. The control center was taken over, and the prison was completely controlled by prisoners in approximately one-half hour. The riot lasted for the next 36 hours. (See the accompanying box, "Corrections in the Real World.")

Prison administrators have learned from such uprisings. In many of the riots, staff and inmates could sense that "something was about to happen" days or even weeks before the actual event. It is possible that correctional officials are now quicker to act on the possibility of a riot before larger problems occur. Despite the efforts of correctional officials, riots still occur. In February 2006, for instance, a number of riots, involving thousands of inmates, occurred in the Los Angeles County Jail system. These riots occurred over a period of several days. One inmate was killed, many others injured, and the units involved received extensive property damage. Administrators attributed the cause of the riots to racial tension between black and Hispanic inmates.

Corrections in the Real World

During the New Mexico riot, 33 inmates were killed, but the death toll alone does not capture the brutality of this riot. Roger Morris's study of the riot describes how welding torches and other tools were used to torture inmates, especially "snitches."[26] Although none of the correctional officers who were taken hostage died in this event, many were injured, and some were allegedly sexually assaulted by the prisoners.

Institutional Programs

One important aspect of security, control, and successful reentry (see Chapter 13) is having prisoners constructively occupy their time. However, there has been growing concern that, because of crowding and resource limitations, prison systems have not been providing enough opportunities for inmates to participate in rehabilitation programs. Some prisons have been described as "warehousing" prisoners.[27]

Although a wide variety of prison programs exist, most focus on three basic areas: (1) education, (2) counseling, particularly substance abuse treatment and crisis intervention, and (3) work and vocational training (see **Figure 12–2**). These programs respond to the characteristics of most prisoners: undereducated, drug- and/or alcohol-addicted, and poor histories of employment.

The earliest attempts of correctional systems at providing rehabilitative programs were hindered by the fact that practitioners did not understand the psychological factors that help individuals make successful rehabilitative changes. Most rehabilitative programs, even in the twentieth century, were based primarily on knowledge or skill development such as education or participation in work experience programs.

Figure 12–2 A correctional counselor works with a small group of prison inmates.
Source: *© Thinkstock/age fotostock.*

Basing correctional education or treatment on skill development alone neglected the fact that offenders often had more extensive problems, such as drug or alcohol addictions, involvement in negative associations (such as memberships in gangs), attitudes and values associated with criminal behavior (such as a lack of empathy for others), and often poor social and interpersonal skills. By neglecting to help offenders change these factors, prison education programs might have been guilty of "setting offenders up to fail." Ex-prisoners are released to their home communities, with the same temptations of alcohol, drugs, or gang life. If they do not resist these negative influences, there is a greater likelihood of recidivism. In addition, some correctional systems have failed to provide appropriate aftercare or prisoner reentry programs (refer to Chapter 13), so these ex-prisoners had little formal structure in their lives.

Some programs offered in prisons today respond to the limitations listed earlier by helping prisoners develop better interpersonal and social skills. There is evidence to suggest that interventions should be targeted at prisoners with the highest levels of risk.[28] Moreover, interventions should also be guided by a careful assessment of the offender's needs (e.g., antisocial attitudes, substance abuse, or negative associations—such as membership in a gang). Last, treatments should be responsive to the unique characteristics of the prisoner, such as the inmate's intellectual abilities, motivation to change, and readiness for treatment.[29] Educational and treatment staff members today are more likely to be aware of these issues and tailor their interventions with prisoners more carefully to maximize their efforts.

Educational Programs

The Georgia Department of Corrections reported that in April 2007, more than 69 percent of their inmates had not earned their grade 12 or general equivalency diploma.[30] This is not an unusual finding; jail and prison inmates often have poor histories of educational achievement. But the grade 12 diploma is the minimum requirement for most jobs today, and persons who do not have this level of education are at a disadvantage. As a result, educational programs are offered in most prisons, and they range from adult literacy to college courses. The following are examples of educational programs offered in Florida prisons:[31]

- *Mandatory Literacy*—A 150-hour classroom program that is mandatory for inmates with less than ninth-grade reading level.
- *Adult Basic Education*—Helps inmates develop their reading, writing, and mathematics skills.
- *General Education Development*—A high school equivalency program.[31]

One of the challenging aspects of providing rehabilitative services to correctional populations is that many inmates have histories of failure with school systems and are sometimes resistant to classroom learning. As a result, the educational and treatment staff members who work with prisoners must have a lot of patience, excellent interpersonal skills, and the ability to encourage and motivate their students. Staff must resist the urge to neglect prisoners who are least motivated and showing little progress.

Earlier it was said that some inmates attend college classes while incarcerated, and approximately 3 percent attend postsecondary courses.[32] Several decades ago more students attended, but the federal government's elimination of Pell Grants for postsecondary correctional education in 1994 has made it difficult for prisoners to obtain college courses. Although only a small percentage of inmates is ready for college-level instruction, those who complete degrees have generally remained crime-free.[33] Despite these successes, such programs are controversial because many working citizens in the community who could not afford to attend college were funding the education of these prisoners. Concern over this issue led to the elimination of federal funding for postsecondary correctional education, although some colleges provide instruction free of charge to prisoners.

Counseling Programs: Substance Abuse Treatment and Crisis Intervention

A broad array of individual and group counseling techniques is used in prisons. To a large extent, the techniques used at any given facility depend on both the training and background of the treatment personnel employed there and the characteristics of the inmate population (e.g., sex offenders, mentally ill offenders, and persons with histories of violence). Although an in-depth consideration of counseling and treatment strategies is beyond this book's scope, we illustrate two of the most common areas of treatment, substance abuse intervention and crisis intervention.

There is a strong relationship between drug use and crime. A recent federal study showed that 56 percent of state prisoners and 50 percent of federal prisoners had used drugs in the month prior to their offense, while 32 percent of state inmates were using drugs at the time of their offense compared to 26 percent of their federal counterparts.[34] Rates of alcohol dependency were also high. Given these facts, it is important that a range of substance abuse programs be offered in prisons. Mumola and Karberg report that approximately 40 percent of state prisoners and 49 percent of federal inmates had participated in treatment since their admission to prison, but that only 15 percent of state prisoners had received treatment with a professional, and it appears that most inmates are in peer education and counseling sessions, such as Alcoholics Anonymous or Narcotics Anonymous (AA and NA) programs.

The need for education or treatment in prison settings for addictions has long been recognized. There are a number of different approaches to the delivery of these programs. Some prison-based substance abuse treatment programs are delivered in a separate prison setting (e.g., a living unit) and all of the participants have addiction problems. One example of this is the **Therapeutic Community** (TC) approach, which is a group-based treatment model that relies upon participants to support each other's recovery. Prendergast and colleagues describe how one prison-based community has a three-phase treatment approach that assesses the offender's needs, allows participation in education and counseling programs that target problem areas, and prepares the inmate for life in the community.[35] These prison-based programs are a long-term intervention and require about a year or so to complete. So far, research has shown that these programs are effective at reducing recidivism.[36]

Although long-term inpatient or residential treatment programs have high rates of effectiveness, they are not always easy to implement because they require some expertise, separation from the general prison population, and commitment from the prison management to supply enough resources. Accordingly, most prisons offer less-expensive interventions. Some provide classroom-based addiction education programs. Mumola and Karberg recently found that 17.8 percent of state prisoners and 30.2 percent of federal inmates who were drug dependent had participated in these types of interventions.[37]

Perhaps the most commonly encountered substance abuse treatment programs in jails or prisons are AA and NA groups. In addition to being inexpensive and easy to implement (the programs are operated entirely by volunteers), these self-help programs assist prisoners to maintain a sober lifestyle. Another advantage is that inmates can participate in these groups at any point in their imprisonment and prisoners might not be eligible for placement in more intensive addictions programs, such as TC, until they are close to their release date.

In reality, mental health professionals working in prison can expect to spend a considerable portion of time performing crisis intervention. This means working with prisoners who are in the midst of some immediate life crisis that may have erupted only a short time ago (e.g., in reaction to rejection by a spouse or child during a visit). It is not uncommon for treatment staff to be called to deal with prisoners who are suicidal, experiencing a breakdown in mental health, or behaving belligerently toward other prisoners or staff. Crisis intervention involves working to defuse stress, restore emotional calm, and achieve stability so that problems can be approached rationally in a manner that does not harm the prisoner or others.

Work and Vocational Programs

Prisons tend to be self-contained operations; to run the facility, the staff often rely upon inmate labor. As a result, many inmates are assigned jobs ranging from facility or grounds maintenance to working in the prison library, laundry, or kitchen. Depending on the security level of the facility, others work in prison factories that

Corrections and Policy

Prison administrators are criticized when inmates are idle, yet they are also criticized when prisoners are producing goods that compete against private manufacturers or when prisoners are hired by prison–private enterprise partnerships. Under these circumstances, what is the best option?

manufacture various products, some work at construction, some work in prison agriculture, and still others work at data processing. Some prisoners are paid prevailing wages for their efforts, although many receive only a few cents per hour.

Over the past 100 years, inmates have built furniture for state offices, contributed to war efforts, constructed roads, and, in many states, highway signs and license plates are manufactured by prison labor. In fact, former U.S. Supreme Court Justice Warren Burger advocated that prisons be turned into "factories within fences."[38] Most people support the fact that inmates are learning a vocational skill while in prison, although some businesses and labor organizations have protested that **prison industries** (prison operations that manufacture items for sale in the community) undercut fair competition.

One alternative to prison industries is the partnership of private corporations and the prison work force, and a number of enterprising companies have actually developed assembly lines within prisons where they put inmates to work. Although seemingly a win-win solution (the inmates enhance their skills, and the corporation has a large supply of labor), these partnerships have been criticized because some argue that inmates are exploited.[39] This example illustrates one of the challenges of operating a prison: Sometimes there are so many stakeholders that solutions with which all participants are happy are difficult to achieve.

Most prison systems offer inmates opportunities for work or rehabilitation, although in some places those programs may be limited. One of the problems is that security and control (e.g., no escapes, murders, riots or major incidents within the facility) is the primary goal of most prison administrators, and rehabilitation is a secondary priority. Over the past several years, however, there has been more political interest in reducing recidivism and trying to introduce rehabilitative programs that have been demonstrated to work with prisoners.

Evidence-Based Corrections

Increasing attention has been placed on **evidence-based corrections**, which means using rehabilitation strategies that have been demonstrated by research to be effective in reducing recidivism. Many policymakers are critical that we spend a lot of public funds on incarcerating offenders, but we do not understand what approaches are effective in reducing offender recidivism, or what type of offender is most amenable to treatment.[40] Latessa and colleagues argue that much of what has occurred in correctional rehabilitation over the past several decades has been "quackery," unproven interventions often based on our stereotypes of offender behavior, untested principles, or popular fads.[41] The boot camp (see Chapter 9) was an example of an intervention that was presumed to be effective, appealed to popular beliefs about offenders, and was intended to reduce jail or prison overcrowding. The problem, however, was that boot camps did not reduce recidivism or facility overcrowding, and did not save taxpayers money. If boot camps had been effective with a broad range of adult offenders, there would have been more than 95 of these facilities by 2000 (84 confinement and 11 community-based camps).[42]

Latessa describes four key elements of effective correctional programs. Effective correctional programs must:

- Assess offenders' needs and risks.
- Use treatment models that are known to be effective, such as cognitive behavioral intervention.
- Employ the "3 Cs" of effective corrections (use credentialed staff, ensure that the facility is credentialed by seeing that it is founded on fairness and improvement of lives, and base interventions on credentialed knowledge).
- Understand the principles of effective intervention.[43]

These goals are fairly straightforward, and we provide a brief review.

As mentioned earlier, an assessment of needs and risks is often accomplished when an offender is classified for security levels upon admission to a prison system. The second goal is to use interventions that have been demonstrated to be effective, and these programs generally involve using a number of approaches that confront the offender's criminal thinking, increase the offender's social skills, and attack specific problems known to be related to crime (e.g., addictions). Increasing the professionalism of the staff who work in rehabilitative programs and using interventions that have been demonstrated to be effective are the bases of the third element of effective correctional programs. The last goal is associated with the first and links interventions that research has demonstrated to be effective with the offender's needs and risks.

It is difficult, however, to deliver effective correctional programs. One significant barrier is financial. Wilhelm and Turner outline how nine states cut funding to education, vocational, and substance abuse treatment programs during the budget crises of 2001–2002.[44] Even in normal economic conditions, there has been little political backing for funding rehabilitative interventions, although the public typically supports the rehabilitation of offenders.[45] Sometimes correctional staff members are reluctant to introduce new programs, and even when some new change is adopted, in some cases the correctional staff has undermined the reform.[46]

Altogether, there are many challenges in providing effective correctional rehabilitation programs, but the benefits seem to far outweigh the costs—especially when they are guided by research that demonstrates their effectiveness and is tailored to the individual prisoner's risks and needs.

Institutional Services and Health Care

Modern prisons are much like small self contained cities. Accordingly, many human services available in the free community are made available or are replicated for prisoners. Some major examples include these:

- Food services that satisfy minimal dietary standards
- Clothing and laundry services
- Correspondence services (e.g., mail, phone, and visitation)
- Commissary services where prisoners may purchase food and supplies
- Recreational services ranging from television viewing areas to areas for physical exercise and games
- Religious services and accommodations for prisoners of varying recognized faiths
- Legal services and resources
- Health and dental care services

As in the case of counseling services (discussed earlier), coverage of each of these services is beyond the scope of this book. Instead, we focus in greater detail on one complex service area, health care.

We noted earlier that the government has an obligation to provide adequate medical care for incarcerated offenders. The increased population of jail and prison inmates has created a corresponding need for health care, and this has become a significant challenge in many places. Prisoners tend to be a rather unhealthy population, and rates of illness in correctional populations tend to be much higher than those among people living in the community. In Chapters 7 and 11, we outline how there are high populations of persons with mental illness residing in correctional facilities. In addition, life-threatening communicable diseases such as HIV/AIDS and hepatitis are found at higher rates among correctional populations. Other serious communicable diseases such as TB or sexually transmitted infections are encountered by medical care staff in correctional facilities, as are infections such as methicillin-resistant staphylococcus aureus. We have already discussed how inmates with severe health problems are creating a demand for high levels of correctional medical services, as are inmates who are aging and those with mental illness. One of the problems that correctional administrators have to overcome is that all of these seriously ill prisoners require medical and psychiatric services at higher levels than inmates in the general population.

Because jails often hold inmates for only a few days before they are able to make bail, these inmates historically needed little access to health care, and most discussion of health care in corrections focused on health care in prisons. Yet, we are finding that some inmates are held in jail for years while their cases are processed or when they are serving a period of jail incarceration after a lengthy detention. A federal study of jail inmates found that some were actually incarcerated for a year or more,[47] and some state offenders who should have been sent to prison served their entire prison sentence in the local jail. These inmates are likely to require a greater amount of health care.

Because inmate populations have grown so rapidly in many places and spending for inmate medical care has not increased at the same rate, some critics suggest that prison medical care has been rationed.[48] Often prison health care is delivered in an environment that was not designed for modern medical services, and jail or prison health care facilities are often cramped and lack basic fixtures. It is also difficult for jails or prisons to attract medical staff. Many of these professionals would rather not work in the austere environment of prison with inmates for patients and in professional isolation (there are few other nursing staff or physicians with whom to interact).

Although not correctional health care in the strictest sense, many correctional facilities provide therapeutic programs using their health care staff. In addition to providing counseling, for example, some psychologists provide sex offender counseling and treatment (including facilitating sex offender groups). Other programs that might be delivered by therapeutic staff include parenting programs (e.g., how to care for children) and life or social skills courses. Nurses, by contrast, might deliver prenatal classes, addictions education programs, HIV/AIDS awareness and prevention programs, as well as sexuality classes.

Jail and prison administrators have responded to the challenge of delivering better correctional health care. Larger jails and prison systems hire (or contract with) physicians to deliver care. Facilities constructed in the past few decades have more appropriate treatment facilities, including equipment for medical and dental care. Larger prison systems have introduced specialized care facilities for infants and their mothers and for persons with mental illness, and some have introduced hospices to provide specialized care for dying inmates. Even though the delivery of medical care is still the largest source of litigation in jails or prisons, correctional administrators have made significant progress in delivering better correctional health care in many jurisdictions. As a result, rates of mortality in some correctional populations are actually less than rates of mortality in the community.[49]

Conclusion

Correctional systems have displayed considerable success in reducing prison murders. We reviewed that these changes were a consequence of improved inmate classification, changes in prison design, and comprehensive prison security programs. Moreover, the current group of correctional officers and managers is better trained, better educated, and more professional than prior generations of correctional staff were. Given these successes, there is increasing attention being paid to the problem of offender recidivism and using rehabilitative programs to increase offender success when offenders return to the community. Many taxpayers support the use of imprisonment but also want offenders rehabilitated.[50]

The problem is that some offenders are resistant to rehabilitation, and even when they have the best of intentions, they often fail in their first few months of living in the community. Prisoners require an investment in their health, education, and skill development while in prison. Many offenders do not receive much in the way of rehabilitative programs, and Petersilia reports, "Of the entire California prison population, an estimated 20% of inmates are altogether idle, never participating in any prison program during their entire stay."[51] Funding is a key reason why more inmates do not receive better correctional rehabilitation and treatment. A recent Florida study, for instance, reports,

> *"In Fiscal Year 2006–07, the Legislature appropriated over $2.1 billion to the Department of Corrections; of this, $36.4 million or 1.7% of the agency budget, was allocated to correctional education and substance abuse programs."*[52]

If correctional systems are spending less than 2 percent of their entire budget on rehabilitation, it is difficult to imagine that there will be a significant reduction in recidivism rates in the near future.

READY FOR REVIEW

- To increase safety, most facilities use objective classification instruments that develop a score that will aid in the decision to place an inmate in the living unit and level of security that best matches the offender's needs.
- The rapid growth in prison populations has been associated with a number of negative consequences, including increases in violence, the spread of communicable diseases, and levels of institutional stress. Moreover, the creative energy of the prison managers and staff is directed at solving the overcrowding problem rather than at other challenges (e.g., inmate rehabilitation).
- Security is the foundation of all activities in a prison and is based on a number of factors, including clear agency goals, support from headquarters, staff training and development, using appropriate equipment, keeping inmates constructively occupied, and security practices to prevent violence or escapes and control contraband.
- The prison murder rate decreased 93 percent from 1980 and 2002. This decrease has been associated with improved conditions in the prison environment, fewer violent offenders imprisoned, changes in the architectural design of prisons that enable better supervision, and improved prison management (e.g., holding inmates more accountable for disruptive behaviors).
- Prison riots have been attributed to two causes: Expressive riots are spontaneous acts that occur without much planning, whereas instrumental riots are planned and have some form of goal, such as facilitating an escape.
- Prison populations tend to have higher rates of communicable diseases and illnesses than do the general population. Increased attention has been paid to the relationships between prison and community health and the idea that solving these problems in the prison will improve community health.
- Evidence-based corrections are correctional rehabilitation programs that are demonstrated to be successful in making a significant change with the prisoner.
- Effective correctional programs assess the offender's needs and risks, use treatment approaches that have been demonstrated to be effective by research, employ the "3 Cs" of effective corrections, and understand the principles of effective intervention.

KEY TERMS

evidence-based corrections Using correctional rehabilitation strategies that have been demonstrated by research to be effective

expressive riot A spontaneous or unplanned riot that typically occurs as a prisoner reaction to an unpopular decision or in response to an incident

external classification Determines a prisoner's security or custody classification and where the prisoner will be placed in the facility

instrumental riot A prison riot that has another motive, such as aiding in an escape, or as part of a collective protest (or uprising)

internal classification Method of inmate classification used to make housing (e.g., with whom an inmate will be placed) and programming assignments

isolation cell A single-room "hardened" cell (e.g., bare concrete walls and stainless steel fixtures) for the temporary control of disruptive inmates

objective classification Method of inmate classification that relies on numerical scores—typically, the higher the score, the higher the level of security that the inmate requires

prison industries Prison-based manufacturing industries that use inmate labor and then sell the products in the community (e.g., office furniture sold to state government offices)

Prison Rape Elimination Act of 2003 A federal law that provides grants for the collection of national statistics about prison sexual violence, develops strategies to prevent these crimes, and provides funding so that correctional agencies can reduce or eliminate prison rape

reclassification In many prison systems an inmate's classification rating is evaluated annually to determine whether the inmate is in a facility or housing unit with the appropriate level of security

segregation unit A block of isolation cells used to hold temporarily prisoners who are out of control

subjective classification Basing security and housing decisions on the judgment of the classification officers

therapeutic community A group-based treatment model often used for substance abuse (but also for other offenses such as sexual abuse), where participants support each other's treatment or recovery

YOU ARE THE CORRECTIONS PROFESSIONAL SUMMARY

1. Should you get involved in this situation? If you decided to intervene, how would you approach these officers? It is difficult to intervene in

the private matters of consenting adults, but workplace relationships can lead to problems in correctional facilities. Inmates who know about an officer's personal relationships are empowered by this knowledge (they may use that information to manipulate the officer). Even though fraternization may not be addressed in policy, you might advise these two staff members to keep their relationship discreet.

2. What are some possible negative outcomes of this relationship? In addition to empowering inmates, workplace relationships make it hard for the persons involved to be remain objective (e.g., in case one of these officers was held hostage).
3. How will "stepping up to supervisor" affect the relationships with your former coworkers? It is often a difficult transition to move from a rank-and-file staff member to supervisor. In addition to the new expectations of the job, many of your coworkers will remind you of times that you violated the same policies you now must enforce when you were "one of them."

NOTES

1. A. M. Durham, *Crisis and Reform: Current Issues in American Corrections* (Boston: Little, Brown, 1994).
2. J. Austin, *Findings in Prison Classification and Risk Assessment* (Longmont, CO: National Institute of Corrections, 2003).
3. J. Austin and P. L. Hardyman, *Objective Prison Classification: A Guide for Correctional Agencies* (Longmont, CO: National Institute of Corrections, 2004), x.
4. Austin, *Findings in Prison Classification,* 1.
5. T. R. Clear, "Inmate Classification," *Criminology and Public Policy* 2(2003): 213.
6. Austin, *Findings in Prison Classification,* 5.
7. P. M. Harrison and A. J. Beck, *Prisoners in 2005* (Washington, DC: Bureau of Justice Statistics, 2006), 7.
8. Office of the Inspector General, *Neither Safe nor Secure: An Assessment of Indian Detention Facilities* (Washington, DC: Office of the Inspector General, 2004), 50.
9. V. Hassine, "Prison Violence from Where I Stand," in *Turnstile Justice,* eds. R. L. Gido and T. Alleman, 38–56 (Upper Saddle River, NJ: Prentice Hall, 2003).

10. L. B. Gadkowski and J. E. Stout, "Tuberculosis in Corrections," *Infectious Diseases in Corrections Report* 9, no. 2 (2006): 1–2, 4, 6.
11. C. J. Mumola, *Suicide and Homicide in State Prisons and Local Jails* (Washington, DC: Bureau of Justice Statistics, 2005), 1.
12. E. E. Atherton and R. L. Phillips, *Guidelines for the Development of a Security Program* (Alexandria, VA: American Correctional Association, 2007).
13. B. Useem and A. M. Piehl, "Prison Buildup and Disorder," *Punishment & Society* 8(2006): 87–115.
14. R. Wener, "Effectiveness of the Direct Supervision System of Correctional Design and Management," *Criminal Justice and Behavior* 33(2006): 392–410.
15. Hassine, "Prison Violence from Where I Stand," 38.
16. Mumola, *Suicide and Homicide,* 1.
17. Federal Bureau of Investigation, "Crime in the United States," 2003, www.fbi.gov/ucr/cius_02/pdf/02crime2.pdf.
18. R. W. Dumond, "Inmate Sexual Assault: The Plague That Persists," *Prison Journal* 80(2000): 407–414.
19. T. W. Franklin and C. A. Franklin, "Violence Reduction: Incarceration Shouldn't Be a Death Sentence," in *Issues in Correctional Health Care,* eds. R. Ruddell and M. Tomita, 103–122 (Richmond, KY: Newgate Press, 2008).
20. Human Rights Watch, "Prisoners' Voices," 2006, www.hrw.org/reports/2001/prison/voices.html.
21. A. Hochstetler, D. S. Murphy, and R. L. Simons, "Damaged Goods: Exploring Predictors of Stress in Prison Inmates," *Crime and Delinquency* 50(2004): 436–457.
22. Human Rights Watch, "No Escape: Male Rape in U.S. Prisons," 2001, www.hrw.org/reports/2001/prison/report.html.
23. J. J. Stephan and J. C. Karberg, *Census of State and Federal Correctional Facilities, 2000* (Washington, DC: Bureau of Justice Statistics, 2003), 9.
24. Stephan and Karberg, *Census,* 9.
25. Federal Bureau of Investigation, *Law Enforcement Officers Killed and Assaulted, 2004* (Washington, DC: FBI, 2005), 3.
26. R. Morris, *The Devil's Butcher Shop: The New Mexico Prison Uprising* (Albuquerque: New Mexico University Press, 1988).
27. J. Irwin, *The Warehouse Prison: Disposal of the New Dangerous Class* (Los Angeles: Roxbury, 2005).
28. D. A. Andrews and C. Dowden, "Risk Principle of Case Classification in Correctional Treatment: A Meta-analytic Investigation," *International*

Journal of Offender Therapy and Comparative Criminology 50(2006): 88–100.

29. C. Di Placido, T. L. Simon, T. D. Witte, D. Gu, and C. P. Wong, "Treatment of Gang Members Can Reduce Recidivism and Institutional Misconduct," *Law and Human Behavior* 30(2006): 95–96.
30. Georgia Department of Corrections, "Inmate Statistical Profile," 2007, www.dcor.state.ga.us/.
31. Office of Program Policy Analysis and Government Accountability, *Corrections Rehabilitative Programs Effective, but Serve Only a Portion of the Eligible Population* (Tallahassee, FL: Office of Program Policy Analysis and Government Accountability, 2007), 2.
32. J. Page, "Eliminating the Enemy: The Import of Denying Prisoners Access to Education in Clinton's America," *Punishment and Society* 6(2004): 357–378.
33. M. E. Batiuk, K. F. Lahm, M. McKeever, N. Wilcox, and P. Wilcox, "Disentangling the Effects of Correctional Education: Are Current Policies Misguided? An Event History Analysis," *Criminal Justice* 5(2005): 55–74.
34. C. Mumola and J. C. Karberg, *Drug Use and Dependence, State and Federal Prisoners, 2004* (Washington, DC: Bureau of Justice Statistics, 2006), 1.
35. M. L. Prendergast, E. A. Hall, H. K. Wexler, G. Melnick, and Y. Cao, "Amity Prison-Based Therapeutic Community: 5-Year Outcomes," *Prison Journal* 84(2004): 36–60.
36. Prendergast et al., "Amity Prison-Based Therapeutic Community."
37. Mumola and Karberg, *Drug Use and Dependence,* 9.
38. W. E. Burger, "Prison Industries: Turning Warehouses into Factories with Fences," *Public Administration Review* 45(1985): 754–757.
39. D. Burton-Rose, D. Pens, and P. Wright, *The Celling of America: An Inside Look at the U.S. Prison Industry* (Monroe, ME: Common Courage Press, 1998).
40. D. L. MacKenzie, "The Importance of Using Scientific Evidence to Make Decisions About Correctional Programming," *Criminology and Public Policy* 4(2005): 249–257.
41. E. J. Latessa, F. J. Cullen, and P. Gendreau, "Beyond Correctional Quackery: Professionalism and the Possibility of Effective Treatment," *Federal Probation* 66(2002): 43–59.
42. Stephan and Karberg, *Census,* 12.
43. E. J. Latessa, "Promoting Public Safety through Effective Correctional Interventions: What Works and What Doesn't?" in *Corrections Policy: Can States Cut Costs and Still Curb Crime?* eds. E. Gross, B. Friese, and

K. Bogen-Schneider, 31–36 (Madison: University of Wisconsin Center for Excellence in Family Studies, 2003), 33–34.

44. D. F. Wilhelm and N. R. Turner, *Is the Budget Crisis Changing the Way We Look at Sentencing and Incarceration?* (New York: Vera Institute of Justice, 2002).
45. B. Applegate and R. King Davis, "Examining Public Support for 'Correcting' Offenders," *Corrections Today* 67(2005): 94, 102.
46. A. C. Lin, *Reform in the Making: The Implementation of Social Policy in Prison* (Princeton, NJ: Princeton University Press, 2000).
47. D. J. James, *Profile of Jail Inmates, 2002* (Washington, DC: Bureau of Justice Statistics, 2004).
48. D. S. Murphy, "Health Care in the Federal Bureau of Prisons: Fact or Fiction," *Californian Journal of Health Promotion* 3, no. 2 (2005): 23–37.
49. C. Mumola, *Medical Causes of Death in State Prisons, 2001–2004* (Washington, DC: Bureau of Justice Statistics, 2007), 3.
50. F. Cullen, "The Twelve People Who Saved Rehabilitation: How the Science of Criminology Made a Difference," *Criminology* 43(2005): 1–42.
51. J. Petersilia, *Understanding California Corrections* (Berkley, CA: California Policy Research Center, 2006), 39.
52. Office of Program Policy Analysis and Government Accountability, *Corrections Rehabilitative Programs Effective, but Serve Only a Portion of the Eligible Population* (Tallahassee, FL: Office of Program Policy Analysis and Government Accountability, 2007), 2.

13

Release from Prison, Parole, and Prisoner Reentry

Chapter Objectives

- Describe why the topics of release and reentry are so important to corrections and criminal justice.
- Differentiate the various methods of release from prison, and describe trends in the use of these methods.
- Differentiate prerelease and temporary release, and identify types of temporary release.
- Define parole, and distinguish it from probation.
- Describe the historical evolution of parole as well as the manner in which parole is organized and administered today.
- Describe the key phases of the parole release process, including consideration of the factors that influence release decisions.
- Describe the parole supervision and termination process.
- Critically analyze the major issues in parole, including the following:
  - The movement to abolish parole release
  - Rights of persons on parole
  - Liability of parole officials
  - The lack of resources for parole supervision and services
- Discuss what research shows about patterns of recidivism among those released from prison.
- Demonstrate understanding of (1) the challenges faced by prisoners upon reentry into the community, (2) approaches to improving reentry,

and (3) what research shows about the effectiveness of reentry programs.

- Critically analyze the controversies surrounding prisoner reentry, including the following:
  - Employment of ex-prisoners
  - The impact on families and communities
  - The reentry of sex offenders
  - The role of crime victims in reentry

CASE STUDY

You are a newly appointed member of a five-person parole board. Your panel has just completed a hearing to determine whether a 30-year-old male prisoner should receive parole release. This prisoner has now served 6 years of a 5- to 10-year sentence and has accumulated 2 years of good time. He was denied parole release at the 4-year point because the board did not believe that he had served enough of the sentence given the nature of the offense; the members stated that he should serve at least 2 additional years.

The prisoner was convicted of aggravated assault as a result of a domestic violence incident involving his spouse. Although he did not use a weapon, he seriously assaulted his wife with his fists during an argument, and the assault resulted in a broken jaw, three facial lacerations requiring stitches, and multiple facial contusions. Though eventually making a full physical recovery, the victim required more than a week of hospitalization and extensive neurological tests, and she was treated for blackouts. The prisoner was severely intoxicated at the time of the assault and had a history of alcoholism dating to his juvenile years. He was adjudicated for simple assault at the age of 17 and successfully completed a probation sentence. At the age of 21, he was convicted of simple assault against his spouse (who was then his girlfriend) and sentenced to probation, with the first six months served in jail. He was on probation for this offense at the time of the crime that resulted in his imprisonment.

The prisoner has an excellent record of conduct in prison. There is no evidence of disciplinary infractions, and he has successfully completed the prison treatment program for alcoholics. He has held a job in the prison industry program throughout the incarceration period and has completed his general equivalency diploma (GED). A local factory owner is willing to hire the prisoner if he is released, and the prisoner plans to return home to live with and support his spouse and now

7-year-old daughter. The prisoner expressed deep remorse and reported to the parole board that he assumes full responsibility for his crime and views the crime as an outcome of alcoholism. He reports that he has not consumed alcohol since the day of the crime, that he intends to be involved in Alcoholics Anonymous upon release, and that he believes, with help, he can continue to overcome his drinking problem.

Though the victim lives more than 200 miles from the prison, she has been visiting the prisoner as often as possible. She strongly supports parole and wants her family reunited. Prison officials are also recommending that parole be granted. However, the prosecutor and judge in the small city to which the prisoner would return have issued strong statements arguing against early release. You are aware that there is a political campaign in the city meant to crack down on domestic violence incidents. This is in reaction to heavy media coverage of two recent cases of men murdering their spouses. Earlier this week, a local headline read "Wife Beater Being Considered for Early Prison Release." Should you vote for or against parole release?

1. What are the advantages and disadvantages of the prisoner serving the full 10-year term?
2. If the prisoner is paroled at this time, what should be included in the special conditions of supervision?

Introduction

The topic of incarceration was considered in Chapter 7 (jails) as well as in Chapters 10 through 12 (prisons). This chapter addresses the process of people coming out of correctional facilities and transitioning back into society, commonly referred to as **reentry**. Incarcerated populations have steadily increased in size over the last three decades. Thus, reentry has emerged as one of the most significant issues facing criminal justice.

Most people who are placed in correctional institutions will reenter society at some point. Each year, hundreds of thousands of people exit jails, prisons, or other correctional facilities to return to life in the community.[1] Almost 650,000 people were released from adult prisons in the United States in 2005.[2] However, as will become evident in this chapter, there are formidable barriers to successful reentry. Overcoming these barriers is one of the keys to improving the criminal justice system. The reason for this is simple. No system of any kind, be it a technological, educational, governmental, or other system, can function well if too many of its resources are consumed dealing with its failures.

Release from Incarceration

Release from incarceration usually involves more than a simple two-step process of (1) the offender serving the amount of time stipulated by the sentence, or a portion thereof, and then (2) being released directly into the community from the facility at which the sentence was served. The amount of time a person actually serves can be affected by several considerations. Depending on the case, some considerations could include the following:

- The extent to which the sentencing system used in the jurisdiction is determinate or indeterminate (see Chapter 6)
- Any special law applicable to a particular case (e.g., "three strikes" or mandatory minimum legislation)
- Reductions in the sentence because of time previously served in jail or because of good/meritorious behavior in prison (sometimes known as "**good time**"—see Chapter 6)
- The discretion of criminal justice officials involved with the case (e.g., prosecutors, judges, correctional officials)
- Availability of space and other resources in the prison system
- The wishes of the community, the victim, and/or the victim's family

In addition, many times prisoners are not released into the community directly from the facility or facilities at which they served their sentence. It is common for there to be a period of transition during which the person is exposed to progressive increments of freedom and gradually reintroduced to life outside. Prisoners are often transitioned from higher to lower-security facilities and released from low-security ones. Also, as described later, many jurisdictions use temporary and/or prerelease programming to ease the transition. This is important because the culture to which a person is exposed in prison is usually quite different from the one to which he or she will return on the streets.

Methods of Release

Although terminology varies from one jurisdiction to the next, there are three basic methods by which people are released from prison. First, an individual may be released at the expiration of the maximum amount of time allowed by the sentence (minus any good time that might apply), without any supervision in the community following release. Known as **unconditional mandatory release** or "maxing out," this type of release is less common than are the other two types, conditional mandatory release and discretionary parole release, because unconditional release permits less discretion regarding the length of the prison stay and allows for less control over the ex-prisoner in the community. With **conditional mandatory release**, the person is released under the provision of law after serving a legally stipulated portion of the sentence (minus good time) with a period of supervision in the community following release; that is, release is conditional upon behavior in the community. By contrast, with **discretionary parole release** (hereafter referred to as parole release), the offender is released into community supervision at the discretion of a parole board or similar authority, rather than under the provision of law. Parole release takes place prior to the expiration of the sentence, but offenders are usually not eligible to be considered for parole until a certain minimum period of time has been served.

Conditional mandatory release and discretionary parole release are similar, except that in the case of the former, release is required by law. With the latter, release is at the discretion of paroling officials. In contrast to unconditional mandatory release, both involve community supervision after release. This is why unconditional mandatory release is sometimes called *mandatory* parole in contrast to *discretionary* parole.

In the not too distant past, the majority of people exiting prison did so through discretionary parole release. This was before indeterminate sentencing systems and parole came under widespread criticism during the 1970s and 1980s. Since then, conditional mandatory release has replaced discretionary parole as the most commonly used release method. According to the U.S. Department of Justice, discretionary parole accounted for 55 percent of all persons released from state prisons in 1980, compared to only 22 percent in 2004. By contrast, conditional mandatory release accounted for only 19 percent of releases in 1980, compared with 39 percent by 2004.[3] In fact, the federal government and some states have abolished discretionary parole release (see Chapter 6).

Pre- and Temporary Release Programming

Irrespective of the method of release used, most jurisdictions employ temporary and/or prerelease programming. **Prerelease programs** are implemented shortly (e.g., six months) before the scheduled release date with the intent of preparing the prisoner for reentry and testing how he or she will respond to greater freedom. These programs typically emphasize some combination of education, employment, life skills training, and abstinence from substance abuse and other illegal activities. A prerelease program can operate at a separate low-security facility with structured living, such as a prerelease center or halfway house. Alternatively, some prisons have separate housing units for inmates participating in these programs.

Temporary release programs allow persons confined to a correctional facility to leave the facility for a set period to participate in approved activities in the community, such as working at a job. One goal of these programs is to help the offender establish or maintain social ties that will ultimately enhance reentry.

Another goal is to give prisoners incentives for desired behavior while incarcerated, because participation usually is reserved for carefully screened cases. Temporary release is sometimes incorporated with prerelease programming, such as when a person at a prerelease center or halfway house holds a job in the community. In other cases, temporary release may be granted to select prisoners well in advance of the prerelease period.

Common types of temporary release include work release, study release, and furlough. Work release programs allow the prisoner to leave the facility for a specified time frame (e.g., 7:00 A.M. to 5:00 P.M. Monday through Friday) to hold a job in the community. Although work release participants are paid prevailing wages, most such jobs are minimum wage or only slightly greater. Typically, correctional officials deduct some of the pay for such things as rent, restitution, and child support, and the balance is then placed in prisoners' institutional accounts. In study release programs, prisoners are permitted to leave the facility to complete vocational training or further their formal educations (e.g., completing a GED or college courses). With furlough programming, prisoners are granted brief (e.g., 24- to 48-hour) leaves from the facility to participate in such preapproved activities as searching for employment or housing, or reuniting with family. Furloughs may be granted on a regular basis, such as bimonthly, or on an irregular as-needed basis, such as when the prisoner needs to attend a family member's funeral.[4]

Corrections in the Real World

During the 1988 presidential election campaign, the Democratic candidate, Massachusetts Governor Michael Dukakis, was depicted in a series of ads by the Republican candidate, Vice President George Bush, as a soft-on-crime liberal and a threat to public safety. The Bush campaign noted that Dukakis had permitted a prison furlough for Willie Horton, an African American man serving life for murder in Massachusetts. During the furlough, Horton committed a number of crimes, including the rape of a white woman. Many observers believe this strategy helped Bush win the election. About this same time, a study was underway of temporary release programming in Massachusetts. The study was published in 1991 and, based on a careful scientific analysis of recidivism involving many cases, concluded that "prerelease programs following prison furloughs, and prison furloughs alone, appear to reduce dramatically the risks to public safety after release."[5]

Parole

The term **parole** has two distinct meanings. First, parole refers to one type of release from prison (see the previous description of discretionary parole release). Second, parole refers to a period of structured supervision and programming to which the former prisoner is exposed in the community following release from incarceration. Obviously, persons who are granted parole release receive supervision in the community from parole officers following release. Persons receiving conditional mandatory release also usually undergo a period of community supervision, often from probation or parole agencies.

Parole and Probation: A Comparison/Contrast

Parole is frequently confused with probation. This is understandable because the two have overlapping goals. Recall from Chapter 8 that the goals of probation include protection of the community, assisting the offender to better himself or herself in the community, and gathering information upon which officials can base decisions about the case. These same goals apply to parole supervision, but because parole has another dimension (parole release, as noted earlier), additional goals are involved. One such goal is the control and management of prisoners. Parole means potential for early release and can give prisoners incentives to comply with prison rules and make constructive use of their time, such as working, furthering their education, or benefiting from counseling. Although not always formally acknowledged, another goal of parole is to function as a sort of "safety valve" to help relieve prison crowding. When a facility grows too crowded: (1) more prisoners than usual can be paroled, and/or (2) prisoners can be paroled earlier than they otherwise would.

Probation and parole supervision are quite similar. Even though the offender's legal status differs (probationer vs. parolee), both are exposed to supervision and programming in the community. Also, both are expected to follow rules and conditions to avoid revocation. The same agency (and even the same officer) may supervise both probationers and parolees. However, parole supervision is not a community-based sentence like probation is. Instead, parole supervision is a continuation or extension of a prison sentence. For this reason, parole supervision is usually stricter than probation supervision is, with tighter rules, closer enforcement, and less tolerance for violations of any kind. Also, parolees often face more pronounced adjustment problems than probationers do. Parolees have the stigma of being ex-cons, which can restrict their civil rights and opportunities for conventional behavior, and they are less accustomed to living outside an institution.

Historical Overview of Parole

Parole has a lengthy history, and only the major highlights of this history are mentioned here.[6] The history of parole is filled with controversy. This is not surprising because parole involves freeing offenders from prison before they have completed their terms.

Like probation, parole evolved out of earlier practices. One major predecessor of parole was penal **transportation** (or transportation—outlined in Chapter 7), which was authorized by English law in the sixteenth century and was also practiced by other countries. In lieu of another punishment (commonly execution), select convicts were sent to the English colonies for a specified period of labor. These convicts could earn a pardon for their offenses by following conditions associated with the transportation contract and by staying out of trouble after the period of indentured labor. Transportation evolved out of the ancient practice of banishing offenders to the wilderness, as the Egyptians did with the Hebrew leader Moses during biblical times. As populations and cities grew and there were fewer places to which people could be banished, transportation became popular in some European countries that possessed colonies; there was also demand for cheap and free labor in European colonies.

A more direct forerunner of parole was the **ticket of leave system** pioneered during the 1840s by Alexander Machonochie at the Norfolk Island penal colony off the coast of Australia. This system was implemented slightly later by Sir Walter Crofton in Ireland.[7] The basic idea was that prisoners who made good progress while in prison, by moving through a series of phases that gradually provided more freedom and responsibility, could earn early release via a ticket of leave; ex-prisoners were then supervised in the community by the police, other government officials, or civilian volunteers. Continued release was conditional on good behavior in the community. Machonochie was relieved from his post by the British government because his practices were viewed as too lenient on criminals, but Crofton's use of the same basic ideas in Ireland helped promote the rise of indeterminate sentences and parole in the United States. Crofton's efforts were a major point of focus at the first meeting of the National Prison Association (the forerunner of the American Correctional Association) held in Cincinnati, Ohio, in 1870.

In the United States, the concept of releasing select prisoners early in exchange for desired conduct in prison, and then following such release with community supervision, was popularized in the 1870s by Zebulon Brockway. Brockway was the warden at New York's Elmira Reformatory, and in his system, prisoners served their sentences around an indeterminate range of time and were then released when officials deemed them ready for community supervision. This system facilitated prison management and also proved to be an effective way to control prison crowding. Although there was much criticism and debate, the system spread rapidly. By the mid-1900s, parole had become a standard aspect of criminal justice and was widely available across the United States.

The 1940s and 1950s saw the rise of a **medical model** of corrections. During this era, penologists looked to the field of medicine as a protocol for their work. Under the medical model, crime was viewed as pathology to be (1) assessed, diagnosed, and classified in relation to its causes; (2) treated based on the diagnosis; and finally (3) followed up to monitor the effects of treatment and encourage the desired outcomes. The model gave correctional officials wide discretion to tailor the specifics of a sentence to the needs of a given offender, and parole fit exceptionally well with the model's emphasis on follow-up. Hence, this has been described as the era of "clinical parole" and is conventionally regarded as parole's heyday.[8]

Starting in the 1960s, the correctional medical model in general and parole in particular became targets of intense criticism from both conservative and liberal sectors.[9] Conservatives had long voiced concern that, although parole may bolster prison management and help control crowding, it is soft on crime because it undermines retribution and deterrence by shortening stays in prison. Similarly, critics pointed out that parole can compromise public safety unless three doubtful conditions are met: (1) prison programs must promote rehabilitation well enough to make early release tenable, (2) paroling authorities must accurately distinguish and predict which offenders are ready for early release, and (3) supervision and programming in the community must be adequate to prevent recidivism. Critics pointed to a now famous review of 231 research studies by Robert Martinson and his colleagues that had concluded that correctional rehabilitation efforts frequently do not have the desired effects on recidivism.[10]

On the other hand, more liberal critics such as David Fogel expressed concern that parole is unjust for offenders.[11] For instance, parole can result in unfair disparities in time served. Two offenders from the same jurisdiction, having highly similar backgrounds, convicted of the same crime, and displaying similar adjustment to prison could still serve vastly different amounts of time; one might be released after 3 years, and the other might serve 10 years. Likewise, the possibility for early parole can subtly pressure prisoners to participate in rehabilitation programs that are of questionable effectiveness. Prisoners may feel they have little choice but to participate in these programs if they are to achieve early release. But some people may view such participation as simply another con game, translating into a no-win situation for prisoners.

The last two decades of the twentieth century were dominated by conservative policies in the United States. Conservative criticisms of criminal justice and correctional practices helped propel many jurisdictions into a crackdown on crime that continued into the present century. Some jurisdictions moved from indeterminate to determinate sentencing and also either eliminated or sharply curtailed parole release. By the end of 2000, 16 states and the federal government had abolished parole release for all or most categories of offenders, and 4 states had abolished it for select violent crimes.[12]

The trend across the country over the last three decades has been (1) to restrict the ability of both judges and parole authorities to reduce the use of imprisonment (or the severity of prison terms), and simultaneously, (2) to empower lawmakers and prosecutors to send more people to prison and keep them there for longer terms. A major result has been unprecedented levels of incarceration and institutional crowding, as well as serious prison management problems.[13] Ironically, the prison management and crowding problems exacerbated by the attack on indeterminate sentencing and parole release probably helped to sustain and expand parole supervision. More than 784,400 adults were on parole in the United States at the end of 2005, a 1.6 percent increase over 2004.[14]

The Organization and Administration of Parole Today

Parole release decision making is administered by a parole board, commission, or similar paroling authority as established by law. Where discretionary parole release is used today, it is generally a function of the executive branch of government. On the other hand, parole supervision is administered by the field services agency authorized by law in a given jurisdiction to supervise persons exiting prison. Depending on the jurisdiction, parole release and supervision can be jointly administered under the same agency or administered separately. Similarly, probation and parole field services can be administered separately or jointly. In Kentucky, for instance, field services are combined into the Division of Probation and Parole within the Department of Corrections, which in turn, is part of the Justice and Public Safety Cabinet; the Office of the Parole Board is a separate unit of this cabinet. Regardless of the particular setup, it is important that parole release authorities be familiar enough with prison operations to make well-informed decisions, yet independent enough that they are not overly influenced by institutional objectives. This can be a tricky balance.

Parole board members play a key role in establishing parole policy. They can make release decisions, set conditions for community supervision, and make revocation decisions. In most places, they are appointed by the governor for a specific term (e.g., four years) that can be renewed. Although there is a lot of variation, it is fairly common for states to have between five and seven board members, and some states include part-time members. In Kentucky, for example, there are seven full-time and two part-time members, plus an executive director; members are appointed by the governor to four-year terms.

Keep in mind that even jurisdictions that have abolished or seriously restricted discretionary parole release typically practice community supervision with persons coming out of prison, although it can be called something other than parole supervision (e.g., supervised release). As mentioned earlier, prisoners receiving conditional mandatory release under law still undergo a period of community supervision. For instance, at the federal level, even though offenders convicted after 1987 cannot be paroled, many receive conditional mandatory release and are supervised in the community by federal probation officers.

The Parole Release Process

Where parole release is permitted, prisoners must become eligible for parole. Eligibility means only that early release will be considered, and eligibility is usually attained by serving a required portion of the sentence, minus good time accrued as well as any time served in jail before imprisonment. In recent decades, some jurisdictions have prolonged eligibility for parole by passing mandatory minimum laws, which require a specified minimum term to be served in prison. A similar strategy is truth-in-sentencing legislation, which requires that relatively high percentages (e.g., 85%) of prison terms must be completed. These laws target certain categories of offenders (e.g., career criminals) or those convicted of certain crimes (e.g., sex or drug-dealing offenses).

When prisoners approach eligibility, they must work with staff at the institution to prepare a **parole plan** that describes postrelease plans for issues such as housing, employment, avoidance of trouble, and pursuit of rehabilitation opportunities. Additionally, institutional staff write a **preparole report**. The report describes the prisoner and his or her crime and reviews his or her behavior in prison, as well as progress toward becoming rehabilitated. Such a report may or may not contain a recommendation about whether parole should be granted.

The next step in the process is a **parole grant hearing** where the prisoner appears before the parole board, a subset of the board, or officers who report to the parole board. Depending on the case and the jurisdiction, prison officials, prosecutors, victims (or victims' families), and other community stakeholders may be present. It has become common for communities to be given advance notification of parole hearings as well as opportunities to voice concerns. Because parole officials can review the parole plan and preparole report in advance, the usual parole grant hearing is brief and revolves around an interview with the prisoner. The focus is on what the prisoner has done to better himself or herself while incarcerated and why the person believes he or she can make it in the community without returning to crime. Prisoners may be notified of the decision at the end of the hearing or considerably later, such as three weeks. When parole is denied, the prisoner will be eligible to see the board again at a future date. If parole is granted, the prisoner will transition to the period of parole supervision, as described in the next section.

The U.S. Supreme Court has considered the topic of parole release decision making.[15] The Court ruled that the level of due process required during the decision to revoke probation or parole (see Chapter 8) is lower when release decisions are made. Parole release is an act of grace, so jurisdictions have discretion to decide what constitutes acceptable due process at grant hearings.

Researchers have studied the factors that influence parole release decisions. Parole officials ordinarily assign much weight to the nature and seriousness of the offense for which the person was imprisoned, including the sentence for that offense and the time served. They also assign weight to the person's prior legal history, especially as regards prior felonies and violence. Also, officials may give attention to the person's responsiveness to previous correctional efforts, their behavior in prison and efforts toward rehabilitation, considerations pertaining to the victim, and scores on tests of readiness for parole.[16]

Some jurisdictions have tried to bring greater structure, consistency, and objectivity to the release decision-making process by using presumptive parole dates and parole guidelines. With **presumptive parole dates**, prisoners are told upon (or shortly after) their admission to prison the date they can expect to receive parole if they follow prison rules and participate in programs as directed. **Parole guidelines** are similar to sentencing guidelines and risk/needs assessment instruments. These guidelines allow officials to rate persons being considered for parole on variables believed to predict recidivism, such as prior legal history and age at the time of the offense. Prisoners with total scores falling into certain categories are considered worse risks. The idea is to reduce unfair disparity in decision making and to help promote community safety. An example of such an instrument, developed by the U.S. Parole Commission and known as the Salient Factor Instrument, appears in **Figure 13–1**.

Parole Supervision and Termination

The number of adults under parole supervision increased at an average rate of 1.4 percent per year between 1995 and 2005. As discussed in Chapter 2, of the more than 784,400 persons under parole supervision at the end of 2005, only about a quarter had served time for violent crimes and another quarter for property crimes. More parolees (37%) had served time for drug crimes.[17] (See **Figure 13–2**.) Refer to Chapter 2 (Table 2–4) for other parolee characteristics.

In Chapter 8, we pointed out that there are different types of probation. There are also different types of parole. **Straight parole** means releasing people directly from prison onto the streets under supervision of the parole field services agency. By contrast, **residential parole** involves paroling people from prison into a community residential facility (sometimes called a halfway house) where they undergo a temporary period of structured and supported living designed to ease the transition to independent living in the community. As a last measure to avoid revocation and reincarceration, residential parole can also be used with people who have violated the conditions of parole in the community.

Parole supervision is meant to control ex-prisoners in the community and to assist them in improving their lives and avoiding a return to crime. In this sense, parole supervision is similar to probation supervision. After assessing risk and needs, parole officers determine how much and what kind of supervision and services are warranted by a given case. They investigate and monitor the parolee's

Salient Factor Score Calculator – Parole

Please fill in the following information. When you have completed the form, please click **"Calculate Score"** to calculate the Salient Factor Score.

Item A. PRIOR CONVICTIONS/ADJUDICATIONS (ADULT OR JUVENILE) 0

None = 3; One = 2; Two or three = 1; Four or more = 0

Item B. PRIOR COMMITMENT(S) OF MORE THAN 30 DAYS (ADULT/JUVENILE) 0

None = 2; One or two = 1; Three or more = 0

Item C. AGE AT CURRENT OFFENSE/PRIOR COMMITMENTS 0

26 years or more	+3 or less prior commitments = 3 +4 prior commitments = 2 +6 or more commitments = 1
22–26 years	+3 or less prior commitments = 2 +4 prior commitments = 1 +6 or more commitments = 0
20–21 years	+3 or more prior commitments = 1 +4 or more prior commitments = 0
19 years or less	+ any number prior commitments = 0

Item D. RECENT COMMITMENT FREE PERIOD (THREE YEARS) 0

No prior commitment of more than 30 days (adult or juvenile) or release to the community from last such commitment at least 3 years prior to the commencement of the current offense = 1; Otherwise = 0

Item E. PROBATION/PAROLE/CONFINEMENT/ESCAPE STATUS VIOLATOR THIS TIME 0

Neither on probation, parole, confinement, or escape status at the time of the current offense, nor committed as a probation, parole, confinement, or escape status violator this time = 1; Otherwise = 0

Item F. OLDER OFFENDERS 0

If the offender was 41 years of age or more at the commencement of the current offense (and the total score from Items A – E above is 9 or less) = 1; Otherwise = 0

Calculate Score | Reset | **Total Salient Factor Score:**

Figure 13–1 Salient Factor Parole Guideline Instrument.
Source: *Reproduced from the Public Defender Service of the District of Columbia, www.pdsdc.org/CriminalLawDatabase/salientfactorscore.asp.*

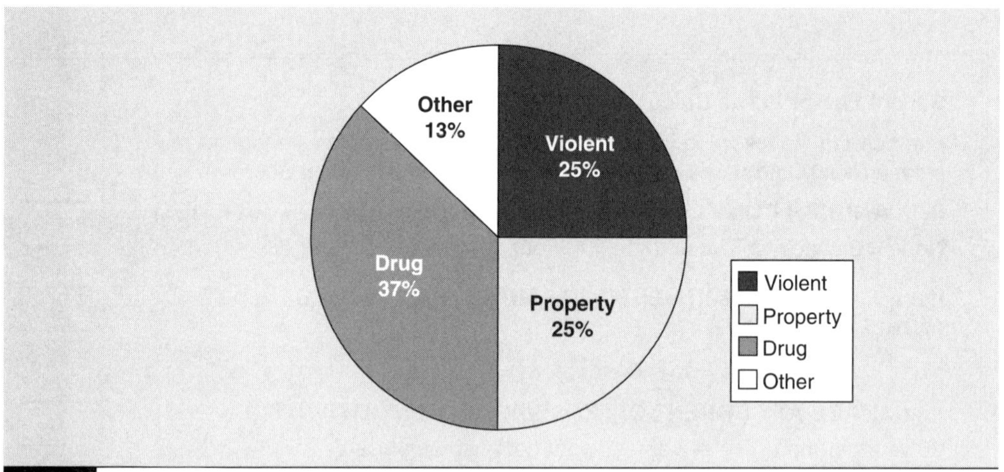

Figure 13–2 Type of offense among parole population, 2005.
Source: *Adapted from L. E. Glaze and T. P. Bonczar, (2006, November).* Probation and Parole in the United States, 2005 *(Washington, DC: U.S. Department of Justice, Bureau of Justice Statistics Bulletin).*

activities and report back to parole board officials who will ultimately decide how the case should be terminated. In addition, officers provide services such as counseling directly to parolees and arrange for parolees to receive services from community agencies. In some places (usually larger jurisdictions), parole officers have specialized caseloads, consisting, for instance, of all sex offenders or offenders with drug problems. This specialization enables parole officers to provide better supervision, because officers are able to better understand the characteristics of these offenders and the community resources available to help them.

Parolees are expected to abide by the standard and special conditions of their parole (which are very similar to probation conditions) and to participate in programs meant to help them avoid recidivism. Key conditions usually include avoiding law violations (including firearm and drug restrictions), cooperating with the parole agency (e.g., reporting as required), and maintaining gainful legitimate employment so as to meet financial obligations such as child support.

Richard McCleary conducted a classic study of parole supervision that demonstrates the importance of agency bureaucracy and politics in shaping parole supervision. McCleary found that parole agencies are often preoccupied with avoiding "trouble" from parolees, defined as anything that threatens to upset the status quo of the parole bureaucracy. Trouble can tarnish the agency's image in the eyes of the public and bring undesired scrutiny and pressure from political officials to change practices. During the parole supervision process, parole agents try to maintain the status quo and circumvent trouble by exposing parolees to incentives for desired conduct and disincentives for undesired behavior. The parolees McCleary called "dangerous men" are not necessarily violent or prone to violence, but they fail to respond predictably to the system of incentives and disincentives. This failure implies the potential for trouble, so agents try to identify and label dangerous men as quickly as possible. Although the label has to be used sparingly because it requires more supervision and documentation for the parolee, it can help forestall or soften criticism if trouble later occurs. The agency can point out that the particular parolee was identified as a problem early on, thereby reducing the element of surprise. Likewise, agents are quick to seek revocation of a parolee with the dangerous man label, thus preempting trouble. McCleary concludes that "parole outcomes are shaped in the main by the likes and dislikes of bureaucrats."[18]

Parole supervision can end in one of three ways. First, the maximum term of correctional supervision permitted under the sentence originally imposed by the judge may expire. Using a simple example, an offender sentenced to prison for an indeterminate term of 5 to 10 years may serve 6 years in prison followed by 4 years on parole, with all correctional supervision ending at the 10-year point. Second, parole authorities may discharge the ex-offender from parole before expiration of the maximum term. This is usually done with a recommendation from the field services agency because of the ex-offender's favorable progress on parole. Third, parole may be revoked because of violations of conditions. Revocation decisions are made by parole authorities or, in jurisdictions that use mandatory conditional release, by the court.

As with probation revocation, parole revocation can occur in response to new crimes or **technical parole violations** (more commonly known as a technical violation, which is a violation of a condition of parole). And as with probation, new crimes or violations by parolees do not necessarily result in revocation; parole conditions may be tightened, jail time may be assigned, or a period of residential parole may be used. When parole revocation is sought, it is similar to probation revocation and governed by the same case law due process requirements.[19] These requirements include a probable cause inquiry followed by a formal revocation hearing (see Chapter 8).

Parole Issues

Earlier it was mentioned that a combination of conservative and liberal criticisms led to the abolition or serious curtailment of parole release in some jurisdictions. However, defenders of rehabilitation in general, and of parole in particular, have countered these criticisms with arguments of their own.[20] Examples of these arguments include the following.

- The incentive for prisoners to earn early release is essential for prison management and the control of prison crowding.
- Rather than increasing disparities in time served, parole release can actually reduce sentencing disparity by ironing out inconsistencies introduced by plea negotiations and judicial discretion.

- Taking discretion away from parole authorities and judges does not make discretion disappear; lawmakers and prosecutors simply gain more of it.
- Parole authorities sometimes promote community protection by keeping dangerous prisoners incarcerated for longer periods than would occur with mandatory release.
- Some correctional programs have been shown to promote rehabilitation, and people sometimes need inducements to begin participation before they can benefit from these programs.

Based on counterarguments like these, parole release has withstood the challenge of its critics in the majority of jurisdictions. Nonetheless, there are other pressing issues confronting parole today. Some of these include (1) parolees' rights, (2) the liabilities of parole officials, and (3) the crucial importance to criminal justice of adequate resources for parole supervision and services.

Parolee Rights

When people are convicted of a felony, they often lose certain civil liberties. Later, when they leave prison to be supervised in the community (whether through discretionary parole release or conditional mandatory release), these liberties are generally not restored immediately. The person may have to wait a certain amount of time before rights related to factors such as voting, housing, and holding certain jobs or licenses can be restored. Also, depending on the jurisdiction, it may be necessary for the person to petition an executive official (such as a governor), the paroling authority, or a court to restore rights.[21]

In addition, the U.S. Supreme Court has made it clear that people under correctional supervision in the community are not entitled to the same constitutional protections as free citizens. Recall from Chapter 8 that a warrantless search can be conducted of a probationer's residence based only on reasonable grounds versus the stricter standard of probable cause.[22] In 2006, the U.S. Supreme Court went further still for parolees, upholding a California statute that allowed parole agents or the police to conduct a "suspicionless" search of parolees. The majority decision stated that "parolees have fewer expectations of privacy than probationers, because parole is more akin to imprisonment than probation is to imprisonment."[23] Likewise, the exclusionary rule (discussed in Chapter 3) does not apply at parole revocation hearings; any evidence obtained in violation of Constitutional protections can be considered at these hearings.[24] These rulings illustrate the "get tough" directions criminal justice policy has taken over the last three decades or so.

People who support imposing limitations on ex-felons' rights argue that doing so is necessary to protect the community and to maintain moral and ethical standards that would be compromised by granting former criminals the full spectrum of rights. Many supporters also believe that part of the price for crime should be restrictions on rights. Some believe these restrictions help deter crime.

On the other hand, critics point out that imposing restrictions on rights rubs against the grain of rehabilitation. We expect ex-offenders to become independent and responsible citizens, yet we place restrictions on their ability to do so. Critics also believe that people coming out of prison have paid their debts and should not continue to be penalized. One commentator refers to restrictions on ex-prisoners' rights as "invisible punishment."[25]

Parole Official Liability

Most people would agree that, like other public servants, parole board and field services officials should be accountable for their decisions. In recent years, efforts have increased to hold these officials legally liable when, for example, an individual on parole murders someone. Traditionally, however, most parole board officials have had full immunity from liability, while field services officers usually had partial immunity. Immunity has been upheld by the courts,[26] based on arguments that officials need latitude to perform their jobs effectively and that officials ultimately are not responsible for the choices parolees make. But critics charge that immunity can encourage irresponsible decisions, thus threatening public safety. Also, Joan Petersilia has expressed concern that a trend toward more liability for parole officials will result in greater proportions of people having parole revoked for technical violations, followed by return to prison. In other words, parole officials may believe it is better to be safe than sorry and, in the process, may fuel prison crowding. As a general rule, parole officials can be held liable when it is demonstrated that they failed to follow procedures required by law or administrative policy (e.g., ignoring parole release guidelines they were required by law to consider, or neglecting to follow agency requirements when supervising a parolee in the community).[27]

The Shortage of Resources for Parole Supervision and Services

Many citizens do not appreciate that community supervision and services for released prisoners represent one of the most crucial phases of criminal justice processing. Unfortunately, because parole supervision and services occur at the back end of the criminal justice process and are relatively unpublicized in the vast majority of cases, resource allocations have traditionally been meager. This has helped place the criminal justice system in a self-defeating cycle.

The 784,408 adults under parole supervision in the United States at the end of 2005 represent an increase of almost 105,000 parolees (more than 15%) since 1995.[28] The record high parole population necessitates that officers frequently have high caseloads, meaning that they may be unable to provide as much supervision and services to each parolee as needed. To worsen matters, the parole populations of many jurisdictions consist increasingly of ex-prisoners who, as a result of prison crowding, have not received much meaningful programming while incarcerated and have been released early. These people may need heightened supervision and services to be successful in the community. However, because so many resources are being devoted to prison construction and expansion, resources are often lacking to fund parole.

Most people agree that recidivism among people released from prison is too high (see the next section). Undoubtedly, high recidivism is partially caused by the shortage of resources available for community supervision and services. However, the normal public and political response to recidivism is not to call for more resources but instead for a "no-nonsense," "zero tolerance" approach that will lead more ex-prisoners to be returned to prison, often for technical violations. Of course, such a reaction simply intensifies both institutional crowding and the tendency to use early release to control crowding. This, in turn, results in further strain on community resources, seemingly intractable recidivism problems, and ever-increasing reliance on incarceration. Breaking this cycle, which has been described by Stephen Richards and his colleagues as a "perpetual incarceration machine," is a key challenge facing efforts to achieve successful prisoner reentry.[29]

Recidivism and Reentry

Throughout this text, we have discussed the high rates of recidivism for the various correctional sanctions. In the following section, we discuss the recidivism of released prisoners.

Recidivism of Released Prisoners

Recall from Chapter 8 (Table 8–1) that recidivism can be defined in different ways. Recall further that a lack of recidivism, though important, is only one indication of correctional effectiveness. As with probation, however, most studies of the effectiveness of prisons and parole have focused on recidivism.

The U.S. Justice Department estimated that a total of 503,800 ex-prisoners were discharged from parole supervision during 2005. Of these, just more than 45 percent were classified as successfully completing supervision. Slightly more than 38 percent were returned to incarceration facilities for new offenses or technical violations, and another 14.8 percent received unsuccessful discharges for absconding or something else that did not result in reincarceration. This means that almost 53 percent of those exiting parole in 2005 received some type of unsuccessful discharge.[30]

A study of persons discharged from parole between 1983 and 2000 found that persons receiving discretionary parole release were more likely than those receiving conditional mandatory release (mandatory parole) to be discharged successfully from community supervision (see **Figure 13–3**).[31] This is true even though those receiving conditional mandatory release generally served greater percentages of their

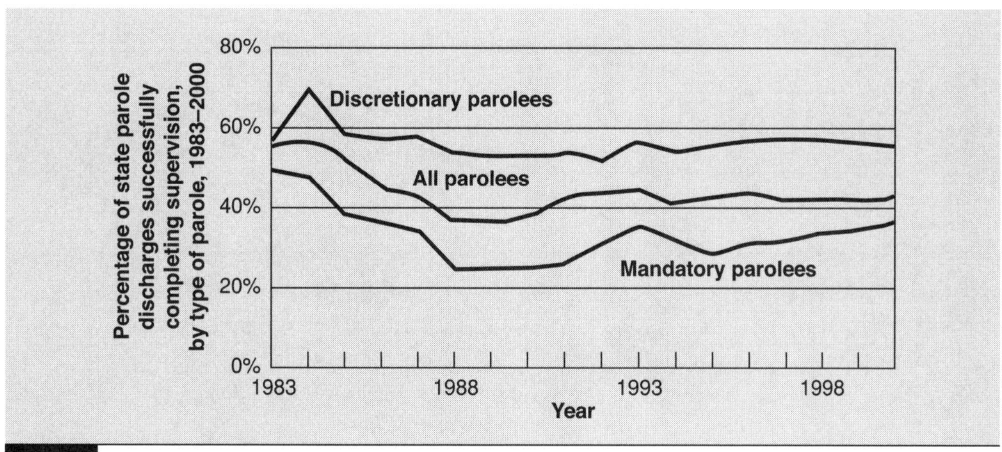

Figure 13–3 Parole success by release type.
Source: *Reproduced from T. Hughes and D. J. Wilson.* Reentry Trends in the United States, *www.ojp.usdoj.gov/bjs/reentry/reentry.htm.*

sentences in prison. Also, persons being released from prison for the first time were considerably more likely to have successful discharges than those who had served prison time prior to the current term of incarceration.[32]

In 2002, the U.S. Justice Department published a follow-up study of 272,111 ex-prisoners released from prisons in 15 states during 1994; this constituted two-thirds of all prisoners released in the nation that year. Ex-prisoners were tracked for three years, and recidivism was defined in terms of rearrests, reconvictions, or reincarcerations.[33] The study shows that by 1997, 67.5 percent of the ex-prisoners had been rearrested for new crimes (mostly felonies and serious misdemeanors), 46.9 percent had been reconvicted of a new crime, and 51.8 percent had been returned to prison for either new crimes or technical violations. The likelihood of rearrest was especially high during the first year after release. Nearly 30 percent of the ex-prisoners were rearrested in the first six months of release, and by the one-year mark, the figure had climbed to more than 44 percent (see **Figure 13–4**). This suggests that reentry challenges are especially difficult early in the period following release.

Rearrest rates were higher among those released after serving time for property crimes (73.8%) compared to those who had served time for drug offenses (66.7%) or violent crimes (61.7%). Rearrest rates were also disproportionately high among men, African Americans, younger persons, and persons with longer prior legal records. Of the 272,111 released prisoners, 21.6 percent were rearrested for a violent crime over the three-year period.[34]

A study of recidivism in Tennessee reported that the proportion of persons who were released and subsequently returned to prison within a three-year period increased from 49 percent in 1993 to 55 percent in 1997. The study found that parole release was being used to control prison crowding in Tennessee.

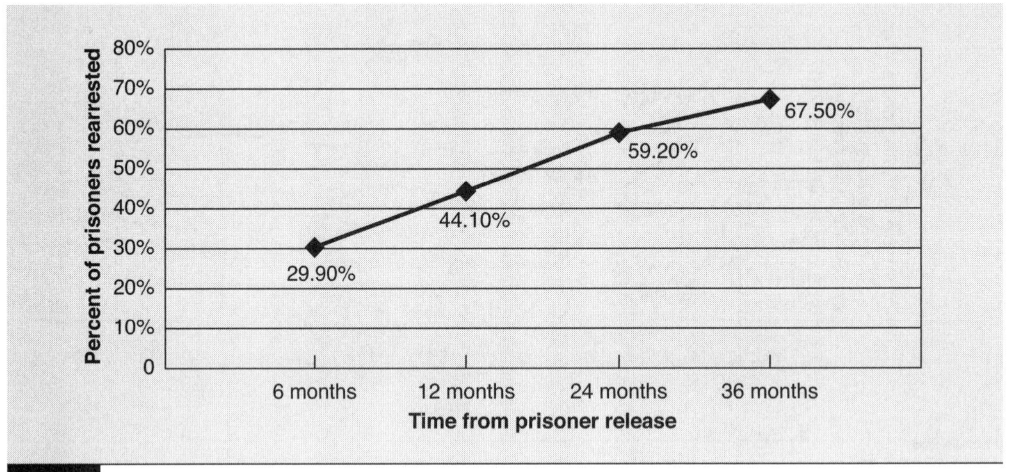

Figure 13–4 Time to rearrest among prisoners released in 1994.
Source: *Adapted from P. A. Langan and D. J. Levin, (June 2002).* Recidivism of Prisoners Released in 1994 *(Washington, DC: U.S. Department of Justice, Bureau of Justice Statistics Special Report).*

However, there was little evidence that the increase in returns to prison took place because of higher-risk prisoners getting parole or because of a major increase in new criminal behavior among paroled offenders. Instead, most of the increase was caused by a greater tendency of parole officials to revoke parole for technical violations.[35] When parole officials have increased caseloads and demands placed upon them, they may become less tolerant of parolees who are not following the technical conditions of community supervision.

Prisoner Reentry

In previous chapters, we have discussed the challenges individuals face upon entry into jail and prison. In the following section we discuss the challenges faced by those upon completion of their incarceration in prison.

Challenges

People who leave prison to reenter the community undergo a dramatic and sometimes abrupt change in living environments. This change makes successful reentry a major challenge. Former prisoners are leaving a structured and depriving environment, one in which their choices have been severely constrained. In many instances, dependency and unquestioning compliance have been rewarded by prison staff. Though minimal subsistence (food, shelter, clothing, and the like) has been provided during incarceration, many prisoners find it difficult or impossible to earn and save money adequate for release. Furthermore, among peers in the prison culture, behaving civilly toward others has often been equated with weakness, whereas distrust, cunning, violence, and exploitation of others have been valued and rewarded.

In some respects, then, the prison environment is a contradiction to the community environment to which most prisoners return.[36] In the free world environment, former prisoners are expected to "make it"—to impose structure and organization on life activities, to exercise responsibility and independence, and to interact with others in a civil way. Failure to do these things is commonly taken as evidence that the person was never really ready for release. Perhaps it is really not so ironic that one classic study found that people who were most successful at "adjusting" to prison were the least successful at adjusting to the community upon release, while those who showed difficulty adjusting to prison often adjusted better upon release.[37]

In addition to facing formidable financial challenges, ex-prisoners often find upon release that many or all of the prosocial ties they had in the community upon entering prison have been severed. Ties to family, law-abiding friends, work, and other important social institutions may be weak or broken.[38] The stigma of being an "ex-con" contributes greatly to this problem, especially as regards obtaining housing and adequate work. Many prisoners return to economically disadvantaged neighborhoods that display a range of social problems, including dilapidated housing, poor health care, and high crime. Opportunities for building prosocial ties may seem blocked, and consequently, the person may default into a criminal subculture on the street.

When considering reentry challenges, it is a mistake to believe that the problems and needs confronted by women are identical to those faced by men. Many women offenders have the following histories:

- Living in extreme poverty
- Neglected physical and mental health care
- Being traumatized by abuse (as juveniles and/or as adults)
- Being dependent on drugs

Upon leaving prison, women confront not only stereotypes about ex-convicts, but also gender stereotypes that can limit their opportunities for successful reentry, especially as regards adequate employment. The situation is further compounded for racial minority women because these women must confront racial stereotypes as well. Additionally, many women must assume responsibility for dependent children as single parents, and separation during the period of incarceration has often strained the parent–child relationship.[39]

Improving Reentry

In view of the challenges just described, a critical aspect of successful reentry is **reintegration**, or reestablishing prosocial ties that have been weakened or broken and building new ties to conventional society. Given the record high number of prisoners returning to communities across the country, there has been a great deal of focus in recent years on how to accomplish this.

Corrections professionals who specialize in the topic of reentry have identified certain key areas for effective reentry programming. For example, recommendations developed by Joan Petersilia are summarized in **Table 13–1**.

Another set of recommendations has been offered by Jeremy Travis, who identified the following principles of effective reentry.

Table 13–1 Petersilia's Areas for Reforming Reentry Practices

1. Alter the prison environment to teach skills (work, education, etc.) needed in the community rather than violence.
2. Use discretionary parole release, basing decisions on parole guidelines that predict recidivism.
3. Improve parole supervision and services, targeting ex-offenders who pose the greatest risks and needs.
4. Build partnerships between parole and the community, including both formal and formal social control agents (e.g., police, family members, other ex-offenders, victim advocates).

Source: J. Petersilia. *When Prisoners Come Home: Parole and Prisoner Reentry* (New York: Oxford University Press, 2003), chap. 9.

1. Prepare for reentry. Planning for reentry should be central to prison operations. This includes
   a. Preparing the prisoner for meaningful work in the community
   b. Attending to health care needs
   c. Strengthening prisoners' family ties
   d. Involving community organizations in facilitating reentry
   e. Helping prisoners develop realistic expectations for the reentry process
2. Establish links between prisoners and their communities, including links with
   a. Criminal justice agencies, including the courts and the police
   b. Private organizations and community groups, such as health care providers and faith-based groups
3. Maximize the moment of release instead of treating it as routine. This could involve having exit orientation classes and a welcoming ceremony with the prisoner's family.
4. Strengthen the supports surrounding the ex-prisoner. This includes building prosocial ties between the ex-prisoner and family, peers, community organizations (such as those related to housing and work), social service agencies (such as vocational education and drug treatment agencies), and criminal justice agencies—especially parole and police agencies.
5. Promote reintegration by fostering a long-term successful transition back into society. This includes recognizing and rewarding the ex-prisoner's positive achievements rather than being preoccupied with punishing failures.[40]

Given that women face special challenges, it is important that reentry programming be responsive to these needs. One study concludes that programming for women should focus on survival skills and building self-confidence and self-reliance. This means providing assistance with housing, employment, ties to others in the community, and problem solving.[41] In addition, Stephanie Covington suggests that programming for women should include child care assistance, parenting education, and mentoring by positive female role models.[42]

One interesting approach to managing prisoner reentry is the use of **reentry courts**. These courts are similar to the drug courts mentioned in earlier chapters except that instead of being used as diversion to keep offenders out of jail or prison from the outset, they are set up to help people transition from prison back into the community. Some courts are staffed by the judicial branch of government (judges and other court officials), whereas others are staffed primarily by parole officials. Prisoners are released from institutions under the supervision of the court, which monitors and coordinates transition to community life. Services include drug and mental health treatment as well as assistance with education, jobs, and housing. Regular court hearings are conducted to monitor the ex-prisoner's progress and compliance with conditions.[43]

Something that has not received much attention is the way reentry is subjectively experienced by the ex-prisoner and what ex-prisoners need to do to be successful. In this regard, Shadd Maruna studied a group of former prisoners who were, by all counts, at high risk of recidivism, but who still managed to refrain from crime in the community. Maruna compares this group with a group of former prisoners who did recidivate.[44] He found that the key to abstaining from crime was for ex-prisoners to make sense of their lives—to construct a life story that both accounted for the negative past and highlighted personal change toward a positive future. This story needs to be credible not only to others but also

Race and Gender in Corrections

There has not been much research on special reentry challenges facing minorities and the implications of these challenges for programs. However, an article by Aretha Marbley and Ralph Ferguson calls for the African American and Hispanic communities to build partnerships with the criminal justice system. The goal of these partnerships is to improve reentry among ex-prisoners of color and to mitigate the negative effects of high incarceration rates on minority communities and families because these communities are affected disproportionately by high levels of imprisonment.[45]

to the ex-prisoner. For this to take place, the ex-prisoner must adopt three related attitudes:

1. Start perceiving that he or she is contributing something positive to society, such as through a rewarding job and/or marriage
2. Be able to separate his or her true self from the criminal past, such as making a distinction between committing illegal behavior in the past and being a bad person
3. Begin to sense that he or she is in control of the future and capable of overcoming difficulties

Obviously, the policies recommended by people like Travis (reviewed earlier) could help accomplish what Maruna suggests.

Research on Reentry Programming

Many studies have evaluated the effectiveness of reentry programs, and illustrations are described in this section. As will be apparent, the research has produced mixed findings. Although some studies show that reentry programs can have favorable outcomes, other programs appear to have few positive effects.

It has long been argued that meaningful and adequately paying work is important to successful reentry, not only because of concrete financial rewards, but also because of more intangible psychological benefits (refer back to Maruna's three points earlier).[46] Although some research has shown that programs designed to improve ex-offenders' job skills and employment statuses are associated with lessened recidivism, other studies have not confirmed this.[47] One program that has produced some favorable preliminary results is the Ready4Work initiative, a collaborative initiative between the U.S. Department of Labor and public/private ventures. The program focuses on building partnerships between faith, criminal justice, business, and social service organizations in the community to provide ex-offenders with job training and placement. There is additional emphasis on assistance with housing, health care, and substance abuse treatment. Ex-offenders are also provided mentors. Early findings showed that only 1.9 percent of program participants had been returned to prison within six months of release, while only 5 percent has been returned within a year.[48]

Other reentry programs have addressed the substance abuse needs of ex-prisoners. One study reports that 22 percent of offenders with a history of substance abuse who were paroled to community-based substance abuse treatment facilities were reconvicted of a crime over a two-year period after release; this compared with 34 percent of those with substance abuse histories who were paroled without such treatment.[49] However, an evaluation of Project Greenlight, a New York prison-based reentry program, produced less encouraging findings.

This project targeted the substance abuse and other needs of persons nearing release by trying to alter patterns of criminal thinking and behavior. It was based on principles other research has shown to be key to reducing recidivism. Yet one year after release, project participants fared worse on various indicators of recidivism than did a group of nonparticipants.[50]

The Preventing Parolee Crime Program (PPCP) was initiated by the California Department of Corrections in the late 1990s as a statewide, multifaceted effort to reduce parole recidivism. PPCP focused on education (basic math and literacy), employment, housing, and substance abuse. An evaluation demonstrated that the program was associated with reductions both in parolees being returned to prison within one year of release and in rates of absconding.[51]

As mentioned earlier, women making the transition from prison to the community often display some special problems and needs. One study reports that correctional officials tend to underestimate women's needs in the areas of substance abuse treatment, employment, and housing. The study finds that initiatives to meet these needs were associated with lower chances of failure on parole.[52] Altogether, the evidence seems to support the proposition that when parolees are given a broad range of supportive community-based services, they tend to have better success at community reentry.

Key Controversies Surrounding Prisoner Reentry

Many controversies surround the topic of prisoner reentry. Some of the key ones are described in this section.

Corrections and Policy

President George W. Bush called for increased attention to prisoner reentry in his 2004 State of the Union speech. In 2005, Congress enacted legislation aimed at improving reentry and reducing barriers, including the Second Chance Act and the Re-Entry Enhancement Act. The Reentry Initiative has been launched by the U.S. Department of Justice in collaboration with the U.S. Departments of Education, Health and Human Services, Housing and Urban Development, and Labor. The initiative provides funds for reentry programs that will help promote community safety and reduce serious recidivism. In 2002–2003, the federal government provided more than $150 million to fund 89 reentry programs across the nation as part of the Serious and Violent Offender Reentry Initiative.[53]

Offender Employability As mentioned previously in this chapter, people leaving prison often face certain legal restrictions as a result of their status as ex-felons. Given that meaningful employment is viewed as being essential to successful reentry, one of the most controversial issues involves restrictions that limit ex-prisoners' employability. Because of fears and stereotypes, many employers are reluctant to hire former prisoners. In many places, there are laws keeping former prisoners from obtaining licenses to enter certain jobs, such as teaching, health care, barbering, real estate, or insurance sales. Also, background checks are becoming a routine part of hiring decisions, a trend greatly facilitated by computer automation and by laws and policies making it easier for employers to obtain information from the government about criminal history. Of course, the presence/absence of a criminal record is the major focus of these checks, and a record is all the more stigmatizing when it includes a prison term.[54]

Proponents of imposing restrictions on ex-prisoner employability argue that the restrictions are needed to protect the public, both physically and financially. Some proponents believe that the restrictions are part of the long-term price the former offender must pay for criminal behavior and that restrictions might help deter other people from choosing crime. However, opponents counter that many of the restrictions are irrelevant to crime prevention and filled with shortcomings.[55] Opponents say that the restrictions pose unnecessary barriers to reintegration by closing off access to legitimate income and conventional social ties.

It remains to be seen which of these sides will prevail. At present, there are contradictory trends in law and policy. On the one hand, laws restricting licensure and employment have been on the increase, and it has become easier for employers to get criminal background data. On the other hand, federal laws such as the Second Chance Act and the Re-Entry Enhancement Act are aimed at reducing barriers to reentry.

Controversies

How great a risk does an employer take by hiring an ex-offender? Are people who have committed a crime in the past quite likely to do so in the future? A study addressing this question found that people with records of previous arrests were indeed more likely to offend than those without such records. However, this likelihood was only greater in the time frame right after the previous crime was committed. The risk of future crime declined dramatically and quickly thereafter. By the time six or seven years had passed, people with a prior record had virtually the same risk of offending as people without a prior record.[56]

The Impact of Incarceration and Reentry on Families and Communities A second controversy relates to the impact of incarceration and reentry on families and local communities. Traditionally, the logic has been that criminal offenders need to be removed from their families and communities, in many cases for prolonged periods, so as to (1) protect family and community members from criminal victimization and criminal influences, and (2) punish and attempt to rehabilitate the offender in a controlled environment. Advocates of this logic suggest that regardless of financial costs and other negative effects of imprisonment, the benefits to families and communities are worth it.

Although it is clear that some offenders pose a danger to the community and must be removed, lessons from the mass incarceration and mass reentry eras in corrections have led some to question the logic just outlined. It has become evident that incarceration (especially for long periods), followed by weak or nonexistent reentry programming, can do much long-term harm to families and communities. When used on such a mass scale, incarceration has aggregate or cumulative effects. It can remove large numbers of wage earners from homes, destabilize marriages, and negatively affect child development by leaving children without parents (most typically, fathers). It can remove taxpayers from communities, often disadvantaged communities in desperate need of tax revenue. When followed by little or no attention to reentry services, it can contribute to the disorganization and deterioration of communities. In effect, then, mass incarceration and lack of attention to reentry can weaken what informal social control might have existed in families, schools, and neighborhoods. Economically disadvantaged and predominately minority communities are affected most.[57]

The product can be the kind of self-perpetuating cycle described earlier in this chapter when discussing parole resources (see **Figure 13–5**). When families and communities become destabilized and the informal social control they should be providing grows weaker, two consequences can be high recidivism among reentering prisoners as well as high crime among youth growing up in the community.[58] High recidivism and crime affect neighborhood culture by prompting anger, fear, and distrust among residents. A cultural context of this nature is ripe to prompt calls to "get tough" or "crack down" on criminals—in essence, intensification of the very calls that gave rise to mass incarceration policies from the outset. When the get tough response to fear and anger involves greater use of incarceration (and it typically does), fewer resources are available for reentry; now the cycle is set to repeat.

Sex Offenders In recent years, probably no issue in criminal justice has sparked greater public fear and outrage than the return to communities of people who have been incarcerated for sex crimes, especially sexual crimes against children. Much of this fear and anger is in response to isolated and highly publicized incidents such as the murder of Megan Kanka in New Jersey by a parolee who had served time for sex offending. This tragic incident resulted in Megan's Law, and

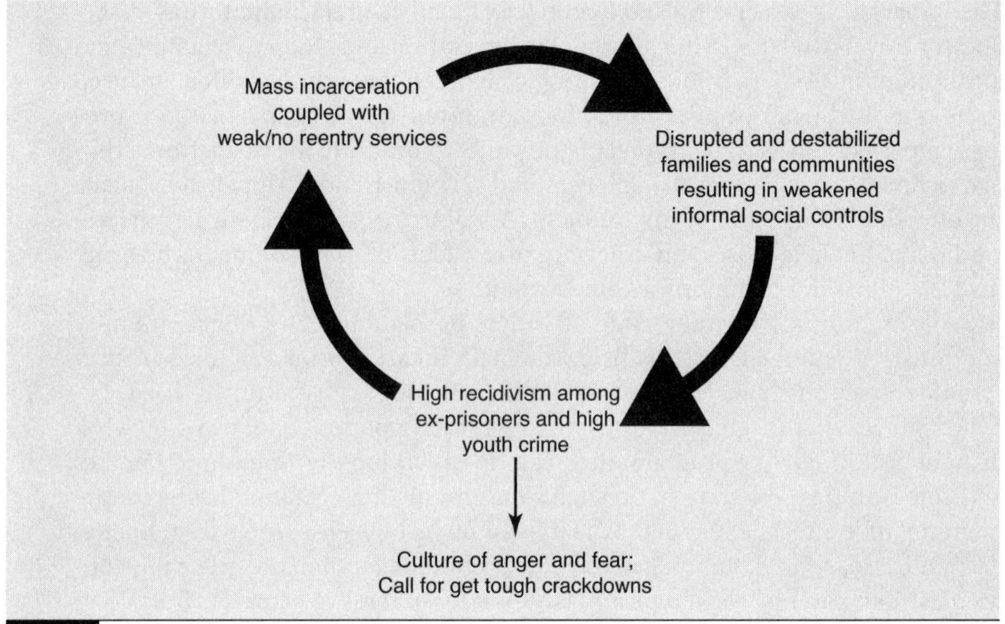

Figure 13–5 Self-perpetuating cycle of mass incarceration.

such laws (described later) now exist in some form in every state. Public fear and anger has also been fueled by media productions such as Dateline NBC's popular *To Catch a Predator*, where the objective during each episode is to employ undercover techniques to identify and apprehend potential sex offenders who are interacting with children over the Internet.

Every state requires sex offenders to register themselves with local authorities in a **sex offender registry**, a compilation of convicted sex offenders in the jurisdiction that can be accessed to identify these offenders and monitor their whereabouts. Further, more and more jurisdictions have passed **sex offender notification laws** requiring the local public to be notified (e.g., through Web sites, newspapers, special bulletins) when sex offenders are reentering the community from prison.[59] In addition, some jurisdictions have laws that prohibit these offenders from living or loitering specified distances from schools, day care facilities, and parks. These laws can make large sectors of urban areas off limits to sex offenders for housing. In addition, when such laws go into effect, sex offenders may need to move to be in compliance. The result can be that these offenders colonize, for instance, in run-down motels.[60]

Critics charge that laws of this nature do little more than fuel political rhetoric and media hype. Far from improving public safety, the laws may create a false sense of security by creating an appearance that "something" is being done. In

Controversies

In July 2006, Kentucky began enforcement of a law that prohibited registered sex offenders from residing within 1,000 feet of a school, day care center, or playground. Here are two reactions that demonstrate the controversy over the law.[61]

"It does raise some civil liberties concerns and that's especially troubling when it doesn't appear that these sorts of restrictions work."—American Civil Liberties Union Representative

"Ideally, you would like to put them in a boat and send them to an island. The problem is, they are a part of society and we have to deal with them. Nobody wants sex offenders living next to them, but they have to live someplace."—County Sheriff

fact, the real effect may be to further stigmatize and disenfranchise offenders, impair their reintegration, and increase the likelihood of recidivism. Nevertheless, proponents counter that communities have a right to know when sex offenders are living in the neighborhood so that offender activities can be monitored and so those citizens who wish to can take precautions. Some proponents also claim that the stigma and disenfranchisement resulting from these laws are fitting punishment for sex crimes and might help deter such offenses in the future. At the same time, research fails to support the contention that community notification efforts reduce recidivism among sex offenders.[62] Although it is not realistic to expect communities to be unsuspicious and highly supportive of reentering sex offenders, there is also a need to form realistic expectations of what can be accomplished with registry, notification, and housing laws.

On May 17, 2007, U.S. Attorney General Alberto Gonzales released proposed national guidelines for sex offender registration and notification to set minimum standards and provide guidance for states in setting up systems. Gonzales also announced the availability of $25 million in federal money to assist communities with implementing the guidelines. The proposed guidelines are available at www.ojp.usdoj.gov/smart/guidelines.htm.

Reentry and Crime Victims The efforts of victims' rights advocates have resulted in most jurisdictions giving victims (or their families) some type of opportunity for input during parole release decision making. In addition, it is now commonplace for restitution to the victim to be a condition of parole supervision. More recently, there has been debate over whether victims should play a greater role in the reentry process.

Corrections and Policy

The Bush administration aggressively promulgated various policies meant to protect children from sex offenders (as well as other forms of victimization).

- The PROTECT Act (signed in 2003) expands funding for Amber Alerts so that the public can become alerted more quickly about missing children.
- The Adam Walsh Child Protection Act (signed in 2006) allowed integration of state sex offender registry systems to establish a national registry (see the following for guidelines), established mandatory sentences at the federal level for serious crimes against children, granted money to the states to help incarcerate sex offenders, and established regional task forces to address Internet crimes against children.
- Operation Predator was launched to help police track sex offenders targeting children.
- Project Safe Childhood was initiated to help police at all levels of government investigate and prosecute Internet crimes against children.[63]

Some claim that giving victims and victim organizations an expanded role in reentry could work against offender reintegration. For example, expanding the emphasis on victims could make the reentry process too emotionally charged and render the best interests of the offender a distant second to those of the victim. When offenders leave prison, they have paid their "debt to society," and it is time to focus on turning them into productive citizens.

However, Susan Herman and Cressida Wasserman argue that giving victims a greater role in reentry is compatible with offender reintegration. Both victims and offenders ultimately have a stake in reducing the likelihood of recidivism. "Victims and victims' organizations can positively assist in . . . reintegration . . . by providing decision makers with important and relevant information; offering experience and expertise; encouraging offender accountability; and furthering the goals of victim empowerment, safety, restitution, and reintegration."[64] Herman and Wasserman discuss three general ways by which victims can be meaningfully involved in reentry.

1. Participation during key stages of criminal justice processing, such as providing victim impact statements during sentencing, participating in the parole release decision process, and communicating with parole officials about their interactions with ex-offenders (e.g., as regards child support payment)

2. Participation in programs meant to educate offenders about the impact of the crime, such as participation in restorative justice or mediation conferences with the offender during and/or after incarceration
3. Involvement in developing and implementing reentry programs, such as designing and delivering class sessions on the impact of crime or participating in reentry court efforts—especially efforts related to supervision of the ex-offender and monitoring of his or her compliance with conditions

There is no doubt that offender reentry will continue to be surrounded by these and a host of other controversies in future years. But as discussed throughout this chapter, reentry is a critical aspect of the criminal justice system. Improving it is one key to improving the system.

Application to Criminal Justice and Corrections

The material in this chapter has clear application to criminal justice in general and other aspects of the correctional system in particular. Illustrations of this applicability are described in this section.

Laws guide and sometimes dictate decisions about which offenders will go to prison and how much time they will serve. Furthermore, parole and other methods of release are established and regulated by law, and the same is true for parole supervision agencies. Law also regulates the reentry process by specifying what restrictions can and cannot be placed on former prisoners, as well as conditions under which criminal justice officials may be liable for persons under supervision. As with other elements of criminal justice, then, law is the foundation upon which release, parole, and reentry operate.

Police agencies both affect, and are affected by, release decisions and reentry processes. For instance, the release of prisoners who pose a threat to the community affects the workload of the police, so police agencies have an interest in shaping release policies through discussions with other criminal justice officials and elected representatives. On the other hand, if prisoners are held excessively long in a manner that fuels institutional crowding, a prisoner backlog can develop in local jails, thus affecting the local police. After prisoners are released, the police work with parole officers to monitor former prisoners and apprehend those who have reoffended. Because the police are the front line between the criminal justice system and the community, police–community relations can be adversely affected by serious recidivism on the part of ex-prisoners, especially for sex crimes and violence. By the same token, the police must deal directly with the community and social problems associated with high rates of imprisonment and/or inattention to reentry services.

The courts are closely linked to release and reentry as well. Obviously, judges shape release dates through their authority to sentence the guilty, but so do prosecutors and defense attorneys when they negotiate sentence recommendations during plea bargaining. Judges, prosecutors, and defenders may also have input into discretionary release decisions. When ex-prisoners return to the community, judges may have some say over what restrictions apply. Traditionally, in some jurisdictions, field supervision functions have been administered by the judicial branch of government. And in jurisdictions where reentry courts have been established, the judicial branch is playing an unprecedented role in reentry. If ex-prisoners reoffend, the trial court may process the case as a new crime. Finally, appellate courts have rendered decisions stipulating the due process rights of prisoners seeking parole release as well as ex-prisoners under supervision in the community.

Other components of corrections are closely related to release and reentry. Indeed, many of the people reentering society from prison have a history of participating in diversion programs, being on probation, and/or serving time in local jails. As mentioned at different points in this chapter, correctional officials are often heavily reliant on parole and other release methods to manage the behavior of prisoners and to keep institutional crowding in check. However, there is potential for diverging interests in this area. Although the reality of institutional crowding and resource limitations make the early release of some prisoners a virtual necessity, many police, court and legislative officials, as well as citizen groups express frustration that prisoners do not serve enough of their sentences.

When correctional officials are trying to determine which prisoners should receive parole, they tend to draw heavily on case file information contributed by police, court, and probation agencies. People released from prison into community supervision are frequently supervised by agencies that supervise probationers as well. The same officer may have both probationers and parolees on his or her caseload. Parolees may participate in the same community programs as probationers and even those who have been diverted from the system. Rather than being returned to prison, someone under parole supervision who violates sentence conditions may receive a jail term and serve that term alongside pretrial detainees, probation violators, and offenders serving jail terms as the primary sentence of their crimes.

READY FOR REVIEW

- Most people sentenced to prison will reenter society at some point, and most will face formidable challenges when doing so. Thus, release and reentry are critically important topics in criminal justice and corrections.
- Release from prison is affected by several factors, and the process is more complex than many people realize. Prisoners are typically released from programs designed to make the transition to the community gradual. These include prerelease and temporary release programs. Types of temporary release include work release, study release, and furlough.
- There are three principal methods of release from prison, including unconditional mandatory release, conditional mandatory release, and discretionary parole release. The latter used to be the most common method, but as jurisdictions either abolished or restricted parole release, conditional mandatory release became the most common method.
- The term *parole* encompasses both parole release (as one method of prison release) and parole supervision, which occurs in the community after release. People can be placed under parole supervision following either parole release or conditional mandatory release.
- Though the two programs have similar goals, parole supervision is different from probation in that parole is used with people exiting prison, whereas probation is used in lieu of a prison term. Additionally, parolees often face greater adjustment problems than probationers do.
- Parole evolved out of the English practice of transportation in the sixteenth century and the ticket of leave system used off the coast of Australia and in Ireland during the mid-1800s. In the United States, it was pioneered at Elmira Reformatory in New York along with indeterminate sentencing.
- In the mid-1900s, parole was dominated by a medical model, which conceptualized the treatment of crime as analogous to the treatment of disease. However, this model came under heavy criticisms for being too soft on crime and for being unfair to offenders. Much of the criticism was based on the premise that correctional rehabilitation efforts had not been very effective.
- From the latter 1970s on, the trend across the United States was for jurisdictions to either abolish or curtail parole release. Nevertheless, most jurisdictions did not abolish it, and record numbers of people are under parole supervision today.
- Parole release decisions are administered by a parole board (or similar authority), whereas parole supervision is administered by a field services agency. States have different ways of organizing and administering their parole and probation services.

- Before parole release can be considered, prisoners must become eligible for parole, and eligibility is governed by law. When a prisoner becomes eligible, he or she will develop a parole plan, and institutional officials will prepare a preparole report to assist with decision making. The parole board will then conduct a parole grant hearing to decide whether parole is to be granted. Boards give a lot of consideration to the nature of the crime and the time served as well as to the offender's legal history.
- The number of persons under parole supervision across the country has been on the rise primarily because of massive increases in incarceration rates. More parolees have served time for drug crimes than for any other crime category.
- Prisoners may be paroled directly to the community (straight parole) or into a community residential facility (residential parole). Parole supervision is quite similar to probation supervision in that parolees are monitored for compliance with conditions and provided with services to promote their betterment. Research shows that the dynamics of the parole bureaucracy can affect which parolees are successful.
- People may be discharged from parole at the expiration of the sentence or earlier than that for making good progress, or they may have parole revoked for violations or new crimes. Parole revocation is governed by the same case law and due process requirements as probation revocation.
- Defenders of parole mount rebuttals to the arguments of those who seek to abolish or restrict parole. There are numerous controversial issues surrounding parole. These include the following:
  - To what extent is it necessary and fair to place restrictions on the civil liberties of parolees?
  - To what extent should parole officials be held liable for the behavior of parolees?
  - How can we increase resources for parole supervision and better control recidivism at a time when we expend so many resources on imprisonment?
- Research by the U.S. Justice Department has demonstrated that just more than half of the people released from prison return to prison within three years. The chances of recidivism are highest early in the period following release. There is some evidence that parole release is associated with a greater likelihood of successful discharge from supervision than conditional mandatory release. There is concern among some observers that excessive numbers of parolees are returned to prison for technical violations rather than for having committed new crimes.
- Reentry to the community from prison presents a host of challenges. The living environment of the free community differs drastically from the prison environment. Ex-prisoners usually face financial difficulties as well as difficulties establishing or reestablishing ties to conventional social

institutions. Women leaving prison often face special challenges, including being single parents and facing barriers to adequate employment.

- Corrections professionals have offered a number of suggestions for improving reentry. These include greater attention to preparation for reentry while in prison, improving collaboration among community agencies and the criminal justice system, and improving the supports available to former prisoners in the community. Some jurisdictions have started using reentry courts to better coordinate services and monitor ex-prisoners in the community. Some commentators have suggested that women need programs tailored especially to their needs because their needs are often different from those of men. Research has shown that it is also important to consider the way ex-prisoners subjectively experience reentry.
- Reentry programs often target the housing, employment, substance abuse, and other needs of ex-prisoners. However, research on the effectiveness of such programs has produced mixed results. Some programs are associated with lower recidivism, whereas others are not.
- Reentry is a highly controversial topic. Illustrations of controversies are as follows:
  - Some argue that restrictions must be placed on the employment of ex-prisoners to protect the community. Others counter that these restrictions are unnecessary and encourage recidivism. Policies and laws are contradictory on this point.
  - The traditional logic that many or most criminals need to be removed from their communities and families and incarcerated is being questioned. It is becoming evident that mass imprisonment and subsequent mass reentry can have destabilizing effects on those families and communities that are disproportionately affected by high levels of incarceration (i.e., economically disadvantaged minorities in urban areas). The destabilization of families and communities can increase the chances of both recidivism among reentering prisoners and crime among youth. In turn, such increases can exacerbate high levels of incarceration and reentry, thereby leading the cycle to repeat.
  - Recent years have witnessed stronger efforts to monitor sex offenders returning to the community from prison, as evidenced by sex offender registry and notification laws. Although advocates espouse community safety and the right of the community to be informed, proponents argue that these laws do little good and, in fact, may make communities less safe by disenfranchising ex-prisoners.
  - Traditionally, arguments have been made that victims and victim organizations should play a minimal role in reentry so that focus can be maintained on successful adjustment among offenders. More recently, arguments have been made that victims and victim organizations should have their role in reentry expanded, because they can actually facilitate reentry processes.

KEY TERMS

conditional mandatory release A type of prison release in which the person is released under the provision of law after having served a legally stipulated portion of the sentence (minus good time) with a period of supervision in the community following release; sometimes referred to as "mandatory parole"

discretionary parole release A type of prison release in which the offender is released into community supervision at the discretion of a parole board or similar authority instead of under the provision of law; offenders are generally not eligible to be considered for parole until a certain minimum period of time has been served

good time A provision whereby prisoners can earn reductions in the sentence for good/meritorious behavior in prison

medical model An era starting in the 1940s and 1950s when crime was viewed and treated as pathology, analogous to the manner in which the medical profession treats illness

parole Can refer to a type of release from prison (*see* discretionary parole release) or a period of supervision and programming in the community subsequent to release

parole grant hearing A proceeding involving the prisoner, parole board officials, and possibly others where consideration is given to whether a prisoner should be paroled

parole guidelines Instruments that allow officials to rate persons being considered for parole on variables believed to predict recidivism

parole plan A document describing a prisoner's plan for such factors as housing, employment, and counseling should parole be granted

preparole report A document prepared by prison officials that describes the prisoner and his or her crime and reviews his or her behavior in prison as well as progress toward rehabilitation

prerelease programs Programs, focused on areas such as employment and life skills, that are implemented shortly before the scheduled release date with the intent of preparing the prisoner for reentry to society

presumptive parole dates An arrangement whereby prisoners are told upon (or shortly after) their incarceration the date they can expect to receive parole if they follow prison rules and participate in programs as directed

reentry The process of people coming out of correctional institutions and transitioning back into society

reentry courts Courts (similar to the drug courts) that are set up to help people transition from prison back into the community

reintegration Involves a former prisoner reestablishing prosocial ties that have been weakened or broken and building new ties upon release

residential parole A variety of parole whereby people are paroled from prison into a community residential facility (halfway house) where they undergo a temporary period of structured and supported living designed to ease the transition to independent living in the community
sex offender notification laws Laws that require the public to be notified when sex offenders are reentering the community from prison
sex offender registry A compilation of convicted sex offenders in a jurisdiction that can be accessed to identify these offenders and monitor their whereabouts
straight parole The direct release of people from prison onto the streets under supervision of the parole field services agency
technical parole violation A violation of a condition of parole (most often known as a technical violation)
temporary release programs Programs allowing persons confined to a correctional facility to leave the facility for a temporary period to participate in approved activities in the community geared toward establishing or maintaining social ties; types include work release, study release, and furlough
ticket of leave system An antecedent of parole in which prisoners who exhibited increments of favorable progress while incarcerated could earn early release and be supervised in the community contingent on good behavior
transportation An antecedent to parole in which the English sent convicts to their colonies for a specified period of labor and permitted the convicts to earn a pardon by following certain rules and avoiding trouble with the law
unconditional mandatory release A type of prison release whereby the person is released upon having served the maximum amount of time allowed by the sentence without any supervision in the community following release; this is also known as "maxing out"

YOU ARE THE CORRECTIONS PROFESSIONAL SUMMARY

1. What are the advantages and disadvantages of the prisoner serving the full 10-year term? The main advantage in this case is political; the prosecutor and judge in the local jurisdiction would be satisfied by the message sent and, therefore, not likely to direct criticisms at the parole board. If the prisoner is going to reoffend (which, of course, is unknown at this time), then continued incarceration would prevent, or at least delay, reoffending. However, additional prison time is unlikely to benefit the prisoner or his family at this point. Because the prisoner has accumulated two years of good time and is likely to accumulate more if parole is denied, he would probably serve less than two additional years before maxing out his sentence. If he is let out through unconditional

mandatory release, correctional officials would have no authority to supervise him in the community or provide support services. This could make recidivism more likely.

2. If the prisoner is paroled at this time, what should be included in the special conditions of supervision? If the prisoner is paroled, the special conditions of parole supervision (which would supplement the general conditions) should probably include mandatory alcohol treatment in addition to the self-help the prisoner has planned. This might be combined with family counseling to assist with reunification of the family as well as counseling to help the prisoner deal appropriately with anger, frustration, and stress. Finally, rather than the prisoner going directly home from the institution, it would probably be wise for reentry to be gradual. You might consider prerelease programming with family furloughs and/or a brief stay at a halfway house on residential parole.

NOTES

1. See J. Petersilia, *When Prisoners Come Home: Parole and Prisoner Reentry* (New York: Oxford University Press, 2003); J. Travis, *But They All Come Back: Facing the Challenges of Prisoner Reentry* (Washington, DC: Urban Institute Press, 2005).
2. U.S. Department of Justice, Office of Justice Programs, "Reentry" home page, www.reentry.gov/.
3. L. E. Glaze and T. P. Bonczar, *Probation and Parole in the United States, 2005* (Washington, DC: U.S. Department of Justice, Bureau of Justice Statistics Bulletin, November 2006).
4. For additional information about temporary release programming, see B. R. McCarthy, B. J. McCarthy Jr., and M. C. Leone, *Community-Based Corrections*, 4th ed. (Belmont, CA: Wadsworth/Thompson, 2001).
5. D. P. LeClair and S. Guarino-Ghezzi, "Does Incapacitation Guarantee Public Safety? Lessons from the Massachusetts Furlough and Prerelease Programs," *Justice Quarterly* 8(1991): 26.
6. For a few extended treatments of the history of parole, see G. Cavender, *Parole: A Critical Analysis* (Port Washington, NY: Kennikat Press, 1982); D. J. Rothman, *Conscience and Convenience: The Asylum and Its Alternatives in Progressive America* (Boston: Little, Brown, 1980); J. Simon, *Poor Discipline: Parole and the Social Control of the Underclass, 1890–1990* (Chicago: University of Chicago Press, 1993).

7. See Cavender, *Parole*. Also: N. Morris, *Maconochie's Gentlemen: The Story of Norfolk Island and the Roots of Modern Prison Reform* (New York: Oxford University Press, 2002).
8. Simon, *Poor Discipline.*
9. F. T. Cullen and K. E. Gilbert, *Reaffirming Rehabilitation* (Cincinnati, OH: Anderson, 1982).
10. D. Lipton, R. Martinson, and J. Wilks, *The Effectiveness of Correctional Treatment: A Survey of Treatment Evaluation Studies* (New York: Praeger, 1975). See also: A. von Hirsch, *Doing Justice: The Choice of Punishments* (New York: Hill and Wang, 1976).
11. D. Fogel, *"We Are the Living Proof": The Justice Model for Corrections*, 2nd ed. (Cincinnati, OH: Anderson, 1979). See too: The American Friends Service Committee, *Struggle for Justice: A Report on Crime and Punishment in America* (New York: Hill and Wang, 1971).
12. T. A. Hughes, D. J. Wilson, and A. J. Beck, *Trends in State Parole, 1900–2000* (Washington, DC: U.S. Department of Justice, Bureau of Justice Statistics Special Report, October 2001).
13. J. Austin and J. Irwin, *It's About Time: America's Imprisonment Binge,* 3rd ed. (Belmont, CA: Wadsworth/Thompson Learning, 2001). Also see: J. Irwin, *The Warehouse Prison: Disposal of the New Dangerous Class* (Los Angeles: Roxbury, 2005).
14. Glaze and Bonczar, *Probation and Parole.*
15. *Greenholtz v. Inmates of the Nebraska Penal and Correctional Complex*, 442 U.S. 1 (1979).
16. See the following for examples of studies: R. Burns, P. Kindade, M. C. Leone, and S. Phillips, "Perspectives on Parole: The Board Members' Viewpoint," *Federal Probation* 63(1999): 16–22; B. M. Huebner and T. S. Bynum, "An Analysis of Parole Decision Making Using a Sample of Sex Offenders: A Focal Concerns Perspective," *Criminology* 44(2006): 961–992; K. D. Morgan and B. Smith, "Parole Release Decisions Revisited: An Analysis of Parole Release Decisions for Violent Inmates in a Southeastern State," *Journal of Criminal Justice* 33(2005): 277–287; J. C. Runda, E. E. Rhine, and R. E. Wetter, *The Practice of Parole Boards* (Lexington, KY: Host Communications Printing, 1994).
17. Glaze and Bonczar, *Probation and Parole.*
18. R. McCleary, *Dangerous Men: The Sociology of Parole,* 2nd ed. (New York: Harrow and Heston, 1992), 140.
19. *Morrissey v. Brewer*, 408 U.S. 271 (1972); *Gagnon v. Scarpelli*, 411 U.S. 778 (1973).

20. For excellent coverage on this point, see Cullen and Gilbert, *Reaffirming Rehabilitation*.
21. C. A. Cripe and M. G. Pearlman, *Legal Aspects of Corrections Management*, 2nd ed. (Boston: Jones and Bartlett, 2005).
22. *Griffin v. Wisconsin*, 483 U.S. 868 (1997).
23. *Samson v. California*, 126 S. Ct. 2193 (2006).
24. *Pennsylvania Board of Probation and Parole v. Scott*, U.S. 97-581 (1998).
25. Travis, *But They All Come Back*.
26. *Martinez v. California*, 444 U.S. 277 (1980).
27. Petersilia, *When Prisoners Come Home*. And see: J. C. Watkins Jr., "Probation and Parole Malpractice in a Noninstitutional Setting: A Contemporary Analysis," *Federal Probation* 53(1989): 29–34.
28. Glaze and Bonczar, *Probation and Parole*.
29. S. C. Richards, J. Austin, and R. S. Jones, "Kentucky's Perpetual Prisoner Machine: It's About Money," *Review of Policy Research* 21(2004): 93–106.
30. Glaze and Bonczar, *Probation and Parole*.
31. T. Hughes and D. J. Wilson, "Reentry Trends in the United States," www.ojp.usdoj.gov/bjs/reentry/reentry.htm.
32. Hughes et al., *Trends in State Parole*.
33. P. A. Langan and D. J. Levin, *Recidivism of Prisoners Released in 1994* (Washington, DC: U.S. Department of Justice, Bureau of Justice Statistics Special Report, June 2002).
34. Langan and Levin, *Recidivism of Prisoners*.
35. J. A. Wilson, "Bad Behavior or Bad Policy? An Examination of Tennessee Release Cohorts, 1993–2001," *Criminology and Public Policy* 4(2005): 485–518.
36. A. Cordilia, *The Making of an Inmate: Prison as a Way of Life* (Cambridge, MA, 1983); E. Goffman, *Asylums* (New York: Doubleday, 1961); C. Haney, "The Psychological Impact of Incarceration: Implications for Postprison Adjustment," in *Prisoners Once Removed: The Impact of Incarceration and Reentry on Children, Families, and Communities*, eds. J. Travis and M. Waul, 33–66 (Washington, DC: Urban Institute Press, 2003); R. A. Schill and D. K. Marcus, "Incarceration and Learned Helplessness," *International Journal of Offender Therapy and Comparative Criminology* 42(1998): 224–232.
37. L. Goodstein, "Inmate Adjustment to Prison and the Transition to Community Life," in *Correctional Institutions*, eds. R. M. Carter, D. Glaser, and L. T. Wilkins, 3rd ed., 285–302 (New York: Harper & Row, 1985).

38. B. M. Huebner, "The Effect of Incarceration on Marriage and Work over the Life Course," *Justice Quarterly* 22(2005): 281–303.

39. For an excellent discussion of the issues former women prisoners face, see S. S. Covington, "A Woman's Journey Home: Challenges for Female Offenders," in *Prisoners Once Removed: The Impact of Incarceration and Reentry on Children, Families, and Communities,* eds. J. Travis and M. Waul, 67–103 (Washington, DC: Urban Institute Press, 2003); also see B. E. Richie, "Challenges Incarcerated Women Face as They Return to Their Communities: Findings from Life History Interviews," *Crime and Delinquency* 47(2001): 368–389.

40. Travis (2005), *But They All Come Back*, chap. 12.

41. P. O'Brien, " 'Just Like Baking a Cake': Women Describe the Necessary Ingredients for Successful Reentry After Incarceration," *Families in Society* 82(2001): 287–295. See also: McCarthy et al., *Community-Based Corrections.*

42. Covington, "A Woman's Journey Home."

43. C. Lindquist, J. Hardison, and P. K. Lattimore, *Reentry Courts Process Evaluation (Phase 1): Final Report* (Washington, DC: National Institute of Justice, April 2003).

44. S. Maruna, *Making Good: How Ex-convicts Reform and Rebuild Their Lives* (Washington, DC: American Psychological Association, 2001).

45. A. F. Marbley and R. Ferguson, "Responding to Prisoner Reentry, Recidivism, and Incarceration of Inmates of Color," *Journal of Black Studies* 35(2005): 633–649.

46. J. K. Liker, "Wage and Status Effects of Employment on Affective Well-Being among Ex-felons," *American Sociological Review* 47(1982): 264–283; P. H. Rossi, R. A. Berk, and K. J. Lenihan, *Money, Work, and Crime: Experimental Evidence* New York: Academic Press, 1980).

47. For some positive findings, see B. Morrell, "Can Ex-offenders Ever Become Productive Citizens? Barriers to and Policies toward Reentry," *Criminal Justice Research Reports* 7(2006): 86–88. Also see R. P. Seiter and K. R. Kadela, "Prisoner Reentry: What Works, What Does Not, and What Is Promising," *Crime and Delinquency* 49(2003): 360–388. For less-encouraging findings, see C. A. Visher, L. Winerfield, and M. B. Coggeshall, "Ex-offender Employment Programs and Recidivism: A Meta-analysis," *Journal of Experimental Criminology* 1(2005): 295–315.

48. C. Farley and S. Hackman, *Ready4Work in Brief—Interim Outcomes Are In: Recidivism at Half the National Average* (Philadelphia: Public/Private Ventures, September 2006), www.ppv.org. Also see: L. Jucovy, *Just Out:*

Early Lessons from the Ready4Work Prisoner Reentry Initiative (Philadelphia: Public/Private Ventures, February 2006).

49. D. A. Zanis, F. Mulvaney, D. Coviello, A. I. Alterman, B. Savitz, and W. Thompson, "The Effectiveness of Early Parole to Substance Abuse Treatment Facilities on 24-Month Criminal Recidivism," *Journal of Drug Issues* 33(2003): 223–235.
50. J. A. Wilson and R. C. Davis, "Good Intentions Meet Hard Realities: An Evaluation of the Project Greenlight Reentry Program," *Criminology and Public Policy* 5(2006): 303–338.
51. S. Zhang, R. E. L. Roberts, and V. J. Callanan, "Preventing Parolees from Returning to Prison through Community-Based Reintegration," *Crime and Delinquency* 52(2006): 551–571.
52. P. J. Schram, B. A. Koons-Witt, F. P. Williams III, and M. D. McShane, "Supervision Strategies and Approaches for Female Parolees: Examining the Link between Unmet Needs and Parolee Outcome," *Crime and Delinquency* 52(2006): 450–471.
53. P. K. Lattimore, C. A. Visher, L. Winterfield, C. Lindquist, and S. Brumbaugh, "Implementation of Prisoner Reentry Programs: Findings from the Serious and Violent Offender Reentry Initiative Multi-site Evaluation," *Justice Research and Policy* 7(2005): 87–109.
54. See Petersilia, *When Prisoners Come Home,* for a good summary of restrictions.
55. P. M. Harris and K. S. Keller, "Ex-offenders Need Not Apply: The Criminal Background Check in Hiring Decisions," *Journal of Contemporary Criminal Justice* 21(2005): 6–30.
56. M. C. Kurlychek, R. Brame, and S. D. Bushway, "Scarlet Letters and Recidivism: Does an Old Criminal Record Predict Future Offending?" *Criminology and Public Policy* 5(2006): 483–504.
57. For excellent treatment of these issues, see Travis, *But They All Come Back*, chap. 6. Also see J. Travis and M. Waul (eds.), *Prisoners Once Removed: The Impact of Incarceration and Reentry on Children, Families, and Communities* (Washington, DC: Urban Institute Press, 2003), particularly Chapters 5–8 and 10.
58. C. E. Kubrin and E. A. Stewart, "Predicting Who Reoffends: The Neglected Role of Neighborhood Context in Recidivism Studies," *Criminology* 44(2006): 165–198.
59. P. Cromwell, L. F. Alarid, and R. V. del Carmen, *Community-Based Corrections*, 6th ed. (Belmont, CA: Thompson Wadsworth, 2005).
60. C. Kirby, "Relocating Sex Criminals Can Backfire," *Lexington Herald-Leader*, July 18, 2006, A1, A13.

61. Kirby, "Relocating Sex Criminals," A13.
62. Civic Research Institute, "Study of Wisconsin Sex Offenders Finds Extensive Community Notification Efforts Have No Effect on Recidivism," *Criminal Justice Research Reports* (September/October 2006): 11–12.
63. Office of the Press Secretary, "President Signs H.R. 4472, the Adam Walsh Child Protection and Safety Act of 2006" (fact sheet), July 27, 2006, www.whitehouse.gov/news/releases/2006/07/20060727-7.html.
64. S. Herman and C. Wasserman, "A Role for Victims in Offender Reentry," *Crime and Delinquency* 47(2001): 429.

Juvenile Justice and Corrections

14

Chapter Objectives

- Discuss the limitations of the legal definition of delinquency.
- Describe the major patterns of delinquency.
- Discuss the potential causes of delinquency.
- Use the risk-protection framework to discuss a youth's likelihood of delinquency.
- Describe the major juvenile justice reforms designed to control delinquency throughout history.
- Understand the principle of *parens patriae*.
- Identify two key differences between early juvenile courts and adult courts.
- Explain three ways in which contemporary juvenile courts differ from the earlier model.
- Understand the major stages through which a delinquency case proceeds.
- Describe correctional options available to juvenile courts.
- List the characteristics of effective correctional intervention.
- Discuss potential causes and solutions to disproportionate minority contact.
- Discuss the rationale for maintaining a separate juvenile court.

CASE STUDY

During the past six months, there have been several incidents in your county that involved juvenile offenders in serious crimes including rape, arson, and homicide. Newspapers are filled with scathing editorials that attack the juvenile justice system and challenge what are perceived as protective and lenient juvenile court policies. A civil demonstration was held in protest of a decision to retain one of the cases in juvenile court. As the juvenile court judge, you have a public relations nightmare on your hands. You have been asked to respond to the attacks during a city council meeting. As you are preparing your remarks, you ask yourself the following questions:

1. What is the basis for having a separate court for juvenile offenders?
2. What protections are juveniles afforded and why?
3. Is juvenile court effective in controlling delinquency?
4. How are juvenile offenders held accountable for their actions?

Introduction

As early as 1825, different philosophies have guided the treatment of adult and juvenile offenders. The inception of the first juvenile court in 1899 attested to the special status accorded to youthful offenders. Although since that time many juvenile justice reforms have come and gone, there is still a commitment to maintaining a separate justice system for youth. Arguments supporting this separate system have revolved around three key beliefs: Juvenile crime stems from youths' immaturity and lack of capacity to understand the long-term consequences of their actions; juvenile offenders are more malleable because they are not as entrenched in antisocial behavior as adult offenders are; and it is the duty of the state to protect youth from themselves and others and to guide them toward a prosocial pathway. Throughout time, these beliefs have been challenged, and the juvenile justice system, although still separate, has grown more similar to its adult counterpart. The purpose of this chapter is to provide students with an overview of juvenile justice philosophy and practice throughout history. Students will be introduced to the concept of delinquency and its potential causes, historical approaches to juvenile justice, and contemporary juvenile justice and correctional practices.

Defining Delinquency

For the purposes of this chapter, we adopt the legal definition of delinquency as set forth by Elrod and Ryder: "**Delinquency** consists of those behaviors that are prohibited by the family or juvenile code of the state and that subject minors to the jurisdiction of juvenile court."[1] There are two types of behaviors that are prohibited by family or juvenile codes: (1) criminal offenses, or behaviors that violate a local, state, or federal law and that would be a crime if committed by an adult (e.g., robbery or theft); and (2) **status offenses**, or behaviors that violate laws or regulations that apply only to children (e.g., curfew violation, truancy, beyond control). A **minor** is defined as "a youth at or below the upper age of the original jurisdiction in a State."[2] State legal codes establish upper age limits ranging from 15 to 17 years, with age 17 being the most common (38 states). Lower age limits for juvenile court jurisdiction range from 6 to 10 years; however, in 35 states no lower age limit is specified.[3]

Legal definitions of delinquency, like the one used here, suffer from two key problems. First, they exclude delinquent behavior that never comes to the attention of officials and, thus, provide a false characterization of the true nature and extent of delinquent behavior that occurs. Second, they cover an overly broad range of behaviors. The number of behaviors categorized as "status offenses" and the vague statutory definitions of these offenses allow for broad interpretations that could result in almost any youth being subjected to state sanctions. Recognizing this, the Uniform Juvenile Court Act of 1968 and the majority of state codes have opted for a more narrow definition of delinquency that focuses only on criminal offenses and excludes status offenses.[4] Although only criminal offenses are labeled as "delinquent" in most states, it should be noted that in all states both types of acts can trigger intervention by the juvenile court.

Another troubling aspect of the legal definition of delinquency has to do with the arbitrary age cut-offs for juvenile court jurisdiction. The age limitations for juvenile court jurisdiction are based on the legal concept of *mens rea*, or criminal intent (see Chapter 3). Youth are deemed less able to understand the difference between right and wrong, or to understand the consequences of their actions, and thus, are less able to formulate criminal intent. What accounts for the variation across states? Are youth in some states able to distinguish right from wrong at an earlier age?

In many cases, the age cut-offs for juvenile court jurisdiction are not in concert with age limitations on other "adult" behaviors. For example, in all states, 18-year-olds are deemed able to formulate criminal intent and be prosecuted in adult courts, but in most states they are not able to legally purchase or drink an alcoholic beverage. Moreover, the age cut-offs ignore variations in the social and psychological development of youth.[5] Major areas of adolescent development include cognitive, social, moral, physical, and competency development. Adolescent development is gradual and nonlinear, is rarely in lock-step agreement with chronological age, and cannot be assessed based on a single trait.[6] Thus, although a physically developed 14-year-old boy may look capable of formulating criminal intent, he may not have the cognitive capacities necessary to determine right from wrong or consider the consequences of his actions.

The Nature and Extent of Juvenile Crime and Victimization

In Chapter 2, we discussed data sources for criminal behavior among adults. In the following section, we provide a detailed discussion of the various data sources through which we learn about the extent of juvenile delinquency.

Measuring Delinquency

What do we know about juvenile crime? Before answering this question, it is important to consider the common methods for measuring crime. As you will recall from Chapter 2, the most common methods for measuring crime trends in the United States are the Uniform Crime Reports (UCR) and the National Crime Victimization Survey (NCVS). There are three other data sources that are particularly relevant to measuring the nature and extent of juvenile crime—self-report, panel, and cohort studies.

Self-report surveys, discussed in Chapter 2, examine a broader range of delinquent behavior than those captured in the UCR. Youths are asked to report their involvement in behaviors ranging from status offenses to serious crimes. As such, these studies uncover delinquent behavior that is not detected by or reported to the police and reveal different patterns of delinquency than those revealed in official statistics. A **panel study** is a special type of self-report study that uses a research design that follows a representative sample of youth (typically surveying them in early adolescence and following them into adulthood with repeated self-report measures) throughout the time period under study. By following the youth throughout adolescence, these designs allow researchers to determine both the onset and course of an individual's delinquent career. Additionally, panel designs allow the researcher to identify when a number of factors that are known to influence delinquency (e.g., gang membership, poverty, family disruption, delinquent peers) appear and disappear and thus unravel the causal time order of delinquency. A number of panel studies have been conducted and have provided tremendous insight into the development of theories of delinquency, including the identification of the **chronic offender**.

The Monitoring the Future Study

The **Monitoring the Future (MTF) Project** began in 1975 to (1) study changes in the attitudes, beliefs, and behaviors of young people in the United States and (2) monitor trends in substance use and abuse among adolescents and young adults. The MTF project is conducted by the University of Michigan Survey Research Center under a series of research grants from the National Institute of Drug Abuse. Self-report questionnaires are administered annually to approximately 50,000 randomly selected students selected by multistage random sampling from the 8th, 10th, and 12th grades in about 420 public and private schools. Biannual follow-up surveys are conducted by mail with selected respondents who have graduated from each senior class since 1976. Results from this study are available at www.monitoringthefuture.org/. These results are widely reported and regularly used in the White House Strategy on Drug Abuse.

Youth Risk Behavior Surveillance System

The **Youth Risk Behavior Surveillance System (YRBSS)** was created by the Centers for Disease Control and Prevention in 1992 to examine categories of behavior that contribute substantially to the leading causes of death among people in the United States. These six categories include (1) tobacco use, (2) alcohol and other drug use, (3) sexual behaviors, (4) dietary behaviors, (5) behaviors that contribute to unintentional injuries and violence, and (6) physical inactivity.

The YRBSS collects data biennially from February to May in odd-numbered years (e.g., May 2005) from a nationally representative sample of approximately 15,000 high school students at 80 schools from 10 to 15 states. Results from this study are widely reported to describe unhealthy and risky behaviors among adolescents throughout the United States and are available at www.cdc.gov/HealthyYouth/yrbs/index.htm.

Cohort studies are similar to panel studies, but they follow a group of people who have something in common over a period of time. For example, the cohort studied by Wolfgang, Figlio, and Sellin (1972) included boys who were born in Philadelphia in 1945 and lived there until they were 18, and data from police, school, and court records were collected on the boys over a 10-year period.[7] Cohort studies are limited in number because of the expense associated with their methods. The primary strength of cohort studies is that they permit the study of patterns of behavior over the life course. Although each method for measuring delinquency has its own set of strengths and weaknesses, the methods are complementary. Taken together, they provide a relatively accurate picture of juvenile crime in the United States.

Patterns of Delinquency

Figure 14–1 reflects the number of juvenile arrests for property crimes per 100,000 juveniles in the population between 1980 and 2005. As depicted, juvenile arrest

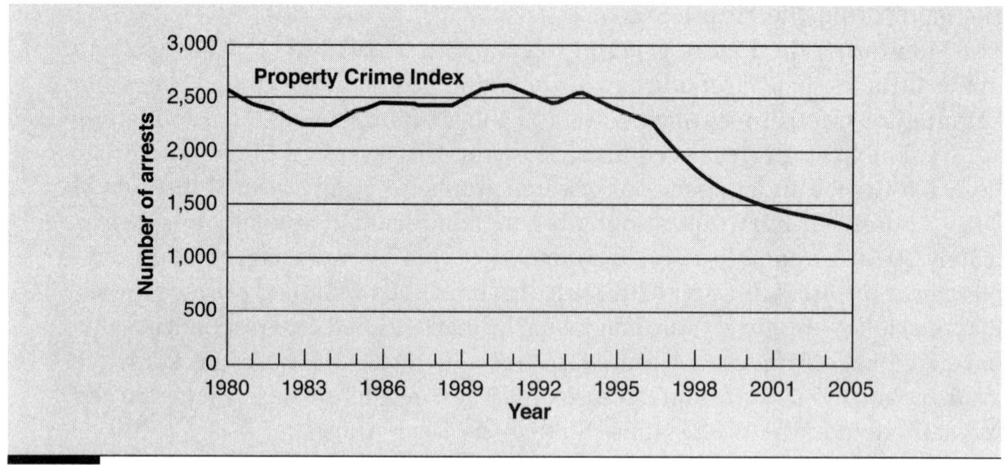

Figure 14–1 Number of arrests per 100,000 juveniles ages 10–17, 1980–2005.
Source: *Juvenile Arrest Rates for Property Crime Offenses. Office of Juvenile Justice and Delinquency Prevention, Statistical Briefing Book. Available electronically at: http://ojjdp.ncjrs.org/ojstatbb/crime/JAR_Display.asp?ID=qa05206.* Accessed October 10, 2007.

rates for property crimes remained relatively stable until 1994, at which time they began a steady decline. In 2005, the juvenile arrest rate for property crime reached an all-time low at 1,246 per 100,000 juveniles.[8]

Greater fluctuations were seen in juvenile arrests for violent crime during this time period with dramatic increases occurring between 1988 and 1994 (**Figure 14–2**). This peak in juvenile violence has been attributed to a number of social factors, including the breakdown of the family, declining morals, violence in the media, and

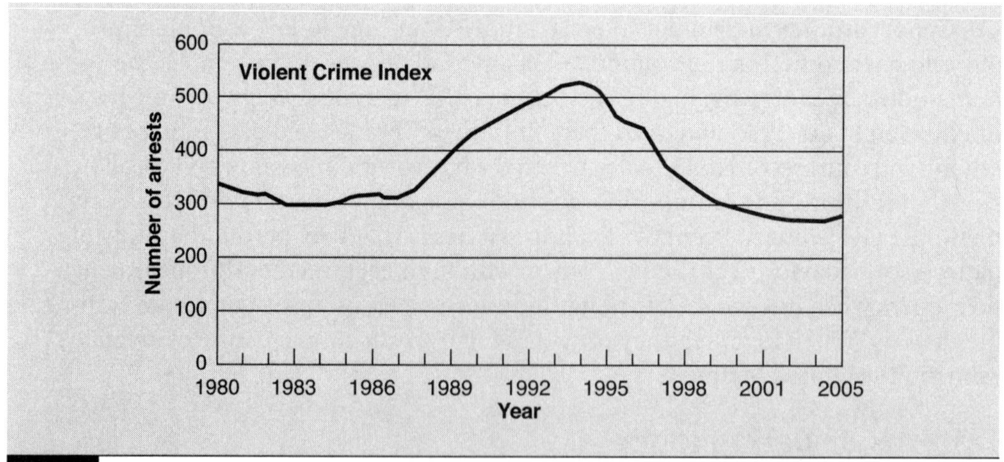

Figure 14–2 Number of arrests per 100,000 juveniles ages 10–17, 1980–2005.
Source: *Juvenile Arrest Rate Trends. Office of Juvenile Justice and Delinquency Prevention, Statistical Briefing Book. Available electronically at: http://ojjdp.ncjrs.org/ojstatbb/crime/JAR_Display.asp?ID=qa05201.* Accessed October 10, 2007.

the crack epidemic. A steady decline in juvenile violence was observed between 1994 and 2004 with the arrest rate for violent crime reaching its lowest level since 1980. In 2005, however, the arrest rate for violent crime increased to 283 per 100,000 juveniles, an increase of 5 percent since 2004.[9]

Although talks of juvenile crime waves may be overstated, it is a well-documented fact that, throughout history, juveniles have always committed a greater proportion of crime relative to their proportion in the general population.[10] In 2004, youth ages 10 to 17 years accounted for 11.4 percent of the general population and 15.8 percent of all arrests (15.5 percent of arrests for violent crime and 27.5 percent of arrests for property crimes).[11]

Official statistics, such as those cited previously, reflect the number and rates of juvenile arrests. They do not, however, provide information regarding the proportion of youth involved in delinquency. How widespread is delinquent behavior? According to many self-report studies, virtually every youth reports engaging in at least one delinquent act during his or her lifetime.[12] Most youths, however, report involvement in minor, less serious offenses that never come to the attention of officials. When considering more serious offenses, the proportion of youths involved is much lower. For example, 10 percent of respondents in one self-report study reported stealing from a store in the past year, and 11 percent reported getting in a serious fight.[13] Several self-report studies have revealed that the frequency and severity of delinquency tends to escalate in the late teens.[14] Most youths are not repeat offenders and are not involved in any type of serious delinquency. Studies show, however, that there is a small proportion of youths who engage in repeated delinquency. For example, two cohort studies revealed that approximately 6 percent of the youths studied were responsible for more than half of the offenses attributed to members of the cohort.[15]

All data sources suggest that the nature and extent of delinquency varies across race and gender (see **Table 14–1**).[16] Arrest data suggest that black youths are involved in a disproportionate amount of juvenile crime; although they accounted for only 16 percent of the population between the ages of 10 and 17 years in 2003, they accounted for 27 percent of all juvenile arrests. Self-report studies have revealed that, overall, delinquency is more proportionately distributed across races. These same studies and cohort studies, however, indicate that a higher proportion of black youths is involved in the types of serious delinquency captured by the UCR.[17]

In 2003, males between the ages of 10 and 17 years accounted for 71 percent of juvenile arrests.[18] According to official statistics, however, the rate and severity of female delinquency is on the rise. From 1980 to 2000, the juvenile arrest rate for females increased 35 percent.[19] Most of this increase was attributable to increases in arrests for assaults (see **Figure 14–3**).[20] When assaults were omitted from the violent crime index, the female delinquency trends were fairly stable; girls accounted for only 10 percent of violent juvenile crime.[21] A comprehensive

Table 14–1 Variations in Juvenile Arrest Rates Across Gender and Race

		Percent of Total Juvenile Arrests					
Most Serious Offense	2003 Juvenile Arrest Estimates	Female	Ages 16–17	White	Black	American Indian	Asian
Total	2,220,300	29%	68%	71%	27%	1%	2%
Violent Crime Index	92,300	18	67	53	45	1	1
Murder and nonnegligent manslaughter	1,130	9	89	49	48	1	2
Forcible rape	4,240	2	63	64	33	2	1
Robbery	25,440	9	75	35	63	0	2
Aggravated assault	61,490	24	64	59	38	1	1
Property Crime Index	463,300	32	63	69	28	1	2
Burglary	85,100	12	65	71	26	1	1
Larceny-theft	325,600	39	62	70	27	1	2
Motor vehicle theft	44,500	17	75	56	40	1	2
Arson	8,200	12	39	81	17	1	1
Other (simple) assault	241,900	32	57	61	36	1	1
Forgery and counterfeiting	4,700	35	87	77	20	1	2
Fraud	8,100	33	82	66	32	1	1
Embezzlement	1,200	40	94	68	30	0	2
Stolen property (buying, receiving, possessing)	24,300	15	73	57	41	1	1
Vandalism	107,700	14	56	80	18	1	1
Weapons (carrying, possessing, etc.)	39,200	11	64	66	32	1	2
Prostitution and commercialized vice	1,400	69	86	51	47	0	1

Sex offense (except forcible rape and prostitution)	18,300	9	49	71	26	1	1
Drug abuse violation	197,100	16	83	72	26	1	1
Gambling	1,700	2	85	12	86	0	2
Offenses against family and children	7,000	39	65	77	20	2	2
Driving under the influence	21,000	20	98	94	4	2	1
Liquor laws	136,900	35	90	92	4	3	1
Drunkenness	17,600	23	87	89	8	2	1
Disorderly conduct	193,000	31	59	64	34	1	1
Vagrancy	2,300	25	75	62	37	1	1
All other offenses (except traffic)	379,800	27	72	74	23	1	2
Suspicion	1,500	24	74	66	33	1	0
Curfew and loitering law violation	136,500	30	71	68	30	1	1
Runaway	123,600	59	64	73	20	2	5
U.S. population ages 10–17	33,499,000	49	24	78	16	1	4

Source: H. N. Snyder and M. Sickmund, *Juvenile Offenders and Victims: 2006 National Report* (Washington, DC: Office of Juvenile Justice and Delinquency Prevention, March 2006).

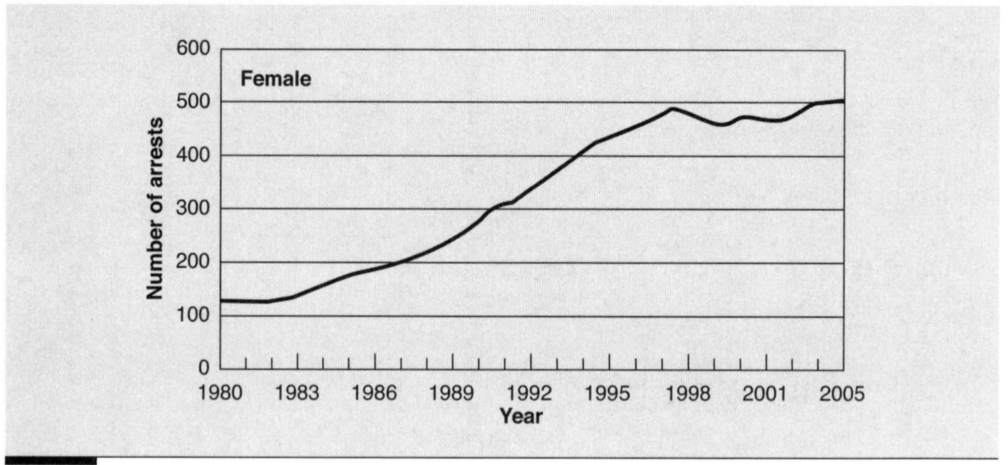

Figure 14–3 Arrests per 100,000 females ages 10–17 for simple assault, 1980–2005.
Source: *Juvenile Arrest Rates for Simple Assault by Sex, 1980–2005. Office of Juvenile Justice and Delinquency Prevention, Statistical Briefing Book. Available electronically at: http://ojjdp.ncjrs.org/ojstatbb/crime/JAR_Display.asp?ID=qa05241*. Accessed October 10, 2007.

analysis of official and self-report data on female delinquency led the Office of Juvenile Justice and Delinquency Prevention (OJJDP) Girls' Study Group to conclude that the increases in girls' violent delinquency reported in the UCR stem more from changes in the laws and the actions of officials rather than changes in the behavior of girls. Indeed, what were once considered normal fights between family members are now classified as assaults that attract formal police intervention and more frequently result in arrest.[22]

In sum, the available methods for measuring delinquency suggest the following:

1. Levels of juvenile crime have been relatively stable over the past 25 years, with the exception of a peak in violent crime between 1988 and 1994.
2. Juvenile arrests account for approximately 16 percent of all arrests.
3. Although delinquent behavior is widespread, most delinquency involves minor offenses that do not come to the attention of authorities.
4. A small percentage of youths (approximately 6%) are responsible for more than half of juvenile crime.
5. Black youths are disproportionately involved in serious crime.
6. Delinquency is primarily a male domain.

Juvenile Victimization

As discussed in Chapter 2, the NCVS collects data every six months from approximately 50,000 people 12 years of age and older regarding their experiences with criminal victimization. Researchers with the Bureau of Justice Statistics regularly compile reports focusing on criminal victimization among juveniles (except for homicide, where juvenile victims are those 17 years of age and younger, juveniles are defined as those between the ages of 12 and 18 as a result of the fact that the NCVS collects data only about those 12 years of age and older) based on those data. The most recent report available, *Juvenile Victimization and Offending, 1993–2003*, uses data from the NCVS and homicide data from the Federal Bureau of Investigation's Supplementary Homicide Reports contained in the UCR to describe 10-year trends in victimization among juveniles.[23]

Between 1993 and 2003, victimization for all nonfatal crimes measured (rape/sexual assault, robbery, aggravated assault, and simple assault) declined 54 percent, from about 130 victimizations per 1,000 teenagers in 1993 to about 60 in 2003. Nevertheless, juveniles 12 to 18 years of age were more than twice as likely as adults to be victimized by nonfatal violent crime (83.8 vs. 31.7 per 1,000), particularly for simple (57.3 per 1,000) and aggravated assault (15.5 per 1,000). These victimization rates were higher for males than for females; both white and black juveniles were equally likely to be victimized by violent crime. Urban youth were more likely than suburban and rural youth to be victimized by violent crime. More than two in three violent crimes against juveniles were committed with a weapon.

Although school was the most common location for violent victimization for both groups, the proportion of victimizations occurring at school was higher for youths 12 to 14 years old than for youths 15 to 17 years old (53.4% vs. 32.0%). After adjusting for the hours of exposure to criminal victimization, juveniles were most vulnerable to victimization during the hours immediately after school (3 to 6 P.M.). Older teens had higher levels of victimization at night than did younger teens.

Data sources reveal similar patterns in juvenile delinquency and victimization. That is, those individuals and groups more likely to be involved in delinquency are also at an increased risk of victimization. Thus, studying the causes of delinquency is a promising avenue for reducing victimization among juveniles.

Causes of Delinquency

What accounts for this widespread involvement in delinquency? Much of the behavior exhibited by adolescents is risky behavior (e.g., binge drinking, fighting, skipping school) that may be maladaptive in the larger social context but seems logical in adolescents' context.[24] Other adolescent behavior is more serious, however, and requires explanation that can guide intervention. This section of the chapter explores potential causes of delinquency through a brief discussion of four prominent theories and a description of a risk-protection framework that guides contemporary juvenile justice practice. It also discusses two factors that seem particularly prominent in the lives of delinquent youth: substance abuse and gangs.

Theories of Delinquency

A complete discussion of all of the theories of delinquency is beyond the scope of this text. Thus, this discussion is limited to four prominent theories of delinquency that have shaped juvenile justice research and practice: social learning, strain, social bond, and labeling theories.

Social learning theories have their roots in the Chicago school of criminology. The basic assumption is that criminal behavior is learned by observing the behavior of others. Edwin Sutherland introduced the theory of differential association in 1947. He asserted that the techniques and definitions (e.g., motives, rationalizations, attitudes) that support criminal behavior are learned through interaction that occurs within intimate personal groups (e.g., family, peer groups). The concept of "differential association" refers to the idea that youth are exposed to some people who have attitudes that support conformity and to some people whose attitudes support law-breaking behavior; a youth becomes delinquent because of an excess of definitions favorable to violation of law over definitions unfavorable to violation of law.

Later social learning theories put forth by Burgess and Akers (1966) and Akers (1996) build on Sutherland's theory by emphasizing the concept of modeling and incorporating the principles of operant conditioning from the field of learning psychology. These theorists assert that youth imitate the behavior of persons in their environment who are similar to them, likeable and respected, and rewarded for their behavior. Then, whether or not the behavior is repeated is determined by the balance of anticipated or actual rewards and punishments that follow a behavior; the behavior associated with the most net rewards is likely to persist.

Social learning theory has garnered substantial empirical support. For example, several studies have demonstrated that children imitate the behavior they view on television;[25] and having criminal parents and associating with delinquent peers are among the strongest correlates of crime.[26] Juvenile justice agencies frequently employ interventions aimed at increasing youths' exposure to positive role models by connecting them with prosocial peer groups or adult mentors. Also popular are cognitive intervention programs that teach youth problem-solving skills through the use of social learning techniques of modeling and performance feedback.

Strain theory represents one of the most dominant schools of criminological thought. The basic idea of strain theory, as set forth by Robert Merton, is that the inability to achieve desired goals through legitimate means motivates people to engage in crime. For example, Cohen argued that the source of strain for lower-class males was their inability to achieve middle-class status. They weren't socialized to delay gratification, to develop self-control, or to be courteous and respectful. Thus, these lower-class boys weren't successful in achieving "status" in schools governed by middle-class values. This led to "status frustration," or strain. As a way of coping with this status frustration, these lower-class boys joined together in a delinquent subculture and engaged in negativistic and nonutilitarian behaviors. In the context of the delinquent subculture, they could establish their own criteria for status and acceptance.

Robert Agnew extended traditional strain theory to include two other sources of strain: (1) the anticipated or actual removal of positively valued stimuli (e.g., loss of privileges, getting kicked off of an athletic team, breaking up with girl/boyfriend); and (2) the anticipated or actual presentation of negative stimuli (e.g., sexual or physical abuse, family conflict, curfew). Additionally, Agnew does not believe that delinquency is an automatic response to strain; instead, Agnew asserts that whether or not someone responds to strain with delinquency depends on his or her interpretation of the stressors, the ability to cope with adversity, and the level of social support available. For example, a youth who views the strain as deserved or fair will likely ignore it and move on, whereas a youth who views it as unfair and becomes angry will likely retaliate or rebel.

There is weak empirical support for the idea that people are pressured to deviate because of unachievable aspirations or desires. Some researchers report that people with high aspirations are driven to conformity and that it is low aspirations that are associated with deviance. There is more support for the link between delinquency and the types of strain identified by Agnew. The policy implications associated with Agnew's strain theory also are more relevant to contemporary juvenile justice practice. For example, the cognitive skills training programs that are so popular in correctional settings work with youth to help them identify their stressors and develop problem-solving skills that will help them respond in more prosocial, constructive ways.

Travis Hirschi's **social bond theory** is a derivative of control theory. Control theory assumes that everyone is equally motivated to engage in delinquency, and that the variation in social and personal controls is what explains why some people commit crime and others conform. For Hirschi, it is the strength of the social bond that controls whether or not youth engage in delinquency. Hirschi describes the social bond as a connection between youth and society that includes four elements:

- Attachment refers to the psychological and emotional connection one feels toward other persons or groups. The logic is that if a person does not care about the wishes and expectations of others, he or she is free to deviate.
- Commitment is demonstrated by the degree to which the adolescent has developed a stake in conformity by investing in certain activities, particularly school. The logic here is that whenever the individual considers deviant behavior, he or she will also consider the cost of the behavior, such as losing those stakes (i.e., being suspended from school, losing a job).

- Involvement refers to time spent in conventional activities such as studying or athletics. The more youth are involved in conventional activities, the less time there is for delinquency or other antisocial behaviors.
- Belief refers to belief in conventional values and norms. The more a person believes in the law, the more likely the person is to conform.

Studies have found general support for social bond theory. According to the research, attachment is the most important element of the social bond. Many contemporary interventions focus on improving communication and supervision within the family as a means of enhancing the attachment between the youths and their parents. Other practices attempt to involve youths in conventional activities such as sports or work.

Labeling theory, also known as societal reaction theory, was the leading theory of the 1960s when concerns about unfairness and discrimination in the juvenile justice system were at their peak. Rooted in symbolic interaction theory in the field of sociology, labeling theory emphasizes the possible negative consequences of labeling a youth as delinquent. The basic assumption is that society's negative reaction to initial deviance may alter youths' self-image and increase their likelihood of engaging in subsequent delinquency. It does this by promoting a self-fulfilling prophecy whereby youth begin to perceive themselves as delinquent and behave accordingly. Additionally, this new label of "delinquent" puts strains on interpersonal relationships and blocks youths' opportunities for engaging in conventional activities, both of which further elevate their likelihood of future delinquency. Another key concern of labeling theorists is that certain youth, by virtue of their more advantaged status within the community, are able to avoid labeling whereas poor, minority youth who lack the resources to hire a good attorney or obtain services outside of the juvenile justice system are most likely to be labeled delinquent. In other words, the nature and extent of juvenile justice processing can be affected by the social status characteristics of youth.

The results of studies that have tested the validity of labeling theory are mixed. Some studies have found no significant differences in the likelihood of subsequent delinquency for youth who are labeled delinquent and formally processed through the juvenile justice system, compared to youth who are able to avoid the label through more informal responses.[27] The hypotheses regarding the effects of status characteristics have also received little support; most studies show that legal factors play a more pronounced role in determining who and what gets labeled than do status characteristics. Despite this, the diversion programs that are rooted in labeling theory are a mainstay in most juvenile justice systems.

Risk-Protection Framework

A large body of research has been dedicated to identifying factors associated with delinquency. Findings from this research have contributed to a **risk-protection framework** that guides juvenile justice intervention. This framework includes three important concepts—risk factors, protective factors, and resiliency.

Risk factors are characteristics that increase youths' likelihood of engaging in delinquency. They fall within five major domains, including individual, family, school, peers, and community (see **Table 14–2**). Among the factors that have been shown to have the strongest relationship with delinquency are lack of

Table 14–2 Risk and Protective Factors

	Risk Factors	Protective Factors
Individual	• Antisocial attitudes • Rebelliousness	• Positive expectations for the future • High self-esteem • Intolerance toward deviance • Positive social orientation
Family	• Sexual/physical/emotional abuse • Conflict • Lack of supervision • Lack of discipline • Antisocial parents/sibling • Lack of parental involvement	• Good relationship with parents • Good family communication • High levels of parental supervision
School	• Poor academic performance • Underachievement • Lack of involvement in extracurricular activities • Poor relationships with teachers/students • Dislike for school	• Reading and math skills • Commitment to school • Strong school motivation • Doing well in school • Positive attitude toward school
Peer group	• Lack of friendship network • Antisocial friends • Victimization by peers	• Nondelinquent friends
Community	• Availability of drugs/firearms • High mobility • Economic deprivation • Lack of informal social control/collective efficacy	• Nondisadvantaged neighborhool • Low neighborhood crime • Prosocial community norms • Informal social control/collective efficacy

Sources: Hawkins, J. D. (1999). Preventing crime and violence through communities that care. *European Journal on Criminal Policy and Research, 7*(4), 443–458.

Huizinga, D. (1996). The influence of delinquent peers, gangs, and co-offending on violence. Fact Sheet. Washington, DC: Office of Juvenile Justice and Delinquency Prevention.

Thornberry, T. P., and Krohn, M. D. (2003). *Taking Stock of Delinquency: An Overview of Findings from Contemporary Longitudinal Studies*. New York, NY: Kluwer/Plenum Publisher.

Thornberry, T. P., Huizinga, D., Loeber, R. (2004). Causes and correlates studies: Findings and policy implications. *Juvenile Justice, 9*(1), 3–19.

supervision and discipline, family conflict, academic failure, lack of commitment to school, antisocial attitudes, having antisocial peers, and substance abuse.[28]

Protective factors are characteristics that decrease youths' likelihood of engaging in delinquency. They moderate the effects of adversity and help youths counteract risk factors.[29] Three categories of protective factors have been identified in the literature: (1) positive personality and social orientation; (2) warm and supportive relationships with family members or other adults; and (3) prosocial family and community norms.[30]

Resiliency refers to the capacity for successful adaptation to disruptive, stressful, or challenging circumstances.[31] There are many children who, despite their exposure to multiple risk factors, are able to beat the odds and grow into productive, law-abiding citizens.[32] The difference for these children typically lies in some factor in their lives, either internal or external, that makes them resilient and enables them to overcome adversity.

Programs with the best chance of preventing delinquency are those that try to reduce known risk factors and, simultaneously, implement strategies designed to enhance these identified protective factors.[33] Applying this risk-protection framework, many juvenile justice agencies use assessment practices that help them identify a youth's risk and protective factors upon intake. Then, individualized treatment plans are developed to address these factors and reduce the youth's likelihood of engaging in subsequent delinquency.

There is considerable congruence between the four delinquency theories that were previously discussed and the risk-protection framework. For example, the risk factor of family conflict undermines the parent–child attachment that is an important element of Hirschi's social bond; the presence of antisocial peers reflects the theoretical assumptions of social learning theory; and the positive coping skills that are found to be a source of resiliency are an important component of Agnew's strain theory. Knowledge of these theories and the factors associated with delinquency will help readers understand many of the contemporary juvenile justice practices that are discussed later in this chapter.

Substance Abuse

According to the 2004 National Survey on Drug Use and Health, 30 percent of youth ages 12 to 17 reported having used an illicit drug at least once during their lifetime, 21 percent reported drug use during the past year, and 10.6 percent of youth reported having used an illicit drug during the past month.[34] Among 12 to 17-year-olds surveyed, 17.6 percent reported using alcohol during the past month, and 11.1 reported being binge drinkers. Consequences of substance abuse among youth include poor school performance, absenteeism from school, cognitive problems, injury and disease, developmental lags, and alienation from peers.[35] There are numerous family and community consequences as well, including family crises and dysfunction, the high cost of medical and other treatment services that are needed to address the problem, and increased rates of delinquency.

Although it is not clear from research that substance abuse directly causes other kinds of delinquency, there is a strong correlation between the two.[36] Studies suggest that compared to the general population, substance abuse is more prevalent among youth involved in the juvenile justice system. Data from the Arrestee Drug Abuse Monitoring program revealed that, of the sample of juvenile detainees drug tested in 2000, 42 to 55 percent of males and 17 to 58 percent of females tested positive for at least one illicit drug.[37] No such estimates are available for alcohol use among juvenile offenders, but extrapolating from the data on illicit drug use, it is expected that the prevalence of alcohol use among juvenile offenders is quite high.

Recognition of the serious consequences of juvenile substance abuse has spawned the implementation of numerous prevention programs in schools and other community-based settings. Among the most popular approaches are the "just say no" campaign and the Drug Abuse Resistance Education program, neither of which has proved very successful in reducing substance abuse among youths.[38] The most effective approach to substance abuse prevention appears to be skills training through which youths are taught to recognize high-risk situations, cope with stress, resist peer pressure, and make positive choices.

In the juvenile justice system, a significant level of resources is allocated to the treatment of substance abuse. An increasing number of specialized juvenile drug courts have been implemented in the past 10 years. Youths on probation often are required to participate in some type of treatment program as a condition of their probation. Many probation departments provide in-house educational programs for groups of youths on their caseloads. Most commonly, however, youths are referred to local agencies for treatment ranging from standard outpatient treatment to long-term residential treatment. Sixty-six percent of the residential facilities responding to the 2002 Juvenile Residential Facility Census indicated that they provide on-site substance abuse services.[39] These services include assessment, treatment planning, and individual and group therapy.

Youth Gangs

There is plenty of evidence to suggest that the choices youths make about peer groups have a strong influence on whether they choose a delinquent or prosocial pathway. Studies revealing that 62 to 93 percent of delinquent acts occur in group settings[40] suggest that there is more to it than simply having delinquent peers. What is it about the group context that leads to delinquency?

Peer groups vary in their level of organization and commitment to the delinquent lifestyle. The typical size of a delinquent peer group is 2 to 4 youths; these youths are usually a subset of a larger peer group. The membership in most peer groups is unstable, with youths constantly shifting between peer groups. In these peer groups, it appears that the motivation for delinquency arises after the peer group is assembled and occurs as the result of "contagious reciprocity of behaviors and actions."[41] For example, one study found that 79 percent of the offenses self-reported by youth were conceived of within a peer group less than half an hour before they occurred.[42] At the other end of the spectrum, however, are more organized, cohesive youth gangs that assemble for the purpose of crime. It is these organized gangs that are of grave concern to juvenile justice professionals.

There is no agreed-upon definition of a gang. One of the most popular definitions is offered by Klein, who defines a **gang** as any identifiable group of youngsters who (1) are perceived by others as a distinct aggregation in their neighborhood, (2) recognize themselves, usually through a group name, as a denotable group, and (3) have been involved in a sufficient number of delinquent incidents that call attention to themselves.[43] According to the most recent National Youth Gang Survey, there are 21,600 youth gangs consisting of 731,500 gang members nationwide.[44] Every city with a population more than 250,000 reports gang activity, as do some smaller cities.

Studies have shown that gang membership intensifies delinquent behavior. For example, a study of Colorado youths found that although only 14 percent were reported gang members, they accounted for more than 70 percent of serious crime.[45] Other studies have reported that gang members commit violent offenses at a rate that is seven times higher than the rate of youths who are not gang members.[46]

Several factors have been found to increase a youth's likelihood of joining a gang.[47] Individual risk factors that contribute to gang involvement include a history of physical aggression, early involvement in delinquency and drug use, mental health problems, and a high prevalence of stressful life events. Family factors that contribute to gang membership include coming from a broken home, family poverty, poor family management, child abuse/neglect, and gang involvement of family members. Community factors that contribute to gang involvement include the availability of drugs, the presence of many youth who are in trouble, feeling unsafe in the neighborhood, low neighborhood attachment, a low level of neighborhood integration, poverty, and neighborhood disorganization.

Given these multiple risk factors, a multidimensional, comprehensive model of prevention, intervention, and suppression is needed to reduce gang activity.[48] School-based prevention programs inform youths of the dangers of gangs and help youths develop the problem-solving and social skills that are needed to resist gang involvement. Interventions with youths who are already involved with gangs focus on providing alternative leisure-time activities, legitimate employment opportunities, and mentoring relationships with prosocial role models. Last, intensive police surveillance, arrest, and prosecution are used to suppress gang activities in high-risk communities.

Corrections in the Real World

Jeremy is 16 years old and has just been placed on your probation caseload for possession of methamphetamine. He has one prior adjudication for stealing a bike when he was 14. Jeremy currently lives with his aunt; his mom is in prison on a similar drug charge and he never knew his father. He and his aunt appear to have a good relationship. She works nights as a nurse, and Jeremy spends most of the evenings alone. Jeremy attends school regularly and earns mostly Cs. He plays on his high school football team. Jeremy spends a lot of time with three friends, all of whom have been in trouble with the court. He reports drinking and smoking marijuana regularly. He stated that he does not use other drugs and that he was just holding the meth for a friend when he got pulled over for speeding. Jeremy likes to work on cars and after high school plans to go to a vocational school to study mechanics. Jeremy has been very cooperative with the court but does not seem to understand the serious nature of his charge. What is Jeremy's likelihood of future delinquency? What are some risk and protective factors that affect his likelihood of delinquency? How would you use this information to alter his likelihood of delinquency?

Controlling Delinquency Throughout History

How to control the behavior of the young has challenged generations across time and place. This section of the chapter highlights the key juvenile justice reforms that have been designed to control delinquency throughout history.

Model of Family Government

In the early years of colonization in the United States, the family was responsible for maintaining social control over children.[49] The father had absolute authority over all family affairs; wives and children were expected to abide by the father's wishes. The primary type of discipline was corporal punishment, and one early Massachusetts law prescribed the death penalty for children who disobeyed their parents. Additionally, laws held parents legally responsible for providing their children with moral education and for teaching them how to read and write.

This **model of family government** was dismantled by the Industrial Revolution at the end of the eighteenth century.[50] An increase in child labor and long hours at the factories weakened family ties. Families were displaced from their land and were without work. At the same time, Irish immigrants were pouring into the country. Wealthy Americans were concerned that this impoverished class was threatening their safety. New methods of social control were developed.

Houses of Refuge

Incarceration in jails and workhouses became the preferred mechanism of control for adults and children alike. Concerned about the negative influence of the adult offenders, members of the Society for the Prevention of Pauperism pushed for the separation of adult and juvenile offenders and established the first house of refuge in New York City in 1825.[51] **Houses of refuge** were schools of instruction characterized by education, labor, discipline, and a moral regimen. Because the focus was on prevention rather than punishment, houses of refuge were authorized to house not only delinquent children but also dependent and neglected children. Institutionalization was viewed as necessary for saving these children from pauperism and the evil influences of the city. Many of the youths were released to apprenticeships; others stayed in the houses of refuge until legal age of majority.

This unfettered power to intervene in the lives of children was first challenged in *Ex Parte Crouse* (1838). The father of a child placed in the Philadelphia House of Refuge by her mother for a noncriminal offense filed a writ of *habeas corpus* requesting her release. The state Supreme Court denied the motion, declaring that the rights of natural parents who are "unequal to the task of education, or unworthy of it, [should] be superceded by the *parens patriae*, or common guardian of the community."[52] Moreover, the court ruled that the youth's incarceration was justifiable because it was for the purpose of reformation and assistance rather than punishment. It was this case that established the principle of ***parens patriae*** as the basis for state intervention with youths.[53]

Placing Out

In the mid-1800s, concerns about overcrowding, reports of abuse, enforced silence, and hard labor in the institutions caught the attention of the Child Savers, a group of reformers whose mission was to prevent delinquency through a range of community-based programs.[54] Early initiatives of the Child Savers included the distribution of food and clothing, the provision of temporary shelters for homeless youths, and the establishment of playground programs in an attempt to reach children of the streets. Later, the practice of **placing out** was introduced as a way to rescue vagrant and poor youths from the instability of the cities. Introduced by Charles Loring Brace and the Children's Aid Society, this program involved transporting vagrant and poor children out West where they would be placed with farm families. These families, it was believed, could provide these youths with a warm and stable home environment. Critics of the practice argued that many of these children were exploited by the families with whom they were placed and rarely were accepted as part of the family.

Probation

Probation was another community-based initiative that was developed in the latter half of the eighteenth century as an alternative to incarceration. As pointed out in Chapter 8, John Augustus, a Boston bootmaker, introduced the idea of probation in 1841 when he suggested to a judge that a young boy be given the opportunity to avoid incarceration and prove himself worthy of remaining in the community. The judge approved the request and appointed Augustus as the overseer of the young man's progress. Augustus served in this capacity in more than 2,000 cases throughout his life. In 1869, the state of Massachusetts enacted a statute appointing official probation officers to gather information on youths and recommend the best course of action for state intervention. Many states soon followed suit and enacted statutes authorizing some type of juvenile probation.

Reform Schools

With a continued rise in delinquency in the late 1800s, government began to take over the administration of juvenile institutions. Newer institutions, called **reform schools**, were to emphasize formal schooling and reformation for delinquent, dependent, and neglected youths. By 1890, almost every state outside the South had a reform school. In contrast to the harsh institutional environment of the houses of refuge, youths were housed as small groups in cottages designed to emulate family living. As with earlier efforts at institutionalization, the reform school became overcrowded and dangerous. The focus on reformation gave way to custodial concerns, abuse, and exploitation through hard labor.

The First Juvenile Court

The state's right to intervene in the lives of children who had committed no criminal offense was again challenged in *People v. Turner* (1870). This case involved a youth, Daniel O'Conell, who was committed to Chicago House of Refuge against his parents' wishes because he was perceived to be in danger of becoming a pauper. In response to an appeal filed by the parents, the Illinois Supreme Court ruled that Daniel was being punished in the House of Refuge; as such, due process protections were necessary.

This ruling, and other similar rulings, erected barriers to what was deemed by the Child Savers as necessary intervention with delinquent, unruly, and impoverished youth. It also raised concerns regarding the court's leniency with youth. A campaign ensued that pushed for juvenile justice reform and led to the creation of the Illinois Juvenile Court Act of 1899 authorizing the first separate court for juveniles. The court was established based on the principle of *parens patriae*. Under this principle, the state would assume guardianship for youths whose parents were deemed unfit. By 1925, all but two states had established separate juvenile courts.

The new juvenile courts differed from adult courts in two key ways. First, whereas adult courts were interested in justice for the victim and punishment for the offender, juvenile courts were designed to rehabilitate youthful offenders; that is, their overriding purpose was to provide guidance and assistance to children to help them become law-abiding citizens. Second, in contrast to adult courts, juvenile courts were characterized by informal courtroom procedures and greater discretion. Because the court was designed to serve in the best interests of the children, formal due process protections were deemed unnecessary and seen as possible obstructions to individualized treatment. Attorneys for the state or the youths were not considered essential to the courts' operation.

A review of juvenile justice reforms leads many to question whether they were motivated by benevolence or a desire for control.[55] Many of the reforms were spearheaded by the Progressive reformers who were known for their humanitarian concerns. Many historians, however, view juvenile justice reforms as a means of controlling social unrest and protecting the class privileges of wealthy do-gooders. By creating a separate juvenile court dedicated to helping rather than punishing, the state of Illinois was able to sidestep due process protections erected by *People v. Turner* (1870) and resume the institutionalization of youth for noncriminal behavior.

Contemporary Juvenile Courts

Contemporary juvenile courts differ from the earlier model in three key ways. First, juveniles are now accorded many of the due process rights granted to adult offenders. Between 1966 and 1971, the rulings in several Supreme Court cases expanded the due process rights of juvenile offenders (**Figure 14–4**). Prior to this, the court had ruled in favor of fewer due process rights for juveniles based on the idea that youths were receiving a "compensating benefit" under the principle of *parens patriae*.[56] The rulings in *Kent v. United States* (1966), *In re Gault* (1967), and *In re Winship* (1970) questioned the extent to which this compensating benefit existed; they rejected the argument that youths were being protected rather than punished and, accordingly, afforded juveniles many of the due process rights granted to adults. The Supreme Court drew the line, however, in *McKeiver v. Pennsylvania* (1971) when it ruled against granting juveniles the right to a trial by jury. In this ruling, the Court argued that the adversarial nature of jury trials would destroy the "traditional character of juvenile proceedings" and provide no guarantee of increased accuracy in the juvenile court process.

Second, congressional acts have placed some restrictions on the use of incarceration for youth.[57] In 1968, Congress passed the Juvenile Delinquency Prevention and Control Act that recommends that status offenders be handled outside of the juvenile court system. In later acts (Juvenile Justice and Delinquency

U.S. Supreme Court Cases

Kent v. United States (1966)
Courts must provide the "essentials of due process" in transferring juveniles to the adult system.

In re Gault (1967)
In hearings that could result in commitment to an institution, juveniles have four basic constitutional rights.

In re Winship (1970)
In delinquency matters, the state must prove its case beyond a reasonable doubt.

McKeiver v. Pennsylvania (1971)
Jury trials are not constitutionally required in juvenile court hearings.

Figure 14–4 Landmark cases expand rights to juveniles.
Source: *Reproduced from H. N. Snyder and M. Sickmund,* Juvenile Offenders and Victims: 2006 National Report *(Washington, DC: U.S. Department of Justice, Office of Juvenile Justice and Delinquency Prevention, 2006).*

Prevention Act of 1974, 1980, 2002), the deinstitutionalization of status offenders and nonoffenders was required for participation in the Formula Grants Program that allocated funds to states for the development of community-based programs. The 1974 act required that youths be separated by sight and sound from adult offenders, and the 1980 act required the removal of youths from adult jails and lockups.

Third, in response to the increase in juvenile crime between 1988 and 1994, many states have enacted laws and other policy changes making their juvenile justice systems more punitive.[58] Forty-five states either expanded the type of crimes or lowered the age for which youth can be transferred to the adult system; 31 states have expanded sentencing options available to juvenile courts; and 47 states have removed confidentiality requirements that were designed to protect youths and made juvenile records and proceedings more open to the public. Additionally, the missions of juvenile courts have shifted from a sole focus on "serving the best interest of the youth" to a more balanced approach designed to provide more rights to victims of juvenile crime and hold offenders more accountable for their behaviors.[59]

Processing Delinquency Cases in Juvenile Court

There is great variation in the operation of juvenile courts across the nation. Moreover, within each juvenile court, the manner in which a case is handled can vary from case to case. **Figure 14–5** presents a simplified version of the major

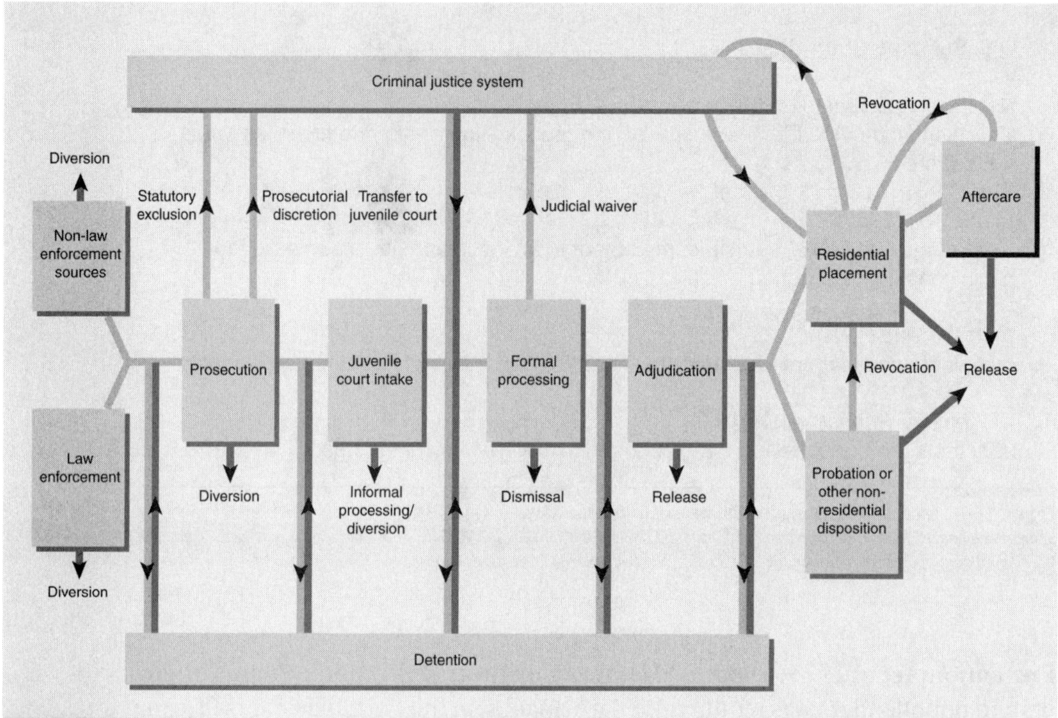

Figure 14–5 Simplified view of case flow through the juvenile justice system.
Source: *Reproduced from H. N. Snyder and M. Sickmund,* Juvenile Offenders and Victims: 2006 National Report *(Washington, DC: Office of Juvenile Justice and Delinquency Prevention, March 2006).*

stages through which a delinquency case proceeds. In this section of the chapter, the typical case proceedings are described, followed by discussions of detention, diversion, and transfer to adult court, occurrences that can happen at many points along the way.

Typical Case Proceedings

The police are the primary gatekeepers to the juvenile justice system.[60] In 2004, 81 percent of the cases referred to juvenile court were the result of an arrest or other encounter with law enforcement.[61] Once a youth is taken into custody, a **preliminary screening** is conducted to ascertain the most appropriate resolution to the situation. This screening generally involves a review of the youth's past record and discussions with the youth, parents, and victim. The police have several options available, including releasing the youth on the spot or to parents with a warning only; releasing youth to parents with a referral to other community resources; filing formal charges and releasing the youth to parents; or filing formal charges and detaining the youth. If the youth is to be detained, a **detention**

hearing must be held within 24 hours to determine if the youth poses a threat to the community or self and the likelihood of the youth appearing in court. About one in five cases is detained.

Next, **intake screening** is conducted to determine the most appropriate way to handle a case. A juvenile probation officer, or some other designated officer of the juvenile court, is typically responsible for intake screening. The factors considered by this officer include type and seriousness of the offense, harm to victim, age, legal history, parental response to offense, school performance, and the youth's reaction to the incident. Delinquency cases also require prosecutorial review. The role of the prosecutor is to establish jurisdictional authority over the case (e.g., residence, age, location of offense) and to determine if there is probable cause for filing charges. At this point, the prosecutor may decide to transfer the youth to adult court based on either statutory requirements for automatic transfer or prosecutorial discretion.

Upon conclusion of the intake process, a determination is made as to whether the case will be formally or informally handled. In 2004, 56 percent of the juvenile cases resulted in a petition requesting formal intervention, an increase of 80 percent since 1985.[62] In 1 percent of the cases in 2002, the judge waived the case to the adult court. Although no petition was filed in the other cases, only a portion of them was dismissed outright. The others were referred to some other community service or diversion program (i.e., a program that addresses the youth's needs but allows the youth to avoid official processing through the juvenile justice system) or to a sanction such as community service, restitution, or informal probation.

Adjudication is the next stage in the process for those cases in which a **petition** for formal juvenile court intervention is requested. An **adjudicatory hearing** is similar to the trial in adult courts; facts of the case are reviewed and in some cases disputed by the defense counsel. In contrast to adult courts, however, judges rather than juries hear the cases and determine the outcomes. As in adult courts, youths can enter a **guilty plea** at this point in the process. In this case, having agreed to the facts outlined in the petition, they waive their rights to an adjudicatory hearing. The end result of a guilty plea or an adjudicatory hearing is either an adjudication or an acquittal. An adjudication is similar to a conviction in adult court and means that a judge has determined that the facts outlined in the petition are supported by the evidence. An **acquittal** means that the judge determined that the facts in the petition are not supported, and the petition is dismissed. Of the petitioned cases, approximately 67 percent result in an adjudication.

Once adjudicated, a youth appears in juvenile court for a **dispositional hearing**. During this hearing, a decision is made as to the most appropriate course of action for the youth. The decision is based on a **predisposition report** prepared by a probation officer that includes a comprehensive review of the youth's behavioral and social history and a recommendation. This report is much like the presentence report in adult court (see Chapter 8). The judge has a variety of options available ranging from release to residential placement. In 2004, of the cases resulting in adjudication, 15 percent received a sanction such as community service or restitution; 63 percent were placed on probation; and 22 percent were ordered into residential placement.[63] Whereas sentences in adult courts are typically imposed for the purpose of punishment, juvenile court dispositions are typically more focused on promoting rehabilitation and serving the best interests of the child.

Juvenile Detention

Until the 1980s, it was common practice to book youths accused of delinquent acts into local adult jails. Although this practice continues in some rural areas of the country, it has been greatly curtailed by the Juvenile Justice and Delinquency Act of 1974. This act requires that "juveniles alleged to be or found to be delinquent shall not be detained or confined in any institution in which they have contact with adult inmates."[64]

Amendments to the act in 1980 specify that youths be removed from adult jails and lockup facilities, with the exception of a six-hour grace period while awaiting transfer to a juvenile facility or while awaiting a court hearing. It also required that youths be separated from the adult inmates by sight and sound when housed in a facility that also houses adults. Because of these requirements, many jurisdictions built separate **juvenile detention** facilities. These facilities are secure facilities that house youths prior to and following adjudication for delinquent acts.[65]

A juvenile offender may be held in detention during the court process when deemed necessary to ensure his or her appearance in court, protect the community, or protect the juvenile.[66] In 2004, 21 percent of juveniles were placed in detention at some point prior to their case dispositions; this represents an increase in the use of detention of 47 percent since 1985.[67]

Youths also may be placed in juvenile detention facilities as a short-term post-adjudication sanction. Courts try to avoid long-term placement in juvenile detention facilities; the facilities are intended to be temporary holding facilities and typically have few services available to the youths.[68]

Juvenile Diversion

The labeling theory, as described earlier, contributed to a proliferation of diversion programs. Some argue that the minute a youth has any formal contact (e.g., with the police), the potential for labeling exists. Thus, true diversion occurs when police caution and release youth with no report, no counseling, and no sanction. Edwin Schur refers to this type of diversion as "radical nonintervention."[69] The do-nothing approach is based on the argument that delinquent behavior is a normal response to adolescent development and experiences.

A more accurate and encompassing definition of juvenile **diversion** is any action that minimizes further penetration into the juvenile justice system.[70] Using this definition, diversion can occur at any stage, including arrest, intake, adjudication, or disposition. The primary purposes of diversion programs are to avoid or minimize the consequences of labeling and reduce the workload and cost of the juvenile court and related components of the juvenile justice system. Some of the most common forms of juvenile diversion include these:

- *Community-based case management and counseling*—Youths are referred to a social service or mental health agency for assessment and counseling. These programs typically include a range of services, including individual, group, and family counseling for a range of problem areas. Additionally, youths and their families may be referred to other community-based programs that can address a specific need (e.g., employment training).
- *Teen court*—Rather than participating in the formal juvenile court process, youths are referred to a teen court where local youths assume the roles of attorneys, jurors, and judges who review the facts of the case and choose an appropriate sanction. Common teen court sanctions include a letter of apology, restitution, and community service (see Chapter 5).
- *Drug court*—Many youths are diverted from the traditional juvenile court to specialized drug courts that hear only the cases of drug-involved youth. The court, attorneys, and local treatment agencies work collaboratively to address the treatment needs of youths. Youths appear frequently before the drug court judge and receive immediate rewards or sanctions for their progress.
- *Victim–offender mediation*—These programs, rooted in a restorative justice philosophy, attempt to repair the harm caused by crime by bringing together the youth, victim, and members of the community for the purpose of mediation and conflict resolution. Youths are frequently required to apologize and pay restitution to the victim.
- *Alternatives to detention*—Instead of pre- or postadjudication detention, youths are placed on house arrest or electronic monitoring in the community.

Many studies question the extent to which diversion programs are meeting their goals. In most cases, youths are given the opportunity to participate in a diversion program in lieu of formal adjudication. If youths do not agree to participate, or agree initially but fail to follow through with the program requirements, they are referred for formal processing. Some argue that because of this coercive nature, diversion programs can be just as stigmatizing as formal juvenile court intervention. Two studies confirm this assertion, having found no significant differences between the self-concepts of youths involved in the juvenile justice system and those who participated in a diversion program.[71]

Other studies have reported that, because of net widening, diversion programs have diminished capacity to reduce the cost of delinquency intervention.[72]

Additionally, many diversion programs are more costly than are formal juvenile justice programs.

The true test, of course, is whether or not diversion programs are successful in reducing the subsequent delinquency of youth. Studies comparing the recidivism rates of diversion participants with a matched group of youths involved in another type of juvenile justice program have produced mixed results, with most of the studies reporting no significant differences between groups.[73] Palmer and Lewis (1980) argue that despite these discouraging results, some youths have benefited from diversion programs; the key seems to be in matching the right youth to the right programs.

Transfer to Criminal Court

As mentioned previously, most states have enacted laws making it easier to transfer youths to adult courts. These laws are based on the belief that more punitive responses than those imposed in juvenile court are needed to deter and rehabilitate the new breed of violent juvenile offenders. The number of cases transferred to criminal court increased by 6,000 from 1985 to 1994. Since that time, however, the number of cases waived to adult court steadily declined, and by 2002 it had declined by 46 percent.

Three mechanisms are used to transfer youths to adult courts:

- *Judicial waivers*—The juvenile court judge transfers the case to criminal court after considering the merits of the case and determining that the youth is no longer amenable to rehabilitation.
- *Prosecutorial waivers*—Based on the gravity of the offense, the prosecutor decides to file the case in criminal court.
- *Legislative exclusions*—State statutes exclude certain classes of crimes, usually serious and violent offenses, from juvenile court jurisdiction. These statutes also specify the age at which youths can be transferred to adult court for each type of offense.

Blended sentencing is a fourth option that has been introduced in 20 states. Although the models of blended sentencing vary from state to state, the basic idea is that youths receive both juvenile and adult sentences with the adult sentence being enacted only when the youth fails to meet the conditions of the juvenile sentence.[74]

Available research challenges the wisdom of juvenile waivers in three key ways. First, transfer to adult court does not ensure more certain and severe sanctions. In about half of all transfer cases, youth received sentences comparable to or more lenient than what they would have received in juvenile court.[75] Second, juveniles who are transferred to criminal courts are more likely to reoffend.[76] Moreover, they have been found to reoffend more quickly and with more serious offenses. Third, youths incarcerated in adult prisons experience higher rates of victimization than do youths incarcerated in juvenile facilities.

Juvenile Corrections

At the dispositional hearing, a judge must determine the most appropriate setting for the youth's treatment to occur. In addition to the predispositional report, juvenile courts are increasingly using **risk assessment** (see Chapter 8) to help guide this decision. The purpose of risk assessment is to estimate the likelihood of the youth engaging in subsequent delinquency. Standardized tools that include the risk factors known to be associated with delinquency are used for this purpose. The youth receives a score and is classified as low, medium, or high risk of delinquency. The court uses this information to apply the least restrictive sanction necessary to treat and control the youth's behavior.

The dispositional options available to juvenile courts are many and varied. The National Council of Juvenile and Family Court Judges recommends that all jurisdictions have the following options available:

- Probation
- Restitution
- Community service
- Mental health, substance abuse, and sex offender evaluation and treatment services in both community-based and residential settings
- Educational interventions including alternative educational environments
- Day and evening treatment centers
- Wrap-around and coordinated case management services for multiple youth needs
- Placement resources (e.g., foster care, group homes, secure facilities)[77]

Although the majority of adjudicated delinquents were placed in some type of community-based option, 22 percent were placed in some type of residential care in 2004.[78] The most common correctional options for juveniles are described in the following subsections.

Probation

Probation is the conditional release of youths into the community where they are provided with supervision and treatment. It is the most common form of disposition for juvenile offenders; approximately 55 percent of adjudicated youth nationwide are placed on probation.[79] Juvenile probation operates in much the same way as adult probation: Youths are required to abide by conditions designed to control and treat their behavior, the intensity of probation varies based on the risk level of the youth, and probation officers serve investigative, supervisory, and service provider functions. Key differences include smaller caseload sizes (e.g., 41 juveniles vs. 114 adults), a stronger focus on the individualized treatment of youth, and greater collaboration with schools, families, and other community contexts in which the juvenile lives.

The probation experience can vary from youth to youth. Low-risk youth may be placed under a minimum level of supervision that requires only that they visit with a probation officer once a month to review behavior and progress. In contrast, higher-risk youths are required to meet more frequently with their probation officer and participate in various types of intervention programs. Additionally, these youths may be subjected to various types of surveillance including drug testing, curfew checks, and electronic monitoring. Many youths are required to pay restitution or perform some type of community service as a condition of their probation.

Probation officers work with the youths and their families to develop individualized treatment plans that are designed to reduce the youths' likelihood of subsequent delinquency. These plans outline specific goals and tasks to be completed as part of the youths' probation. Common treatment plans require youths to participate in substance abuse treatment, mental health counseling, and educational programs. Some intervention programs are provided in-house by probation officers; more often, youths are referred to local service providers. In some jurisdictions, youths with multiple needs are referred to case management or wrap-around services that work collaboratively with the probation department, schools, and other community-based services in an effort to surround the youths with the type of treatment services and support needed to encourage positive behavioral change.

Youths who fail to comply with their treatment plan or with other rules and regulations of probation can be charged with technical violations and returned to juvenile court for contempt. If the youth is found to be in contempt, the judge can continue the youth on probation (perhaps with stricter conditions) or revoke the youth's probation and impose another, more severe sanction. Youths continued on probation after being found in contempt are often placed in a higher level of supervision and may be ordered to serve a brief term in detention.

Because of the variation in probation both within and across jurisdictions, it is difficult to gauge its effectiveness. Most research suggests that probation is at least as effective at reducing delinquency as is the institutional treatment of youths.[80] Given this, and the lower costs of probation, it remains an appealing correctional option for juvenile courts.

Day Treatment Centers

Day treatment centers have emerged as an alternative to incarceration for high-risk youths and are often used in conjunction with probation. These centers are similar to day reporting centers in the adult system (see Chapter 9) in that they are designed to provide treatment and supervision in one common setting. Day treatment centers typically hold both day and evening hours. Youths who have been suspended or expelled from school may be required to spend the school day at the center where they receive a range of services, including educational services, substance abuse treatment, individual and group counseling, and life skills training. These centers also offer after-school programs for youths who are attending school but who require additional structure and services in the evening hours. Day treatment centers can be used as a stand-alone sanction or as a condition of probation. They are typically staffed by a combination of educational, counseling, and court personnel.

Foster Care

Foster care is a form of out-of-home placement in which youths are removed from their parents and temporarily placed with foster parents who, after being licensed, are paid a daily rate by the state to provide care for youths. Foster care is designed to keep the youth in a family-like setting within the community. It is typically used in cases where there is documented evidence of abuse or neglect. It is also used for delinquent youths who are unable to achieve their treatment goals in the context of an unstable and sometimes chaotic family life.

A more recent form of foster care, **therapeutic foster care** (TFC; also called treatment foster care), was introduced in the mid-1980s as an alternative for youths who had treatment needs so extensive as to require institutionalization.[81] TFC is a highly structured program. Foster parents must participate in preservice training on youths and adolescent development and behavioral modification techniques. They also receive ongoing monitoring and daily consultation with a case manager. The foster parents serve as members of a treatment team working with schools, juvenile courts, and other social service agencies to provide comprehensive services and support to the youths and their families.

The benefits of TFC include allowing youths to stay in the community and attend local schools, providing more individualized treatment, and minimizing their exposure to the antisocial peers that they would encounter if institutionalized.[82] Studies of the TFC model developed by the Oregon Social Learning Center reveal positive outcomes when compared to comparison groups in traditional foster care or other community-based programs. These outcomes included a reduction in the number of days that the youth was incarcerated,[83] fewer arrests, and higher rates of program completion.[84] A recent review of studies on TFC reveals significant reductions in violence among chronic delinquents targeted for program participation.[85]

Group Homes

Group homes are similar to adult halfway houses (see Chapter 9). They are small, nonsecure, community-based settings where juvenile offenders are placed as an alternative to incarceration or as a step-down program after a period of incarceration. According to the most recent Juvenile Residential Facility Census, 38 percent of the facilities that hold juvenile offenders identify themselves as group homes, and approximately 12 percent of the youths in residential placement are in group homes.[86] Group homes are typically small, family-like settings; 59 percent of group homes house 1 to 10 residents and most (82%) are privately run.[87]

Group home residents are usually required to attend school or work. They receive a variety of services, including individual and group counseling, life skills training, and educational support. Although some of these services are provided in-house, youths are typically referred to services in the community.

Long-standing problems associated with group homes include the inadequate pay and benefits for staff and a lack of staff training and support.[88] Very little research is available on the effectiveness of group homes in reducing recidivism.

Boot camps

Following the lead of the adult system, juvenile corrections agencies have developed boot camps nationwide.[89] As with the adult models, juvenile **boot camps** are designed for first-time, nonviolent offenders and emphasize strict regimens, military drill, and hard labor. Despite the widespread appeal of boot camps, there is considerable debate about their merits.[90] Proponents of juvenile boot camps argue that the military model provides the structure and control needed for a safe environment, builds camaraderie and discipline, and fosters respect for staff. Opponents argue that the military model undermines the type of therapeutic relationships and atmosphere needed for positive development, fails to address youths' individual needs, and creates an environment that is antithetical to youths' home environment. Surveys of staff and juveniles from 27 boot camps and 22 traditional residential facilities reveal that perceptions regarding boot camps were more favorable except that the juveniles in boot camps more frequently reported perceptions of danger from staff.[91] This same study found that when compared to youths in traditional residential facilities, boot camp participants experienced a greater reduction in anxiety and reported decreased impulsivity and increased prosocial attitudes. Despite these positive results, outcome studies have found no significant differences in the recidivism rates of boot camp participants and control groups of offenders in other correctional programs.[92]

Secure Care Facilities

The most restrictive correctional options for youth are **secure care facilities**, which are the option most similar to adult prisons. These institutions also are commonly called "training schools," reflecting their emphasis on education and other types of rehabilitation. According to the most recent Juvenile Residential Facility Census, there are 389 of these facilities across the nation.[93] The type and level of security vary across facilities. Many are "staff secure" facilities in which security is accomplished by the staff through close monitoring of youth and visitors. Others have a variety of security options, including locked sleeping rooms for youths, locked doors or gates, and perimeter razor wire.

The more-secure facilities are reserved for the most serious and chronic juvenile offenders. Approximately 35 percent of juveniles committed to secure facilities have been adjudicated for a violent offense.[94] Most have a prior history of delinquency and have undergone various types of community-based intervention before being incarcerated. The median time spent by juveniles in state-level secure facilities is 113 days.[95] The Juvenile Justice and Delinquency Prevention Act of 1974 called for the deinstitutionalization of status offenders; statistics suggest that this has largely been achieved with only 2 percent of the juveniles in state facilities having been incarcerated for a status offense in 2003.[96]

Secure care facilities share many common features. They are characterized by a high level of structure throughout the day. The typical schedule requires youths to participate in a variety of planned activities from 6:00 A.M. until 9:00 P.M. Youths enjoy about one hour of free time per day. Common activities include educational programming, individual and group counseling, and recreational activities. Although educational programs are more individualized in institutions, most of the youths' day is spent in a classroom setting. Additionally, a range of interventions is provided to youths to address their individual risks and needs. Most institutions have established mental health, substance abuse, and sex offender treatment programs. Some institutions specialize in one type of intervention.

Phase systems and token economies are common behavioral modification techniques used to encourage participation and compliance among the youths. Youths typically enter into the facility on an orientation phase during which they learn the rules and get acclimated to the institution. After they achieve the requirements of this phase, they can progress through the next phases that involve more responsibility and more privileges. Youths who violate the rules can be moved back a phase as a consequence of their behavior. In token economies, youths earn "tokens" or points for accomplishing specific tasks (e.g., maintaining good hygiene, cleaning their room). These points can then be exchanged for some type of reward (e.g., an extra privilege, a snack). Although these types of behavioral modification systems are associated with increased compliance and lower rates of institutional misconduct, they do little to promote long-term behavioral change after a youth leaves the institution.[97]

Because of the variation in recidivism measures and follow-up periods, it is difficult to estimate a national rate of recidivism for youths released from correctional institutions. Three states using the measure "any new arrest" reported recidivism rates of 55 percent; 8 states using the measure "any new conviction/adjudication" reported rates of 33 percent.[98] Many juvenile justice professionals attribute high rates of recidivism to a lack of transitional or aftercare services for youths upon their release from the institution.[99]

Aftercare

The period of transition from a highly structured institutional program back to the community has been characterized as disorienting, alienating, and stressful.[100]

Aftercare services help youths navigate this transition by providing a range of services and supervision. Effective aftercare begins as the youth enters the institution and involves formal (e.g., juvenile court, schools) and informal (e.g., family, neighbors) networks of social control in prerelease planning and postrelease supervision.[101] Aftercare stands in contrast to traditional practices that compartmentalize agencies' work with juvenile offenders and instead promotes interagency collaboration among all of the agencies that work with youths.

One of the most prominent aftercare models is the Intensive Aftercare Program (IAP) designed for high-risk juvenile offenders. This model includes "three distinct, yet overlapping segments: (1) prerelease and preparatory planning during incarceration; (2) structured transition that requires the participation of institutional and aftercare staff prior to and following community reentry; and (3) long-term, reintegrative activities that ensure adequate service delivery and the necessary level of social control."[102] This model has been tested in three sites. The findings of a five-year experimental research study suggest that although each site successfully implemented the core components of the IAP model, there were no significant differences between the recidivism rates of the IAP and control groups.[103] Other less stringent studies of aftercare services have produced more promising outcomes, with programs contributing to lower rates and less serious recidivism.[104]

Effective Correctional Interventions for Youth

As evidenced in the previous descriptions of correctional options, there is no conclusive evidence regarding what programs are effective in reducing delinquency. Evaluation results are mixed; one evaluation might report positive results and the next evaluation of the same type of program may report it to be a dismal failure. How can this be? These mixed results are a direct reflection of the variation in program design and implementation. Although similar in name and purpose, two programs can have significant differences in terms of funding levels, staff qualifications, community resources, services actually delivered, and types of youths served. The problem is that these contradictory findings impede the development of evidence-based practices capable of reducing delinquency.

Over the past 20 years, several researchers have devoted their work to identifying programs and practices that promote positive outcomes for youth. This work has entailed the systematic review and synthesis of available studies and has produced a body of literature on effective correctional interventions. One review of 200 studies found that, overall, participation in treatment contributed to a 12 percent reduction in recidivism among serious delinquents.[105] This same review found that some programs reduced recidivism by as much as 40 percent. The best programs were behavioral programs that targeted more serious offenders and provided interpersonal skills training and individual counseling.

Several reviews have been able to isolate other programmatic factors that contribute to improved outcomes. For example, they have revealed that noninstitutional correctional programs contribute to lower rates of recidivism among youth than do institutional programs.[106] It is speculated that this increased effectiveness stems from the opportunity that community-based programs provide for "reality testing." That is, youths are able to test the skills learned in their natural environment while continued feedback and support are still available. Other characteristics of effective programs include these:

- Matching offenders to program intensity and duration based on their level of risk for relapse or delinquency
- Matching offenders to programs that will best accommodate their personal characteristics (e.g., age, race, gender) and learning styles (e.g., level of cognitive functioning)
- Addressing offenders' criminogenic needs, or those needs most strongly associated with substance abuse and delinquency
- Providing multiple types of services to address the constellation of risks and needs
- Using a cognitive-behavioral approach
- Involving family members in the treatment process
- Providing aftercare or follow-up services in the community[107]

Although more research is needed, this work has contributed greatly to the development of evidence-based practices for juvenile corrections.

Race and Gender in Corrections

Because girls make up such a small proportion of juvenile offenders, they have been neglected in delinquency research and programming. Startling patterns in official rates of female delinquency are now making the development of effective girls' programming a priority within juvenile justice agencies. Although the major risk factors for delinquency appear to be similar across gender, girls have some unique needs that require gender-specific programs and practices. For example, mental health problems such as depression and post-traumatic stress disorder are more prevalent among girls. Additionally, there is some evidence to suggest that although boys get into trouble as the result of risk-taking behaviors and antisocial attitudes, girls are more apt to get into trouble because of problems with the relationships in their lives. The OJJDP Girls' Study Group is engaging in research to further identify the unique needs of girls and to identify effective models for girls' interventions.

Contemporary Issues in Juvenile Justice

Obviously, the juvenile justice system has evolved into a system that differs slightly (or perhaps dramatically, depending on the jurisdiction) from the juvenile justice system that began in 1899. Nevertheless, the system continues to evolve. In the following section, we discuss three issues faced by the contemporary juvenile justice system. We begin this section with a discussion of disproportionate minority contact.

Disproportionate Minority Contact

Although a longstanding problem, recent attention has been paid to the problem of disproportionate minority contact (DMC) in the juvenile justice system. DMC occurs when minority youths are represented in proportions greater than their representation in the general population.[108] For example, in 2003, blacks constituted 16 percent of the population between the ages of 10 and 17 years, but they represented 27 percent of the delinquency caseload, 29 percent of delinquency cases referred to juvenile court, 36 percent of youth held in detention, and 38 percent of youths in residential placement. Although this overrepresentation occurs with other minority groups, it is most pronounced for black youths.[109]

There are multiple sources of DMC. The Child Welfare League of America asserts that DMC is partially caused by higher rate of black involvement in more serious delinquency; differential treatment of youth stemming from racial bias; "get tough" policies such as expanded waiver laws, the war on drugs, immigration laws, and anti-gang laws; and higher levels of exposure to factors that increase

Corrections and Policy

The Juvenile Justice and Delinquency Prevention Act of 1988 first addressed the issue of DMC by requiring states receiving federal formula grants to submit a strategy for reducing disproportionate minority "confinement." In 2002, the act expanded this mandate to disproportionate minority "contact" based on the recognition that problems regarding the overrepresentation of minority youths existed at each stage of the juvenile justice process and not just at confinement decisions.[110]

youths' risk of delinquency including poverty, substance abuse, mental health problems, and familial incarceration.[111]

The most successful strategies for reducing DMC are those that address disparity at each decision point. Common approaches include cultural awareness and education training for law enforcement and other juvenile justice personnel, community-oriented policing, and the use of objective-based decision-making tools that guide prosecutors' charging decisions and juvenile court dispositions.[112]

Death Penalty for Juveniles

As the "get tough" philosophy crept into juvenile court, and legislation made it easier to transfer youths to criminal courts, more juvenile offenders faced the possibility of adult punishments, including the death penalty. Fervent debates surround the application of death penalty to juvenile offenders. Supporters believe that in the case of particularly heinous crimes, the offender's age is irrelevant. Opponents believe that the death penalty constitutes cruel and unusual punishment, particularly as it applies to juveniles who are less mature, and therefore less culpable, than adult offenders are.

Despite repeated requests, it was not until 1987 that the Supreme Court addressed the constitutionality of the death penalty for juveniles.[113] In *Thompson v. Oklahoma* (1987),[114] the Court reversed the death sentence of Thompson who, at the age of 15, had participated in the murder of his brother-in-law on the basis that executing a 15-year-old defendent constituted cruel and unusual punishment. In *Stanford v. Kentucky* (1989),[115] the court upheld the death penalty for persons who commit murder at 16 or 17 years of age, on the basis that it did not appear to offend society standards of decency and, thus, did not constitute cruel and unusual punishment under the Eighth Amendment.

The constitutionality of the death penalty for juveniles was not revisited until 2005 when, in *Roper v. Simmons* (2005), the Supreme Court ruled that the death sentence for persons under 18 years of age at the time of the crime constituted cruel and unusual punishment. In this ruling, the Supreme Court cited the infrequent use of the death penalty for juveniles and the abolition of the death penalty for juveniles in several states as evidence that society views juveniles as "categorically less culpable than the average criminal."[116]

From 1973, when the death penalty was reinstated in the United States, until the 2005 ruling in *Roper v. Simmons*, a total of 226 death sentences were imposed and 22 offenders were executed for capital offenses they committed as juveniles.[117] The Supreme Court ruling in *Roper v. Simmons* (2005) brought the United States in line with international law that had identified the death penalty for juveniles as a human rights issue.

Controversies

On January 16, 1995, at age 17, Eddie Capetillo agreed to join four friends in carrying out a robbery. Eddie carried a firearm into the home of Matt and Allison Vickers and, in the course of trying to gather the four people in the house together and force them to hand over money, shot Kimberly Williamson, a friend of Allison's who was living in the house. Bullets entered Kimberly's neck and chest, causing her death. Eddie was indicted for capital murder on December 21, 1995. Eddie was a student at Klein High School in Texas and had no criminal record. At the trial, the state introduced into evidence a chronicle of disciplinary problems at school and in his neighborhood. Sixteen witnesses testified on Eddie's behalf at his trial, reporting that Eddie was quiet and respectful and spoke about his love for playing soccer. His Navy recruitment officer testified that Eddie had been a highly motivated and respectful member of their delayed entry program. A friend testified that he had seen Eddie on the day of the offense and observed that Eddie had been under the influence of LSD and cocaine. A physician spoke as an expert witness about the dangerous effects those drugs have on a person's perceptions and actions. Eddie was convicted and sentenced to death in the Court of Harris County, Texas, on January 31, 1996. He faced an execution date of March 30, 2004. The execution was stayed awaiting the decision in *Roper v. Simmons* and has since been commuted to a sentence of life in prison. Did the Supreme Court make the correct ruling?
Source: Excerpted from American Bar Association, Juvenile Death Penalty Cases, www.abanet.org/crimjust/juvjus/juvcases.html.

Abolish Juvenile Court?

In many ways, the practices and philosophies of juvenile courts have converged with those of adult courts: Juvenile court procedures have become more complex as due process rights have been expanded; the severity of sanctions has increased for juvenile offenders; and new laws increasing the public's access to court proceedings have detracted from the court's protective role. Given this, recent discourse has centered on the wisdom of keeping a separate court for handling juvenile offenders.[118]

The push for abolishing juvenile court was driven by the increases in the rates of juvenile arrests for violent crime. Three basic points are used in the argument. First, most proponents of abolishing the juvenile court argue that it is too lenient. They invoke the old adage "if you are old enough to do the crime you are old enough to do the time" and assert that juvenile courts do a disservice to our youth by not holding them accountable. Second, proponents of abolition argue that victims of crime are ignored in juvenile court. They assert that victims do not care about the age of the offenders who harmed them—they want just deserts. Third, they cite high recidivism rates and equivocal program evaluation results and argue that the juvenile court has not been effective in rehabilitating youth under its care.

There are also those who argue for abolition of juvenile court on grounds that it allows excessive discretion and abuse of due process under the guise of rehabilitation. Despite recent court cases that enhance youths' due process rights, juvenile court is not as protective of these rights as adult court is.

Three counterarguments are offered by juvenile court advocates. First, they argue that the push for change has been fueled by distorted information about the level of serious juvenile crime. The increase in the rate of juvenile violence was overblown by the media and politicians; this led to a moral panic about crime and grave predictions about a new generation of young superpredators.[119] Second, they vehemently disagree with the contention that juvenile court is ineffective in the rehabilitation of youths, citing programs that have lowered recidivism by 20 to 30 percent. Third, they argue that juvenile cases that have been transferred to adult court have not experienced better results. That is, they are not necessarily punished more harshly, nor are they deterred from further delinquency.

How can the same data lead to such oppositional viewpoints? Clearly, perceptions of data are colored by the philosophical lens of the observer. Although the strength of the argument is likely to ebb and flow with each rise in juvenile crime, it is unlikely to result in the abolishment of juvenile court given its strong tradition in the United States.

READY FOR REVIEW

- The legal definition of delinquency includes those behaviors that are prohibited by the family or juvenile code of the state and that subject minors to the jurisdiction of juvenile court.
- There are two types of behaviors that are prohibited by family or juvenile codes: criminal offenses and status offenses.
- A minor is defined as "a youth at or below the upper age of the original jurisdiction in a state." The upper age varies from state to state.
- Self-report, panel, and cohort studies are important sources of information about juvenile crime. Combined with UCR and NCVS, they provide a relatively accurate picture of juvenile crime in the United States.
- The available methods for measuring delinquency suggest that levels of juvenile crime have been relatively stable over the past 25 years with the exception of a peak in violent crime between 1988 and 1994.
- Although juvenile victimization for all nonfatal crimes declined 54 percent between 1993 and 2003, juveniles 12 to 18 years of age were still more than twice as likely as adults to be victimized by nonfatal violent crime.
- Four prominent theories of delinquency include social learning, strain, social bond, and labeling theories. Understanding the main assumptions of these theories can help students understand many of the contemporary juvenile justice practices.
- Many factors associated with an increase or decrease in youths' likelihood of delinquency have been identified through research. Two of the strongest risk factors for delinquency include substance abuse and gang membership.
- Formalized, government-run efforts to control delinquency began with the house of refuge, the first separate institution for the care of delinquent, dependent, and neglected children.
- Early juvenile justice reforms, including the first juvenile court, were rooted in the principle of *parens patriae*. Under this principle, the state was obligated to protect youths from themselves and others and to provide care and guidance for youths whose parents were absent or deemed unfit.
- Although juvenile courts today continue to operate under the principle of *parens patriae*, their mission has shifted from a sole focus on serving the best interests of the child to a more balanced effort to meet the needs of the offender, victims, and the community.
- Whereas juvenile courts were once characterized by informality, juveniles are now accorded many of the same due process rights accorded adult offenders.

- In response to the peak in juvenile violence between 1988 and 1994, many states enacted laws and other policy changes to make their juvenile justice systems more punitive.
- The major stages of juvenile court include arrest, intake, adjudication, and disposition. The manner in which a case is handled varies within and across jurisdictions.
- At any point in the case processing, youths can be diverted from further penetration into the juvenile justice system to minimize the consequences of labeling. Diversion can range from "doing nothing" to drug court.
- Prosecutors and judges can use their discretion to transfer a case to the adult court. Additionally, legislation has been passed that automatically excludes juveniles charged with certain offenses from juvenile court jurisdiction.
- Predispositional reports and risk assessment are used to determine the least intrusive sanction necessary to treat and control youths' behavior.
- A variety of correctional options is available to juvenile courts. Probation is the most common form of intervention for juvenile offenders. Other community-based interventions include restitution, community service, and day treatment centers.
- Out-of-home placements include foster care, group homes, boot camps, and other secure care facilities. Secure care facilities are the most restrictive correctional options available for youths and are the option most similar to adult prisons.
- Aftercare has been identified as an essential element of any system of corrections. Prerelease planning and postrelease supervision reduce the likelihood of offender recidivism.
- Recent research has focused on identifying programmatic elements and practices that promote positive outcomes for youths. Programs that implement these evidence-based practices have achieved reductions in recidivism up to 40 percent.
- Disproportionate minority contact occurs at every decision point in the juvenile justice system. Federal mandates to reduce DMC have led to a broad array of initiatives with varying degrees of success.
- A 2005 Supreme Court case ruled that the death penalty for juveniles was unconstitutional. This ruling brought the United States in line with international law.
- The convergence in the philosophies and practices of juvenile and adult courts has prompted debate about the need for separate juvenile courts. The dominant sentiment is that a separate court that understands and is sensitive to childhood and adolescent development is needed to ensure appropriate intervention.

KEY TERMS

acquittal A determination by the judge that the facts in the petition are not supported and the petition should be dismissed

adjudication A determination by the judge that the facts outlined in the petition are supported by evidence; similar to a conviction in adult court

adjudicatory hearing A hearing during which the judge reviews the facts of the case. Similar to the trial in adult courts but without a jury

aftercare Prerelease planning and postrelease supervision designed to help youth navigate the transition from secure care back to the community

blended sentencing Youths receive both juvenile and adult sentences with the adult sentence being enacted only when the youth has failed to meet the conditions of the juvenile sentence

boot camp Institutional program designed for first-time, nonviolent offenders; emphasizes strict regimens, military drill, and hard labor

chronic offender A youth who engages in high rates of delinquent activity throughout his or her delinquent career. Panel studies regularly demonstrate that a small percentage of chronic offenders (typically less then 8 percent of a sample of youth) are responsible for the majority of delinquent activity among that sample of youth

cohort studies Studies in which a group of people who have something in common are followed over a period of time

day treatment centers An intensive, highly structured alternative to incarceration that provides youths with treatment and supervision in one common setting during day and evening hours and allows youths to return home at night

delinquency Those behaviors that are prohibited by the family or juvenile code of the state and that subject minors to the jurisdiction of juvenile court

detention hearing A hearing held within 24 hours of arrest to determine if further detention is warranted

dispositional hearing A formal court hearing during which a decision is made as to the most appropriate course of action for the youth

diversion Any action that minimizes further penetration into the juvenile justice system

foster care A form of out-of-home placement in which youths are removed from their parents and temporarily placed with foster parents who, after being licensed, are paid a daily rate by the state to provide care for youth

gang Any identifiable group of youngsters who are perceived by others as a distinct aggregation in their neighborhood, recognize themselves as a denotable group, and have been involved in a sufficient number of delinquent incidents that call attention to themselves

group homes Small, nonsecure, community-based settings where juvenile offenders are placed as an alternative to incarceration or as a step-down program after a period of incarceration

guilty plea Formal agreement by youths to the facts outlined in the petition
house of refuge The first separate institution for delinquent, dependent, and neglected youths; established in New York in 1825
intake screening A review of a youth's social and behavioral history, usually conducted by a probation officer or some other arm of the court, to determine whether a case should be referred for formal juvenile court intervention
judicial waivers The juvenile court judge transfers the case to criminal court after considering the merits of the case and determining that the youth is no longer amenable to rehabilitation
juvenile detention Secure facilities that house youths prior to and following adjudication for delinquent acts
labeling theory A theory of delinquency that assumes that society's reaction to initial deviance alters youths' self-image and increases their likelihood of engaging in subsequent delinquency
legislative exclusions State statutes that exclude certain classes of crimes from juvenile court jurisdiction or that specify the age at which youths can be transferred to adult court for each type of offense
mens rea The capacity for formulating criminal intent
minor A youth at or below the upper age of the original jurisdiction in a state
model of family government During colonization in the United States, the family was responsible for maintaining social control over children
monitoring the future (MTF) project A project conducted by the University of Michigan Survey Research Center under a series of grants from the National Institute on Drug Abuse since 1975 to (1) study changes in the attitudes, beliefs, and behaviors of young people in the United States; and (2) monitor trends in substance use and abuse among adolescents and young adults
panel studies Longitudinal studies where a representative sample of youths is followed over a period of time (typically several years) to examine their involvement in delinquent activity and the circumstances surrounding their involvement in this activity
parens patriae The obligation of the state to assume guardianship for youths whose parents were absent or deemed unfit
petition A request for formal juvenile court intervention that outlines the accusations against the youth
placing out Vagrant and poor youth from the cities were transported out West to live with farm families who would provide them with a warm and stable home environment
predisposition report A report prepared by a probation officer that includes a comprehensive review of the youth's behavioral and social history; serves as the basis for decisions made during the disposition hearing
preliminary screening Conducted by police to ascertain the most appropriate resolution to police–youth contact. Generally involves a review of the youth's past record and discussions with the youth, parents, and victim

probation The conditional release of youths into the community where they are provided with supervision and treatment

prosecutorial waivers Based on the gravity of the offense, the prosecutor decides to file the case in criminal court

protective factors Characteristics that decrease youths' likelihood of engaging in delinquency

reform schools Cottage-like institutions developed in the late 1800s designed to emulate family living. Emphasize formal schooling and reformation for delinquent, dependent, and neglected youth

resiliency The capacity for overcoming adversity

risk assessment A systematic case review conducted for the purpose of estimating the likelihood of the youth engaging in subsequent delinquency

risk factors Characteristics that increase youths' likelihood of engaging in delinquency

risk-protection framework A framework used by juvenile justice professionals to estimate the likelihood of future delinquency and to develop intervention plans

secure care facilities State-run institutions providing the most restrictive correctional option for youths

self-report studies Studies that ask youths to report their involvement in a range of delinquent behaviors

social bond theory A theory of delinquency that assumes that the likelihood of delinquency increases as youths' connections to society weaken

social learning theory A theory of delinquency that assumes criminal behavior is learned by observing the behavior of others

status offenses Behaviors that violate laws or regulations that apply only to children

strain theory A theory of delinquency that assumes delinquency is motivated by strain experienced as the result of an inability to achieve desired goals through legitimate means, the presentation of negative stimuli, or the removal of positive stimuli

therapeutic foster care Enhanced form of foster care for high-risk youth; foster parents must participate in preservice training and daily consultation with a case manager and serve as members of a treatment team

Youth Risk Behavior Surveillance System A project conducted biennially since 1992 by the Centers for Disease Control and Prevention to examine categories of behavior that contribute substantially to the leading causes of death among people in the United States. These six categories include (1) tobacco use; (2) alcohol and other drug use; (3) sexual behaviors; (4) dietary behaviors; (5) behaviors that contribute to unintentional injuries and violence; and (6) physical inactivity

YOU ARE THE CORRECTIONS PROFESSIONAL SUMMARY

1. What is the basis for having a separate court for juvenile offenders? Because youths lack maturity and the capacity to fully understand the consequences of their actions, they are less culpable. Therefore, instead of punishment, they require protection and guidance that can best be provided by a separate court that understands and is sensitive to their stage of development.
2. What protections are juveniles afforded and why? Despite new laws giving the public more access to juvenile proceedings, much of the information in juvenile cases is kept confidential to help minimize the effects of labeling. Additionally, as a result of a recent Supreme Court ruling, juvenile offenders can no longer receive the death penalty.
3. Is juvenile court effective in controlling delinquency? Programs that use best practices have achieved reductions in recidivism of up to 40 percent. Research has shown that community-based interventions are more effective in reducing recidivism. There are lower recidivism rates associated with juvenile courts than with adult courts.
4. How are juvenile offenders held accountable for their actions? In recent years, most juvenile courts have modified their mission statements to place more emphasis on offender accountability. Juvenile courts have also become more punitive with more intensive sanctioning options available. Some methods for holding juveniles accountable include restitution, community service, court-ordered treatment, and various types of sanctions including detention, probation, electronic monitoring, and long-term incarceration.

NOTES

1. P. Elrod and S. Ryder, *Juvenile Justice: A Social, Historical, and Legal Perspective*, 2nd ed. (Boston: Jones and Bartlett, 2005).
2. National Center for Juvenile Justice, home page, www.ncjj.org.
3. M. King and L. Szymanski, "National Overviews," *State Juvenile Justice Profiles* (Pittsburgh, PA: National Center for Juvenile Justice, 2006), www.ncjj.org/stateprofiles/.
4. J. T. Whitehead and S. P. Lab, *Juvenile Justice: An Introduction*, 5th ed. (Cincinnati, OH: Anderson Publishing, 2006).

5. Elrod and Ryder, *Juvenile Justice.*
6. American Bar Association, *Adolescent Development: An Overview* (Washington, DC: American Bar Association, n.d.).
7. M. E. Wolfgang, R. M. Figlio, and T. Sellin, *Delinquency in a Birth Cohort* (Chicago: University of Chicago Press, 1972).
8. H. Snyder, *Juvenile Arrests 2005* (Washington, DC: Office of Juvenile Justice and Delinquency Prevention, forthcoming).
9. Snyder, *Juvenile Arrests 2005.*
10. T. J. Bernard, *The Cycle of Juvenile Justice* (New York: Oxford University Press, 1992).
11. Bureau of Justice Statistics, "Crime and Justice Data Online," July 27, 2006, http://bjsdata.ojp.usdoj.gov/dataonline/.
12. D. S. Elliott, S. S. Ageton, D. Huizinga, B. A. Knowles, and R. J. Canter, *The Prevalence and Incidence of Delinquent Behavior: 1976–1980* (Boulder, CO: Behavioral Research Institute, 1983); K. Maguire and A. L. Pastore, *Sourcebook of Criminal Justice Statistics 1996* (Washington, DC: U.S. Department of Justice, 1996).
13. Maquire and Pastore, *Sourcebook.*
14. Elliott et al., *Prevalence and Incidence of Delinquent Behavior.*
15. Wolfgang et al., *Delinquency in a Birth Cohort*; P. E. Tracy, M. E. Wolfgang, and R. M. Figlio, *Delinquency Careers in Two Birth Cohorts* (New York: Plenum, 1990).
16. H. N. Snyder and M. Sickmund, "Juvenile Offenders and Victims: 2006 National Report (Washington, DC: Office of Juvenile Justice and Delinquency Prevention, March 2006).
17. D. S. Elliott and S. S. Ageton, "Reconciling Race and Class Differences in Self-Reported and Official Estimates of Delinquency," *American Sociological Review* 45(1980): 95–110; Wolfgang et al., *Delinquency in a Birth Cohort.*
18. Snyder and Sickmund, "Juvenile Offenders and Victims."
19. Child Welfare League of America, "The CWLA National Girls Initiative: Growing Girls for Greatness" (paper presented at the National Girls Initiative Symposium, Washington, DC: October 2003).
20. H. N. Snyder, "Juvenile Arrests 2003," *Juvenile Justice Bulletin* (Washington, DC: U.S. Department of Justice, 2005).
21. S. Steffensmeier, J. Schwartz, and H. Zhong, "An Assessment of Recent Trends in Girls' Violence Using Diverse Longitudinal Sources: Is the Gender Gap Closing?" *Criminology* 43, no. 2 (2005): 355–405.

22. M. Chesney-Lind, "Gender and Justice: What about Girls?" (paper presented at the National Girls' Initiative Symposium, Washington, DC: October 2003). Snyder and Sickmund, "Juvenile Offenders and Victims."
23. H. N. Snyder, *Juvenile Arrests 2003*, NCJ 209735 (Washington, DC: U.S. Department of Justice, Office of Juvenile Justice and Delinquency Prevention, August 2005).
24. American Bar Association, *Adolescent Development*.
25. A. Bandura, *Social Foundation of Thought and Action: A Social Cognitive Theory* (Englewood Cliffs, NJ: Prentice Hall, 1986).
26. D. J. Simourd and D. L. Andrews, *Correlates of Delinquency: A Look at Gender Differences* (Ottawa, Ontario: Department of Psychology, Carleton University, 1994).
27. R. Paternoster, G. P. Waldo, T. G. Chiricos, and L. S. Anderson, "The Stigma of Diversion: Labeling in the Juvenile Justice System," in *Courts and Diversion: Policy and Operations Studies*, eds. P. L. Brantingham and T. G. Blomberg (Beverly Hills, CA: Sage, 1979), 127–142; S. Rausch, "Court Processing versus Diversion of Status Offenders: A Test of Deterrence and Labeling Theories." *Journal of Research in Crime and Delinquency* 20(1983): 39–54.
28. D. Huizinga, R. Loeber, and T. P. Thornberry, *Recent Findings from the Program of Research on Causes and Correlates of Delinquency* (Washington, DC: Office of Juvenile Justice and Delinquency Prevention, 1995); Simourd and Andrews, *Correlates of Delinquency*.
29. J. D. Hawkins, R. F. Catalano, and J. Y. Miller, "Risk and Protective Factors for Alcohol and Other Drug Problems in Adolescence and Early Adulthood: Implications for Substance Abuse Prevention," *Psychological Bulletin* 112(1992): 64–105.
30. N. Garmezy, "Stress-Resistant Children: The Search for Protective Factors," in *Recent Research in Developmental Psychopathology*, ed. J. E. Stevenson, (New York: Pergamon, 1985), 213–233; J. C. Howell, *Juvenile Justice and Youth Violence* (Thousand Oaks, CA: Sage, 1997).
31. G. E. Richardson, B. L. Neiger, S. Jensen, and K. L. Kumpfer, "The Resiliency Model," *Health Education* 216(1990): 33–39; D. Cicchetti and N. Garmezy, "Prospects and Promises in the Study of Resilience," *Development and Psychopathology* 5(1993): 497–502.
32. Richardson et al., "The Resiliency Model."
33. Hawkins et al., "Risk and Protective Factors."
34. Substance Abuse and Mental Health Services Administration, *Findings from the 2004 National Survey on Drug Use and Health* (Rockville, MD: Department of Health and Human Services, 2005).

35. A. Crowe, *Working with Substance Abusing Youths: Knowledge and Skills for Juvenile Probation and Parole Professionals* (Lexington, KY: American Probation and Parole Association, 1999).
36. Office of National Drug Control Policy (2004).
37. National Institute of Justice, *2000 Arrestee Drug Abuse Monitoring: Annual Report* (Washington, DC: National Institute of Justice, 2003).
38. Whitehead and Lab, *Juvenile Justice.*
39. M. Sickmund, "Juvenile Residential Facility Census 2002: Selected Findings," *Juvenile Offenders and Victims National Report Series* (Washington, DC: Office of Juvenile Justice and Delinquency Prevention, June 2006).
40. Elrod and Ryder, *Juvenile Justice.*
41. M. Gold, *Delinquent Behavior in an American City* (Pacific Grove, CA: Brooks/Cole Publishing, 1970).
42. Gold, *Delinquent Behavior.*
43. M. W. Klein, *The American Street Gang* (New York: Oxford University Press, 1995).
44. A. Egley, J. Howell, and A. K. Major, *National Youth Gang Survey: 1999–2001* (Washington, DC: Office of Juvenile Justice and Delinquency Prevention, July 2006).
45. T. P. Thornberry, "Membership in Youth Gangs and Involvement in Serious and Violent Offending," in *Serious and Violent Juvenile Offenders: Risk Factors and Successful Interventions,* eds. R. Loeber and D.P. Farrington (Thousand Oaks, CA: Sage, 1998), 147–166.
46. F. Esbensen and D. Huizinga, "Gangs, Drugs, and Delinquency in a Survey of Urban Youth," *Criminology* 31 (1993): 565–589; T. P. Thornberry, M. D. Krohn, A. J. Lizotte, and D. Chard-Wiershem, "The Role of Juvenile Gangs in Facilitating Delinquent Behavior," *Journal of Research in Crime and Delinquency* 30(1993): 55–87.
47. J. Howell, *Preventing and Reducing Juvenile Delinquency: A Comprehensive Framework* (Thousand Oaks, CA: Sage, 2003).
48. I. A. Spergel and G. D. Curry, "Strategies and Perceived Agency Effectiveness in Dealing with the Youth Gang Problem," in *Gangs in America,* ed. C. R. Huff (Newbury Park, CA: Sage, 1990), 288–309.
49. Elrod and Ryder, *Juvenile Justice*; B. Krisberg and J. Austin, *Reinventing Juvenile Justice* (Newbury Park, CA: Sage, 1993).
50. Krisberg and Austin, *Reinventing Juvenile Justice.*
51. Elrod and Ryder, *Juvenile Justice.*

52. *Ex Parte Crouse*, 4 Wheaton (PA) 9 (1838).
53. Whitehead and Lab, *Juvenile Justice*.
54. Krisberg and Austin, *Reinventing Juvenile Justice*.
55. Krisberg and Austin, *Reinventing Juvenile Justice*; A. M. Platt, *Child Savers—The Invention of Delinquency* (Chicago: University of Chicago Press, 1969).
56. Snyder and Sickmund, "Juvenile Offenders and Victims."
57. Snyder and Sickmund, "Juvenile Offenders and Victims."
58. Snyder and Sickmund, "Juvenile Offenders and Victims."
59. G. Bazemore, "On Mission Statements and Reform in Juvenile Justice: The Case of the Balanced Approach," *Federal Probation* 56, no. 3 (1992): 54–70.
60. Unless otherwise noted, the information in this section is taken from Snyder and Sickmund, "Juvenile Offenders and Victims."
61. Office of Juvenile Justice and Delinquency Prevention (OJJDP), "Statistical Briefing Book," 2007, http://ojjdp.ncjrs.gov/ojstatbb/court/qa06203.asp?qaDate=2004.
62. OJJDP, "Statistical Briefing Book."
63. OJJDP, "Statistical Briefing Book."
64. Snyder and Sickmund, "Juvenile Offenders and Victims."
65. K. M. Hess and R. W. Down, *Juvenile Justice,* 4th ed. (Belmont, CA: Wadsworth, 2004).
66. OJJDP, "Statistical Briefing Book."
67. OJJDP, "Statistical Briefing Book."
68. Hess and Down, *Juvenile Justice*.
69. E. Schur, *Radical Nonintervention: Rethinking the Delinquency Problem* (Englewood Cliffs, NJ: Prentice Hall, 1973).
70. M. W. Klein, "Issues in Police Diversion of Juvenile Offenders," in *Back on the Street: The Diversion of Juvenile Offenders,* eds. R. M. Carter and M. W. Klein (Englewood Cliffs, NJ: Prentice Hall, 1976), 73–104; E. W. Vorenberg and J. Vorenberg, "Early Diversion from the Criminal Justice System: Practice in Search of a Theory," in *Prisoners in America,* ed. L. E. Ohlin (Englewood Cliffs, NJ: Prentice Hall, 1973), 151–183.
71. Paternoster et al., "The Stigma of Diversion"; Rausch, "Court Processing versus Diversion."
72. T. G. Blomberg, "Diversion from Juvenile Court: A Review of the Evidence," in *Juvenile Justice Philosophy: Readings, Cases and Comments,* 2nd ed., eds. F. L. Faust and P. J. Brantingham (St. Paul, MN: West, 1979), 415–429;

T. Palmer and R. V. Lewis, *An Evaluation of Juvenile Diversion* (Cambridge, MA: Oelgeschlager, Gunn, and Hain, 1980).

73. Klein, "Issues in Police Diversion of Juvenile Offenders"; Palmer and Lewis, *An Evaluation of Juvenile Diversion.*
74. Howell, *Preventing and Reducing Juvenile Delinquency.*
75. M. A. Bortner, "Traditional Rhetoric, Organizational Realities: Remand of Juveniles to Adult Court," *Crime and Delinquency* 32(1986): 53–73.
76. D. M. Bishop, C. E. Frazier, L. Lanza-Kaduce, and L. Winner, "The Transfer of Juveniles to Criminal Court: Does It Make a Difference?" *Crime and Delinquency* 42(1996): 171–191.
77. National Council of Juvenile and Family Court Judges, *Juvenile Delinquency Guidelines: Improving Court Practice in Juvenile Delinquency Cases* (Reno, NV: National Council of Juvenile and Family Court Judges, 2005).
78. OJJDP, "Statistical Briefing Book."
79. Snyder and Sickmund, "Juvenile Offenders and Victims."
80. J. D. Wooldridge, "Differentiating the Effects of Juvenile Court Sentences on Eliminating Recidivism," *Journal of Research on Crime and Delinquency* 25(1988): 264–300.
81. P. Chamberlain, "Treatment Foster Care," *Juvenile Justice Bulletin* (Washington, DC: Office of Juvenile Justice and Delinquency Prevention, December 1998); P. Meadowcroft and B. A. Trout, *Troubled Youth in Treatment Homes: A Handbook of Therapeutic Foster Care* (Washington, DC: Child Welfare League of America, 1990).
82. Chamberlain, "Treatment Foster Care."
83. P. Chamberlain, "Comparative Evaluation of Specialized Foster Care for Seriously Delinquent Youths: A First Step," *Community Alternatives: International Journal of Family Care* 2(1990): 21–36.
84. P. Chamberlain and J. B. Reid, "Comparison of Two Community Alternatives to Incarceration for Chronic Juvenile Offenders," *Journal of Consulting and Clinical Psychology* 66(1998): 624–633.
85. R. A. Hahn, O. Bulikha, J. Lowy, A. Crosby, M. T. Fullilove, A. Liberman, E. Moscicki, S. Snyder, F. Tuma, P. Corso, and A. Schoefield, "Effectiveness of Therapeutic Foster Care for the Prevention of Violence: A Systematic Review," *American Journal of Preventive Medicine* 28, no. 251 (2005): 72–90.
86. Sickmund, "Juvenile Residential Facility Census 2002."
87. Sickmund, "Juvenile Residential Facility Census 2002."
88. Elrod and Ryder, *Juvenile Justice.*

89. Sickmund, "Juvenile Residential Facility Census 2002."

90. D. L. MacKenzie, D. B. Wilson, G. S. Armstrong, and A. R. Gover, "Impact of Boot Camps and Traditional Institutions on Juvenile Residents: Perceptions, Adjustment, and Change," *Journal of Research in Crime and Delinquency* 38, no. 3 (2001): 279–313.

91. MacKenzie et al., "Impact of Boot Camps and Traditional Institutions on Juvenile Residents."

92. D. L. MacKenzie, D. B. Wilson, and S. B. Kider, "Effects of Correctional Boot Camps on Offending," in *Annals of the American Academy of Political and Social Science*, vol. 578, eds. D. P. Farrington and B. C. Welsh (Place: Publisher, 2001), 126–143.

93. Sickmund, "Juvenile Residential Facility Census 2002."

94. Snyder and Sickmund, "Juvenile Offenders and Victims."

95. Office of Juvenile Justice and Delinquency Prevention, "Juveniles in Corrections," http://ojjdp.ncjrs.org/ojstatbb/index.html.

96. OJJDP, "Juveniles in Corrections."

97. D. Lester, M. Braswell, and P. Van Voorhis, "Radical Behavioral Interventions," in *Correctional Counseling and Rehabilitation*, 4th ed., eds. P. Van Voorhis, D. Lester, and M. Braswell (Cincinnati, OH: Anderson, 2000), 129–148.

98. Virginia Department of Juvenile Justice, "Juvenile Recidivism in Virginia," *DJJ Research Quarterly* 3(April 2005): 1–12.

99. D. M. Altschuler and T. L. Armstrong, "Intensive Interventions with High-Risk Youths," in *Intensive Interventions with High-Risk Youths: Promising Approaches in Juvenile Probation and Parole*, ed. T. L. Armstrong (New York: Criminal Justice Press, 1991), 45–85.

100. J. Irwin, *The Felon* (Englewood Cliffs, NJ: Prentice Hall, 1970); V. McArthur, *Coming Out Cold* (Lexington, MA: Lexington Books, 1974).

101. S. V. Geis, "Aftercare Services," *Juvenile Justice Bulletin* (Washington, DC: Office of Juvenile Justice and Delinquency Prevention, September 2003).

102. Altschuler and Armstrong, "Intensive Interventions with High-Risk Youths," 15.

103. R. G. Wiebush, D. Wagner, B. McNulty, Y. Wang, and T. N. Le, *Implementation and Outcome Evaluation of the Intensive Aftercare Program: Final Report* (Washington, DC: Office of Juvenile Justice and Delinquency Prevention, 2005).

104. Geis, "Aftercare Services."

105. M. W. Lipsey, D. B. Wilson, and L. Cothern, "Effective Intervention for Serious Juvenile Offenders," *Juvenile Justice Bulletin* (Washington, DC: Office of Juvenile Justice and Delinquency Prevention, April 2000).

106. Lipsey et al., "Effective Intervention for Serious Juvenile Offenders"; P. Gendreau, "The Principles of Effective Intervention with Offenders," in *Choosing Correctional Options That Work,* ed. A. T. Harland (Thousand Oaks, CA: Sage, 1996), 117–130.

107. Gendreau, "The Principles of Effective Intervention."

108. Snyder and Sickmund, "Juvenile Offenders and Victims."

109. J. Short and C. Sharp, *Disproportionate Minority Contact in the Juvenile Justice System* (Washington, DC: Child Welfare League of America, 2005).

110. Snyder and Sickmund, "Juvenile Offenders and Victims."

111. Short and Sharp, *Disproportionate Minority Contact.*

112. Short and Sharp, *Disproportionate Minority Contact.*

113. L. Cothern, "Juveniles and the Death Penalty," *Coordinating Council on Juvenile Justice and Delinquency Prevention* (Washington, DC: Office of Juvenile Justice and Delinquency Prevention, November 2000).

114. *Thompson v. Oklahoma*, 487 U.S. 815 (1988).

115. *Stanford v. Kentucky*, 492 U.S. 361 (1989).

116. Snyder and Sickmund, "Juvenile Offenders and Victims."

117. V. Streib, "The Juvenile Death Penalty Today: Death Sentences and Executions for Juvenile Crime January 1, 1973–February 28, 2005," 2005, www.law.onu.edu/faculty/streib/streib.htm.

118. B. C. Feld, "Abolish the Juvenile Court: Affirming Personal Responsibility," *Juvenile Justice Update* 2, no. 4 (1996); J. C. Howell, "Abolish the Juvenile Court? Nonsense!" *Juvenile Justice Update* 4, no. 1 (1998).

119. Howell, *Preventing and Reducing Juvenile Delinquency.*

15

The Future of Corrections and Juvenile Justice

Chapter Objectives

- Understand the challenges facing corrections and juvenile justice in the twenty-first century in the following areas:
  - Sentencing practices
  - Overcrowding
  - Aging prison population
  - Communicable diseases in corrections
  - Drugs and drug policies
  - Privatization
  - Faith-based initiatives
  - Technology
  - Specialty courts
  - Evidence-based corrections
  - Community reentry
  - Juvenile justice (including increased emphases on aftercare, collaboration, school resource officers, and adultization of juvenile justice)
- Present a futurist analysis of corrections and juvenile justice.

CASE STUDY

You are a deputy commissioner in charge of research, development, and planning for a state department of corrections. The governor, the commissioner of the Department of Juvenile Justice, and the commissioner of the Department of Corrections have asked that you meet with them regarding issues that will be facing both corrections and juvenile justice in the twenty-first century. They want you to develop a report in which you forecast which areas in corrections and juvenile justice will require a significant change in funding over the next 10-year period. They would like you to conclude the report by developing a list of funding strategies with priority given to what you anticipate will be the emerging trends in corrections and juvenile justice.

1. What kind of factors would need to be considered to ensure that you made the most accurate report to the state leadership?
2. What will be the most challenging areas in corrections and juvenile justice in your state in the twenty-first century?
3. After you have identified a list of challenges facing the Departments of Corrections and Juvenile Justice, how will you decide what is the most important issue for funding? How will you decide where funding can be reduced?

Introduction

In the preceding chapters, we have discussed the paths that someone who is arrested for a crime can take through the criminal justice system, including key considerations along the way. The path that any given case takes is rarely efficient and often has a number of turns, detours, and roadblocks. Nevertheless, as the twenty-first century unfolds, there are a number of trends and challenges emerging in corrections and juvenile justice that will alter the path even more in the future. In this chapter, we discuss some of the main trends and challenges.

Trends and Challenges Facing Corrections in the Twenty-First Century

In the near future, the fields of corrections and juvenile justice will face a number of challenges that have the potential to dramatically change both fields. A few of the most pressing challenges are discussed in the following section.

Retribution and Sentence Severity

The first trend in corrections and juvenile justice for the twenty-first century has actually been underway since the 1980s. That trend involves increased sentencing severity for individuals who engage in crime. As we have discussed throughout the book (particularly in Chapter 6), the movement to incarcerate greater numbers of offenders for longer periods of time had dramatic implications for all areas of corrections and juvenile justice and, indeed, for the system as a whole. Despite significant reductions in the crime rate during the 1990s, the number of prisoners incarcerated in the United States passed 1.5 million for the first time in 2005. In fact, 491 of every 100,000 U.S. citizens were in a federal or state prison at the end of 2005, a 19 percent increase over the rate of 411 per 100,000 in 1995, just after the crime rate began to fall in the United States.[1]

Two of the most controversial changes in sentencing have involved the widespread increased use of mandatory sentencing and habitual offender statutes (discussed in Chapter 6). These have increased both the number of offenders going to prison and the number of prisoners incarcerated for longer periods of time. Given that the federal government and many states are moving toward determinate sentencing where convicted offenders must serve 85 percent of their sentence prior to release, the number of prisoners will probably continue to grow in the future, regardless of whether crime rises sharply over the next decade. This will lead to further overcrowding and an increasingly older population of prisoners. Each of these issues is discussed in further detail in the following subsections.

Overcrowding in Correctional Facilities

Partly because of the increased focus on harsher sentencing policies that began in the 1980s, the United States now has the highest incarceration rate of any country in the world (industrialized or nonindustrialized). This incarceration rate is five to eight times higher than the incarceration rates of England, Canada, Germany, and France.[2] As such, the most immediate consequence of "get-tough" policies discussed earlier is overcrowding in both prisons and jails (discussed in Chapters 7 and 11). Overcrowding poses numerous problems for both prisons and jails. As discussed previously, these problems include double-bunking, reduced availability of programming, health and sanitary issues, and increased tensions inside institutions. Pressing issues such as these mean that the creative energies of the correctional staff are directed into managing the problem of overcrowding rather than focusing on other key issues, such as offender rehabilitation.

Although the percentage of state prisons operating over capacity has declined in the last decade (from 125% in 1995 to 114% in 2005), the percentage of federal facilities over capacity has actually increased (from 125% in 1996 to 134% in 2005).[3] The growth in federal imprisonment has outpaced growth at the state level for the past decade, and at midyear 2007 this trend continued as the population of immigration detainees and sentenced offenders increased.[4] The response to overcrowding at the federal level and in most state jurisdictions has been to expand space in existing facilities, to construct new facilities, or both.

In Chapter 7, we mentioned how jail incarceration rates had also increased significantly in the past two decades, and this growth has caused sheriffs and jail administrators to implement a number of solutions to reduce overcrowding. Most jurisdictions have expanded their capacity by building new jails or additions to existing structures. Some sheriffs have even erected tents to enable them to hold a larger number of inmates. (Although Sheriff Joe Arpaio's "tent city" in Maricopa County, Arizona, is the most well known, other jurisdictions in Colorado, Ohio, and Texas are also contemplating using these tents for jail inmates.) Other sheriffs have placed jail inmates in the community, either in pretrial programs or by creating activities and programs that enable jail inmates to be supervised (e.g., sheriff's work alternative programs) or housed on an intermittent basis (e.g., "weekender" programs).

Given that the federal, state, and local governments continue to adopt stricter enforcement and stringent sentencing policies in the twenty-first century (regardless of the crime rate), institutional crowding is not a problem that appears to have any immediate resolution. The logical alternatives are to (1) sentence far fewer people to incarceration, (2) drastically reduce the average time served for many offenses, and (3) dramatically expand space by expanding existing facilities and building new ones. However, none of these alternatives appear politically palatable, at least not at a level that would be required to reduce the crowding problem significantly.

Aging Prison Population

Along with overcrowding, increasingly punitive sentencing policies have also created an older prison population. On December 31, 2005 (the most recent year for which data were available at the time of this writing), there were 63,500 state and federal prisoners 55 years of age and older.[5] This number represents almost 5 percent of the total prison population in the United States and equates to a rate of 208 per 100,000. This rate is almost four times that of only 10 years ago—69 per 100,000 in 1996 and 49 per 100,000 population in 1990.[6]

With increasing numbers of aging prisoners come a wide variety of challenges. Perhaps the most important challenge presented by a graying prison population is that of health care for aged offenders. Older prisoners face a wide variety of chronic illnesses (e.g., hypertension, arthritis, pulmonary disease, diabetes, emphysema) that require constant medical attention. Estimates put the cost of keeping an inmate older than the age of 60 incarcerated at $70,000 per year compared with approximately $22,000 per year for the "typical" adult prisoner. As health care costs continue to rise, the chronic medical needs of older inmates will present a heavier financial burden for both state and federal correctional systems.[7]

Elderly inmates also present a number of mental health challenges as well. Correctional facilities in the United States currently house more mentally ill individuals than do either hospitals or mental institutions, and more than half of the individuals incarcerated in prisons and jails in the United States have a mental health problem. More than two in five prisoners displayed symptoms that met the criteria for mania, while about one in four prisoners reported symptoms of major depression. The problem of mental illness is even more pervasive in jails than in prisons because jails often serve not only as a holding area for offenders for a short period of time but also as a holding area for mentally ill persons pending their transfers to mental health facilities.[8]

Despite the fact that mental illness is more prevalent among younger prisoners (62.6 percent of those 24 years and younger have mental health problems compared with 39.6 percent of those 55 years and older), certain mental illnesses are more problematic for elderly offenders than they are for their younger counterparts. Dementia, depression, and psychiatric disorders appear to be far more prevalent among offenders over the age of 50 than among their younger counterparts. The noisy, physically strenuous environment where older offenders experience threats from younger offenders often creates differential levels of stress for elderly offenders and exacerbates the symptoms of mental and physical illness among this group.[9] As the prison population ages through the twenty-first century, the need for mental health treatment among older offenders will continue to grow as well.

Communicable Diseases in Corrections

At the end of 2004, there were more than 23,000 individuals who had been diagnosed with the human immunodeficiency virus (HIV) incarcerated in U.S. prisons. This number represents almost 2 percent of the state prison inmate population and slightly more than 1 percent of the federal prison population. The percentage of prisoners who are HIV-positive is higher for females (2.4 percent) than for males (1.7 percent).[10] Although this number has decreased from 25,680 (2.2 percent of the prison population) in 1998, the overall rate of acquired immune deficiency syndrome (AIDS) in the prison population remains more than three times the rate in the general population. As such, management of HIV-positive prisoners and prisoners with AIDS will continue to be a challenge for twenty-first-century corrections managers. A major concern is the high cost of caring for an inmate infected with HIV/AIDS. Although the death rate has dropped significantly in recent years, this is a result of medications that have an annual cost of $13,900 to $36,500 per inmate.[11]

Although 42 percent of newly diagnosed HIV/AIDS cases involved men who have sex with men, more than one in six cases involved men who acquired their infection through injection drug use. Given that a disproportionate number of individuals sentenced to prison are regular drug users, this finding may increase the problem of managing offenders with HIV/AIDS in the future.

Controversies

A controversy in corrections related to HIV/AIDS is the practice in some European nations of providing prisoners with access to clean needles for intravenous drug use and bleach to clean those needles. This practice is controversial because providing these supplies gives tacit approval to this illegal behavior. In addition, both bleach and needles can be used as weapons. Advocates of these policies argue that preventing a single case of HIV reduces the taxes and increases life expectancy.

Additionally, a number of myths and misconceptions about HIV/AIDS have surfaced in corrections. One of the most common fears among correctional officers is the fear of contracting HIV/AIDS from a prisoner supervised during the course of their duties. Despite the fact that we were unable to identify a single incident where a correctional officer had developed an HIV or AIDS infection as part of job-related duties, many officers are extremely fearful of this occurrence. Another myth among officers is that HIV/AIDS are spreading rapidly in the prison population; as the numbers reported earlier suggest, this is also not the case.[12]

Nevertheless, given the limited space available in prisons for an ever-increasing number of prisoners, correctional systems have increasingly begun to house offenders with HIV (80 percent of the systems) and AIDS (60 percent of the systems) in the general population with no restrictions. Although about one in three systems use mandatory testing for AIDS, less than one in five systems (17 percent) inform correctional staff of the identity of prisoners with AIDS, and only two states make condoms available to inmates on a "targeted" basis (to vulnerable populations only) or through a "sick call" procedure.[13] Given the exaggerated concern among correctional staff regarding AIDS, these divergent facts will continue to make HIV/AIDS in prison a source of controversy during the twenty-first century.

Although HIV/AIDS is the most serious communicable disease that affects corrections, rates of hepatitis, sexually transmitted infections, and tuberculosis are also much higher in correctional facilities than they are in the general population outside institutions. To reduce the spread of these diseases (both inside jails and prisons and in the larger society), correctional health practitioners are more likely to take a proactive role in assessment, detection, and treatment than they were in previous years. Moreover, these correctional health professionals may be more likely to collaborate with community-based public health officials in the future because a large proportion of inmates with communicable diseases routinely is released from jail or prison to the community.

Drugs and Drug Policies

As mentioned earlier in this book, the "war on drugs" is considered by many to be the most important cause of tremendous increases in the prison population over approximately the last three decades. Approximately one in five prisoners is incarcerated for drug offenses, and these percentages are almost 50 percent higher for blacks and Hispanics than they are for whites. In the federal system, more than half of sentenced inmates are incarcerated for drug offenses.[14] This increase at the end of the twentieth century, and the fairly stable rates of drug offenses that have continued to exist, suggest that the tougher drug penalties enacted in the 1990s have done little to decrease the use and distribution of drugs in the United States.

Welch suggests four elements that are important in any discussion of the impact of drug control policy on corrections, regardless of the school of thought one adopts when considering drug control strategies.[15] These four elements are discussed here.

The **supply reduction** element is important for those who favor using law enforcement strategies to reduce drug use. Advocates of this philosophy argue that if law enforcement can reduce the supply of illegal drugs, the consumption of illegal drugs (and thus the number of individuals incarcerated for both distribution and possession of illegal drugs) will then decline. Those who oppose the proliferation of law enforcement strategies in the effort to reduce the problem of drugs often favor **treatment** instead of supply reduction. This group suggests that treatment for substance abuse in the criminal justice system is ineffective because substance abuse treatment programs are often superseded by the coercive nature of incarceration and the mandatory nature of these programs in that incarcerative setting.

Another important position in the drug control policy debate revolves around **prevention/education**. No matter what philosophy one follows in trying to reduce illegal drugs in the United States, people generally agree that prevention and education are important. Nevertheless, the debate hinges on whether these prevention/education messages should be primarily (perhaps deceivingly so) negative or factual. The final factor involved in the drug control debate centers around **decriminalization**. Proponents of decriminalization suggest that removing the criminal nature of drug use reduces the stigma surrounding drug use and encourages an open discussion of the harm caused by legal drugs (e.g., alcohol, tobacco, prescription drugs). On the other hand, opponents suggest that removing the legal prohibition of drug use will encourage widespread use of illegal drugs. These opponents suggest that widespread use will serve to increase the harmful actions brought about by drug use, and this strategy is strongly opposed by the Office of National Drug Control Policy.

During times of economic stress or budget crises, there is increased attention paid to the costs and benefits of incarcerating nonviolent offenders. Steve

Aos, for instance, calculates that imprisoning violent and property offenders makes good economic sense. Every dollar spent imprisoning a violent offender saves taxpayers $2.74, whereas the economic benefit of imprisoning a property offender is actually greater, saving taxpayers $2.84 for each dollar spent on corrections. In terms of drug offenders, though, Aos found that each dollar spent for incarcerating these offenders produced only a $0.37 return.[16] Although economic analyses of crime are often difficult to conduct (e.g., we have a good idea of the costs of incarceration but can only estimate the economic benefits of an offender's incapacitation), they provide us with a useful starting point when discussing these issues.

Some jurisdictions have enacted laws that mandate treatment for drug offenders rather than just punishment. The public seems to support these initiatives, and in 2000, more than 60 percent of Californians voted for Proposition 36, a program that uses the coercive power of the courts to force nonviolent drug offenders to participate in treatment. These treatment programs are still in their infancy, and despite their "growing pains"—including a lack of funding to the county agencies that must deliver these programs—initial evaluations show promising results.

The University of California at Los Angeles has conducted four evaluations of the Substance Abuse and Crime Prevention Act (the formal name of the legislation). A five-year evaluation released in April 2007 revealed that slightly more than one-third of participants completed treatment, and their rates of reoffending were much less than the rates of offenders who did not complete treatment.[17] Further, a cost-benefit analysis revealed that for every dollar spent in these programs, there was a return of $2.50 in reduced incarceration costs (and a $4.00 return for participants who completed treatment). Given that most of the participants in California were long-term drug users whose drugs of preference were methamphetamine and cocaine, these are promising results. Other states are monitoring the success of this California initiative and may introduce similar programs.

Despite the various schools of thought regarding the approach the United States should take to drug control policy, there is little argument that the approach to control illegal drugs remains largely legalistic and punitive. Furthermore, as we highlighted in Chapter 6, despite a report from the United States Sentencing Commission (USSC) in May 2007 that current drug policies (particularly those centered on crack cocaine) were discriminatory and overwhelmingly used against nonwhites, there have been no changes to federal sentencing laws.[18] As such, there is little available evidence that suggests that philosophies toward illegal drugs, or policies based on those philosophies, will change in the near future.

Privatization

As we outlined in Chapter 7, the 1990s saw an increasing trend toward privatization of corrections in the United States. Despite the original excitement and discussion generated by this move, however, the vast majority of prisoners continue to be housed in government-operated facilities. Nevertheless, the number of inmates incarcerated in private facilities has grown 74 percent since 2000 in the federal system and 7 percent in the state systems.[19]

Proponents of privatization argue that (1) private industries can deliver equal care and custody for less money; (2) private prisons can reduce recidivism of offenders who served their time in private correctional facilities; and (3) private prisons can be built much faster and operated much more efficiently than can typical state-run institutions.[20] Opponents counter these arguments by suggesting that: (1) profiting from the punishment of offenders is controversial, even unethical; and (2) private prisons deliver substandard care and are entirely driven by the profit motive.

Both sides in the debate over privatization have some merit. Privatization has some positive aspects, such as financing, construction, and flexibility. It also has downfalls, however, including the potential to reduce officer professionalism or provide fewer goods and services to inmates to save money. Nevertheless, as proponents have suggested, if organizations such as the American Correctional Association can develop standards of care that all private institutions must meet prior to receiving a contract, privatization may be more attractive in the future than it has been in the past.[21] Regardless of one's stance on this issue, privatizing corrections tends to generate considerable debate, and this debate will continue well into the twenty-first century.

Faith-Based Initiatives

Before people used the term *rehabilitation* to describe efforts to prepare incarcerated individuals for release from prison, the term *reformation* was in vogue. This term is grounded directly in religion. Despite the fact that the link between correctional programming and religion can be traced to the eighteenth century and the important role played by the Quakers in the formation of American penology, until recently little scientific attention has been paid to the role of religion in reducing recidivism.

Nevertheless, under President George W. Bush, the role of **faith-based programs** has received increased attention. In 2001, President Bush launched his "faith-based" initiative by creating a federal office called the White House Office of Faith-Based and Community Initiatives to assist and encourage faith-based organizations in seeking federal funds to combat social problems, including reentry and rehabilitation programs for prisoners.[22] Despite the criticism that this initiative violated the separation of church and state, a number of grant programs have now become available to community organizations largely operated as faith-based initiatives.

Perhaps the most well known of these programs in corrections is the **Prisoner Reentry Initiative** that President Bush introduced in his state of the union address in January 2004 (refer also to Chapter 13). According to a press release describing the program, the goal of this initiative was to "assist ex-prisoners re-entering the community through an employment-based program that incorporates housing, mentoring, job training, and other services."[23] Examples of programs receiving this type of funding include the Exodus Transitional Community in East Harlem, New York, and the city of Memphis Second-Chance Program.

The limited empirical evidence that exists regarding the effectiveness of these programs suggests that although religion operates as a multidimensional method in correctional programming, religious programming may play a role in reducing recidivism for offenders who choose to take advantage of those initiatives.[24] Nevertheless, as a number of authors observe, the limited number of evaluations of these programs often fail to formally document the impact of religion on recidivism and often ignore factors that might lead individuals to volunteer to participate in these programs.[25] Correctional programs requiring offenders to participate voluntarily are often not effective for more general populations.

Despite criticism of faith-based initiatives, the federal office President Bush created has been ruled constitutional (with some limitations on funding) in several district and federal court decisions. As such, the issue of funding of faith-based programming in corrections will continue to be a topic of discussion in corrections in the early part of the twenty-first century.

Technology

Of all the twentieth-century changes that affected corrections, the one that has the most potential influence to effect change in the twenty-first century revolves around **criminal justice technology**. Modern innovations involving electronic sensing devices, computers, weapons, and other technologically sophisticated devices will no doubt continue to both (1) assist correctional and juvenile justice officials in the performance of their daily activities of confinement and control, and (2) present challenges to the protection of individual liberties of those being confined and controlled.[26]

A major technological innovation involves advances in data storage that began at the end of the twentieth century and continues today. With advances in computer technology, criminal justice data can now readily be stored, accessed, and shared by law enforcement, court, corrections, and juvenile justice officials. This information has the potential to benefit these officials by providing easy access to data used to make accurate and timely decisions. Nevertheless, this increased technology also has the potential to provide inaccurate information (particularly resulting from data coding or entry errors) that could adversely affect innocent people. Further, such access to data about so many individuals raises important privacy questions. Criminal justice agencies routinely face questions regarding whether appropriate safeguards are in place to prevent both inaccurate data and abuses of accurate data (e.g., unauthorized access and use of this confidential information). In the worst-case scenario, clerical errors can lead to deaths of innocent people (e.g., homeowners who defend their homes against unexpected entry of strangers, law enforcement or probation/parole officers acting on inaccurate information).[27]

New technologies for surveillance present important benefits and controversies as well. These technologies include **face-recognition technology**, **surveillance cameras**, **radio-frequency identification tags**, and **body cavity screening systems**. Although each of these innovations may be beneficial to the law enforcement and corrections communities, each also has the potential to violate constitutional rights of the individuals upon whom they are used.

Weapons and control devices are another area in which technological innovations have important implications. Technologies such as **stun belts**, **TASER stun devices**, and the **active denial system** all provide both corrections and law enforcement officers advanced weaponry they can use to more effectively subdue individuals they are trying to control. Nevertheless, each of these weapons has the potential for abuse. Several of the more technologically sophisticated weapons may cause fatal injury to people with heart conditions or pacemakers, neither of which are immediately obvious to those officials using these weapons in a potential conflict.[28] Despite the potential hazards or misuse associated with these stun devices, some physicians have actually reported that their use has decreased injuries. Bozeman, an emergency duty physician, notes, "Use of the devices appears to drastically reduce overall injuries and deaths, largely from a reduction in the use of other more dangerous methods available to police officers, such as striking violent suspects with a nightstick or flashlight or shooting them with a firearm."[29]

Perhaps the most well known technological advance in criminal justice has been the continued development of DNA technology. With DNA technology, law enforcement and corrections officers can obtain evidence from small bits of genetic material (as few as six cells or a fingerprint). DNA evidence has been used to facilitate investigations and prosecutions; it is particularly useful for prosecuting crimes committed inside prisons, given that strong norms against "snitching" restrict the use of eyewitness testimony. It has also been used to gain the release of a number of wrongfully convicted persons incarcerated for crimes they did not commit.

As with any technological innovation, however, legal and constitutional issues also surround the use of DNA. The most controversial issue regarding DNA evidence is the use of "John Doe" warrants, where officials have DNA evidence but no suspects to whom they can match the DNA. These warrants have been used to extend indictments past the statute of limitations. Additional issues concerning DNA revolve around whose DNA should be collected and maintained in computer records. Some officials have suggested that DNA evidence should be obtained from all arrestees and this evidence should be logged and stored for future use. Critics question this action, however, because thousands of individuals are arrested only to have charges dropped for lack of evidence each year. Regardless of one's political or philosophical view, a number of controversies will continue to surround DNA evidence in the twenty-first century.[30]

More correctional facilities are also using video technology to substitute for in-person court appearances, inmate visiting, and for telemedicine (linking an inmate with a physician in the community). Video technology reduces the costs and risks of transporting inmates to court appearances. Likewise, it can provide prisoners with access to a medical specialist who may be hundreds of miles away. Further, video visitation is being used in more facilities. It lets inmates and visitors communicate via closed-circuit television. These visits are less intimate than contact visits are, but they offer several advantages: (1) visitors do not have to be subjected to a search, (2) less contraband is brought into facilities by visitors (which in turn increases security), and (3) inmates may have more frequent contact with visitors.

As we have discussed previously, technological innovations are sometimes a double-edged sword. Each technological innovation has the potential to assist law enforcement, court, and corrections officials in the performance of their daily duties. Nevertheless, some innovations also have the potential for abuse or misuse, especially because they are often highly dependent upon human involvement. New innovations are continually being developed that have the potential to change the scope and nature of corrections and juvenile justice in the future. These innovations will face the same (if not greater) challenges, and judges and policymakers will have to decide whether the benefits of the technologies outweigh the potential harm.

Specialty Courts

In previous chapters, we briefly described how some offenders (such as persons with mental illness or drug problems, or ex-prisoners reentering the community) sometimes appear before courts that have a specialized docket. These **specialty courts** are a relatively new innovation and were developed in response to entrenched or seemingly unsolvable problems, such as the high arrest rates involving persons with mental illness or drug use. Police, court officials, and correctional staff became discouraged because the traditional justice system approach was not always effective at stopping the cycle of rearrest, new appearances before the court, and readmission to jail or prison. Part of the problem is that the justice system was not always able to acknowledge and address the underlying problems of these offenders.

Since 1989 (when the first drug court was introduced—see Chapter 5 for further discussion), hundreds of similar courts have been introduced and have expanded throughout the United States to provide services to different types of offenders. In addition to drug or mental health courts, courts have been established for persons convicted of firearms offenses, domestic violence, and driving while intoxicated. Moreover, there is a long history of teen courts (some were established in the 1940s, although they did not become popular until recently).

One of the key features of these specialized courts is that the judge, prosecutor, and defense attorneys have developed expertise working with a specific type of offender population. In addition to developing expert knowledge, these officials are also able to approach the case with some degree of sensitivity toward the main challenges that these offenders confront. Court officials and staff members working in these courts also have a greater knowledge of community resources and programs that are intended to help the offender. These courts use the coercive power of the justice system to ensure that offenders comply with their treatment or rehabilitative programs; if the offender successfully completes treatment, he or she may be rewarded with the expunging of the offender's criminal record. If, by contrast, the offender does not comply with treatment, the offender may be sent to jail or returned to prison (in the case of a reentry court). Many of these courts feature the offender making frequent (e.g., weekly or biweekly) court appearances and receiving support and encouragement from the judge after the offender's progress is reported.

Specialty courts are typically community-based programs, and many of them are used as a form of diversion from further processing (see Chapter 8). As such, these courts will probably have a greater long-term impact upon community corrections than other correctional services will. Probation and parole officers may play a significant role in the future of these courts.

Initial evaluations show that specialty courts are more effective than some other approaches are, but there are a number of concerns.[31] First, some are critical that these courts force the offender to participate in treatment. Second, if there is not an array of community-based resources to which offenders can be referred, these programs tend to be less successful. Third, some critics suggest that the rights of the participants in these courts are not always protected.[32] Regardless of such concerns, specialty courts appear to be a strategy of responding to difficult-to-manage offender groups and are likely to expand as time goes on.

Evidence-Based Corrections

In Chapter 12, we mentioned that increased attention is being directed to better understand the types of therapeutic, skills-based, and educational programs that research has demonstrated to be effective at helping offenders avoid criminal behavior. Interestingly, much of this research has shown that offenders with the highest levels of risk for reoffending have shown the greatest benefits, but only when rehabilitative programs are tailored to the individual. Although interest in correctional rehabilitation has historically been limited to practitioners and academics, this concern has become more broad-based in the past decade.

Today, politicians, policy analysts, and program administrators are more likely to have a better understanding of the factors that contribute to lower recidivism rates. This interest has translated into a greater degree of funding for university-based research centers that study this issue (e.g., the Center for Evidence-Based Corrections at the University of California in Irvine). Thus, the movement toward evidence-based programming is likely to continue.

Community Reentry

There is growing awareness that one way to increase public safety is to develop effective community reentry strategies to reduce recidivism. In Chapter 13, we stated that there has been a movement to better understand the challenges that ex-prisoners face when returning to the community after serving a prison term (e.g., overcoming the stigma of a criminal conviction, the lack of a recent work history, and poor educational achievement). In addition to traditional programs, such as halfway houses or reentry centers operated by correctional systems, there has been increased interest in developing collaborative programs that bridge the prison and community. A focus of many of these programs is to provide an array of supports to the ex-prisoner (whether those supports are financial, employment-related, therapeutic, or educational) to make the transition from prison to the community more successful. Most evidence suggests this trend will continue and expand in the future, both in adult corrections and juvenile justice (see the following discussion).

Corrections and Policy

In some jurisdictions, concern over the threat posed by certain offenders has prompted state legislators to enact laws that place harsh restrictions on parolees. On April 5, 2007, the Cable News Network reported that Florida sex offenders who had been paroled were living under a bridge on an interstate highway because the conditions of their parole restricted them from living within a half mile of a school or places where children gathered, such as playgrounds or swimming pools. Living under the bridge was the only option that did not violate conditions of parole. On the face, these laws might seem reasonable, but they often have unanticipated consequences. Living under a bridge is not a suitable housing arrangement, and this might actually reduce the parolee's stability and contribute to an offense. This illustrates how laws meant to make us safer can have the unanticipated effect of putting the public at greater risk.

Juvenile Justice

Increased Emphasis on Aftercare

As emphasized in Chapter 14, effective community reentry (aftercare) strategies are no less important in juvenile justice than they are in adult corrections. Traditionally, aftercare has been a neglected component of juvenile justice. Only in recent years has a trend emerged toward giving it some of the attention it warrants, with quite impressive results in some efforts.[33] What must be remembered is that gains made by juveniles during a good residential placement experience can be rendered essentially nil by inadequate or nonexistent aftercare programming upon release.[34]

The typical objection to aftercare in many jurisdictions is that staff and fiscal resources are simply insufficient. Nevertheless, aftercare can actually save resources. If juveniles released with adequate aftercare recidivate at lower rates, taxpayers and funding agencies avoid the cost of the subsequent delinquent act and the cost of reinstitutionalization. If resources are an issue (and they almost always are), the better approach is to reduce the length of incarceration and put the money saved toward aftercare.[35] As Wells and his colleagues suggest, aftercare is at least as important as what takes place during incarceration, and resources traditionally spent on incarceration are often better utilized when effective aftercare programs are in place.[36]

Increased Community and Agency Collaboration

Another noteworthy trend in juvenile justice is increased collaboration between police, probation, prosecutors, and school officials in an effort to control both gangs and illegal firearms among juveniles. These targeted deterrence efforts, such as Operation Ceasefire in Boston and Los Angeles, combine the efforts of school officials, court officials, law enforcement officers, community-based organizations, and juvenile probation officers to (1) reduce gun carrying by offenders on streets; (2) reduce youth involvement with and use of firearms; and (3) keep guns out of places where violence is likely to occur, such as homes with a history of domestic violence and rowdy bars or other "hot spots."

These efforts begin by identifying neighborhoods or areas where gang and gun violence are particularly problematic. After these neighborhoods are identified, high-risk probationers and parolees in the neighborhood are notified that criminal behavior will no longer be tolerated. Probation officers then work together with police officers to arrest and revoke individuals in the identified areas who have outstanding warrants or probation or parole violations. Additionally, prosecutors often file federal firearms charges against offenders who have committed firearm violations, thus increasing the likelihood that these offenders will be removed from the neighborhood for an extended period of time. Such strategies send a clear message to high-risk probationers and parolees and gang members in these areas that violence is not acceptable. These efforts have shown some promise in reducing serious violence in Boston, Indianapolis, Minneapolis, and Los Angeles.[37]

School Resource Officers

On any given day, there are almost 4,000 **school resource officers** (SROs) working in schools throughout the United States (see **Figure 15–1**). SROs are typically city or county law enforcement officers employed by police departments or sheriff's offices and assigned to work specifically in one or more schools. Although SROs may be funded fully or partly by the school district that they serve, they typically remain employees of their respective law enforcement agencies.[38] SROs function as liaisons between the school, community, and police. They also teach law-related education classes, counsel students, and perform law enforcement duties in and around the school.

SRO programs are widely acclaimed and widely used. Despite their popularity, it is difficult to find empirical studies that assess whether this popularity is justified. Research shows that SROs are generally (1) well regarded by the school administrators for which they work, (2) important in influencing perceptions of

Figure 15–1 The use of school resource officers has grown popular.
Source: *© Jim Parkin/ShutterStock, Inc.*

safety among students and staff, and (3) well received by parents and the community.[39] Future research will need to be conducted in an effort to determine the effectiveness of these programs in reducing crime on school campuses.

The "Adultization" of Juvenile Justice

A recent trend concerns the increasing "**adultization**" of juvenile justice, a process by which the juvenile system becomes more like the adult one. In the past 25 years, nearly every state has revised its laws or adopted new legislation to make it easier to transfer adolescents from juvenile courts to adult criminal courts. Additionally, many states have created a "Youthful Offender" category, which gives states a mechanism to impose strict, adult sanctions on juveniles convicted of violent crimes in lieu of waiver to an adult court.

Four other changes provide evidence of an adultization trend. First, the increase in the severity of punishment for juveniles parallels patterns found in the adult criminal justice system. Second, some jurisdictions have implemented fixed sentencing schemes in juvenile courts that are similar to sentencing guidelines in adult courts. These guidelines reduce the amount of discretion traditionally held by juvenile court judges. Third, adult prisons across the United States hold thousands of people who were under age 18 at the time they were admitted to prison to be punished as adults.[40] Fourth, recent legislation is beginning to dismantle the confidential nature of juvenile justice proceedings. For example, in 2006 Kentucky passed House Bill 3, which made felony juvenile convictions open to the public and all juvenile files available to law enforcement.

These trends toward adultization of juvenile justice have led some to argue that separate juvenile courts are no longer necessary (see Chapter 14) and that juvenile courts are more like criminal courts at the beginning of the twenty-first century than at any time since 1899 when the juvenile court first appeared.[41] Furthermore, there is some evidence that the effects of adultization trends may not be entirely negative. One study reports that among persons under 21 years of age, those incarcerated in adult facilities reported greater access to rehabilitative services than those in juvenile facilities.[42]

Given the continuing popularity of "get tough" policies, we expect this more punitive nature of juvenile justice to persist into the foreseeable future. Furthermore, in view of past trends mentioned in Chapter 14, it seems likely that concerns will continue to exist about such punitive trends unfairly affecting racial and ethnic minority youths at disproportionately high rates.[43]

Correctional Futurism

For several years, certain academics, policy analysts, and police administrators have pursued **police futurism** by examining how developments up to 50 years into the future might conceivably affect the field of policing. Police futurists have examined social, political, economic, technical, and crime trends to see whether they can develop proactive (rather than reactive) approaches to these influences. Although some issues cannot be foreseen (e.g., the rapid breakdown of social order and control that occurred after Hurricane Katrina hit New Orleans), futurists try to learn from the past, evaluate the impact of technology, and anticipate the worst-case scenarios and best responses to these scenarios.

Although **correctional futurism** per se does not currently exist, correctional administrators and planners regularly try to gauge the future and its impact upon their organizations. Accurate anticipation is a key to effective preemption of problems. To reduce uncertainty and the possibility of unforeseen negative circumstances, correctional administrators engage in **environmental scanning**, which means looking for wider trends or events that can affect an agency. For example, at the 2003 meeting of the Large Jail Network (an organization for facilities with inmate populations of more than 1,000 prisoners), administrators tried to predict the impact of trends, technology, and global crises on local corrections.

A cumulative theme of this book is that correctional systems are necessarily responsive, and in many ways captive, to decisions made in other justice agencies—political/legislative, police, and court agencies. At the outset of Chapter 1, we suggested that it is useful to think of criminal justice as a social institution of government, both affecting and being affected by the society of which it is part. The same reciprocal relationship exists between corrections and the criminal justice system of which corrections is part. The rest of the system and wider society influence corrections, but correctional agencies also influence the rest of the system and wider society. This insight has significant implications for the future of corrections.

For instance, it is apparent that changing legislative, law enforcement, and court priorities to crack down on drug kingpins or sex offenders will usually have significant and immediate effects on local jails and is also likely to affect the prison system over the longer term. Some of these impacts can be foreseen. As mentioned earlier, increasing the percentage of prisoners with lengthy sentences will result in a greater number of elderly prisoners, and this will in turn result in higher medical care costs. What may be less apparent, however, is that higher medical costs for prisoners (coupled with appellate court orders to provide medical care to elderly prisoners) will eventually become a source of competition with budget requests from trial courts and police agencies. As the saying goes in this regard, "there is no free lunch."

Something that merits attention from criminal justice scholars, policymakers, and practitioners is better understanding how laws or justice system practices sometimes have unanticipated or unforeseen consequences. To illustrate, sending a youngster to an institution for the first time, or waiving a juvenile offender to adult court, may demonstrate a tough, no-nonsense stance toward youth crime. These actions, however, may also reduce the particular juvenile's fear of incarceration, thereby curtailing the deterrent effect of the sanction. The juvenile might also gain status in the eyes of deviant peers for having been "locked up," while simultaneously becoming further stigmatized and disconnected from law-abiding groups. Furthermore, the juvenile is likely to be introduced to offenders he or she would otherwise never have met. These other offenders might try to recruit the juvenile into a gang and might teach him or her about the "best places to party," how to commit a residential burglary, or about some other undesirable act. Although there are clearly times when juveniles need to be confined, this illustration speaks to the importance of policymakers and practitioners working with researchers to learn whether criminal justice practices and legislation are, in fact, having the desired effects while simultaneously minimizing unintended harmful consequences.[44]

Sometimes it is not easy to foresee how the correctional system will be affected by changes in other elements of the justice system, social trends, or global changes that fall outside of the control of corrections or over which the justice system itself has little control. Community and institutional corrections have been profoundly affected by social trends. For instance, the increased use of methamphetamine has created significant challenges for justice systems, especially in some rural areas.[45] In addition to higher levels of crime, drug use trends also lead to other social problems (e.g., a greater need for child protective services when parents become drug-addicted and cannot adequately care for their children). Further, there are numerous other health-related challenges when people abuse this drug, including a painful and disfiguring condition that some dentists have called **meth-mouth** (which is severe tooth loss, damage, and decay brought on by the use of methamphetamine). When these long-term users are booked into jail, they generally require a greater degree of care than do traditional inmates.

Demographic changes will also affect corrections. For instance, the U.S. population is growing older, and the first baby boomers (persons born between 1946 and 1964) have now turned 60 years of age. Some jails have found that the average age of persons admitted has increased over the past 10 years.[46]

Women have also been arrested and imprisoned at higher rates in recent years than previously. This may have a significant long-term impact upon community and institutional corrections. The imprisonment rate for women has grown at a faster rate than that for males over the past decade.[47] If this trend continues, it will increase costs within institutional corrections because women require separate housing units (including the possibility of children's nursery programs) and also need higher levels of psychological and medical health care. Incarcerating more women prisoners may also require that correctional agencies recruit and retain more female correctional officers and implement more adequate rehabilitation programs at women's facilities.

Other changes may be the result of shifting government priorities. State budget crises during the 2000–2002 period affected correctional systems.[48]

Race and Gender in Corrections

In earlier chapters, we noted the disproportionate representation of ethnic and racial minorities in both community and institutional corrections populations. As the number of minorities increases in the general population, how will this affect justice systems and corrections? Some members of these groups, as well as large numbers of disenfranchised whites, feel disconnected from the mainstream community (i.e., they feel they have little stake in the "system"), and this may translate into higher rates of misconduct. We know, for instance, that prison gangs tend to form along racial and ethnic lines, and demographic changes in the prison population may also influence gang populations.

Budget shortfalls caused legislators to rethink severe sentences for nonviolent offenders, cut some prison-based programs (including rehabilitative opportunities for inmates), close some inefficient prisons, delay the opening of other prisons, and cancel the recruitment of new correctional officers. As a result, this one economic change had a direct impact upon prison operations, inmate treatment, as well as career opportunities for correctional officers.

Corrections might also be affected by other forces that are largely beyond the control of justice systems. Some prison systems have plans to respond to a **pandemic** such as influenza or severe acute respiratory syndrome.[49] The global influenza epidemic in 1918–1919, for instance, had an impact on sickness and mortality rates in prisons.[50] In light of these problems, some questions that correctional administrators must ask are: How does a prison best prepare for these outbreaks? How do we best manage and control quarantined or infected inmates (and staff who might not be permitted to leave the institution)?

Correctional agencies also have to respond to global issues, with terrorism being an excellent example. For obvious reasons, certain disenfranchised groups are prime recruitment targets for both domestic and global terrorist organizations, and correctional agencies deal with many individuals who live on the fringe of mainstream society. Because a principal objective of many terrorist organizations is to strike out against "establishment" targets of hatred, such organizations can be attractive to disenfranchised persons. After persons involved in terrorist acts have been convicted and incarcerated, one challenge is keeping them from corresponding with terror cells outside prison to help plot schemes such as the mass transit bombings in Madrid, Spain (2004), and London, England (2005). Another challenge is the day-to-day management of such prisoners. This can be

Corrections in the Real World

Prisons have already started coming to terms with the reality of confining persons both suspected and convicted of terrorist-related acts. The controversial facility established at Guantanamo Bay, Cuba, in 2002 holds *unconvicted* persons suspected by the U.S. government of being "enemy combatants"—al Qaeda and Taliban operatives. Probably the most well known facility that holds *convicted* terrorists is the Administrative Maximum U.S. Prison in Florence, Colorado (known as ADX Florence). The U.S. Bureau of Prisons opened this supermax facility in 1994 as a replacement for the supermax facility at Marion, Illinois (which, in turn, had replaced Alcatraz). ADX Florence holds high-profile federal prisoners and those considered the most extreme risks of violence. The prison holds (or has held) a number of persons convicted in connection with the 1993 World Trade Center bombings (including Ramzi Yousef, a senior al Qaeda member), Richard Reid (the "shoe bomber"), Terry Nichols (who collaborated with Timothy McVeigh in the Oklahoma City bombing), Theodore Kaczynski (the "Unabomber"), and Zacarias Moussaoui (convicted in connection with the September 11, 2001, World Trade Center attacks).

a difficult undertaking when the same prison holds (1) radical Islamic terrorists who justify violence against Americans under the auspices of Jihad, (2) members of the Aryan Brotherhood who despise many nonwhites, and (3) domestic terrorists of the extreme political right who may justify violence under the auspices of Christianity.

Conclusion

Throughout this book, we have portrayed the worlds of corrections and juvenile justice as multifaceted systems influenced by a wide variety of factors. These factors range from decisions made by legislators to enact or repeal criminal and civil laws, to decisions made by prosecutors and police officers to enforce these laws, to decisions made by judges in their choice of sentencing options. These decisions are not the only factors influencing corrections and juvenile justice. External pressures from politicians, the media, the public, or from victims' families, along with internal pressures from corrections professionals and clients, can move justice system leadership forcefully in multiple directions; and changes in the relative strengths of these pressures can abruptly cause directions to change. Corrections and juvenile justice are dynamic fields that often change in unexpected ways.

For most of history, change agents have been people untrained in the fields of corrections and juvenile justice. Some were shoemakers (John Augustus), some were sheriffs (John Howard), some were preachers (William Penn), and others were philanthropists (the Child Savers). All of them had one thing in common: They wanted to improve the lives of offenders and, in turn, increase public safety by reducing crime.

Combined with the zeal of reform-minded people, the fields of corrections and juvenile justice require persons who are educated and skilled to assume a variety of different roles. Many students tend to focus on correctional officer positions and overlook other employment opportunities working within corrections. In some states, for instance, more than one-third of all positions in corrections are not responsible for supervising prisoners. Instead there is a need for program administrators, budget and financial analysts, policy analysts, recruiters, and research specialists. Moreover, for those students who start as correctional officers, there are dozens of opportunities to become canine handlers, correctional counselors, gang investigators, parole officers, trainers, as well as the future correctional supervisors and administrators. Community and institutional corrections have become billion-dollar operations in larger jurisdictions, and many local, state, and federal agencies are actively recruiting college-educated persons to staff and lead these organizations into the future.

We hope that this book, in conjunction with related educational experiences, has increased your interest in working in corrections or juvenile justice. Given the tremendous growth in these fields mentioned at various points throughout this book and the professionalism movement described in Chapter 1, there is strong need for qualified, well-educated people to staff agencies; career opportunities have never been better. We hope that you will take the challenge to become a positive influence in the fields of corrections and juvenile justice.

READY FOR REVIEW

- Sentencing more offenders to longer sentences has become the norm in many jurisdictions. The growth rate in the use of imprisonment in the federal system continues to increase, while state imprisonment rates seem to be stabilizing in most jurisdictions.
- State prison overcrowding reached its peak in 1995 and has eased somewhat since then, but overcrowding in federal prisons has increased to 134 percent of capacity. Jail overcrowding is also a challenge in many jurisdictions, forcing administrators to release some inmates to the community.
- The percentage of elderly prisoners is increasing, and many states are developing specialized housing units for these inmates, as well as reconsidering practices related to compassionate releases.
- The rate of communicable diseases in the corrections population is much higher than the rate among community populations. This has implications for correctional officer safety, costs of imprisonment, and corrections–community partnerships to prevent these diseases from spreading to the community.
- A large number of drug offenders are imprisoned in the United States. More than one-half of the Federal Bureau of Prisons prisoners and approximately 20 percent of state inmates have been convicted of a drug offense. In May 2007, the USSC was critical of sentencing practices for crack cocaine and labeled them discriminatory against nonwhites. Despite this criticism, there appears to be little political willingness to change these laws.
- The involvement of the private sector in the delivery of correctional services is controversial. Many critics believe that punishment should not be delivered by organizations that are interested primarily in profit. Supporters of privatization argue that these corporations can deliver services more efficiently than government organizations can and can reduce taxpayer costs. Most of the growth in private operations is in the area of federal detention, although other operators provide food services, medical care, as well as facility construction and leasing.
- There is a long history of involvement of religious groups in the delivery of correctional programs. In recent years, there has been a renewal of interest in faith-based initiatives.
- Correctional administrators are taking advantage of a number of technological innovations, including methods of surveillance, information management, communication, and nonlethal weapons or control. Although these innovations enable correctional officers to operate more efficiently or safely, there are also criticisms, especially if innovations are misused.

- A number of different types of courts have been developed to deal with specific groups of offenders, such as persons with mental illness, drug abusers, or driving under the influence of alcohol offenders. In addition to having expert knowledge about working with these groups, specialty courts generally monitor the community-based treatment of offenders and sometimes reward first-time offenders by expunging their criminal convictions.
- There is increased interest in developing evidence-based corrections, or rehabilitative interventions that research has demonstrated reduces offender recidivism.
- Politicians and policymakers are increasingly interested in finding better ways to make the transition from prison to community more successful for prisoners. Many of these challenges relate to social skills development (both within the institution and in the community) and finding employment.
- The field of juvenile justice is also experiencing trends that will shape the future. Examples include increased attention to aftercare, interagency collaboration, and school safety. At the same time, there is evidence that the juvenile system is growing more like the adult system, a trend called "adultization."
- Correctional administrators are increasingly interested in predicting how economic, social, demographic, and political trends will affect the delivery of correctional services. These predictions enable correctional administrators to forecast the need for expanding correctional capacity as well as to predict changes in the work force and offender populations.

KEY TERMS

active denial system A device that allows an individual to send a beam of electromagnetic energy through another individual, thus causing tremendous pain but no permanent harm

adultization A trend to make the juvenile justice system more like the adult system

body cavity screening systems Noninvasive systems based on magnetic resonance imaging used to detect contraband, such as very hard plastic or metal weapons or narcotics being hidden and/or transported inside the body

correctional futurism Examining future trends to determine how they might affect correctional agencies

criminal justice technology the use of scientific knowledge to solve practical problems faced by practitioners in criminal justice and corrections

decriminalization A strategy whereby criminal penalties for drug possession, use, and distribution are abolished or reduced

environmental scanning a strategy for examining and forecasting changes, trends, or events that can affect the agency (e.g., how demographic trends will affect the future practice of corrections)

face-recognition technology a technology designed to scan facial characteristics of individuals and match them with a database for recognition
faith-based programs Social service programs conducted by nonprofit religious organizations that offer treatment or rehabilitative services
meth-mouth Severe tooth loss, damage, and decay brought on by the use of methamphetamine
pandemic A worldwide outbreak of an infectious disease, such as influenza or severe acute respiratory syndrome
police futurism Examining future trends to determine how they might affect policing
prevention/education A drug control strategy whereby programs designed to educate the public about the harmful effects of drug use and addiction are used to deter individuals from drug use
prisoner reentry initiative A federal program supported by the U.S. Departments of Justice, Education, Health and Human Services, Housing and Urban Development, and Labor that provides funding to develop or enhance strategies to ease reentry into the community for offenders released from prison
radio-frequency identification tags Small computer tags that are imbedded in items (and potentially people), thereby allowing the items and the movement of the items to be monitored electronically
school resource officers Police officers assigned to work in schools
specialty courts Courts set up to address specialized problems displayed by offenders such as drug abuse, mental illness, and reentry
stun belts Belts containing a battery and control pack that are fastened around an individual's waist or arm that allow another individual to use a remote control to tell the battery pack to give that individual an electric shock; these belts are used in a wide variety of corrections settings
supply reduction A drug control strategy whereby law enforcement seeks to reduce the supply of illegal drugs being transported into the United States in the hope that reduced supply will lead to reduced drug use
surveillance cameras Cameras placed in inconspicuous locations to monitor (and often record) human behavior
TASER stun devices [Thomas A. Swift's Electric Rifle (TASER) stun devices] Devices that fire small, dartlike electrodes with attached metal wires up to 10 meters. Upon contact, these electrodes send a pulse of electricity into the individual to disrupt nerve and muscle functioning. TASERs are used in a wide variety of law enforcement and correctional settings
treatment A drug control strategy whereby individuals with drug abuse and addiction problems participate in programs designed to treat their addictions and thus reduce their use of illegal drugs

YOU ARE THE CORRECTIONS PROFESSIONAL SUMMARY

1. What kind of factors would need to be considered to ensure that you made the most accurate report to the state leadership? You will need to consider all the challenges discussed in this chapter when preparing your report. Additionally, you need to be aware that changes in public opinion or political leadership may dramatically affect each of the challenges that you will face.
2. What will be the most challenging areas in corrections and juvenile justice in your state in the twenty-first century? Each of the areas discussed in this chapter has the potential to be the most challenging area of concern. As such, you should do a careful analysis of these issues in your state prior to developing any sort of list of most challenging problems facing corrections and juvenile justice. Demographic differences (such as the percentage of young or older people in your state) and political differences (such as whether your state leadership is largely conservative or liberal) will structure the order in which you prioritize these challenges.
3. After you have identified a list of challenges facing the Departments of Corrections and Juvenile Justice, how will you decide what is the most important issue for funding? How will you decide where funding can be reduced? The most important decision that you will have to make for your report is the funding recommendations. After completing the careful analysis mentioned in the previous response, you will have a good idea about the challenges that require changes in funding in the upcoming years. Remember, not all challenges can be solved by funding changes; some will need careful legislative or judicial consideration instead.

NOTES

1. P. M. Harrison and A. J. Beck, *Prisoners in 2005*, NCJ 215092 (Washington, DC: U.S. Department of Justice, Bureau of Justice Statistics Bulletin, 2006).
2. A comparison of international incarceration rates is available at the International Centre for Prison studies Web site www.prisonstudies.org/.
3. C. J. Mumola and A. J. Beck, *Prisoners in 1996*, NCJ 164619 (Washington, DC: U.S. Department of Justice, Bureau of Justice Statistics Bulletin, 1997); Harrison and Beck, *Prisoners in 2005*.

4. Federal Bureau of Prisons, *Weekly Population Report*, 2007, www.bop.gov/locations/weekly_report.jsp.
5. Harrison and Beck, *Prisoners in 2005*.
6. D. K. Gilliard and A. J. Beck, *Prisoners in 1997*, NCJ 170014 (Washington, DC: U.S. Department of Justice, Bureau of Justice Statistics Bulletin, 1998).
7. R. H. Aday, *Aging Prisoners: Crisis in American Corrections* (Westport, CT: Praeger, 2003).
8. D. J. James and L. E. Glaze, *Mental Health Problems of Prison and Jail Inmates*, NCJ 213600 (Washington, DC: U.S. Department of Justice, Bureau of Justice Statistics Special Report, 2006).
9. Aday, *Aging Prisoners*.
10. L. M. Maruschak, *HIV in Prisons, 2004*, NCJ 213897 (Washington, DC: U.S. Department of Justice, Bureau of Justice Statistics Bulletin, 2006).
11. "News and Literature Reviews: HIV Costs Have Decreased," *Infectious Diseases in Corrections Report* 9(2006): 8.
12. M. Welch, *Corrections: A Critical Approach*, 2nd ed. (New York: McGraw-Hill, 2004).
13. T. M. Hammet, S. Kennedy, and S. Kuck, *National Survey of Infectious Diseases in Correctional Facilities: HIV and Sexually Transmitted Diseases* (Washington, DC: U.S. Department of Justice, 2007), www.ncjrs.gov/pdffiles1/nij/grants/217736.pdf.
14. Federal Bureau of Prisons, *Weekly Population Report*.
15. Welch, *Corrections*.
16. S. Aos, "Using Taxpayer Dollars Wisely: The Costs and Benefits of Incarceration and Other Crime Control Policies," in *Corrections Policy: Can States Cut Costs and Still Curb Crime?* eds. E. Gross, B. Griese, and K. Bogen-Schneider, (Madison: University of Wisconsin Center for Excellence in Family Studies, 2003).
17. Integrated Substance Abuse Program, *Evaluation of the Substance Abuse and Crime Prevention Act Final Report* (Los Angeles: Integrated Substance Abuse Program, 2007).
18. U.S. Sentencing Commission, *Report to the Congress: Cocaine and Federal Sentencing Policy* (Washington, DC: U.S. Sentencing Commission, 2007).
19. Harrison and Beck, *Prisoners in 2005*, 5.
20. A. Tabarrok, *Changing the Guard: Private Prisons and the Control of Crime* (Oakland, CA: Independent Institute, 2003).
21. Tabarrok, *Changing the Guard*.

22. About.com, "Bush's Faith-Based Initiatives Launched," 2007, http://usgovinfo.about.com/library/weekly/aa012901a.htm.
23. U.S. Department of Labor Employment and Training Administration, home page, www.doleta.gov/pri/.
24. M. T. Sumter and T. R. Clear, *What Works with Religion in the Correctional Setting?* publication 2 (Washington, DC: International Association of Community Corrections Monograph Series, 2002).
25. S. D. Camp, J. Klein-Saffran, O. Kwon, D. M. Daggett and V. Joseph, "An Exploration into Participation in a Faith-Based Prison Program," *Criminology and Public Policy* 5, no. 3 (2006): 529–550.
26. C. E. Smith, M. McCall, and C. P. McCluskey, *Law and Criminal Justice: Emerging Issues in the 21st Century* (New York: Peter Lang, 2005).
27. Smith et al., *Law and Criminal Justice.*
28. Smith et al., *Law and Criminal Justice.*
29. W. P. Bozeman, "Withdrawal of Taser Electroshock Devices: Too Much, Too Soon," *Annals of Emergency Medicine* 46(2005): 301.
30. Smith et al., *Law and Criminal Justice.*
31. M. E. Moore and V. A. Hiday, "Mental Health Court Outcomes: A Comparison of Re-arrest and Re-arrest Severity Between Mental Health Court and Traditional Court Appearances," *Law and Human Behavior* 30(2006): 659–674.
32. A. J. Grudzinskas, J. C. Clayfield, K. Roy-Bujnowski, W. H. Fisher, and M. H. Richardson, "Integrating the Criminal Justice System into Mental Health Service Delivery: The Worcester Diversion Experience," *Behavioral Sciences and the Law* 23(2005): 277–293.
33. D. Josi and D. K. Sechrest, "A Pragmatic Approach to Parole Aftercare: Evaluation of a Community Reintegration Program for High-Risk Youthful Offenders," *Justice Quarterly* 16(1999): 51–80.
34. See G. R. Grissom and W. L. Dubnov, *Without Locks and Bars: Reforming Our Reform Schools* (New York: Praeger, 1989).
35. J. A. Fagan, "Treatment and Reintegration of Violent Juvenile Offenders: Experimental Results," *Justice Quarterly* 7(1990): 233–263.
36. J. B. Wells, K. L. Minor, E. Angel, and K. D. Stearman, "A Quasi-Experimental Evaluation of a Shock Incarceration and Aftercare Program for Juvenile Offenders," *Youth Violence and Juvenile Justice* 4(2006): 219–233.
37. E. F. McGarrel and S. Chermak, *Strategic Approaches to Reducing Firearms Violence: Final Report on the Indianapolis Violence Reduction Partnership*

(Washington, DC: U.S. Department of Justice, 2004), www.ncjrs.gov/pdffiles1/nij/grants/203976.pdf.

38. D. C. May, G. Cordner, and S. D. Fessel, "School Resource Officers as Community Police Officers: Fact or Fiction?" *Law Enforcement Executive Forum* 4, no. 6 (2004): 177–188.
39. May et al., "School Resource Officers."
40. B. Krisberg, *Juvenile Justice: Redeeming Our Children* (Thousand Oaks, CA: Sage, 2005).
41. A. Kupchik, "The Decision to Incarcerate in Juvenile and Criminal Courts," *Criminal Justice Review* 31, no. 4 (2006): 309–336.
42. A. Kupchik, "The Correctional Experiences of Youth in Adult and Juvenile Prisons," *Justice Quarterly* 24(2007): 247–270.
43. Krisberg, *Juvenile Justice*. And see F. E. Zimring, *American Juvenile Justice* (New York: Oxford University Press, 2005).
44. On the general topic of minimizing harm when responding to offenders, see T. R. Clear, *Harm in American Penology: Offenders, Victims, and Their Communities* (Albany: State University of New York Press, 1994).
45. R. W. Weisheit and J. Fuller, "Methamphetamine in the Heartland: A Review and Initial Exploration," *Journal of Crime and Justice* 27(2004): 131–151.
46. R. Kuhlmann and R. Ruddell, "Elderly Jail Inmates: Problems, Prevalence, and Public Health," *Californian Journal of Health Promotion* 3(2005): 50–61.
47. Harrison and Beck, *Prisoners in 2005*.
48. D. F. Wilhelm and N. R. Turner, *Is the Budget Crisis Changing the Way We Look at Sentencing and Incarceration?* (New York: Vera Institute of Justice, 2002).
49. California Department of Corrections and Rehabilitation, "CDCR pandemic preparedness," 2007, www.cdcr.ca.gov/Communications/pandemic_preparedness.html.
50. L. L. Stanley, "Influenza at San Quentin Prison, California," *Public Health Reports* 34(1919): 996–1038.

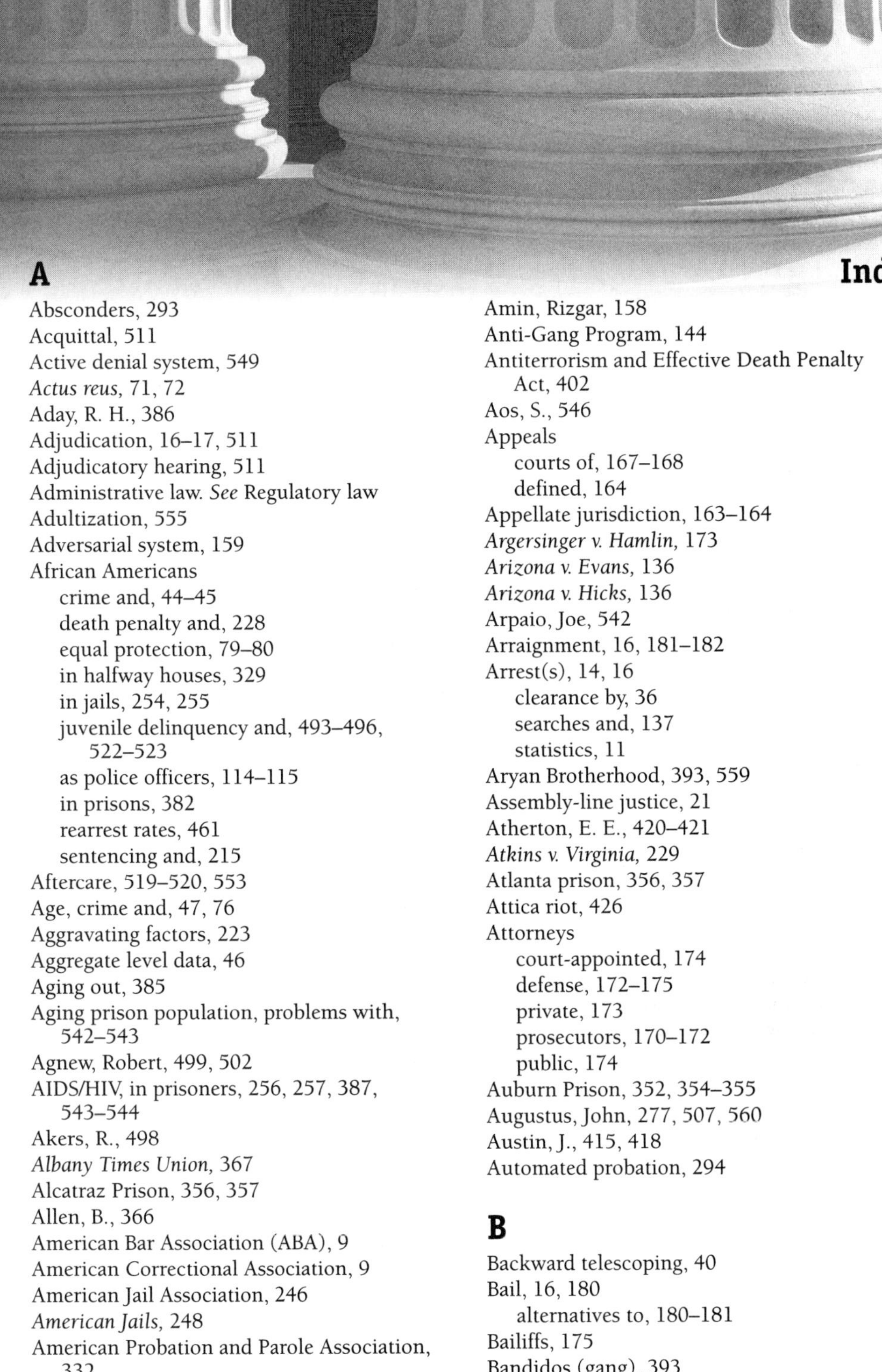

Index

A

Absconders, 293
Acquittal, 511
Active denial system, 549
Actus reus, 71, 72
Aday, R. H., 386
Adjudication, 16–17, 511
Adjudicatory hearing, 511
Administrative law. *See* Regulatory law
Adultization, 555
Adversarial system, 159
African Americans
  crime and, 44–45
  death penalty and, 228
  equal protection, 79–80
  in halfway houses, 329
  in jails, 254, 255
  juvenile delinquency and, 493–496, 522–523
  as police officers, 114–115
  in prisons, 382
  rearrest rates, 461
  sentencing and, 215
Aftercare, 519–520, 553
Age, crime and, 47, 76
Aggravating factors, 223
Aggregate level data, 46
Aging out, 385
Aging prison population, problems with, 542–543
Agnew, Robert, 499, 502
AIDS/HIV, in prisoners, 256, 257, 387, 543–544
Akers, R., 498
Albany Times Union, 367
Alcatraz Prison, 356, 357
Allen, B., 366
American Bar Association (ABA), 9
American Correctional Association, 9
American Jail Association, 246
American Jails, 248
American Probation and Parole Association, 332
Amin, Rizgar, 158
Anti-Gang Program, 144
Antiterrorism and Effective Death Penalty Act, 402
Aos, S., 546
Appeals
  courts of, 167–168
  defined, 164
Appellate jurisdiction, 163–164
Argersinger v. Hamlin, 173
Arizona v. Evans, 136
Arizona v. Hicks, 136
Arpaio, Joe, 542
Arraignment, 16, 181–182
Arrest(s), 14, 16
  clearance by, 36
  searches and, 137
  statistics, 11
Aryan Brotherhood, 393, 559
Assembly-line justice, 21
Atherton, E. E., 420–421
Atkins v. Virginia, 229
Atlanta prison, 356, 357
Attica riot, 426
Attorneys
  court-appointed, 174
  defense, 172–175
  private, 173
  prosecutors, 170–172
  public, 174
Auburn Prison, 352, 354–355
Augustus, John, 277, 507, 560
Austin, J., 415, 418
Automated probation, 294

B

Backward telescoping, 40
Bail, 16, 180
  alternatives to, 180–181
Bailiffs, 175
Bandidos (gang), 393

Barron v. City of Baltimore, 84–85
Bartley-Fox law, 222
Battered woman/spouse syndrome, 78
Bean, Roy, 114
Beardon v. Georgia, 319
Bear Paw Warrior Society, 393
Beccaria, Cesare, 8, 349
Beck, A. J., 253, 382, 389
Beirne, P., 94
Belknap, Joanne, 390
Bell v. Wolfish, 399, 400
Bench trials, 169
Benefit of clergy, 276, 350
Bentham, Jeremy, 8, 349
Bifurcated proceedings, 228
Big house prisons, 351–352
Bill of Rights, 8, 82
  See also individual amendments
Billy the Kid, 8–9
Black Guerilla Family, 393
Blended sentencing, 514
Bloods (gang), 393
Body cavity screening systems, 549
Bondsmen, 180
Bones, 20–21
Bonnie and Clyde, 9
Bontrager, Stephanie, 286
Boot camps, 327–328, 518
Bordenkircher v. Hayes, 214
Bosta, D., 366
Boston Operation Ceasefire, 128–129
Bounds v. Smith, 400
Bozeman, W. P., 549
Brace, Charles Loring, 506
Braithwaite, J., 46
Brockway, Zebulon, 450–451
Broken windows probation, 293–294
Brubaker, 353
Brutalization effect, 230
Bryant, Kobe, 173
Bucqueroux, Bonnie, 130
Building tenders, 353
Bureau of Investigation, 120
Bureau of Prisons, 252, 348, 356–358
Burger, Warren, 87, 431
Burgess, R., 498
Burnout, 371
Burns, William, 122
Burns Agency, 122
Bush, George Herbert Walker, 448
Bush, George W., 91, 170, 424, 467, 472, 547–548

C

Capetillo, Eddie, 524
Capital punishment. *See* Death penalty
Capital Punishment, 52
Capone, Al, 9, 357
Career criminal, 397
Carroll v. United States, 138
Case (precedent) law, 69–70
Causation, 72–73
Celebrated cases, 20
Center for Media and Public Affairs, 19–20
Centralized model of policing, 121
Chain gangs, 352
Challenge of Crime in a Free Society, The (President's Commission on Law Enforcement and Administration of Justice), 116
Challenges for cause, 184
Challenges to the array, 184
Children's Aid Society, 506
Child Savers, 506
Chimel, Ted, 135–136
Chimel v. California, 135–136
Chivalry hypothesis, 43
Cho, Seung-Hui, 66
Chronic offender, 490
Circle sentencing, 210
Circumstantial evidence, 185
Civil law, defined, 65, 66
Civil rights movement, 9
Civil Rights of Institutionalized Persons Act (CRIPA) (1980), 399
Civil War, 8
  policing during the era, 114–115
Classification
  external, 415
  form, 416–417
  internal, 415
  objective, 415
  of prisoners, 262–263, 414–418
  subjective, 415
Clearance by arrest, 36
Clearance by exceptional means, 36
Clemmer, Donald, 394, 395, 397
Clerk of court, 176
Closing statements, 186
Cochrane, Johnnie, 173
Code of Hammurabi, 206
Cohort studies, 491
Cole, D., 79, 80
Colonial America
  court systems in, 161

jails in, 243–244
policing in, 113–114
treatment of crime in, 8
Combined DNA Index, 120
Commission on Law Enforcement and Administration of Justice, 9
Common law, 64
Community-based jail programs, 263–264
Community corrections
defined, 314
history of, 314–317
Community Corrections Acts, 315
Community courts, 166
Community policing, 130
Community service, 320–321
Compassionate release, 387
Compulsion, 77
Computer errors exception to the exclusionary rule, 136
Concurrence, 73
Conditional mandatory release, 446–447
Conditional release, 180
Congregate system, 352
Conjugal visits, 396
Consent decree, 260
Constable, 112
Constitutional law, 69
Constitution of the United States, 3
See also individual amendments
Bill of Rights, 8, 82
due process, 81–88
Continuum of sanctions, 315–316
Contraband, 395
Convict criminologist, 397
Cooper v. Pate, 86, 398–399
Cops, 111
Coroner, 112
Corporate crime, 95–96
Corpus delicti, 70–72
Correctional camps, 360
Correctional futurism, 555–559
Correctional institutions, 355–356
Correctional medical centers, 363
Correctional officers, 364–369
Corrections
See also Community corrections; Juvenile corrections
as a component of the criminal justice system, 12–13
ethnic differences, 19
evidence-based, 431–433, 551–552
role of, 17, 97–99
use of term, 2
Counseling programs for prisoners, 429–430
Court administrators 175–176
Court-appointed counsel, 174
Court of Federal Claims, 167, 168
Court of International Trade, 167, 168
Court reporters, 176
Courts
See also type of
of Appeals, 167–168
community, 166
as a component of the criminal justice system, 12
drug, 187–189
family, 166
of general jurisdiction, 164
impact of, on jails, 259–261
juvenile, 507–514
of last resort, 164
of limited jurisdiction, 164
role of, 12
specialty, 550–551
structure and organization of, 162–168
teen, 189
Court staff
administrators, 175–176
bailiff, 175
clerks, 176
defense counsel, 172–175
judges, 169–170
prosecuting attorney, 170–172
reporter, 176
Court system
federal, 162, 166–168
history of, 159–162
philosophy of, 159
state, 163–166
Court TV, 170
Covington, Stephanie, 465
Coyle, A., 386
Creaming, 283
Cressey, Donald, 397
Crime
age and, 47
categorizing, 88–97
defined, 33
ethnic differences, 43, 44–45
gender differences, 42–44
socioeconomic status and, 46–47
weapon availability and, 47–48
Crime control models, 21–23
Crime data
on death penalty, 52–53

Crime data (*continued*)
on prisoners, 48–50
on probationers and parolees, 50–52
Crime data, sources of
National Crime Victimization Survey, 39–41
self-report surveys, 41–42
Uniform Crime Reports, 33–39
Crimefighters, police as, 124
Crime prevention programs, 124
Criminal justice
as an academic field of study, 6–7
criminology compared to, 7
evolution of, 7–9
professionalism of, 9
as a social and governmental institution, 3–6
training and academic education, distinction between, 7
training versus academic education, argument over, 10
use of term, 2
Criminal justice process
See also under each component
adjudication, 16–17
arrest, 14, 16
corrections, 17
defined, 14
flowchart, 15
funnel, 18
pretrial stage, 16
sentencing, 17
use of term, 2
Criminal justice system
defined, 11
growth of, 9–11
use of term, 2
Criminal justice system, components of
corrections, 12–13
courts, 12
how they work together, 13–14
police, 11
Criminal justice technology, 548–550
Criminal law
defined
procedural, 66–67, 68
substantive, 66
Criminal trial, steps of
closing statements, 186
judge's instructions to jury, 187
jury selection, 184
opening statements, 184–185
presentation of evidence, 185–186
testimony of witnesses, 186
trial initiation, 183–184
Criminogenic, 295
Criminology, criminal justice compared to, 7
Crips (gang), 393
Crofton, Sir Walter, 450
Crouch, B. M., 353
Crucibles of Crime (Fishman), 244
Cruz v. Beto, 400
CSI, 20–21, 111
Cybercrime, 91

D

Dahmer, Jeffrey, 227–228
DARE (Drug Abuse Resistance Education), 503
Dateline NBC, 470
Davis, Katherine Bement, 367
Day fines, 319
Day reporting centers (DRCs), 327
Day treatment centers, 516–517
Deadly force, use of, 138–139
Death penalty, 8
for juveniles, 523–524
methods of, 229
pros and cons, 227–230
statistics on, 52–53, 388
Death row, prisoners on, 388–389
Decentralized model of policing, 121
Decriminalization, 545
Defendants, 178–179
Defense counsel, 172–175
Deferred sentence, 284
Del Carmen, R. V., 325
Delinquency. *See* Juvenile delinquency
Department of Homeland Security, 117, 118
Detention hearing, 510–511
Determinate sentences, 220–221
Deterrence, 207–208
Diallo, Amadou, 133, 134
Diesel therapy, 392
Differential association, 498
Dillinger, John, 9
Directed patrol, 128
Direct evidence, 185
Direct supervision, 261
Discretion, 11
defined, 63
factors influencing, 141–142

Discretionary parole release, 447
Dispositional hearing, 511
Disproportionate minority contact (DMC), 522–523
Dispute resolution centers, 165
District courts, 166–168
Diversion
  defined, 275–276, 513
  effectiveness of, 283
  historical overview of, 276–278
  juvenile programs, 512–514
  process, 281–282
  pros and cons of, 282–283
  staff roles in, 278–279
  types of, 280
DNA
  Combined Index, 120
  technology, 549–550
Double jeopardy, 78
Drug abuse, 93
  juvenile delinquency and, 502–503
Drug courts, 187–189
Drug offenders, 381, 382
Drug policies, 545–546
Due deference phase, 399, 400
Due process, 81–88
  for juveniles, 508–509
  model, 21–22
  revolution, 87
Dukakis, Michael, 448
Durham, A. M., 414
Durham Rule, 75
Durose, M. R., 384
Dyads, 391

E

Earp, Wyatt, 114
Eastern State Penitentiary, 351
Economic sanctions, 317–320
Educational programs for prisoners, 428–429
Edward I, king of England, 112
Edwards v. Arizona, 139
Eighth Amendment, 83, 84
Elderly prisoners, 386–387, 542–543
Electronic monitoring (EM), 324–327
Elliot, B., 395
Elmira Reformatory, 354, 450
Elrod, P., 489
Emergency search, 137
Employment for parolees, 468
England
  crime in, 7, 8, 64, 112–113, 159–160, 349
  gaols, 241–243
Enhanced supervision partnerships, 143
Enlightenment, 7, 8
"Enquiry into the Causes of the Late Increase of Robbers" (Fielding), 113
Entertainment Tonight, 21
Entrapment, 76–77
Environmental scanning, 556
Equal protection, 79–80
Escobedo v. Illinois, 139
Estelle v. Gamble, 400
Ethics
  defense, 174–175
  electronic monitoring and, 325–326
Ethnic and racial differences. *See* Racial differences
Ethnic group, use of term, 44
Evidence
  circumstantial, 185
  defined, 185
  direct, 185
  presentation of, 185–186
  suppressing, 186
Evidence-based corrections, 431–433, 551–552
Evil woman argument, 43
Exclusionary rule, 85–86, 87, 135
Executive branch, defined, 5
Executive clemency, 350
Ex Parte Crouse, 506
Expert witnesses, 177
Expressive riots, 425
External classification, 415
Extradition, 6
Extralegal factors, 206

F

Face-recognition technology, 549
Faith-based programs, 547–548
Families, impact of incarceration and reentry on, 469
Family courts, 166
Family group counseling, 210
Fargo, William, 114
Farkas, Mary Ann, 368–369
Farmer v. Brennan, 399
FBI (Federal Bureau of Investigation), 9
  role of, 120

FBI (*continued*)
  statistics provided by, 11
  Uniform Crime Reports, 33–39, 120
Federal Bureau of Prisons, 252, 348, 356–358
Federal court system, 162, 166–168
Federal jails, 252
Federal law enforcement agencies, 117–120
Federal prisons, 252, 348, 356–358
Federal Transfer Center, 357
Felony, 89
Felony probation, 275
Ferguson, Ralph, 465
Fielding, Henry, 112–113
Fielding, John, 113
Fifth Amendment, 83–84, 139–140
Figlio, R. M., 491
Fines, 319–320
First Instructions (Rowan and Mayne), 113
Fishman, Joseph, 244, 252
Florida v. Bostick, 138
Foot patrols, 130
48 Hours, 111
Forward telescoping, 39–40
Foster care, 517
Fountain, Clayton, 362
Fourteenth Amendment, 83, 84–88, 326
Fourth Amendment, 69, 82, 83, 135–139, 326
France, 8, 349
Franklin, C. A., 424
Franklin, T. W., 424
Frankpledge, 112
Frequent fliers, 256
Fruit of the poisoned tree doctrine, 135
Fugitive apprehension units, 144
Fundamental fairness, 79
Furloughs, 448
Furman v. Georgia, 228–229

G

Gambling, 93
Gamson, W. A., 19
Gangs, 144
  chain, 352
  in jails, 258–259
  in prisons, 391–393
  youth, 503–504
Gangster Disciples, 393
Gaols, 241–243
Garner, Edward, 138–139
Garrett, Pat, 114
Gatekeeper, 240
Gault, In re, 508, 509
Gender differences
  crime committed and, 42–44
  in jails, 254
  juvenile delinquency and, 493–496
  in prisons, 382
  recidivism and, 299
  sentencing and, 216
  women as police officers, 117, 134
General deterrence, 207
General population, 256
Geragos, Mark, 173
Gideon v. Wainwright, 86, 173
Gilbert, Kristin, 227–228
Ginsberg, Ruth Bader, 168
Glaze, L. E., 384
Global Positioning System (GPS), 324
Goals of sentencing, 206
Goffman, Erving, 394
Going rate, 212–213
Goldstein, Herman, 131
Gonzales, Alberto, 471
Good faith exception, 87, 136
Good time, 446
Gordon, M. M., 44
Government
  branches of, 3–6
  law and, 62–63
Grand jury, 171, 181
Great Depression, 8
Greenberg, D. S., 244
Gregg v. Georgia, 229
Grey, Kimberly, 260
Griffiths, A., 243
Group counseling, family, 210
Group homes, 517–518
Guilty beyond a reasonable doubt, 187
Guilty plea, 511

H

Habeas corpus, 398, 402
Habitual misdemeanor offenders, 256
Hagan, John, 14, 43
Halfway houses, 328–329, 363, 454
Hall, J., 72
Hands-off phase, 398–399
Hands-on phase, 399, 400
Haney, C., 229, 363
Hardyman, P. L., 415
Harm, 72
Harrison, P. M., 249, 253, 382, 389
Harrison Act (1919), 356
Harrison Narcotics Act (1914), 93

Harris v. United States, 136
Hassine, Victor, 396, 423
Hawk-Sawyer, Kathleen, 367
Hayes, Paul, 214
Health of prisoners, 256–257, 363, 387–388, 433–434, 543–544
Hearings
  adjudicatory, 511
  detention, 510–511
  dispositional, 511
  parole grant, 453
  preliminary, 181
  sentencing, 204
Hearsay, 186
Hell's Angels, 393
Henry II, king of England, 242
Herman, Susan, 472–473
Hill, Anita, 170
Hilton, Paris, 20, 215
Hinckley, John, 76
Hirschi, Travis, 499–500, 502
Hispanics. *See* Latinos
HIV/AIDS, in prisoners, 256, 257, 387, 543–544
Hochstetler, A., 425
Holland, 8
Holleran, D., 298
Home confinement, 324–327
Homicide, defined, 71
Hoover, Herbert, 9, 115
Hoover, J. Edgar, 116, 120
Horton, Terry Brice, 136
Horton, Willie, 448
Horton v. California, 136
Hospice, 386
Houses of refuge, 506
Howard, John, 349, 560
Hraba, J., 44
Hudson v. McMillian, 399
Hudson v. Palmer, 400
Huff, Ronald, 22–23
Huggins, D. W., 391
Hulks, 242
Human Rights Watch, 424
Hung jury, 187
Hussein, Saddam, 158

I

If It Bleeds, It Leads (Kerbel), 19
Incapacitation, 208
Indeterminate sentences, 220–221
Indian Brotherhood, 393
Indian Posse, 393
Indictment, 171
Individualized justice, 217
Individual level data, 46
Inevitable discovery exception, 87, 136, 140
Informal probation, 284
Informal social control, 243
Information, 171
Information-sharing partnerships, 144
Initial appearance, 179
Inmates. *See* Prisoners
Inquisitorial system, 159
Insanity, 74–76
Instrumental riots, 425–426
Intake screening, 511
Intelligence-led policing, 131–132
Intensive Aftercare Program (IAP), 520
Intensive supervision probation/parole (ISP), 321–323
Interagency problem-solving partnerships, 144–145
Intermediate appellate court level, 164
Intermediate sanctions
  boot camps/shock incarceration programs, 327–328
  community service, 320–321
  costs of, 331–332
  day reporting centers, 327
  defined, 315
  economic, 317–320
  future directions for, 332–334
  goals, 315
  halfway houses, 328–329
  history of, 314–317
  home confinement/electronic monitoring, 324–327
  impact of, 334
  intensive supervision probation/parole, 321–323
  research on, 330–332
  types of, 315, 317–329
Intermittent jail sentence, 264
Intermittent supervision, 261
Internal classification, 415
International Association of Chiefs of Police, 9, 115
Interrogation, 139–140
Intoxication, as a defense, 76
Inverse relationship, 46
Iron law of corrections, 385
Irresistible impulse test, 75
Irwin, John, 355–356, 397
Isolation cells, 423

J

Jails
- community-based programs, 263–264
- difference between prisons and, 12–13, 240–241
- federal, 252
- history of, 241–245
- impact of courts on, 259–261
- inmates, characteristics of, 252–259
- inmates, statistics on, 240
- locally operated, 245–248
- new generation, 261–263
- operated by tribal governments, 248
- private, 249–250
- regional, 250–251
- state, 251
- types of, 245–252

James, D. J., 384
James, Jesse, 8–9
John, king of England, 160
Johnson, Lyndon B., 9, 116, 277
Johnson, Robert, 368, 389
Johnson v. Zerst, 173
Judge Hatchett, 164
Judge Joe Brown, 164
Judge Judy, 164
Judges
- instructions to jury, 187
- role of, 12, 169–170
- selection and qualifications, 170
- traveling, 161

Judicial branch, defined, 5
Judicial reprieve, 276
Judicial review, 168
Judicial waivers, 514
Judiciary Act (1789), 166, 171
Judiciary Act (1925), 166
Juries, early, 160–161
Jurisdiction, original versus appellate, 163–164
Jurisprudence, 64
Jurors, 178
- judge's instructions to, 187
- selection of, 184

Jury nullification, 213
Just deserts, 207
Justice
- by geography, 205
- individualized, 217
- of the Peace, 161

Justice, William Wayne, 353, 401
Justice Prisoner and Alien Transportation System (JPATS), 357
Juvenile corrections
- aftercare, 519–520
- boot camps, 518
- day treatment centers, 516–517
- effectiveness of programs, 520–521
- foster care, 517
- group homes, 517–518
- probation, 515–516
- secure care facilities, 518–519

Juvenile courts
- contemporary, 508–509
- early, 507–508
- processing cases, 509–514

Juvenile delinquency
- causes of, 497–504
- defined, 489–490
- girls and, 521
- measuring, 490–491
- methods used for controlling, 505–508
- risk-protection framework, 500–502
- statistics, 491–496
- theories of, 498–500

Juvenile Delinquency Prevention and Control Act (1968), 508
Juvenile detention, 13, 512
Juvenile Justice and Delinquency Prevention Act (1974), 508 –509, 512, 522
Juvenile justice system
- adjudicatory hearing, 511
- arguments for, 488
- contemporary issues, 522–525, 553–555
- defined, 5–6
- detention, 13, 512
- detention hearing, 510–511
- dispositional hearing, 511
- diversion programs, 512–514
- processing cases, 509–514
- teen courts, 189
- transferring juveniles to adult courts, 514–515
- use of term, 2

Juveniles
- age of, 76, 489–490
- death penalty for, 523–524
- in jails, 255
- as victims, 496–497

Juvenile Victimization and Offending, 1993–2003 (Bureau of Justice Statistics), 497

K

Kaczynski, Theodore ("Unabomber"), 559
Kansas City Gun Experiment, 128
Kansas City Preventive Patrol Experiment, 127–128
Kanka, Megan, 469–470
Karberg, J. C., 367, 425, 429, 430
Keepers, 243
Kelling, G., 293
Kent v. United States, 508, 509
Kerbel, Matthew, 19
Kerle, Ken, 248, 262
Kidd, William ("Captain"), 242
King, N., 350
King, Rodney, 132–133, 134
Kin policing, 111
Klein, M. W., 504
Klopfer decision, 183–184
Klotter, John, 92
Kuhlmann, R., 387

L

Labeling theory, 500
Langan, P. A., 284
La Nuestra Familia, 393
Latessa, E. J., 432
Latin Kings, 393
Latinos
- crime and, 44, 45
- death penalty and, 228
- in jails, 254
- in prisons, 382
- sentencing and, 215

Lattimore, Pamela, 300
Law
- *See also* Civil law; Criminal law
- case, 69–70
- constitutional, 69
- discretion and, 63
- evolution of, 64
- government structure and, 62–63
- procedural, 66–67, 68, 78–88
- regulatory, 70
- rule of, 62, 78–79
- statutory, 67, 69
- substantive, 66, 70–78

Law and Order, 111, 170
Law enforcement agencies
- federal, 117–120
- local, 121–122
- private, 122
- state, 120–121

Law Enforcement Assistance Administration (LEAA), 9, 116, 117, 127, 278
Lay witnesses, 176–177
Leavenworth, 356, 357
Legal issues, policing and, 133–140
Legalistic policing style, 126
Legislative branch, defined, 5
Legislative exclusions, 514
Leonard, V. A., 116
Lewis, R. V., 514
Lewis v. Casey, 400
Lex talionis, 207
Lichtenstein, A., 352
Limits of the Criminal Sanction, The (Packer), 21
Litigation by prisoners, 259–261, 399–403
Living unit, 261
Locke, John, 8
Lockstep, 352
Long-term prisoners, 384–386
Loosely coupled system, 14
Louima, Abner, 133, 134

M

Machonochie, Alexander, 450
Mackey, Pamela, 173
Magistrate's Act (1968), 166
Magna Carta, 160
Mala in se, 88
Mala prohibitum, 88
Malingering, 239
Mandatory minimum sentences, 217, 222–223
Mapp v. Ohio, 85–86, 135
Mara Salvatrucha, 393
Marbley, Aretha, 465
Marbury v. Madison, 69, 70, 168
Marijuana Tax Act (1937), 93
Marion prison, 362
Marquart, J. W., 353
Marshall, John, 84–85, 168
Martinson, Robert, 451
Maruna, Shadd, 465–466
Masterson, Bat, 114
Mayne, Richard, 113
Mays, G. L., 247
McCleary, Richard, 456–457
McKeiver v. Pennsylvania, 508, 509
McVeigh, Timothy, 559

Media,
    portrayal of crime and justice, 19–21
Medical model, 451
Megan's Law, 469–470
Meier, R. F., 46
Mens rea, 71, 73–78, 489
Mental illness, prisoners with, 256, 383–384
Merton, Robert, 498
Messerschmidt, J., 94
Meth-mouth, 557
Metropolitan Police Act (1829) (England), 113
Mexican Mafia, 393
Minneapolis Domestic Violence Experiment, 128
Minor, defined, 489
Minor, K. I., 398
Minorities, as police officers, 117, 134
Minton, T. D., 249
Miranda, Ernesto, 139
Miranda v. Arizona, 87, 139–140
Miranda warning, 87, 139–140
Misdemeanors, 89
Mission statement, 359
Mistake of fact, 76
Mitchell, O., 217
Mitigating factors, 223
M'Naughten, Daniel, 75
M'Naughten Rule, 75–76
Model of family government, 505–506
Monitoring the Future (MTF) project, 491
Montesquieu, Charles-Louis de Secondat, Baron de, 8
Morgan, K., 298
Morris, Roger, 426
Morton, J., 396
Motion
    for change of venue, 186
    for continuance, 186
    defined, 186
    for discovery, 186
    to dismiss charges, 186
    to suppress evidence, 186
Moussaoui, Zacarias, 559
Mueller, Robert, 120
Mullin, Herbert, 227–228
Multidisciplinary, defined, 6
Mumola, C., 386, 429, 430
Municipality police, 121–123,
Murder, defined, 71
Murphy, Daniel S., 397, 425
Murton, Tom, 353

N

Nagel, I. H., 43
National Center for Juvenile Justice, 166
National Chiefs of Police Union, 115
National Commission on Law Observance and Enforcement (Wickersham Commission), 9, 115–116
National Crime Information Center, 120
National Crime Victimization Survey (NCVS), 39–41, 490
    compared with Uniform Crime Reports, 40–41
National Gang Threat Assessment (2005), 393
National Institute of Corrections (NIC), 358
National Prison Association, 9, 450
Native Americans, arrests and jail admissions statistics, 248
Nazi Low Riders, 393
NCIS, 20–21
Necessity, 78
Neighborhood Watch Programs, 127
Net widening, 282–283, 331, 333
Newbold, Greg, 397
New-generation jail, 261–263
New Mexico State Prison riot, 426
New York Times, 262
New York v. Quarles, 140
Nichols, Brian, 175
Nichols, Terry, 559
Nifong, Michael, 19
Nightwatch, 112
Nix v. Williams, 140
No Escape: Male Rape in Prison (Human Rights Watch), 424
Nolo contendre, 181
Nye, F. Ivan, 41

O

Objective classification, 415
Occupational crime, 95
O'Conell, Daniel, 507
O'Connor, Sandra Day, 168
Offenses, types of, 90–93
Office of the Inspector General, 119, 419
Olson, David, 299
Omnibus Crime Control and Safe Streets Act (1968), 9, 116–117
On Crimes and Punishments (Dei delitti e delle pene) (Beccaria), 8
On-view arrest, 124

Opening statements, 184–185
Operation Night Light, 143
Operation Revitalization, 144
Organized crime, 93–95
Original jurisdiction, 163–164
Outlaws (gang), 393
Overcrowding, 541–542
    security issues and, 419–420

P

Packer, Herbert, 21–23
Pagans (gang), 393
Palmer, T., 514
Pandemic, 558
Panel studies, 490
Parens patriae, 506, 508
Parole, 13
    data on, 50–52
    defined, 449
    difference between probation and, 449
    grant hearing, 453
    guidelines, 454
    historical overview of, 450–452
    intensive supervision, 321–323
    issues, 457–460
    official liability, 459
    organization and administration of, 452–453
    plan, 453
    release, 221–222, 447, 453–454
    rights of parolees, 458–459
    supervision and termination, 454–457
    Parole and Community Services Law Enforcement Consortium (PCSLEC), 144–145
Parolee At-Large (PAL), 144
Part I crimes, 33–35
Part II crimes, 36, 37
PATRIOT Act, 22, 87–88, 92
Peel, Robert, 113, 134
Penal transportation, 242, 450
Penitentiary, use of term, 349
Penn, William, 349, 560
People v. Turner, 507, 508
Peremptory challenges, 184
Person offenses, 90
Petersilia, Joan, 298, 334, 397, 435, 459, 463, 464
Peterson, Lacey, 20, 227–228
Peterson, Scott, 173, 227–228
Petition, 511
Phillips, R. L., 420–421
Philosophies of punishment, 206–211
Physical plant, 261
Piehl, A. M., 348, 361, 422
Pinkerton, William, 122
Pinkerton National Detective Agency, 114, 122
Placing out, 506
Plain view doctrine, 136
Plantation-style prisons, 352–353
Plea agreement, 213, 214
Plea bargaining, 181–182, 204, 213–214
Pleas, types of, 181, 511
Plessy v. Ferguson, 115
Police
    as a component of the criminal justice system, 11
    corruption and political patronage, 115
    discretion and, 141–142
    futurism, 555–559
    –probation partnerships, 293
    professionalism, 115
    role of, 11, 123–132
    statistics on, 117
    subculture, 133
    television portrayal of, 111
    working personality, 132–133
Police-community relations, 126–127
Police Departments in Large Cities (U.S. Department of Justice), 134
Policing
    community, 130
    history and evolution of, 111–117
    intelligence-led, 131–132
    kin, 111
    legal aspects of, 133–140
    management, 125–132
    problem-oriented, 131
    reforms, 115–116
    structure and organization of, 117–123
    styles, 125–126
    team, 129
Political crime, 96–97
Posses, 114
Posttrial diversion, 280
Powell v. Alabama, 173
Precedent law. *See* Case law
Predisposition report, 511–512
Preliminary hearing, 181
Preliminary screening, 510
Preparole report, 453

Prerelease centers, 363
Prerelease programs, 447
Presentence investigation (PSI) reports, 204, 218–220, 286–287
President's Commission on Law Enforcement and Administration of Justice, 116
Presumption of innocence, 80–81
Presumptive parole dates, 454
Pretrial activities
  arraignment, 181–182
  bail, 180
  bail, alternatives to, 180–181
  hearing, 181
  initial appearance, 179
  release, 179
Pretrial detainees, 251
Pretrial diversion, 280
Pretrial release, 179
Pretrial stage, 16
Preventing Parolee Crime Program (PPCP), 467
Prevention/education, 545
Prisoners
  characteristics of, 252–259, 381–383
  classification of, 262–263, 414–418
  crimes committed, types of, 255–256, 381–382
  death row, 388–389
  demographics, 253–255, 382–383
  elderly, 386–387, 542–543
  health of, 256–257, 363, 387–388, 433–434, 543–544
  litigation by, 259–261, 399–403
  long-term, 384–386
  Reentry Initiative, 467, 547–548
  rights of, 398–403
  security threat, 391–393
  special-needs, 256–259, 383–389
  statistics on, 48–50, 240
  types of, 397–398
  women, 389–391, 396
  work done by, 352, 353, 430–431
Prisonization, 394
Prison Litigation Reform Act (PLRA) (1995), 69, 401–402
Prison Rape Elimination Act (PREA) (2003), 91, 424
Prisons
  big house, 351–352
  categories of, 360–361
  difference between jails and, 12–13, 240–241
  federal, 356–358
  history of, 348–350
  industries, 431
  plantation-style, 352–353
  reformatories, 354
  riots, 425–426
  security issues, 418–426
  society, 394–397
  specialized, 362–363
  state, 358–359
  types of, 359–362
  violence in, 356, 362, 420, 423–426
  women's, 354–355
Prison staff
  administrative and support, 369
  correctional officers, 364–369
  rehabilitative and treatment experience needed by, 370
  stress and, 370–371
Private attorneys, 173
Private jails, 249–250
Private police agencies, 114, 122
Privatization, 249–250, 546–547
Probable cause, 16, 82, 137
Probation
  automated, 294
  broken windows, 293–294
  changes in, 292–294
  costs of, 296
  data on, 50–52
  defined, 275, 507
  difference between parole and, 449
  effectiveness of, 296–300
  eligibility and placement, 286–287
  fees, 296
  felony, 275
  historical overview of, 276–278
  intensive supervision, 321–323
  issues in, 294–300
  for juveniles, 515–516
  orders, 287, 288, 291–292
  police–probation partnerships, 293
  privatization of, 296
  process, 285–292
  recidivism, 283, 296–300
  risk/needs assessment, 287, 289–291
  for sex offenders, 295
  staff roles in, 278–279
  statistics on, 275
  types of, 284–285
Probation and Parole in the United States, 50

Problem-oriented policing, 131
Procedural criminal law
 defined, 66–67, 68
 due process, 81–88
 equal protection, 79–80
 fundamental fairness, 79
 presumption of innocence, 80–81
 principles of, 78–88
 propriety, 80
Procunier v. Martinez, 400
Professionalism, 9, 115
Progessive Movement, 115
Prohibition, 8, 9
Project Greenlight, 466–467
Project ID, 127
Proof beyond a reasonable doubt, 65
Property offenses, 90–91
Propriety, 80
Prosecutorial waivers, 514
Prosecutors, 170–172
Protective factors, 501, 502
Pseudofamilies, 391
Public defenders, 174
Public safety exception to Miranda, 140
Pugh, R. B., 242
Punishment
 rehabilitation, restitution, and restoration, 209–211
 retribution, just deserts, deterrence, and incapacitation, 206–209

Q

Quakers, 349, 547

R

Race, use of term, 44
Racial differences
 corrections and, 19
 crimes committed and, 44–45
 death penalty and, 228
 equal protection, 79–80
 in halfway houses, 329
 in jails, 254
 juvenile delinquency and, 493–496, 522–523
 in prisons, 382
 sentencing and, 215
Radio-frequency identification tags, 549
Rafter, N., 354–355
RAND Corp., 208, 322–323
Rape, prison, 424–425
Ready4Work program, 466
Reagan, Ronald, 76
Reasonable suspicion, 137
Reception center, 360
Recidivism, 283, 296–300, 460–462
Reclassification, 418
Red Brotherhood, 393
Reentry
 centers, 363
 challenges, 462–463, 552
 controversies regarding, 467–473
 courts, 465
 defined, 445
 improving, 463–466
 research on, 466–467
Re-Entry Enhancement Act, 467, 468
Reentry Initiative, 467, 547–548
Reformation, 547
Reformatories, 354
Reform schools, 507
Regionalization, 250–251
Regulatory (administrative) law, 70
Rehabilitation, 209–211
Rehabilitative programs, 427
 counseling, 429–430
 educational, 428–429
 evidence-based, 431–433
 work and vocational, 430–431
Rehnquist, William, 87
Reid, Richard, 559
Reiman, Jeffrey, 45
Reintegration, 463
Release
 compassionate, 387
 methods of, 446–447
 pre- and temporary, 447–448
 on recognizance (ROR), 180
Residential parole, 454
Residential probation, 285
Resiliency, 502
Restitution, 209–211, 317–318
Restoration, 209–211
Restorative justice, 320
Retained counsel, 173
Retribution, 206–207
Revocation, 292
Richards, Stephen C., 392, 397, 460
Rights of parolees, 458–459
Rights of prisoners, 398–403
Right to counsel, 173–174

Right to speedy trial, 183–184
Riots, reducing prison, 425–426
Risk factors, 501–502
Risk/needs assessment, 287, 289–291, 515
Risk-protection framework, 500–502
Roberts, John, 87
Rodriguez, John, 379–380, 386–387
Roe v. Wade, 70
Roosevelt, Franklin D., 350
Roosevelt, Theodore, 115
Roper v. Simmons, 52, 229, 523, 524
Ross, J. I., 392, 397
Rowan, Charles, 113
Ruddell, R., 247
Ruffin v. Commonwealth, 398
Ruiz v. Estelle, 353, 401
Ruiz v. Johnson, 401
Rule of law, 62, 78–79
Rummel v. Estelle, 385
Rush, Benjamin, 349
Ryder, S., 489

S

Sabol, W. J., 249
Salient Factor Instrument, 454, 455
Santos, Michael, 398
Scheb, J. M, 214
Schlanger, M., 260, 402, 403
Schmalleger, Frank, 91
School resource officers, 554–555
Schur, Edwin, 512
Scientific police management, 127–129
Searches
  arrests and, 137
  deadly force, use of, 138–139
  emergency, 137
  Fourth Amendment and, 135–139
  suspicionless, 138
  vehicle, 138
  warrants, 135–136
Second Chance Act, 467, 468
Section 1983 lawsuits, 399
Secure care facilities, 518–519
Securitas Security Services, 122
Security
  housing units, 362
  issues in prisons, 418–426
  overcrowding and, 419–420
  prison design and, 422–423
  riots, 425–426
  threat groups, 258–259, 391–393
  violence reduction strategies, 423–425
Segregation units, 423
Seiter, R. P., 279
Selective incapacitation, 208–209
Selective incorporation, 85
Self-defense, 77
Self-incrimination, 84
Self-report surveys/studies, 41–42, 490
Sellin, T., 491
Sentencing, 17
  blended, 514
  changes in, 220–226
  circle, 210
  commissions, 216
  consistency problems, 217–218
  disparity, 205–206, 215–217
  goals, 206
  going rate, 212–213
  guidelines, 223–224
  hearing, 204
  mandatory minimum, 217, 222–223
  options, 204–206
  plea bargaining, 181–182, 204, 213–214
  presentence investigations, 218–220
  three strikes laws, 208, 225–226
  truth in, 225
Sentencing Reform Act (1984), 221
Separate system (silent system), 351
Service style of policing, 126
Severance, Theresa, 391
Sex offenders
  notification laws, 470
  probation for, 295
  reentry of, 469–471
  registry, 470
Sex-related offenses, 92
Shapiro, Robert, 173
Sheriff, 112, 122
Shield, The, 111
Shire reeve, 112
Shock incarceration programs, 327–328
Shock probation, 284
Short, James, 41
Silent system (separate system), 351
Silverstein, Thomas, 362
Silverthorne Lumber Co. v. United States, 135
Simons, R. L., 425
Simpson, O. J., 20, 78, 173

Sixth Amendment, 83, 84, 139–140
Skelton, David, 88
Snell, T. L., 388
Social bond theory, 499–500
Social institutions, defined, 3
Social learning theories, 498
Societal reaction theory, 500
Society of Captives, The (Sykes), 395
Socioeconomic status
  crime and, 46–47
  prisoners in jails and, 254–255
  sentencing and, 215
Solitary confinement, 351
Special conditions, 287
Specialized enforcement partnerships, 144
Special-needs prisoners, 256–259, 383–389
Specialty courts, 550–551
Specific deterrence, 207
Speedy Trial Act (1974), 183
Split sentence probation, 284–285
Spohn, C., 298
Square John, 397
Standard conditions, 287
Stanford v. Kentucky, 523
Stare decisis, 70, 88
State court systems, 163–166
State jails, 251
State police agencies, 120–121
State prisons, 358–359
State-raised youth, 397
Status offenses, 89, 489
Statutory law, 67, 69
Steiker, Carol, 212–213
Stephan, J. J., 367, 425
Stewart, Martha, 20, 215
Straight parole, 454
Straight probation, 284
Strain theory, 498–499
Stress, prison employees and, 370–371
Stroud, Robert (Birdman of Alcatraz), 356–357
Study release programs, 448
Stun belts, 549
Stuntz, William, 214
Subculture
  correctional officer, 368–369
  defined, 133, 394
  prison, 394–395
Subjective classification, 415
Subpoenas, 176
Substance abuse, juvenile delinquency and, 502–503
Substance Abuse and Crime Prevention Act, 546
Substantial capacity test, 75
Substantive criminal law
  actus reus, 71, 72
  causation, 72–73
  concurrence, 73
  corpus delicti, 70–72
  defined, 66
  harm, 72
  mens rea, 71, 73–78
  principles of, 70–78
Substantive due process, 69
Supermax prisons, 363
Supervision fees, 318–319
Supply reduction, 545
Supreme Court
  compositions of, 168
  death penalty, 228–229
  death penalty for juveniles, 523–524
  due process, 81–88
  due process for juveniles, 508–509
  interrogations, 139–140
  landmark decisions, 70
  power of judicial review, 69
  rights of prisoners, 398–401
  right to counsel, 173–174
  searches, 135–139
  supervision fees, 319
  women on the, 168
Surveillance cameras, 549
Suspended sentence probation, 284
Suspicionless searches, 138
Sutherland, Edwin, 95, 498
Sykes, Gresham M., 368, 394, 395, 396, 397
Syndicated crime, 93–95

T

TASER stun devices, 549
Team policing, 129
Technical parole violation, 457
Technical violations, 292
Technology, use of, 548–550
Teen courts, 189
Telescoping, 39–40
Television
  portrayal of crime and justice, 19–21
  portrayal of police, 111

Temporary release programs, 447–448
Tennessee v. Garner, 138–139
Terrorism, prison gangs and, 393
Terry, Chuck, 395, 397
Terry v. Ohio, 137
Testimony
  defined, 186
  of witnesses, 186
Texas Rangers, 120
Texas Syndicate, 393
Therapeutic Community, 430
Therapeutic foster care (TFC), 517
Third-party custody, 180–181
Thomas, Clarence, 170
Thompson v. Oklahoma, 523
Three Prisons Act (1891), 356
Three strikes laws, 208, 225–226
Ticket of leave system, 450
Tittle, C. R., 46
To Catch a Predator, 470
Top ten frivolous inmate lawsuits, 401
Torture, 8
Total institution, 394
Tracy, S. J., 368
Transportation, penal, 242, 450
Travis, Jeremy, 385, 463–464, 466
Treatment, drug, 545
Treatment Alternatives to Street Crime (TASC), 277–278, 283
Treatment foster care, 517
Trial by battle, 160
Trial by compurgation, 160
Trial by ordeal, 160
Trial *de novo,* 164
Tribal jails, 248
Trojanowicz, Robert, 130
Tromanhauser, Edward, 397
Truman, Harry, 350
Trustees, 353
Truth-in-sentencing, 225, 453
Turner, N. R., 432
Turner v. Safley, 400
Tyler, Tom, 23

U

Unconditional mandatory release, 446
Uniform Crime Reports (UCRs), 33–39, 120, 490
  compared with National Crime Victimization Survey, 40–41
Uniform Juvenile Court Act (1968), 489
U.S. Attorney General, 171
U.S. Constitution. *See* Constitution of the United States
U.S. Courts of Appeals. *See* Courts of Appeals
U.S. Department of Justice, 118, 119
U.S. district courts. *See* District courts
U.S. Supreme Court. *See* Supreme Court
United States v. Booker, 217, 222, 223
United States v. Robinson, 137
Useem, B., 348, 361, 422

V

Vaughn, J., 325
Vehicle searches, 138
Verdeyen, V., 395
Verdict, 187
Vickers, Allison, 524
Vickers, Matt, 524
Victims
  consent, 77
  juvenile, 496–497
  National Crime Victimization Survey, 39–41
  reentry and impact on, 471–473
  self-report surveys, 41–42
  as witnesses, 177–178
Victim's Assistance Programs, 127
Video technology, 550
Vigilance committees, 114
Vigilante groups, 8
Violence, in prisons, 356, 362, 420, 423–426
Virginia Tech University, 66
Vito, G. F., 298
Vocational programs for prisoners, 430–431
Voir dire, 184
Vollmer, August, 115, 116
Volstead Act (1914), 356
Voltaire, 8

W

Wackenhut, George, 122
Wackenhut Corp., 122
Walker, Samuel, 20, 21, 22, 214
Walnut Street Jail, 349, 351
Walters, G., 394
Warehouse prison, 356
Warlords (gang), 393
Warrants, search, 135–136

Warren, Earl, 85–87, 98, 134, 135–136, 140
Wasserman, Cressida, 472–473
Watchman policing style, 125–126
Weapon availability, crime and, 47–48
Weeks v. United States, 85–86, 135
Welch, M., 545
Welch, Rachel, 355
Wells, Henry, 114
Wells, J. B., 392, 553
Wells Fargo, 122
Welsh, W. N., 260
West, A., 279
Western State Penitentiary, 351
White-collar crime, 95–96
Whites
  crime and, 44
  death penalty and, 228
  in jails, 254
  in prisons, 382
Who's Who in American Jails (American Jail Association), 246
Wickersham, George, 115–116
Wickersham Commission (National Commission on Law Observance and Enforcement), 9, 115–116
Wilhelm, D. F., 432
Williams, Robert, 140
Williamson, Kimberly, 524
Wilson, J. Q., 293
Wilson, O. W., 116
Wilson, Woodrow, 350
Wilson v. Seiter, 400
Winship, In re, 508, 509
Witnesses
  defined, 176–177
  expert, 177
  lay, 176–177
  testimony of, 186
  victims as, 177
Wolff v. McDonnell, 400
Wolfgang, M. E., 491
Women
  See also Gender differences
  battered woman/spouse syndrome, 78
  chivalry hypothesis, 43
  as correctional officers, 367
  evil woman argument, 43
  as police officers, 117, 134
  as prisoners, 354–355, 389–391, 396
  prisons for, 354–355, 389–391
  reentry programs for, 465, 467
  on the Supreme Court, 168
Woodford v. Ngo, 401
Workhouses, 349
Work programs for prisoners, 430–431
Work release programs, 448
Writ of certiorari, 168

Y

Yousef, Ramzi, 559
Youth gangs, 503–504
Youth Risk Behavior Surveillance System (YRBSS), 491